METHODS IN CELL BIOLOGY

VOLUME VIII

Contributors to This Volume

Claudio Basilico
Thomas L. Benjamin
Ronald Berezney
Hollis G. Boren
Carlo M. Croce
L. L. Deaven
Virginia J. Evans
Robert D. Goldman
Earl D. Hanson
Curtis C. Harris
Samuel B. Horowitz
Dieter F. Hülser
Fa-Ten Kao
Hilary Koprowski
Elliot M. Levine
John W. Littlefield
N. G. Maroudas
Harriet K. Meiss
P. M. Naha
D. F. Petersen
Robert Pollack
Theodore T. Puck
C. M. Schmitt
William G. Taylor
Natale Tomassini
A. Vogel
Edith C. Wright
Walid G. Yasmineh
Jorge J. Yunis
H. Ronald Zielke

Methods in Cell Biology

Edited by

DAVID M. PRESCOTT

DEPARTMENT OF MOLECULAR, CELLULAR AND
DEVELOPMENTAL BIOLOGY
UNIVERSITY OF COLORADO
BOULDER, COLORADO

VOLUME VIII

1974

ACADEMIC PRESS • New York and London
A Subsidiary of Harcourt Brace Jovanovich, Publishers

ACADEMIC PRESS, INC.
111 Fifth Avenue, New York, New York 10003

United Kingdom Edition published by
ACADEMIC PRESS, INC. (LONDON) LTD.
24/28 Oval Road, London NW1

LIBRARY OF CONGRESS CATALOG CARD NUMBER: 64-14220

PRINTED IN THE UNITED STATES OF AMERICA

CONTENTS

LIST OF CONTRIBUTORS

Numbers in parentheses indicate the pages on which the authors' contributions begin.

CLAUDIO BASILICO, Department of Pathology, New York University School of Medicine, New York, New York (1)

THOMAS L. BENJAMIN, Department of Pathology, Harvard Medical School, Boston, Massachusetts (367)

RONALD BEREZNEY, Department of Pharmacology and Experimental Therapeutics, The Johns Hopkins University School of Medicine, Baltimore, Maryland (205)

HOLLIS G. BOREN, University of South Florida College of Medicine, Veterans Administration Hospital, Tampa, Florida (277)

CARLO M. CROCE, Wistar Institute of Anatomy and Biology, Philadelphia, Pennsylvania (145)

L. L. DEAVEN, Cellular and Molecular Radiobiology Group. Los Alamos Scientific Laboratory, University of California, Los Alamos, New Mexico (179)

VIRGINIA J. EVANS, Tissue Culture Section, Laboratory of Biology, National Cancer Institute, Bethesda, Maryland (47)

ROBERT D. GOLDMAN,[1] Department of Biological Sciences, Carnegie Mellon University, Mellon Institute, Pittsburgh, Pennsylvania, (123)

EARL D. HANSON, Shanklin Laboratory, Wesleyan University, Middletown, Connecticut (319)

CURTIS C. HARRIS, Lung Cancer Branch, Carcinogenesis Program, National Cancer Institute, Bethesda, Maryland (277)

SAMUEL B. HOROWITZ, Laboratory of Cellular Physiology, Department of Biology, Michigan Cancer Foundation, Detroit, Michigan (249)

DIETER F. HÜLSER, Abteilung Physikalische Biologie, Max-Planck-Institut für Virusforschung, Tübingen, Germany (289)

FA-TEN KAO, Department of Biophysics and Genetics, University of Colorado Medical Center, Denver, Colorado (23)

HILARY KOPROWSKI, Wistar Institute of Anatomy and Biology, Philadelphia, Pennsylvania (145)

ELLIOT M. LEVINE, The Wistar Institute of Anatomy and Biology, Philadelphia, Pennsylvania (229)

JOHN W. LITTLEFIELD, Genetics Unit, Children's Service, Massachusetts General Hospital and Department of Pediatrics, Harvard Medical School, Boston, Massachusetts (107)

N. G. MAROUDAS, Imperial Cancer Research Fund Laboratories, Lincoln's Inn Fields, London, England (93, 101)

HARRIET K. MEISS, Department of Pathology, New York University School of Medicine, New York, New York (1)

P. M. NAHA, National Institute for Medical Research, Mill Hill, London, England (41)

D. F. PETERSEN, Cellular and Molecular Radiobiology Group, Los Alamos Scientific Laboratory, University of California, Los Alamos, New Mexico (179)

ROBERT POLLACK, Cold Spring Harbor Laboratory, Cold Spring Harbor, New York (75, 123)

THEODORE T. PUCK, Department of Biophysics and Genetics, University of Colorado Medical Center, Denver, Colorado (23)

[1]*Present address:* Cold Spring Harbor Laboratory, Cold Spring Harbor, New York.

C. M. SCHMITT, Imperial Cancer Research Fund Laboratories, Lincoln's Inn Fields, London, England (101)

WILLIAM G. TAYLOR, Tissue Culture Section, Laboratory of Biology, National Cancer Institute, Bethesda, Maryland (47)

NATALE TOMASSINI, Division of Experimental Pathology, Childrens Hospital, Philadelphia, Pennsylvania (145)

A. VOGEL, Cold Spring Harbor Laboratory, Cold Spring Harbor, New York (75)

EDITH C. WRIGHT, University of South Florida College of Medicine, Veterans Administration Hospital, Tampa, Florida (277)

WALID G. YASMINEH, Medical Genetics Division, Department of Laboratory Medicine and Pathology, University of Minnesota, Minneapolis, Minnesota (151)

JORGE J. YUNIS, Medical Genetics Division, Department of Laboratory Medicine and Pathology, University of Minnesota, Minneapolis, Minnesota (151)

H. RONALD ZIELKE, Genetics Unit, Children's Service, Massachusetts General Hospital, and Department of Pediatrics, Harvard Medical School, Boston, Massachusetts (107)

PREFACE

In the years since the inception of the multivolume series *Methods in Cell Physiology*, research on the cell has expanded and added major new directions. In contemporary research, analyses of cell structure and function commonly require polytechnic approaches involving methodologies of biochemistry, genetics, cytology, biophysics, as well as physiology. The range of techniques and methods in cell research has expanded steadily, and now the title *Methods in Cell Physiology* no longer seems adequate or accurate. For this reason the series of volumes known as *Methods in Cell Physiology* is now published under the title *Methods in Cell Biology*.

Volume VIII of this series continues to present techniques and methods in cell research that have not been published or have been published in sources that are not readily available. Much of the information on experimental techniques in modern cell biology is scattered in a fragmentary fashion throughout the research literature. In addition, the general practice of condensing to the most abbreviated form materials and methods sections of journal articles has led to descriptions that are frequently inadequate guides to techniques. The aim of this volume is to bring together into one compilation complete and detailed treatment of a number of widely useful techniques which have not been published in full detail elsewhere in the literature.

In the absence of firsthand personal instruction, researchers are often reluctant to adopt new techniques. This hesitancy probably stems chiefly from the fact that descriptions in the literature do not contain sufficient detail concerning methodology; in addition, the information given may not be sufficient to estimate the difficulties or practicality of the technique or to judge whether the method can actually provide a suitable solution to the problem under consideration. The presentations in this volume are designed to overcome these drawbacks. They are comprehensive to the extent that they may serve not only as a practical introduction to experimental procedures but also to provide, to some extent, an evaluation of the limitations, potentialities, and current applications of the methods. Only those theoretical considerations needed for proper use of the method are included.

Finally, special emphasis has been placed on inclusion of much reference material in order to guide readers to early and current pertinent literature.

DAVID M. PRESCOTT

CONTENTS OF PREVIOUS VOLUMES

Volume I

Volume II

Volume III

Volume IV

Volume V

Volume VI

Volume VII

Chapter 1

Methods for Selecting and Studying Temperature-Sensitive Mutants of BHK-21 Cells

CLAUDIO BASILICO AND HARRIET K. MEISS

Department of Pathology,
New York University School of Medicine,
New York, New York

I. Introduction

In order to broaden our knowledge of the genetics and physiology of mammalian cells, a large variety of cells carrying defined genetic markers must be available. While considerable progress along these lines has been made in the last decade, still the only selective procedures that have been fairly well worked out are those for obtaining cell lines with a particular nutritional deficiency (Puck and Kao, 1967; Kao and Puck, 1968) or with resistance to any one of a variety of drugs (Chu, 1971). We thought it would be worthwhile to attempt the selection of conditional lethal temperature-sensitive (*ts*) somatic cell mutants, since this type of mutation has been invaluable for attacking fundamental problems concerning the genetics and physiology of microorganisms and *Drosophila*.

The defect of *ts* mutants is usually attributable to a single amino acid substitution in an essential protein (see review by Drake, 1970), although *ts* mutations resulting from alterations of transfer RNA have also been described (Yamamoto *et al.*, 1972). As a result of such mutations, a protein remains functional at low temperature, but loses its functionality at certain relatively high temperatures, which do not affect the wild-type gene product. Growth is then normal at low temperature, because the affected protein is stable under these conditions, but at higher temperature no growth is observed.

ts mutations have been shown to arise in most organisms studied and to occur all over the genome (Epstein *et al.*, 1963; Edgar and Lielausis, 1964), thus allowing the examination of a wide spectrum of cellular functions. Thus, it is possible to obtain conditional lethal mutations directly affecting indispensable functions such as DNA, RNA, and protein syntheses. Most importantly, the conditional expression of *ts* mutations can provide a unique tool for testing directly the involvement of a specific gene-product in the determination of a specific phenotype. For example, the permanent control exerted by viral gene-product(s) on the phenotype of virally transformed cells could only be demonstrated after the isolation of some classes of *ts* mutants of Rous sarcoma virus. Cells transformed by such mutants have a temperature-dependent expression of the malignant phenotype (Martin, 1970).

It is the purpose of this review to describe and evaluate the procedures we have employed in the last two years for isolating *ts* mutants of the hamster cell line BHK-21 (Meiss and Basilico, 1972).

II. Choice of Cell Lines and Temperature

The choice of the cell line to be used should be mainly determined by the type of studies that are to be carried out with the mutants. For our studies, we chose BHK-21/13, a continuous line of Syrian hamster fibroblasts, which grows in monolayers (Stoker and Macpherson, 1964). The following points are to be considered and were pertinent to our choice of BHK-21:

1. The cell line used should have a good efficiency of plating, since mutants can be isolated only after they have formed a colony originating from a single cell.

2. Selecting a cell line with a short generation time is of advantage for speeding up the procedure, since the multiplication of most cells is slowed down considerably at temperatures below 37°C.

3. The availability of other mutants in the same cell line offers the possibility of obtaining *ts* mutations in cells carrying other genetic markers (see later).

4. The chromosome complement of the cell line to be used is probably of paramount importance, since most *ts* mutations are recessive (Hartwell, 1967; Suzuki and Procunier, 1969; Suzuki, 1970; see also Hayes, 1970). Unfortunately, haploid lines of mammalian cells are not available, but the isolation of recessive *ts* mutations should be easier in a diploid cell line than in cells with a hyperdiploid chromosome complement. Thus, selection of a cell line which is diploid of quasi diploid should be advantageous. In this respect, it is worth mentioning that while we have had good success in the isolation of *ts* mutants of BHK cells, which have a diploid chromosome number, a comparable attempt to obtain *ts* mutants of the mouse cell line 3T3 (Todaro and Green, 1963), which is almost tetraploid (Basilico *et al.*, 1970), has met with very poor results.

5. Since we are interested in studying the cellular functions required for viral multiplication or neoplastic transformation, we chose BHK-21 cells which can be lytically infected or transformed by many viruses.

As for the selection of the permissive and nonpermissive temperatures, the high, nonpermissive temperature is the most important. It must obviously be a temperature at which wild-type cells are not inhibited. The possibility of finding *ts* mutants should increase with increasing temperature, but most mammalian cells do not grow well at temperatures higher than 39° (chick cells, for example, grow well at 41°C); 39°C is the nonpermissive temperature we chose for our studies with BHK. The permissive temperature is probably less critical. Although in theory having a very large temperature differential should increase the probabilities of finding mutants, mammalian cell growth slows down considerably at temperatures below 34°C, so that even a drop of 0.5–1 degree is enough to give markedly increased generation times. Therefore, we chose for our studies 33°–33.5°C, at which temperature BHK cells have a generation time of 16–18 hours as compared to ~12 hours at 37°C and 39°C (in Dulbecco's modified Eagle medium supplemented with 10% calf serum).

It should also be mentioned that it has been our experience that most mutants isolated start to show thermosensitivity at temperatures around 36–37°C. This has also been found with animal virus mutants (Burge and Pfefferkorn, 1966), irrespective of how low the permissive temperature of selection had been. It would therefore appear that choosing too low a permissive temperature has more disadvantages than advantages.

III. Mutagenesis

Temperature-sensitive mutations are believed to be missense mutations. Among the many substances that have been shown to produce base substitutions or missense mutations in microorganisms, *N*-methyl-*N*′-nitro-*N*-nitrosoguanidine (NG), ethyl methane sulfonate (EMS), and several other drugs (see Drake, 1970) may be effective in inducing *ts* mutations in somatic animal cells, as they have been shown to induce formation of other types of recessive mutations (Kao and Puck, 1969; Orkin and Littlefield, 1971; Sato *et al.*, 1972). In bacteria, NG is effective mainly at the DNA replication point (Cerda-Olmedo *et al.*, 1968), but this does not seem to be the case in animal cells (Orkin and Littlefield, 1971). We have used mainly NG, dissolved in H_2O at the time of use. The final concentration in the culture medium was 1 μg/ml, and was left in the culture from 5 to 16 hours at 33°C. A typical inactivation curve under these conditions is shown in Fig. 1. We selected a dose giving a survival of 20–40%, since in bacteria NG is most effective in producing mutations in this range of survival (Adelberg *et al.*, 1965). The time of exposure to the drug is probably not very critical, as Orkin and Littlefield (1971) have shown that 3 hours of exposure to NG are sufficient to produce optimum yield of $HGPRT^-$ mutants in their system, and furthermore, NG is very unstable in solution.

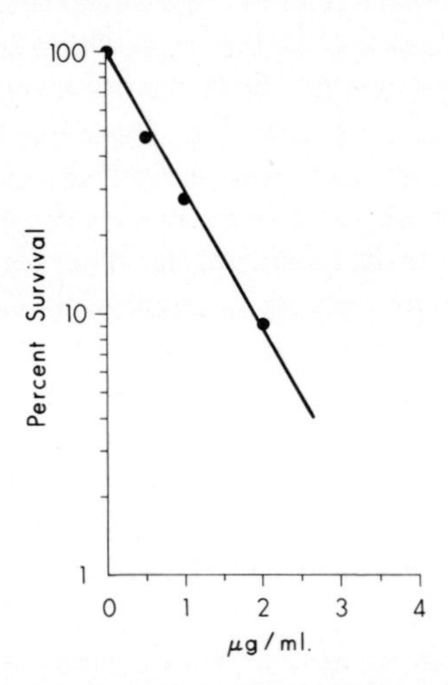

FIG. 1. Petri dishes (60 mm) were seeded with 500 BHK cells, and incubated at 33°C for several hours to allow the cells to attach. At this time, NG was added to the desired concentration. After 16 hours' incubation, NG was removed and the plates were reincubated for about 10 days until colonies could be counted. The number of colonies on the control plate (no NG) represents 100% survival. Each point on the graph was calculated from the average number of colonies on four plates.

After exposure to the mutagen, the cells are washed free of the drug and incubated at the permissive temperature for a time equivalent to at least 3 generations. It has been demonstrated that this is a sufficient time of "fixation" for various types of recessive mutations in somatic animal cells (Chu and Malling, 1968). It should allow for rounds of DNA replication necessary to complete base transitions and segregation, and dilution of pre-existing cellular proteins. In some cases it may allow for possible processes of somatic recombination or chromosomal rearrangements, which could be basic for obtaining *ts* recessive mutations either in a homozygous or in a hemizygous state.

IV. Selection

Selection for *ts* mutants relies on the possibility of finding conditions that are lethal for wild-type (wt) cells, but not for the mutants. However, there are no culture conditions that would give a *ts* mutant selective growth advantage over the wt cells at the permissive temperature. One must thus devise conditions of killing wt cells at the nonpermissive temperature, taking advantage of the fact that under these conditions wt cells multiply normally while the mutants are arrested in some essential step of their metabolism, and consequently in cell division. Preferential killing of wt cells results in enriching the population in mutants. Cells have then to be shifted back to the permissive temperature to allow for growth of the mutant cells.

Basically, the systems which have been used consist of shifting the cells to 39°C, and applying the selective treatment. After a specific incubation time, the cells are shifted back to low temperature, after removal of the killing agent. Such killing agents must have the property of being cytocidal and not simply cytostatic, and of being easily removable from the culture, otherwise they could affect even the mutants, when the cells are shifted to the permissive temperature.

A number of agents can and have been used. These include:

1. Mitotic inhibitors, such as colchicine and vinblastine. Such compounds can be lethal to cells entering mitosis, but do not affect cells in other stages of their life cycle. Such agents could be expected to select generally for growth mutants, since inhibition of division should sooner or later be a characteristic of all conditional lethal *ts* mutants.

2. DNA synthesis inhibitors, or DNA "poisons," such as cytosine arabinoside (Thompson *et al.*, 1970), 5-fluoro-2-deoxyuridine (FUdR) (Meiss and Basilico, 1972), high specific activity thymidine-^{3}H (Thompson *et al.*, 1970), or 5-bromo-2-deoxyuridine (BUdR) and "light" (Naha, 1969). The two

former drugs act by inhibiting DNA synthesis (Graham and Whitmore, 1970; Cohen *et al.*, 1958), and are highly lethal in exponentially growing, DNA-synthesizing cells (Heidelberger, 1965; Pollack *et al.*, 1968; Graham and Whitmore, 1970), but their mechanism of killing is not yet understood (see later). They are probably not incorporated into DNA. The latter two agents are incorporated into DNA, and lethality, in the case of thymidine-^{3}H, is due to the internal radioactive decay of the ^{3}H atoms (Person, 1963); in the case of BUdR, lethality is probably due to the fact that 5-bromouracil-substituted DNA is much more sensitive to UV irradiation than ordinary DNA. Cell death probably results from the formation of strand breaks and chromosomal aberrations (Puck and Kao, 1967; Chu *et al.*, 1972). These four substances have been used with the hope of selecting for DNA synthesis mutants. The rationale here is that they would kill both exponentially growing wt cells and mutants not inhibited in DNA synthesis. However, since many metabolic blocks (e.g., protein synthesis) can suppress DNA synthesis in mammalian cells, these drugs could be used also as a general method of selection.

3. More specific methods of selection, such as incorporation of high specific activity tritiated leucine or uridine (Lubin, 1962; Tocchini-Valentini *et al.*, 1969), to select for mutants with defects in protein or RNA synthesis.

4. A different type of selection, not based on the preferential killing of wt cells, has been applied by Roscoe *et al.* (1973). These authors have used selective detachment of mitotic cells at 39°C to select for *ts* mutants which are blocked at mitosis.

We will describe now in some detail our method of selection, which is based on the use of FUdR and is outlined in Fig. 2. After exposure to the mutagen and intermediate cultivation at 33°C, the cells are replated at the appropriate density and shifted to 39°C. Care must be taken to have exponentially growing cells throughout the procedure, so the number of cells

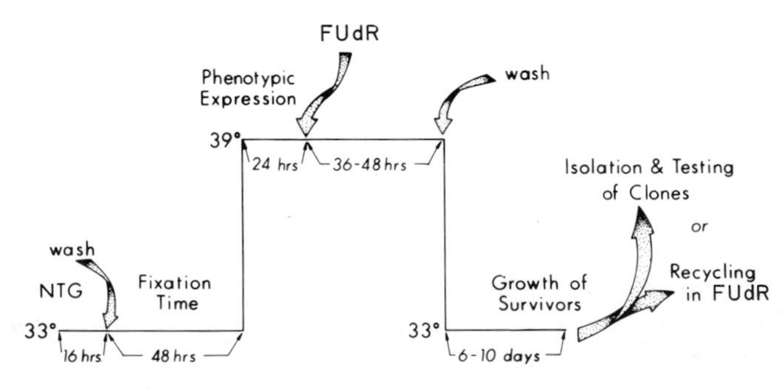

FIG. 2. Schematic representation of the selection procedure for *ts* mutants of BHK.

per plate must not be too high. If the cell number is too high, a large number of cells will not divide because of contact-inhibition, and these will escape the lethal action of FUdR. For the same reason, it is important that the cells are not serum starved. With BHK cells, which grow to over 10^7 cells per 100-mm plate, we try to shift the cells to 39°C at a density of about 5×10^5 cells per 100-mm plate. After shift-up, the cells are incubated at 39°C for about 24 hours, corresponding to about two division times. This time should allow for phenotypic expression of the *ts* defect, which may be delayed for any number of reasons. FUdR is then added to a concentration of 25 μg/ml, in the presence of an excess of cold uridine, to prevent the possible interference of small amounts of fluorouracil, which could be produced by FUdR breakdown, with RNA metabolism.

FUdR is kept in the cultures for approximately 36 hours, after which the cultures are washed free of the drug and shifted to 33°C in the presence of 10 μg of thymidine per milliliter. This treatment generally produces about 99.9% killing of wt cells, while increasing substantially the proportion of mutants in the population, as can be shown by reconstruction experiments. Table I shows the results of two such experiments in which wild-type and mutant cells were mixed in a known ratio and subjected to FUdR treatment at 39°C. It can be seen that one FUdR treatment enriched the cell population in mutants more than 20-fold when AF8, which is a DNA synthesis mutant, was used. When BCH, which is not *ts* for DNA synthesis was tested, the enrichment was about 10-fold. This point will be discussed later.

Since the frequency of mutants we obtained is much lower than 10^{-3}, the cells have to be recycled through FUdR before the isolation of the mutants is feasible. This procedure has the disadvantages of probably killing some classes of *ts* mutants, but it is effective in enriching for "good" mutants (low reversion, low leakiness). Recycling is carried out by growing the cell population at 33°C for 6–10 days before exposing them to another FUdR cycle at 39°C. It is often necessary to trypsinize and dilute the cultures to avoid crowding effects during the FUdR treatment. We have in general submitted our cells to three FUdR treatments (Meiss and Basilico, 1972), except in the case of the *ts* mutants derived from T6a BHK, a hypoxanthine, guanine phosphoribosyl transferase-deficient ($HGPRT^-$) line (Marin and Littlefield, 1968) of BHK cells. For reasons yet unknown, those cells are killed by FUdR much more effectively than regular BHK cells, and therefore the selection procedure is more efficient. After one treatment with FUdR, the surviving fraction is 10^{-4}, of which 10% are good mutants. Another cycle of FUdR treatment increases the proportion of good mutants to over 25%.

As mentioned before, FUdR has been used in the hope that it might select

TABLE I

RECONSTRUCTION OF FUdR SELECTION[a]

Exp. No.	*ts* mutant	Treatment at 39°C	Number of cells forming colonies after treatment		Ratio 39°/33°C	Estimated proportion of *ts* mutants[b] (%)
			39°C	33°C		
I	AF8	FUdR	9.0×10^2	2.2×10^3	0.41	59
		None	1.5×10^5	1.4×10^5	~1	~2.5
II	BCH	FUdR	8.0×10^2	1.2×10^3	0.67	33
		None	1.7×10^5	1.7×10^5	1	~2.5

[a] Wild-type BHK and each of the *ts* mutants were mixed at a 95:5 ratio. After attachment at 33°C, they were incubated at 39°C for 1 day. During that time, presumably wt cells underwent two divisions and the mutants a little less than one division, which yields an estimated proportion of *ts* mutants of 2.5%. Half the cultures were then plated for colony formation at 39°C and 33°C, and the other half received FUdR (25 μg/ml). Two days later, the FUdR was removed and the cells were plated at 39°C and 33°C in the presence of 20 μg of thymidine per milliliter.

[b] Based on the EOP 39°/33°C of 1 for wt BHK and of $< 10^{-5}$ for the mutants.

for mutants affected directly or indirectly in DNA synthesis. It was found, however, that most of the *ts* mutants obtained by this method were not DNA synthesis mutants. It can be asked what was the reason for this result: FUdR is phosphorylated in the cells to FdUMP and then exerts its action by inhibiting the enzyme thymidylate synthetase, which performs the conversion of dUMP to dTMP (Cohen *et al.*, 1958; Hartmann and Heidelberger, 1961; Heidelberger, 1965). FdUMP does not proceed further along the metabolic pathway, and the result is inhibition of DNA replication, since dTTP is not formed. However, the mechanism through which cell death is caused is far from being clear. The mechanism of FUdR killing is probably similar to the mechanism which causes thymineless death in bacteria (see Hayes, 1970). These phenomena have been generally interpreted as being due to unbalanced growth. Although the available evidence shows that cells which are not synthesizing DNA are generally immune to the lethal action of FUdR, while DNA synthesizing cells are not (Pollack *et al.*, 1968), we should remember that in the case of our *ts* mutants the situation is somewhat different. If the killing action of FUdR was due to the phenomena of unbalanced growth mentioned before, it could be thought that cells which are blocked in other steps of their cycle might somewhat correct that imbalance, so that their equilibrium would be less perturbed by DNA synthesis inhibition. This could explain why FUdR does select, although less effectively, also for *ts* mutants not affected in DNA synthesis. (This is shown in the II reconstruction experiment, which involved *ts* mutant BCH, which is not a DNA synthesis mutant.) It is also possible to think that inhibition of some other steps of the cell metabolism may be enough to slow down DNA replication to the point in which the cells escape the lethal effect of FUdR. This, however, does not explain the isolation of *ts* mutant BCH, since the inhibition of DNA synthesis observed in these cells at 39°C is negligible.

It has to be mentioned that other investigators who have used tritiated thymidine as a killing agent have reported similar results (i.e., obtaining large numbers of non-DNA synthesis mutants) (L. Siminovitch, personal communication). Here it may be thought that cells which are not undergoing division have more time to repair DNA damage, but again we have no satisfactory explanation.

In conclusion, the points mentioned above indicate clearly that the selective agents used have proved to be good general selectors, but their specificity for DNA synthesis *ts* mutants has been negated by the experimental results. It is possible that modification of the selection techniques may still prove that these agents are useful also for obtaining mutants with specific defects. The mutants already available should be of help in modifying the selection to such an end, since they can be used for reconstruction experiments.

V. Isolation of Surviving Cells and Testing for *ts* Phenotype

If the last selective treatment is applied on a sufficiently low number of cells per plate, surviving cells can be isolated as clones directly from the plate in which they were subjected to FUdR treatment. Otherwise, after the last FUdR treatment the cells have to be plated out at 33°C in appropriate amounts to permit growth of survivors into single colonies.

Colonies are isolated by the conventional ring method, and the cells are transferred into 35-mm petri dishes. As soon as a sufficient number of cells are obtained, they are trypsinized and plated at 33° and 39°C at a concentration of approximately 5×10^4 cells per 60-mm petri dish. Within a few days, it can be determined whether the cells multiply normally at 39°C. Cells which exhibit fairly normal growth at 33°C and no growth, great delay in growth, or gross morphological alterations at 39°C are propagated further and examined more precisely by plating various cell dilutions at the permissive and nonpermissive temperature. Those which exhibit an efficiency of plating (EOP) ratio 39°/33°C of at most 10^{-3} can be considered *ts* mutants and propagated further. It is advisable to freeze these potentially useful *ts* mutants as soon as possible, to avoid the risk of losing the cells through contaminations or other accidents. These cells should always be recloned at least once before further use, as often the first clonal isolation may have collected also a small number of wt cells. The assay of this mixed population may falsely indicate a mutant with a high reversion frequency.

VI. Frequency and General Behavior of the *ts* BHK Mutants

The fraction of the original cells surviving three FUdR treatments was on the order of 10^{-5}. Upon isolation and testing, about 90% of those displayed *ts* behavior; thus, the overall yield of *ts* mutants was about 9×10^{-6}. However, the mutants isolated fell into five categories, best understood from their efficiency of plating at the nonpermissive temperature (Table II). Mutants of the first type plate normally at the permissive temperature, whereas the yield of colonies at 39°C is 10^{-6} or less, irrespective of the size of the inoculum. Thus, they have low reversion and low leakiness. Mutants of the second type seem to have a high frequency of back-mutants so that their EOP 39°/33° is on the order of 10^{-2} to 10^{-3}. This behavior reflects a high reversion rate, since even after recloning these cells quickly accumulate a high proportion of wt-like cells. They are, therefore, practically useless. Mutants of the third class seem to have some kind of density dependent

TABLE II

EFFICIENCY OF PLATING OF THREE REPRESENTATIVE *ts* BHK MUTANTS AT 33°C AND 39°C[a,b]

	No. of cells seeded						
	5×10^4		5×10^3		5×10^2		
Cells	39°C	33°C	39°C	33°C	39°C	33°C	Ratio 39°/33°C
ts 422E	0	Confluent	0	$>10^3$	0	225	$<1.5 \times 10^{-5}$[c]
ts 13C	150	Confluent	12	$>10^3$	1	220	6×10^{-3}
ts 55X-1	Confluent	Confluent	30	$>10^3$	0	200	—
wt BHK	Confluent	Confluent	$>10^3$	$>10^3$	230	250	~1

[a] From Meiss and Basilico (1972).
[b] Value are expressed as number of colonies per plate. Average of three plates.
[c] Based on the total number of 39°C plates at 5×10^4.

leakiness. They will form no colonies when plated at 39°C at low density, but when plated at densities of 5×10^4 or higher, all the cells will grow, and the cultures will become confluent almost as fast as wt cells. This behavior could be due to a leaky mutation, which is somewhat counteracted by cross-feeding effects. We have not studied these at all, but probably they would not be very useful for studies involving large quantities of cells.

The fourth and fifth categories include cells that do not grow at 39°C, but have also drastically reduced growth rate at 33°C; and cells which have faster growth at 33°C, but grow, albeit slowly, at 39°C. These classes were discarded.

The frequency at which mutants of the first class were obtained was only about 1% of the FUdR survivors. Their overall yield was then $\sim 10^{-7}$, except in the case of the *ts* mutants derived from the BHK T6a (HGPRT$^-$), which was about 10-fold higher. All the studies carried out have been with such mutants, and, although their phenotypes vary, they have some common characteristics: (1) Reversion rate is low, and extended propagations of the cells at 33°C does not result in a substantial accumulation of revertants. This property, however, has to be determined case by case, since if revertants had a considerable selective advantage at the permissive temperature, they would tend to accumulate even if the reversion rate was low. As a general rule, we keep these cells in cultures not longer than 2–3 months. (2) Recloning the cell population yields cells which have essentially the same characteristics of the parental populations. Thus, they are genetically stable and homogeneous. (3) Spontaneous reversion to wt phenotype of all mutants tested can be enhanced 5- to 10-fold by treatment with NG (Meiss and Basilico, 1972).

VII. Complementation Tests and Methods to Test Dominance or Recessiveness of the *ts* Mutations

The methods for genetic analysis of somatic mammalian cells are still limited, since it is not possible to conduct recombination experiments. However, through the technique of somatic cell hybridization, it is possible to attempt complementation studies. Most *ts* mutations are recessive in viruses, bacteria, fungi, and *Drosophila* (Epstein *et al.*, 1963; Hartwell, 1967; Suzuki, 1970; Suzuki and Procunier, 1969). In the case of *ts* mutants of cultured mammalian somatic cells, we are dealing with diploid organisms, and it is reasonable to think that most *ts* mutations obtained may be of the dominant type. This fact would make it difficult to devise a complementation test. The expression of recessive mutations should be rather unlikely, unless: (1) they represent mutations in genes which are normally present in a haploid state (e.g., in the X chromosome, which is always structurally or functionally monosomic) (see review by Lyon, 1972); (2) they represent mutations in genes that, because of deletions of the homologous locus or chromosome, became haploid in the cell in question; (3) they represent heterozygous mutations, that through processes of somatic crossing over, have become homozygous in the mutant cells; (4) a fourth possibility is that somatic animal cells already possess a high degree of heterozygosis, and therefore only a change of the dominant allele should be necessary to change the cells' phenotype.

Therefore, it becomes important to test whether the *ts* mutations obtained are genetically dominant or recessive. The technique of somatic cell hybridization (Ephrussi and Sorieul, 1962; Littlefield, 1966) provides the means of answering this question, and also can be used for testing whether or not mutations affect the same functions. Hybrid cells contain in one nucleus the chromosomes of both parents.

In brief, if two mutants of animal cells are capable of giving rise to a normal frequency of hybrid cells, and these hybrids are capable of sustained multiplication under conditions in which the parental cells are not, it can be deduced that: (a) the two mutations are recessive, as otherwise each mutation would inhibit growth under nonpermissive conditions in the hybrid cell; (b) the two mutants are able to complement, as the gene-product affected in mutant A is provided by mutant B, and the gene-produce affected in mutant B is provided by mutant A.

If the two mutants do not give rise to viable hybrid cells, then it is possible that either they are dominant or they do not complement because both mutants carry the same type of mutation. In the former case, these mutants should not form hybrids with any other mutant, and in the latter they should be perfectly able to give rise to viable somatic hybrids when crossed with mutants of a different genotype.

However, it should be mentioned that even in the most favorable conditions, the frequency of somatic hybrid cells one can obtain is of 10^{-3} to 10^{-4}. This fact makes it necessary to use some system to select for the hybrids, as otherwise they could not be detected among the cell population.

The technique we used for experiments of this type is the following: the pairs of *ts* mutants under test are taken off the plate by trypsinization, and 1×10^6 cells of each are mixed in a glass tube and centrifuged 10 minutes at 1000 rpm at room temperature in an International PRJ centrifuge. The resulting pellet is resuspended in 1 ml of isotonic Tris buffer containing Ca and Mg, pH 7.3. One milliliter of β-propiolactone inactivated Sendai virus (Harris and Watkins, 1965) ($\sim$6000 HA units) is added and the mixture is incubated in ice for 10 minutes, followed by a 30-minute incubation at 37°C. Then the cell mixture is plated at 1.5×10^6 cells per 60-mm petri dish, so that spontaneous as well as Sendai virus-induced fusion can occur. In each experiment, the parent mutant cell lines are also plated individually at equivalent cell densities under the same conditions.

After 2–3 days of cocultivation at 33°C (even without Sendai we have seen that 2 days suffice to obtain an optimum yield of hybrids), the cells are transferred to the nonpermissive temperature of 39°C. The cells are then plated out at 39°C at a density of 1 to 3×10^5 cells per 100-mm petri dish. The plates are then kept at 39°C for 12–14 days, after which time most of the *ts* cells have detached, and either they are stained for colony counting or the colonies are isolated for further analysis. Representative results of such type of experiments are given in Table III. It can be seen that most of the mutants crossed seemed to be able to give rise to viable hybrid cells, capable of growth at 39°C. Crosses of mutants BCH $\times$ BCL, and T23 $\times$ T60 did not produce colonies at 39°C above the frequency in the control population, but each of these mutants was perfectly able to complement with other *ts* mutants. It should also be mentioned that when 20–30 putative hybrid cells were clonally isolated, they were found to be capable of continuous multiplication at 39°C, and their chromosome complement was found to be always in the tetraploid range, which was consistent with their hybrid nature.

The simplest interpretation of these results if the following: (1) These *ts* mutations are recessive, as they can give rise to hybrid cells capable of multiplying at 39°C when crossed with another mutant. (2) Most mutations are not identical, since they complement, with the exception of mutants BCH, BCL, or T23 and T60. This was not surprising because each pair of these noncomplementing mutants appeared to display the same *ts* phenotype. The individual mutants of each pair are likely to be siblings, since they were isolated in the same selection experiment.

However, it is possible to think that the *ts* mutations are dominant, and that the hybrids we obtained are cells that have lost the chromosomes bearing the dominant gene. Chromosome loss is not infrequent in hybrid

TABLE III
HYBRIDIZATION AMONG *ts* MUTANTS OF BHK CELLS

Mutants		Frequency of colonies at 39°C			Complementation[c]
a	b	a[a]	b[a]	a × b[b]	
422E	BCH	$<1.0 \times 10^{-6}$	1.5×10^{-5}	3.7×10^{-4}	+
422E	T23	$<1.0 \times 10^{-6}$	$<1.0 \times 10^{-6}$	$<2.5 \times 10^{-4}$	+
422E	BCL	$<1.0 \times 10^{-6}$	$<1.0 \times 10^{-6}$	3.6×10^{-4}	+
422E	AF8	$<1.0 \times 10^{-6}$	$<1.0 \times 10^{-6}$	1.0×10^{-4}	+
422E	AF6	$<1.0 \times 10^{-6}$	1.0×10^{-5}	1.2×10^{-4}	+
BCH	BCL	1.5×10^{-5}	$<1.0 \times 10^{-6}$	6.0×10^{-6}	−
BCH	AF8	1.5×10^{-5}	$<1.0 \times 10^{-6}$	1.4×10^{-4}	+
AF8	AF6	$<1.0 \times 10^{-6}$	1.0×10^{-5}	1.6×10^{-4}	+
T23	T60	$<1.0 \times 10^{-6}$	$<1.0 \times 10^{-6}$	$<1.0 \times 10^{-6}$	−
T23	T15	$<1.0 \times 10^{-6}$	1.0×10^{-5}	2.2×10^{-4}	+
T60	T15	$<1.0 \times 10^{-6}$	$<1.0 \times 10^{-6}$	0.9×10^{-4}	+

[a] Mutants a or b plated individually at 39°C; values were not affected by Sendai treatment.
[b] Mixed (1:1) culture of mutants a and b plated at 39°C after Sendai assisted fusion.
[c] Based on the ability to form viable hybrid cells capable of growth at the nonpermissive temperature.

cells (Matsuya *et al.*, 1968; Migeon, 1968; Siniscalco, 1970). Since we detect our hybrids by their ability to grow at 39°C, it could be thought that, out of all the possible hybrid cells, we would select for the ones that have lost those specific chromosomes. The main argument against this conclusion is that the frequency of hybridization should have been considerably lower than that obtained using the same technique in known complementing crosses. This was not the case, as the frequency of hybrids we obtained crossing HGPRT$^-$ × thymidine kinase-deficient (TK$^-$) BHK was very similar to that obtained with our *ts* mutants. However, a more rigorous test of the chromosome loss hypothesis is to select for hybrids at 33°C, in the absence of selective pressure for ability to grow at 39°C, and then test the ability of these hybrids to multiply at high temperature. The means to do such an experiment was provided by the fact that we were able to isolate *ts* mutants from BHK T6a, an HGPRT$^-$ derivative of the BHK line. With these cells is was possible to perform a cross with cells wild-type for the *ts* mutations (*ts*$^+$), and isolate the hybrids at the permissive temperature. If the *ts* mutation carried by these cells was dominant, it would have been able to inhibit also the wild-type genome when tested at 39°.

The experiment was carried out as follows: *ts* HGPRT$^-$ BHK were crossed: with *ts*$^+$ TK$^-$ BHK according to the technique described before. After 2 days of cocultivation at 33°C, the cells were trypsinized and plated in medium containing aminopterin 10^{-5} *M*, thymidine 4×10^{-5} *M*, and hypoxanthine 10^{-4} *M* (HAT) (Littlefield, 1965). Aminopterin blocks folic acid reductase, inhibiting endogenous formation of dTMP, and IMP and GMP. Normal cells can overcome this block if supplied with exogenous hypoxanthine and thymidine, which can bypass the block; but TK^- and $HGPRT^-$ cells cannot, because of their inability to utilize one of the exogenous metabolites. By virtue of complementation, the hybrid cells can grow in such a medium while the parent cells are killed. With such a medium, it is possible to select for hybrid cells containing the TK and HGPRT genes; if the plating is done at 33°C, there is no selective pressure for loss of *ts* chromosomes.

The fused cell population was plated in HAT medium at 33° and 39°C, and the frequency of colonies obtained at the two temperatures was determined. In addition, colonies growing at 33°C were shifted to 39°C to detect whether growth continued; hybrid colonies were isolated from plates always kept at 33°C, and their ability to grow at 39°C was tested directly. The results are given in Table IV. It can be seen that: (a) the frequency of colonies arising at 39°C or 33°C was very similar, and most or all of the hybrid cells growing at 33°C continued dividing after shift to 39°C. This implies that the hybrids which were TK^+ and $HGPRT^+$ were also ts^+; (b) hybrid colonies isolated at 33°C were all capable of growing at 39°C, showing that, under conditions that could not have selected for loss of the *ts* genes, their presence did not inhibit the growth of the hybrid cells at 39°C, when in the presence of ts^+ alleles of the same gene.

These experiments provided further evidence that most or all of the *ts* mutations were recessive, at the same time demonstrating the validity of the

TABLE IV

HYBRIDIZATION BETWEEN *ts*, $HGPRT^-$ AND ts^+, TK^- BHK CELLS[a]

ts, $HGPRT^-$ parent	Number of colonies[b] at 39°C in HAT medium	Number of colonies[b] at 33°C in HAT medium	Percent colonies which continued growth upon shift from 33°C to 39°C	Hybrid clones isolated at 33°C and capable of growth at 39°C
T15	416	355	>90	4/4
T23	195	272	>90	3/3
T22	220	280	>90	5/5

[a] Each of the *ts* mutants was fused with thymidine kinase-deficient BHK cells (B1, Littlefield and Basilico, 1966) as described in the text. After 2 days' cultivation at 33°C the cells were plated at 2×10^5/100-mm petri dish at the desired temperature in medium containing HAT (10^{-5} *M* aminopterin, 4×10^{-5} *M* thymidine, 10^{-4} *M* hypoxanthine, 10^{-5} *M* glycine).

[b] Out of 8×10^5 cells plated.

[c] Determined 4–5 days after shift.

complementation test described before. Therefore, one of the hypotheses mentioned before to explain why one can isolate recessive mutants from diploid cell lines must be considered. It is possible that we detect only *ts* mutations whose genes, because of a preexisting or subsequent event, exist in a haploid state, or that all the *ts* mutations we found could be located in genes present on the X chromosome. With the appropriate hybridization experiments, this hypothesis can be tested.

VIII. Nature of the *ts* Mutants

Thus far we have isolated about 30 *ts* mutants using NG as a mutagen and FUdR as the selective killing agent. Recently, we have had success with EMS treatment followed by selection with BUdR and "light" or tritiated leucine. However, these *ts* mutants have not as yet been examined.

Most of the *ts* mutants appear to have originated from alterations in different genetic loci, as evidenced by complementation data (Table V). They display a wide range of morphologies after incubation at 39°C (Fig. 3). Likewise, survival at 39° (as judged by ability to form colonies at 33°C) varies from 1 to 5 days (Meiss and Basilico, 1972). Of the *ts* mutants screened for their ability to incorporate radioactive precursors of DNA, RNA, and protein at 39°C, most do not show a specific inhibition of any one macromolecular synthesis. *ts* mutant AF8 is an exception in that after 12 hours at 39°C, the uptake of thymidine-^{3}H during a short pulse into acid-precipitable material is down to 5% of that observed in the wild-type BHK, while uptake of the other precursors is not drastically affected (or only much later). So far, we have been able to characterize the *ts* defect of one mutant, *ts* 422E (Toniolo *et al.*, 1973). Upon shift to 39°C, these cells, their nuclei and nucleoli become progressively larger (Fig. 3b). The marked enlargement of the nucleoli was suggestive of a defect in rRNA processing which led to an

TABLE V

COMPLEMENTATION GROUPS OF *ts* BHK MUTANTS[a]

1	2	3	4	5	6	7	8	9	10
BCH	AF8	422E	AF6	T22	T15	T23	TD	T14	T2
BCL						T60			
BCB									

[a]This table was constructed on the basis of hybridization experiments such as the ones illustrated in Table III.

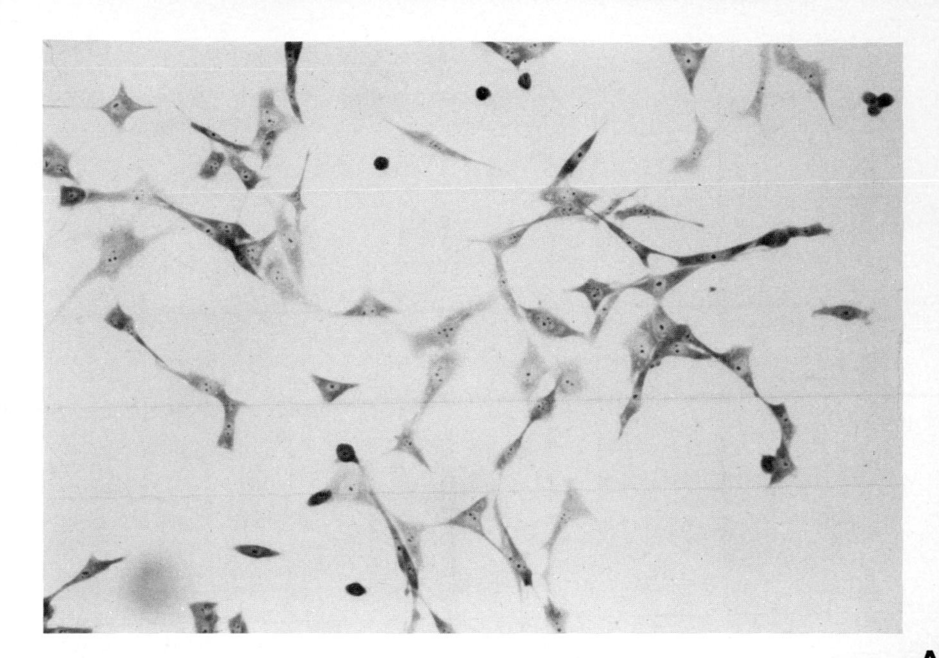

A

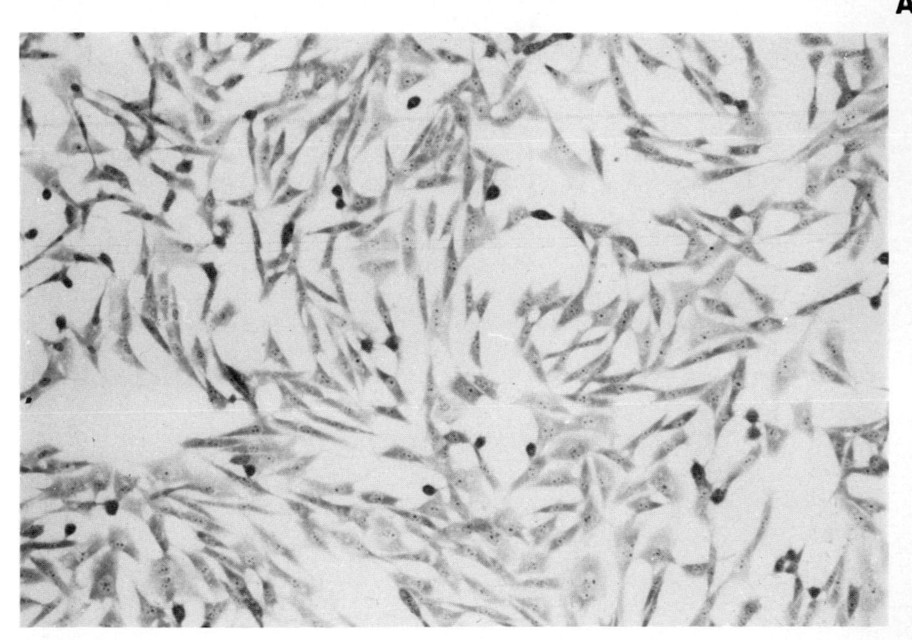

FIG 3. Microphotographs of two *ts* BHK mutants growing at 33°C (bottom) and at 39°C (top) 4–5 days after shiftup, this page, *ts* AF8 cells. On p. 18, *ts* 422E cells. Cells were fixed in 1% glutaraldehyde and stained with Giemsa.

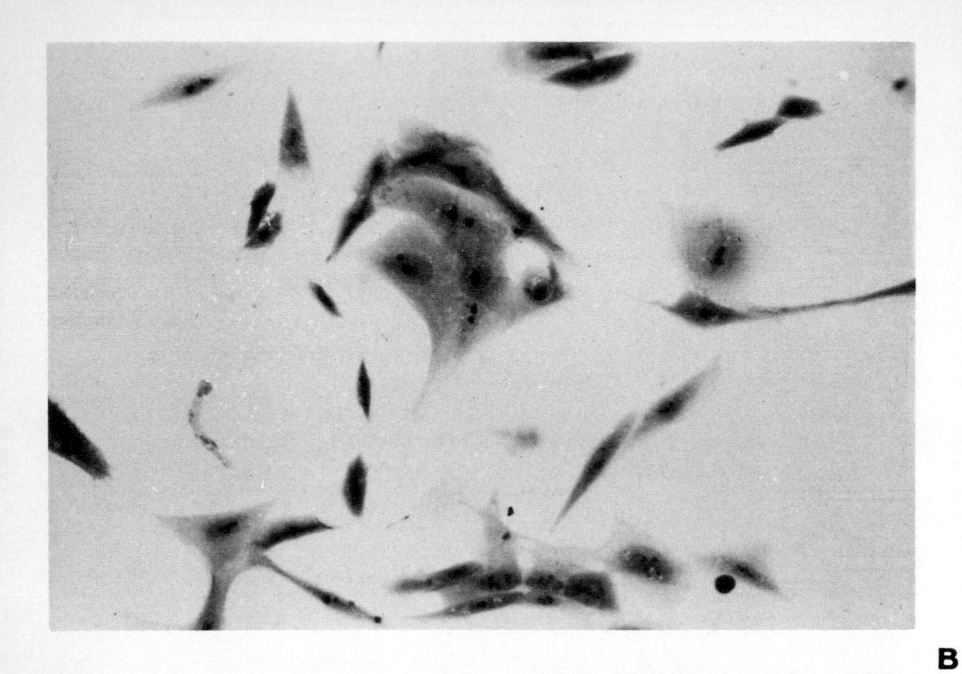

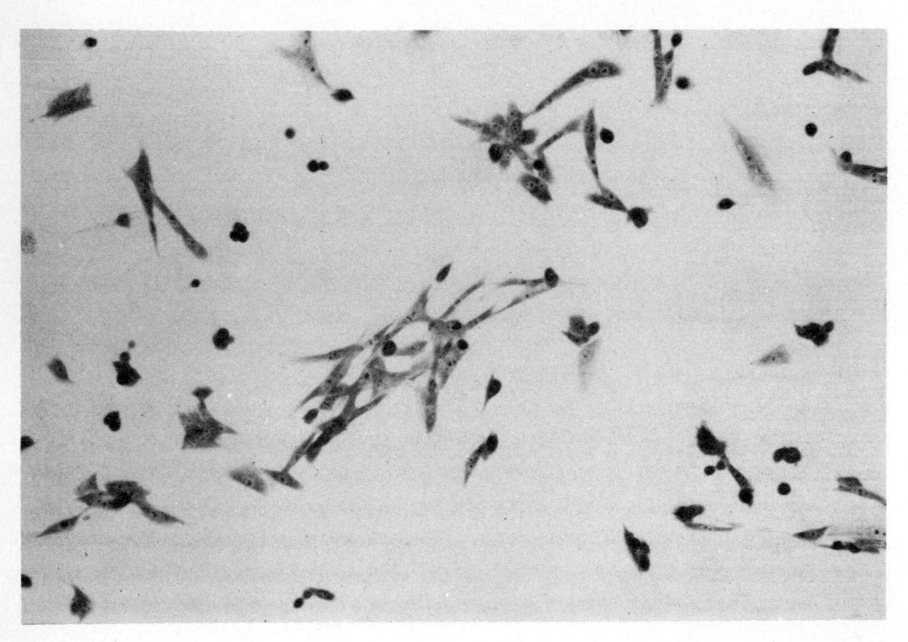

FIG. 3B. For legend see p. 17.

accumulation of rRNA precursors. Thus, we examined the production of 28 S and 18 S rRNA at 39°C and 33°C in *ts* mutant 422E and wt BHK. At 33°C, the mutant behaves like wt cells, while at 39°C, the appearance of 28 S RNA in the cytoplasm is greatly reduced, while production of 18 S RNA and 4–5 S RNA appears to be normal. Further studies revealed that the synthesis of the large 45 S RNA precursor is normal at 39°C, but that processing appears to be arrested after the formation of 32 S rRNA, which can be shown to accumulate to some extent in the nuclear fraction of *ts* 422E.

These data and other data not described, point very strongly to the conclusion that *ts* 422E cannot grow at 39°C because of a defect in rRNA processing. The initial characterization of this mutant points clearly to the potential usefulness of *ts* mutants of somatic animal cells.

IX. Conclusions

From the data discussed above, a number of conclusions can be reached. Most of them suggest that, while the techniques we used were successful in producing a large number of *ts* mutants, there will have to be additional methodology and improvements in the selection procedure if this technique is to be brought to the level of sophistication which has been reached for bacteria.

1. The main thrust of the results we have presented is to demonstrate that *ts* mutants of mammalian cells can be obtained at a reasonable frequency, and that the mutants behave in a reproducible manner. Their reversion frequency is low, but can be increased by mutagens, suggesting that they derive from simple base substitutions (Meiss and Basilico, 1972). In addition, the mutations appear to be recessive, which supports the idea that they derive from alteration of proteins, which are functional at low, but not at high, temperature. Thus, these mutants should be useful for studies of somatic cell genetics and physiology, as postulated.

While this all sounds very positive, we want to caution the reader about the fact that, at present, the isolation of *ts* mutants of mammalian cells is still a large-scale undertaking. The procedure is slow, the frequency of mutants one can obtain is not high, and stringent selective techniques are not available. Sporadic attempts at the isolation of *ts* mutants are probably not going to be very useful.

2. There is still no demonstration that treatment with mutagens is effec-

tive in increasing the yield of *ts* mutants. Although the fact that reversion is increased by mutagens (Meiss and Basilico, 1972) supports this hypothesis, this point will have to be demonstrated conclusively.

3. The selection procedure should be improved. Although our method seemed to be satisfactory for a general selection, the hope that it might have selected for DNA synthesis mutants was not fulfilled by the experimental results. The selection techniques should be rendered more efficient, and mainly more specific. Without such help, the characterization of the mutants is going to be very difficult in most cases.

In addition, the potential mutagenic effect of some of the selective systems will have to be clarified. BUdR and "light," for example, have been long known to be mutagenic in bacteria and viruses (Freese, 1959), and recently have been shown to produce a number of auxotrophic mutations in animal cells (Chu *et al.*, 1972). Even our selection procedure could be mutagenic: in fact, thymine starvation, whether achieved by withdrawing thymine from a thymine-requiring strain or by the use of FUdR is mutagenic in bacteria (Coughlin and Adelberg, 1956; Holmes and Eisenstark, 1968; Drake and Greening, 1970). While in our system FUdR undoubtedly acts as a selective agent (see reconstruction experiments), it cannot be excluded that it might exert also a mutagenic action. However, the high frequency of mutants obtained with T6a cells makes it unlikely that FUdR plays a predominant role in causing the mutations that we isolate. To use a selective treatment which is also mutagenic, while not detrimental for isolating nonspecific *ts* mutants, could make the chances of isolating mutations which affect specific steps of the cell metabolism quite low.

4. The demonstration that our *ts* mutations are recessive, possibly originating from simple base substitutions in genes that are present in a haploid state, suggests that methods capable of provoking monosomies or chromosomal deletions should be of great advantage in increasing the yield of *ts* mutants. The availability of such methods could be of great help, since the yield of *ts* mutants could be probably increased considerably. We would not be forced to rely on the spontaneous occurrence or preexistence of monosomies, which may be very rare, or worse, may be confined only to specific chromosomes in the cell lines used.

Acknowledgments

We are indebted to our colleagues M. Rabinovich, S. Burstin, and H. Renger, for helpful criticism of this manuscript. Most of the work described in this article was supported by PHS contract NCI-E-71-2183. C.B. is a scholar of the Leukemia Society.

REFERENCES

Adelberg, E. A., Mandel, M., and Chen, G. C. C. (1965). *Biochem. Biophys. Res. Commun.* **18**, 788–795.

Basilico, C., Matsuya, Y., and Green, H. (1970). *Virology* **41**, 295–305.

Burge, B. W., and Pfefferkorn, E. R. (1966). *Virology* **30**, 204–214.

Cerda-Olmedo, E., Hanawalt, P. C., and Guerola, N. (1968). *J. Mol. Biol.* **33**, 705–719.

Chu, E. H. Y. (1971). *In* "Chemical Mutagens: Principles and Methods for their Detection" (A. Hollaender, ed.), Vol. 2, pp. 411–444. Plenum, New York.

Chu, E. H. Y., and Malling, H. V. (1968). *Proc. Nat. Acad. Sci. U.S.* **61**, 1306–1312.

Chu, E. H. Y., Sun, N. C., and Chang, C. C. (1972). *Proc. Nat. Acad. Sci. U.S.* **69**, 3459–3463.

Cohen, S. S., Flaks, J. G., Barner, H. D., Loeb, M. R., and Lichtenstein, J. (1958). *Proc. Nat. Acad. Sci. U.S.* **44**, 1004–1012.

Coughlin, C. A., and Adelberg, E. A. (1956). *Nature (London)* **178**, 531–532.

Drake, J. W. (1970). "The Molecular Basis of Mutation." Holden-Day, San Francisco, California.

Drake, J. W., and Greening, E. O. (1970). *Proc. Nat. Acad. Sci. U.S.* **66**, 823–829.

Edgar, R. S., and Lielausis, I. (1964). *Genetics* **49**, 649–662.

Ephrussi, B., and Sorieul, S. (1962). *Univ. Mich. Med. Bull.* **28**, 347–363.

Epstein, R. H., Bolle, A., Steinberg, C. M., Kellenberger, E., Boy de la Tour, E., Chevalley, R., Edgar, R. S., Susman, M., Denhardt, G. H., and Lielausis, I. (1963). *Cold Spring Harbor Symp. Quant. Biol.* **28**, 375–394.

Freese, E. (1959). *Proc. Nat. Acad. Sci. U.S.* **45**, 622–633.

Graham, F. L., and Whitmore, G. F. (1970). *Cancer Res.* **30**, 2627–2635.

Harris, H., and Watkins, J. F. (1965). *Nature* (*London*) **205**, 640–646.

Hartmann, K. V., and Heidelberger, C. (1961). *J. Biol. Chem.* **236**, 3006–3013.

Hartwell, L. H. (1967). *J. Bacteriol.* **93**, 1662–1670.

Hayes, W. (1970). "The Genetics of Bacteria and their Viruses." Wiley, New York.

Heidelberger, C. (1965). *Progr. Nucl. Acid. Res. Mol. Biol.* **4**, 1–50.

Holmes, A. J., and Eisenstark, A. (1968). *Mutat. Res.* **5**, 15–21.

Kao, F. T., and Puck, T. T. (1968). *Proc. Nat. Acad. Sci. U.S.* **60**, 1275–1281.

Kao, F. T., and Puck, T. T. (1969). *J. Cell. Physiol.* **74**, 245–258.

Littlefield, J. W. (1965). *Biochim. Biophys. Acta* **95**, 14–22.

Littlefield, J. W. (1966). *Exp. Cell Res.* **41**, 190–196.

Littlefield, J. W., and Basilico, C. (1966). *Nature* (*London*) **211**, 250–252.

Lubin, M. J. (1962). *J. Bacteriol.* **83**, 696–697.

Lyon, M. F. (1972). *Biol. Rev. Cambridge Phil. Soc.* **47**, 1–35.

Marin, G., and Littlefield, J. W. (1968). *J. Virol.* **2**, 69–77.

Martin, G. S. (1970). *Nature* (*London*) **227**, 1021–1023.

Matsuya, Y., Green, H., and Basilico, C. (1968). *Nature* (*London*) **220**, 1199–1202.

Meiss, H. K., and Basilico, C. (1972). *Nature* (*London*), *New Biol.* **239**, 66–68.

Migeon, B. (1968). *Biochem. Genet.* **1**, 305–322.

Naha, P. M. (1969). *Nature* (*London*) **223**, 1380–1381.

Orkin, S. H., and Littlefield, J. W. (1971). *Exp. Cell Res.* **66**, 69–74.

Person, S. (1963). *Biophys. J.* **3**, 183–187.

Pollack, R. E., Green, H., and Todaro, G. J. (1968). *Proc. Nat. Acad. Sci. U.S.* **69**, 3459–3463.

Puck, T. T., and Kao, F. (1967). *Proc. Nat. Acad. Sci. U.S.* **58**, 1227–1234.

Roscoe, D. H., Read, M., and Robinson, H. (1973). *J. Cell. Physiol.* (in press).

Sato, K., Slesinski, R. S., and Littlefield, J. W. (1972). *Proc. Nat. Acad. Sci. U.S.* **69**, 1244–1248.
Siniscalco, M. (1970). *Excerpta Med. Found. Int. Congr. Ser.* No. 204, pp. 72–83.
Stoker, M., and Macpherson, I. (1964). *Nature (London)* **203**, 1355–1357.
Suzuki, D. T. (1970). *Science* **170**, 695–706.
Suzuki, D. T., and Procunier, D. (1969). *Proc. Nat. Acad. Sci. U.S.* **62**, 369–376.
Thompson, L. H., Mankovitz, R., Baker, R. M., Till, J. E., Siminovitch, L., and Whitmore, G. F. (1970). *Proc. Nat. Acad. Sci. U.S.* **66**, 377–384.
Tocchini-Valentini, G. P., Felicetti, L., and Rinaldi, G. M. (1969). *Cold Spring Harbor Symp. Quant. Biol.* **34**, 463–468.
Todaro, G. J., and Green, H. (1963). *J. Cell Biol.* **17**, 299–313.
Toniolo, D., Meiss, H. K., and Basilico, C. (1973). *Proc. Nat. Acad. Sci. U.S.* **70**, 1273–1277.
Yamamoto, M., Endo. H., and Kuwano, M. (1972). *J. Mol. Biol.* **69**, 387–396.

Chapter 2

Induction and Isolation of Auxotrophic Mutants in Mammalian Cells[1]

FA-TEN KAO AND THEODORE T. PUCK[2]

Department of Biophysics and Genetics,
University of Colorado Medical Center,
Denver, Colorado

I. Introduction

In vitro genetic analysis of somatic mammalian cells requires large numbers of stable, well-characterized mutants (Puck, 1972). In contrast to microbial systems where mutants of many kinds are abundant, the avail-

[1]This investigation is a contribution from the Rosenhaus Laboratory of the Eleanor Roosevelt Institute for Cancer Research and the Department of Biophysics and Genetics (Number 538), and was aided by an American Cancer Society Grant No. VC-81C, and by a grant from the National Science Foundation, No. GB-38664.

[2]American Cancer Society Research Professor.

ability of good genetic markers in mammalian cells has been severely limited. This has, in large part, been due to the lack of effective selection methods comparable to those used in microbial genetics, for example, the penicillin method in bacteria (Davis, 1948; Lederberg and Zinder, 1948).

Using approaches similar in principle to the penicillin method, our laboratory has developed a mutant isolation procedure which selectively kills normal cells but allows mutant forms to survive (Puck and Kao, 1967; Kao and Puck, 1968). The procedure, shown schematically in Fig. 1, involves differential incorporation of 5-bromodeoxyuridine (BUdR) into normal but not mutant cells, followed by illumination with near-visible light which destroys only those cells that contain BUdR-DNA. The mutant cells survive the treatment and are isolated.

Auxotrophic mutants have been extremely useful in microbial genetic analysis. They offer possibilities for many kinds of experiments in which specifically selective media can be used to isolate desired kinds of clones. Many of these mutants have been analyzed biochemically and shown to involve a single gene mutation resulting in absence or malfunction of a specific enzyme (Wagner and Mitchell, 1964). This feature is of particular importance in mammalian cells since various drug-resistant animal cell

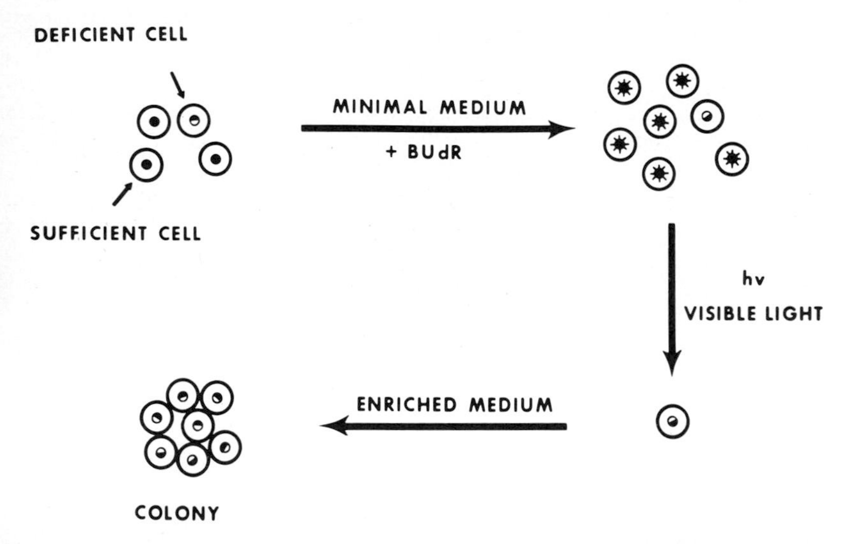

FIG. 1. Schematic representation of the 5-bromodeoxyuridine plus visible light procedure for isolation of nutritionally deficient (auxotrophic) cells. The cells are treated with a mutagen and then exposed to BUdR in a minimal medium in which only the normal, nutritionally sufficient (prototrophic) cells can grow. These alone incorporate BUdR into their DNA and are killed on subsequent exposure to near-visible light. The medium is then changed to a composition lacking BUdR and enriched with various nutrilites, and the deficient cells grow up into colonies.

mutants have been shown to be complex (Harris, 1971; Cass, 1972; Mezger-Freed, 1971, 1972; Morrow *et al.*, 1973).

The BUdR–near-visible light method here described can be used to isolate auxotrophic and other kinds of mutants whose growth and DNA synthesis can be arrested by a set of conditions which does not affect that of the normal cells.

II. Procedures

The procedures described here have been designed for CHO-Kl cells, which have a generation time of 12 hours (Kao and Puck, 1968). This cell culture can be obtained from The American Type Culture Collection Repository, Rockville, Maryland. When other cells are used, appropriate modifications of the procedure will have to be determined. For example, cells with longer generation time may require longer incubation time in the presence of BUdR to ensure maximum incorporation. For each new cell line used, it is important to conduct preliminary experiments to determine the degree of killing of the wild type that can be achieved by the BUdR + visible light procedure under varying conditions. In each case conditions must be arranged to maximize the differential killing of the wild type, but not of the desired mutant cells.

A. Standard Cultivation, and Single-Cell Growth

The standard procedures and media described previously (Ham and Puck, 1962; Ham, 1965) are employed. Important precautionary measures that should be observed are a doubling in the riboflavin concentration of the F12 medium and protection of growth media from excessive exposure to light.

B. Mutagenic Treatment

Treatment with the standard mutagen ethyl methanesulfonate (EMS, Eastman Organic Chemicals) is described as an example. Other mutagens can be employed in a similar manner except that the dose-survival curves for each mutagen must be determined (Kao and Puck, 1969, 1971) in order to select appropriate concentrations (e.g., 10–20% survival level) for mutagenesis experiments.

Day 1 AM Harvest 10^7 actively growing cells and inoculate into four 100-mm Falcon plates at a density of 2.5×10^6 cells each, in complete F12 medium (Ham, 1965) supplemented with 10% fetal calf serum (F12FC10). Incubate for 6 hours.

PM Add EMS to each plate at a final concentration of 400 μg/ml. Incubate for 16 hours, after which about 20% of the cells will have survived.

Day 2 AM Remove the medium from each plate; wash the cells three times with Saline G (see Table II for composition); add fresh F12FC10, and incubate for 5 days to allow mutant expression and cell recovery from mutagenesis treatment. During this period the cells are subcultured when they become confluent. The cells from each plate are divided into two by trypsinization, but the identity of cells from each original mutagenized plate is maintained. Thus it is ensured that mutants arising from different mutagenized plates must represent separate mutagenic events.

C. Starvation, BUdR Incorporation, and Illumination

Day 7 AM The cells have now recovered from the treatment and are undergoing active growth. Trypsinize all duplicate plates and pool cells from each duplicate set into a separate tube containing the deficient medium, F12D (see Section III, E), supplemented with the macromolecular fraction of fetal calf serum, FCM (see Section III, C for its preparation). Count the cells. Inoculate 10^6 cells into 50 plates (60 mm) at a cell density of 2×10^4 cells per plate, a density which ensures minimal cross-feeding among normal and mutant cells in the deficient medium. These plates which contain the deficient medium are incubated for 24 hours in order to achieve starvation of the auxotrophic mutants so that they terminate DNA synthesis.

Day 8 AM Add BUdR (Calbiochem) to each plate at a final concentration of 10^{-5} *M*. Incubation continues for another 24 hours.

Day 9 AM All plates are handled under dim, diffuse light at this step. Remove medium from the plates and add fresh F12 or Saline G to a depth of approximately 2 mm. The plates with their lids in place are placed on the polystyrene shelf of the visible light apparatus and illuminated for 1 hour (see Section III, A). The plates are then removed and complete growth medium, F12 + 10% fetal calf serum, is replaced in the petri dishes. Incubation is resumed for 7 days for colony development from the surviving cells.

After the BUdR and visible light treatment, the great majority of cells die and float up into the medium. It is desirable to change the medium once to remove dead cells in order to provide a more favorable condition for the surviving cells to grow and form colonies.

D. Clone Isolation, Testing, and Auxotrophic Analysis

Day 16 Pick the surviving colonies developed in each plate (see Section III, B for clone isolation technique). With an effective mutagen, about 0–5 colonies should be found per plate. Incubation should continue until each colony contains 200 or more cells. Isolate 100–150 clones, and transfer each to a 35 mm plate containing complete growth medium. These clones are allowed to grow for 3 additional days to obtain enough cells for testing.

Day 19 Each plate containing a clone is trypsinized and 200 cells are plated in each of two 35-mm plates containing F12 and F12D supplemented with 10% FCM, respectively. Incubate for 5 days.

Day 24 Examine microscopically both plates of each clone for cell growth. Discard the clones which grow in both F12 and F12D. Retain those clones which exhibit growth only in F12 but not in F12D. These are mutant clones and are subject to further testing to determine their newly acquired specific growth requirements. Each mutant clone is trypsinized and 200 single cells are inoculated into each of a series of nine 35-mm plates, each containing F12DFCM10 medium to which one of each of the following nine components has been added: glycine, alanine,

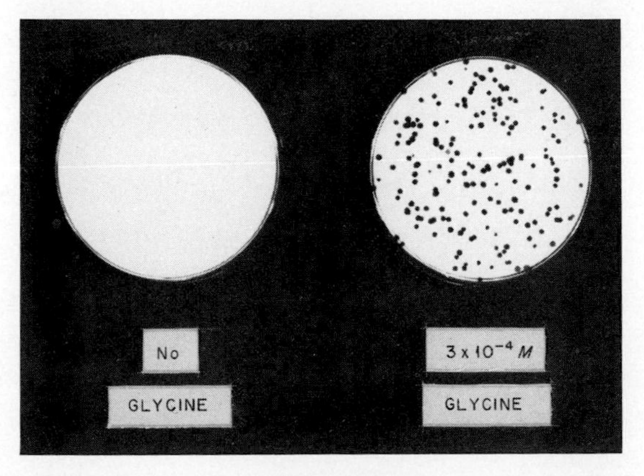

FIG. 2. Demonstration that the glycine-requiring mutant induced by treatment with mutagens yields no colonies in the absence of glycine but 100% plating efficiency in its presence. The original prototroph has 100% plating efficiency in either case.

aspartic acid, glutamic acid, thymidine, hypoxanthine, inositol, vitamin B_{12}, and lipoic acid. These plates are incubated for 4 days.

Day 28 Examine the plates for cell growth in the various deficient media. For example, if a clone can grow only in the plate containing glycine but not in all other plates, this clone is a glycine-requiring mutant (Fig. 2).

III. Special Techniques, Equipment, and Reagents

A. Near-Visible Light Illumination Apparatus

A set of four fluorescent lamps (Westinghouse Cool White, 40 W, 120 cm long) is mounted in a fixture with a parabolic reflector. The lamps are placed 6 cm apart and 16 cm above a transparent polystyrene shelf (25 cm wide, 120 cm long, and 0.5 cm thick). An identical set of four fluorescent lamps is placed under the polystyrene shelf at a distance of 8 cm (Fig. 3), so that

FIG. 3. A fluorescent illumination apparatus for use in the killing of cells containing 5-bromodeoxyuridine (BUdR) in the DNA. The petri dishes, which contain monolayers of cells, are placed on the polystyrene plate and the fluorescent light shines on the cells from both above and below (for details, see text).

the cells are illuminated from both above and below. A maximum of three 60-mm petri dishes containing the cell monolayers per row are placed in the central area of the polystyrene shelf, the edges of which are avoided because of their reduced light intensity. A 30-minute warm-up period for the lamps is provided before the cell exposure in order to achieve reproducible steady-state operating conditions. Excess heat is dissipated by means of a fan directed at the illuminated petri dishes.

B. Clone Isolation Procedure

After the treatment with BUdR, exposure to near-visible light, and subsequent incubation in enriched medium, the colonies which develop from the surviving cells are ready to be picked. The locations of the colonies are first marked with a china-marking pencil on the bottom side of the plate. These marked colonies are examined in an inverted microscope, and those which are sufficiently isolated from other cells are chosen for clone isolation. The medium is removed from the plate, and a stainless steel cloning cylinder (Puck *et al.*, 1956) is placed directly over the center of each colony. The cylinder (6 mm high, 9 mm outer diameter, 6 mm inner diameter) is caused to adhere to the bottom surface of the plate by applying a thin layer of silicone grease to the bottom edge and then firmly pressing it onto the plate so as to enclose the colony. Both the cylinder and the grease have been previously sterilized by autoclaving.

When the cylinder is securely placed on the bottom of the plate, a few drops of trypsin are added to the cylinder. After a few minutes at room temperature, the cells begin to round up. Using a sterile Pasteur pipette with an attached rubber bulb, the cells in the cylinder are released by gentle pipetting and are transferred to a new plate containing F12FC10 medium. An experienced person can easily pick about 60 clones in an hour.

C. Preparation of Macromolecular Fraction of Serum by Sephadex Column

A glass column of 5 cm $\times$ 60 cm is packed with Sephadex G-50, coarse grade, in solution V (Ham and Puck, 1962) diluted to 1:20 with distilled water. The constituents of solution V are listed in Table I.

The serum is detoxified by heating at 60°C for 20 minutes, then tested for its ability to support single cell growth. Any lot of serum which fails to yield growth of CHO-K1 cells with a plating efficiency of 80% and a generation time of 12 hours under these conditions is rejected.

The serum is prefiltered successively through a Millipore filter of pore size 1.2 μm, then with a filter of pore size 0.45 μm, to remove solid materials

TABLE I
COMPOSITION OF SOLUTION V

Chemicals	Concentration (gm/liter)	Preparation
NaCl	148.0	Dissolved in 1 liter (final volume)
KCl	5.7	of triple-distilled water to make
$Na_2HPO_4 \cdot 7H_2O$	5.8	20 × concentrate
KH_2PO_4	1.66	

which are often present in the commercial serum. Fifty milliliters of the pretested and prefiltered serum are added to the top layer of the column, which is then treated continuously with a 1:20 dilution of solution V, as an elution medium. The yellowish band moving downward can readily be seen. When the yellowish front reaches the bottom of the column, a 50-ml sample is collected. Thereafter the column is washed with 200 ml of diluted solution V, and is ready for reuse. This solution of the macromolecular components of serum is used without further fractionation as a standard solution, which is called FCM (Fetal Calf Macro).

After collecting and pooling a large quantity of the FCM, it is tested for its ability to support single-cell growth. Cells are plated in F12 containing several concentrations of the FCM ranging from 2 to 15%. The concentration chosen which usually lies in the neighborhood of 8–10% is a few percent greater than the minimum concentration yielding maximum growth and plating efficiency. The FCM can be prepared by dialysis instead of with Sephadex columns, but we find the latter mode to yield somewhat better and more reliable results.

D. Saline G

This is a balanced salt and glucose solution, containing no $NaHCO_3$, for use outside of the CO_2 incubator (Puck, Cieciura and Robinson, 1958). It is generally employed for washing and other operations with cells. Its composition is listed in Table II.

E. F12D

This is the minimal (or deficient) growth medium and is prepared according to the same procedure used for F12 medium (Ham, 1965), except that the following nine components are omitted: glycine, alanine, aspartic acid, glutamic acid, thymidine, hypoxanthine, inositol, vitamin B_{12}, and lipoic acid. The CHO-Kl cells grow optimally in this medium with the same high plating

TABLE II
COMPOSITION OF SALINE G

Chemicals	Concentration (gm/liter)	Preparation
NaCl	8.0	Dissolved in 1 liter (final volume) of triple-distilled water, sterilized and stored at 4°C
KCl	0.4	
$MgSO_4 \cdot 7H_2O$	0.154	
$CaCl_2 \cdot 2H_2O$	0.016	
$Na_2HPO_4 \cdot 7H_2O$	0.29	
KH_2PO_4	0.150	
Glucose	1.10	
Phenol Red	0.0012	

efficiency as in F12. A few other components, e.g., putrescine, may also be removed from F12, but the resulting medium usually exhibits reduced plating efficiency and growth rate for the CHO-Kl culture.

IV. General Discussion

The procedures described have been adopted to serve in experiments in which the number of cells which have survived mutagenic treatment and which are to be screened for auxotrophs is approximately 10^6. The initial number of cells and the intensity of mutagen treatment must be adjusted accordingly. These particular procedures, in general, will not detect any auxotrophs whose frequency of occurrence is less than 10^{-6}. If the presence of less frequent mutations is to be tested, larger numbers of cells must be screened.

An incubation period of several days has been provided for mutant expression to occur after treatment with the mutagen. This period is also necessary to permit the cells to recover and resume active growth after the mutagenic experience and to ensure that the prototrophic cells are growing vigorously when the BUdR is added. Otherwise they may not incorporate enough BUdR to be killed on the subsequent illumination and such cells may produce colonies which will dilute the number of true auxotrophs and greatly increase the labor required. In the case of treatment with EMS, for example, we find that a 5-day period is generally sufficient for the cells to resume rapid growth after treatment. Cells treated with *N*-methyl-N^1-nitro-*N*-nitrosoguanidine (MNNG) require a somewhat longer time (7–10 days) to recover from the treatment.

The treatment of cells with BUdR and visible light, as described here, reduces the cell survival by a factor of about 10^{-4}. Therefore, approximately 100 cells survive after 10^6 cells are so treated. The surviving clones which have failed to incorporate sufficient BUdR to be subsequently killed on illumination would include: the desired auxotrophic mutants; any thymidine kinase-deficient mutants; cells physiologically or genetically defective in membrane transport, like those described by Breslow and Goldsby (1969); and cells that may have been growing too slowly to incorporate sufficient BUdR to be killed by the subsequent illumination. In typical experiments in which 10^6 CHO-Kl cells are tested after treatment with mutagens like EMS and MNNG, about 10–20 auxotrophic mutants can be isolated by the procedure described.

If the number of surviving clones greatly exceeds that expected on the basis of the foregoing considerations, it is probably due to insufficient opportunity of the cells to recover from the trauma of the mutagenic treatment and incorporate BUdR. Under such circumstances it would be better to discard these clones, and to repeat the procedures with another set of mutagenized cells treated so as better to achieve the actively growing condition between the end of the mutagenesis and the addition of the BUdR.

When other types of cells are used, these considerations should be applied in order to determine the particular conditions required for securing a good yield of mutant clones. It may be necessary to increase the concentration of BUdR or its exposure time, or to augment the BUdR incorporation by the addition of inhibitors like fluorodeoxyuridine which depress endogenous thymidine synthesis.

When a mutant clone has been isolated by virtue of its ability to grow in F12 but not in F12D, its new nutritional requirements must be determined. This can be done by successively adding back to F12D, one at a time, each of the nine components which have been omitted from F12, and testing the mutant cells in each of the resulting solutions. Alternatively, the nine components can be omitted, one at a time, from F12 and the plating efficiency in this series of media determined. The responses of a particular glycine-requiring mutant in these two types of experiments are shown in Table III. It should be remembered, however, that the response of a cell to one added metabolite may be different depending on whether F12D or an enriched version of this mixture is used as the basal medium. A possible example of this behavior has been pointed out (Kao and Puck, 1972a). When such divergencies are found, they indicate interlocking metabolic pathways whose elucidation is always of interest.

The procedure involving exposure of the cells to BUdR plus visible light requires that the induced auxotrophic mutants undergo a 2-day starvation

TABLE III
GROWTH RESPONSES OF A GLYCINE-REQUIRING MUTANT IN VARIOUS DEFICIENT MEDIA[a]

	Growth of a glycine mutant	
Nutrilities which are present in F12 but are omitted in F12D	Individual nutrilites added to the basal medium F12D	Nutrilites omitted individually from the complete F12 medium
None	−	+
Glycine	+	−
Alanine	−	+
Aspartic acid	−	+
Glutamic acid	−	+
Thymidine	−	+
Hypoxanthine	−	+
Inositol	−	+
Vitamin B_{12}	−	+
Lipoic acid	−	+
All the above 9	+	−

[a] + means normal growth; − means no growth.

period, since 1 day is required for incubation in the deficient medium, F12D, and an additional day in F12D plus BUdR. Some of the mutant cells may die during this period of starvation. We have conducted reconstruction experiments in which various mixtures of mutant and wild-type cells have been grown in F12D for 2 days and then subjected to the BUdR-visible light procedure to determine their viability under these conditions. It was found that *pro*$^-$ mutants will survive to the extent of 50% whereas mutants like *gly*$^-$, *ade*$^-$, and *GAT*$^-$ survive only to the extent of 1–10%. In estimating mutation frequencies obtained, therefore, it is necessary to correct for such death due to starvation (Kao and Puck, 1969).

The method here proposed can be used to estimate the frequency of forward mutation induced by various agents. However, such an estimate at this point entails several uncertainties. For example, it is assumed that all the induced mutants will undergo the same number of replications between mutagenesis and application of the BUdR-visible light treatment, and that the loss of mutants by the starvation step is accurately calculated by the correction procedure. On the other hand, the method here proposed has a distinct advantage because frequencies of several different kinds of mutations are simultaneously averaged. This procedure should be more valid in comparing different mutagens by screening procedures, than one which utilizes only a particular kind of mutation.

Apparently some misunderstanding has arisen about the operation of the

BUdR-visible light mutant selection procedure. It is not designed as a mutagenesis procedure, but only to isolate mutants already present. The mutants which are isolated by this procedure are those that cannot synthesize DNA under the conditions of exposure to BUdR in minimal medium. Only the cells destined for death incorporate BUdR, and these will subsequently be killed by exposure to visible light. The auxotrophic mutants which are sought, therefore, never incorporate BUdR and so are unaffected by it mutationally or lethally. Careful control experiments have shown that virtually no mutants are ever recovered when this procedure is applied to normal cells. Cells exposed to standard mutagens, however, yield large numbers of auxotrophic mutants.

The question arises whether auxotrophic mutants obtained by means of the procedures here described are really point mutations in the sense that has come to be accepted in microbial genetics. It is not possible to offer a completely definitive answer to this question in a system like this one, where genetic operations at the molecular level (like mapping of the position in the DNA where the changes have occurred or sequencing of the critical DNA or its resulting protein) have not yet been carried out. Until results of high resolving power are available, a tentative answer to this question can be based on the following considerations: We propose that the term "recessive mutant" be reserved for the present time, for cells which are derived from a given parental form and which obey as many as possible of the following properties: (1) The altered phenotype is derived from its parental form with a very low spontaneous frequency (10^{-6} or less) or as a result of treatment with agents clearly demonstrated to produce true point mutations in other living forms. (2) The rate of spontaneous reversion to the parental form is low (less than or equal to 10^{-6}). (3) When hybridized with a series of other cells, it displays characteristics of recessiveness. (4) When hybridized with cells of the same phenotype, it displays complementation behavior that is consistent with its being a single gene mutant. (5) The pattern of reversion to the parental form produced by different mutagenic agents should be consistent with the hypothesis of a point mutation. (6) Wherever possible, tests for linkage should be carried out and should reveal behavior consistent with a point mutation. (7) Wherever possible, a single, well-defined enzyme deficiency or altered protein structure should be demonstrated in the mutant, but not in the parental cell. (8) Restoration of the specific enzyme or other protein activity should be demonstrable in the revertants. For example, the glyA mutant which has been described has been shown to obey all these criteria (Kao *et al.*, 1969b).

Occasionally auxotrophic mutants isolated by the procedures here described display more than one new nutritional requirement. For example, clones have been described that require the simultaneous addition of glycine, adenine, and thymidine for growth (Kao and Puck, 1968). The simultaneous production of three independent gene mutations would appear to be a most unlikely event. Mutants of this class, which have been studied in our laboratory, have been demonstrated to involve a single mutation. Behavior of this kind can be obtained when the mutation results in loss of an enzyme which produces a key metabolite that is common to several different pathways (Kao and Puck, 1972a). For example, a cell deficient in tetrahydrofolate could exhibit nutritional requirements for glycine, adenine, and thymidine like those which have been found in the mutant we have described.

It will be obvious that these methodologies permit one to isolate auxotrophic mutants with a specific desired growth requirement. For example, if one is interested in obtaining mutants deficient only for adenine, one would add BUdR to the mutagenized cells in a medium containing F12 minus adenine. (Thymidine should also be omitted from this medium to ensure maximum incorporation of BUdR.) Then, after treatment with the visible light, complete F12 would be furnished to the cells. In this manner, only an adenine-deficient mutant would resist incorporation of BUdR in the first phase of the procedure, and would grow out when the adenine was supplied later.

Conversely, one could attempt to find mutants deficient for metabolites not present in F12. In this case one adds a much more enriched medium to the cells after the mutagenic treatment and the BUdR-visible light procedures so that mutants requiring nutrilites not in F12 but in the additionally enriched supplement will also produce colonies. It goes without saying that each metabolite added should first be tested in the wild-type cell to determine the optimal concentration to be used, since it is easy to impair growth of these cells by an excessive concentration of an added metabolite.

The possibility of obtaining new auxotrophic mutants may be difficult because of the diploid nature of mammalian cells. Cox and Puck (1969) have shown how monosomies can be introduced into the cell through the use of carefully regulated treatment with Colcemid. It might also be possible to introduce hemizygous regions by treatment of the cells with X-irradiation or other agents that produce abundant deletions in the genome.

The present mutant isolation procedure has been employed by other laboratories to obtain the following special mutants which appear to be of interest:

1. Temperature-sensitive mutants obtained in the African green monkey kidney cells (Naha, 1969).
2. Isolation of polyoma-transformed BHK cell variants which lack the transformed characteristic of being able to multiply in suspension culture (Wyke, 1971).

3. Auxotrophic mutants in plant somatic cell cultures established from *Nicotiana tabacum* (Carlson, 1970).

4. Hormone-dependent cells isolated from rat mammary tumor culture (Posner *et al.*, 1971).

5. Auxotrophic mutants in *Escherichia coli* obtained with an enrichment factor of 10^5 (Rosner and Yagil, 1970).

6. Additional purine mutants obtained from CHO-Kl cells (Taylor *et al.*, 1971).

7. Isolation of UV-sensitive cells from HeLa S3 (Isomura *et al.*, 1973).

The auxotrophic mutants which have been isolated here and some of their characteristics are summarized in Table IV. All of these except pro$^-$, which is of spontaneous origin, were induced by treatments with a variety of mutagens and isolated as here described. The AT$^-$ mutant was derived from a CHL (Chinese hamster lung culture) and all the rest from the CHO-Kl clone. The *ser*$^-$ mutant was isolated from gly A cells so it possessed two mutant

TABLE IV
AVAILABLE CHINESE HAMSTER CELL AUXOTROPHIC, NONLEAKY MUTANTS

Mutant designation	Growth requirement	Mutagenic agents to produce each mutant class[a]	Number of mutants analyzed
gly A	Glycine	EMS, MNNG, X-ray	4
gly B	Glycine or folinic acid	EMS, MNNG, UV, NMUT	12
gly C	Glycine	EMS, UV	2
gly D	Glycine	EMS, UV	2
ade A	Hypoxanthine, adenine, inosinic acid, their riboside or ribotide, or AIC	ICR-191	3
ade B	Like *ade A*	ICR-191, NDMA, NMUT	20
ade C	Like *ade A*	Sydnone acetamide	1
ade D	Like *ade A*	EMS	1
ade E	Like *ade A*	ICR-191, MNNG	2
ade F	Like *ade A*, but excluding A/C	ICR-191	2
GAT$^-$	Glycine, adenine, and thymidine	EMS, MNNG, UV, NMUT	2
AT$^-$	Adenine and thymidine	EMS	1
ino$^-$	Inositol	MNNG	1
ser$^-$	Serine	EMS	1
pro$^-$	Proline or δ'-pyrroline-5-carboxylic acid	Spontaneous	1

[a] Abbreviations: AIC: 5-aminoimidazole-4-carboxamide; NMUT: nitrosomethylurethane; NDMA: nitrosodimethylamine; ICR-191: an acridine mustard; UV: ultraviolet light.

markers, *ser⁻gly⁻A* (Jones and Puck, 1973). All these mutants exhibit all-or-none growth response with respect to the critical nutrilites (e.g., Fig. 2). Mutants have been obtained which demonstrate "leaky" auxotrophic dependence, but these have not been included in this list. All the mutants listed in Table IV are highly stable, with a spontaneous reversion rate of approximately 10^{-7} or less.

The metabolic defects of some of the mutants listed in Table IV have already been demonstrated, and work on the identification of others is in process. Thus, the *gly A* mutant is deficient in the serine hydroxymethylase enzyme which normally converts serine to glycine (Kao *et al.*, 1969b). The *pro⁻* mutant lacks the enzyme converting glutamic acid to its γ-semi-

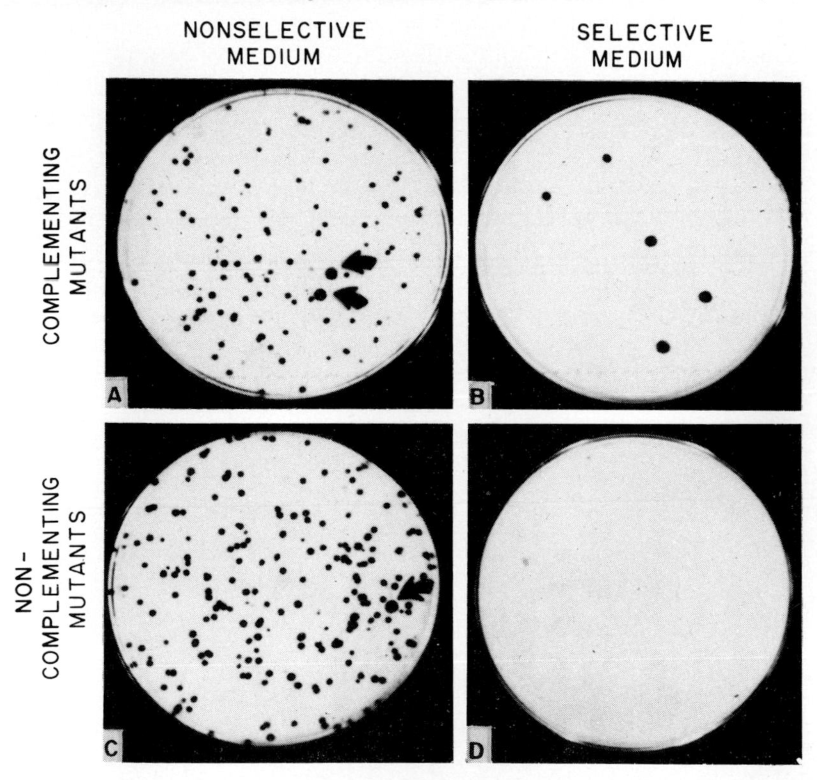

FIG. 4. Plates demonstrating a typical complementation test in various auxotrophic mutants of the same or different phenotypes. After cell fusion, the cell population was inoculated into a series of plates containing either nonselective (A and C) or selective (B and D) medium. Both parental and hybrid cells survived in the nonselective medium regardless of their complementarity (A and C). In the selective medium, however, only hybrid cells from complementing mutants survived (B); the hybrid cells from noncomplementing mutants failed to grow (D). Note the size differences of the colonies developed from the diploid parental cells and the tetraploid hybrid cells (arrows) after a 7-day incubation.

aldehyde in the biosynthetic chain leading to proline (Kao and Puck, 1967). The *ade A* and the *ade B* mutants have been demonstrated to have enzyme defects in the early steps of the *de novo* biosynthesis of inosinic acid, i.e., between 5-phosphorybosyl pyrophosphate and 5-aminoimidazole-4-carboxamide acid ribonucleotide (Kao and Puck, 1972a).

Fifty-five mutants, like those shown in Table IV, have now been analyzed for complementation by conducting pairwise crosses through cell fusion (Kao *et al.*, 1969a; Kao and Puck, 1972a; Fig. 4). These mutants fall into 15 complementation classes, indicating 15 different genes in which mutations have occurred. These mutants have been used for hybridization with human cells and a variety of hybrids have been isolated for use in human linkage and other investigations. The rapid loss of human chromosomes from these hybrids is particularly useful for mapping of human genes (Kao and Puck, 1970; Puck *et al.*, 1971; Jones *et al.*, 1972). The system also offers promise in identifying human regulatory genes and locating them on the human chromosomes (Kao and Puck, 1972b). It is our hope that eventually a sufficient number of Chinese hamster cell auxotrophs can be produced so as to form a series of hybrids which will contain each of the human chromosomes by itself. Growth of these hybrids in a selective medium which depends on the continued presence of the human chromosomes, makes possible cultivation of such clones in the stable state for long periods. If this series of hybrids can be achieved, linkage determination for many human genes will become an extremely simple laboratory operation.

Acknowledgments

It is a pleasure to acknowledge the inclusion of mutants isolated and characterized by Drs. C. Jones and D. Patterson, and of the various contributions to the work described here by J. Brown and J. Hartz.

References

Breslow, R. E., and Goldsby, R. A. (1969). *Exp. Cell Res.* **55**, 339–346.
Carlson, P. S. (1970). *Science* **168**, 487–489.
Cass, A. E. (1972). *J. Cell. Physiol.* **79**, 139–146.
Cox, D. M., and Puck, T. T. (1969). *Cytogenetics* **8**, 158–169.
Davis, B. D. (1948). *J. Amer. Chem. Soc.* **70**, 4267.
Ham, R. G. (1965). *Proc. Nat. Acad. Sci. U.S.* **53**, 288–293.
Ham, R. G., and Puck, T. T. (1962). *In* "Preparation and Assay of Enzymes" (S. P. Colowick and N. O. Kaplan, eds.), Methods in Enzymology, Vol. 5, 90–119. Academic Press, N.Y.
Harris, M. (1971). *J. Cell. Physiol.* **78**, 177–184.
Isomura, K., Nikaido, O., Horikawa, M., and Sugahara, T. (1973). *Radiat. Res.* **53**, 143–152.
Jones, C., and Puck, T. T. (1973). *J. Cell. Physiol.* **81**, 299–304.
Jones, C., Wuthier P., Kao, F. T., and Puck, T. T. (1972). *J. Cell. Physiol.* **80**, 291–298.

Kao, F. T., and Puck, T. T. (1967). *Genetics* **55**, 513–529.
Kao, F. T., and Puck, T. T. (1968). *Proc. Nat. Acad. Sci. U.S.* **60**, 1275–1281.
Kao, F. T., and Puck, T. T. (1969). *J. Cell. Physiol.* **74**, 245–258.
Kao, F. T., and Puck, T. T. (1970). *Nature (London)* **228**, 329–332.
Kao, F. T., and Puck, T. T. (1971). *J. Cell. Physiol.* **78**, 139–144.
Kao, F. T., and Puck, T. T. (1972a). *J. Cell. Physiol.* **80**, 41–50.
Kao, F. T., and Puck, T. T. (1972b). *Proc. Nat. Acad. Sci. U.S.* **69**, 3273–3277.
Kao, F. T., Johnson, R. T., and Puck, T. T. (1969a). *Science* **164**, 312–314.
Kao, F. T., Chasin, L., and Puck, T. T. (1969b). *Proc. Nat. Acad. Sci. U.S.* **64**, 1284–1291.
Lederberg, J., and Zinder, N. (1948). *J. Amer. Chem. Soc.* **70**, 4267–4268.
Mezger-Freed, L. (1971). *J. Cell Biol.* **51**, 742–751.
Mezger-Freed, L. (1972). *Nature (London), New Biol.* **235**, 245–246.
Morrow, J., Colofiore, J., and Rintoul, R. (1973). *J. Cell. Physiol.* **81**, 97–100.
Naha, P. M. (1969). *Nature (London)* **223**, 1380–1381.
Posner, M., Nove, J., and Sato, G. (1971). *In Vitro* **6**, 253–256.
Puck, T. T. (1972). "The Mammalian Cell as a Microorganism—Genetic and Biochemical Studies *in Vitro*." Holden-Day, San Francisco, California.
Puck, T. T., and Kao, F. T. (1967). *Proc. Nat. Acad. Sci. U.S.* **58**, 1227–1234.
Puck, T. T., Cieciura, S. J., and Robinson, A. (1958). *J. Exp. Med.* **108**, 945–956.
Puck, T. T., Marcus, P. I., and Cieciura, S. J. (1956). *J. Exp. Med.* **103**, 653–665.
Puck, T. T., Wuthier, P., Jones, C., and Kao, F. T. (1971). *Proc. Nat. Acad. Sci. U.S.* **68**, 3102–3106.
Rosner, A., and Yagil, E. (1970). *Mol. Gen Genet.* **106**, 254–262.
Taylor, M. W., Souhrada, M., and McCall, J. (1971). *Science* **172**, 162–163.
Wagner, R. P., and Mitchell, H. K. (1964). "Genetics and Metabolism," 2nd Ed. Wiley, New York.
Wyke, J. (1971). *Exp. Cell Res.* **66**, 203–208.

The work reported here shows that animal cells, in the same way as bacterial cells, can be induced to form temperature-sensitive mutants in response to a chemical mutagen. Whether this property is common to all animal cells or restricted to only a few is unknown. An exhaustive study of the effects of various mutagens in inducing cellular mutations has not yet been performed. No attempt would thus be made here to generalize from specific examples.

II. Theoretical Considerations

A. Cell Line

The choice of cell line for studies relating to temperature-sensitive mutation is very crucial for a number of reasons. (1) Established tissue culture cell lines exhibit different degrees of plating efficiency under normal plating conditions; cell lines with a higher efficiency of plating have the obvious advantage that one starts with a larger number of viable cells when exposing to a mutagen. (2) Cell lines differ in the temperature range in which they can grow efficiently; those with a wider range of temperature adaptability are better suited for isolating temperature-sensitive mutants. (3) Some cell lines are easier to manipulate in cloning than others; efficiency of survival of cloned cultures is probably the most important aspect that determines the success or failure in the isolation of a mutant.

Our general experience is that fibroblastic cells (e.g., from kidney, epithelium) are tougher, easier to manipulate in cloning, grow over a wider range of temperatures, and have higher efficiency of plating than nonfibroblastic cells (e.g., lymphocytes, hepatoma cells). Not surprisingly, the temperature-sensitive clones of mammalian cells isolated and reported in the literature (Naha, 1969; Thompson *et al.*, 1970; Meiss and Basilico, 1972) are all fibroblasts.

Chromosomal constitution of a cell line could be an important factor, but at present no data are available for a comparative analysis of gene dosage effect on haploid, diploid, and tetraploid cell lines in the induction of temperature-sensitive mutations of mammalian cells. Mutations rates at different ploidy levels for heat resistance and azaguanine resistance (Harris, 1971) was found to remain constant or decline slightly in cells with increasing numbers of chromosome sets. This trend is not in accordance with expectations based on the assumption of dominant, codominant, or recessive changes of gene or chromosome levels. It has been suggested that at least some variations may arise in somatic cells by stable shifts in phenotypic expression rather than by changes in genetic information (Harris, 1971; Mezgar-Freed, 1971).

B. Temperatures

The temperature range over which the efficiency of plating of a particular cell line remains more or less constant should first be determined with the wild-type cell line. In our hands, the monkey kidney cell line (BSC-1), the human embryo kidney cell line (HEK), and the pig kidney cell line (PK15) all grow equally well between 33°C and 39.5°C; if anything, the efficiency of plating is higher at 39.5°C (Naha, 1969). The plating efficiency of the mutant clones over the temperature range can be taken as an index of the sensitivity of a variant clone to increased temperatures. Ideally, the variant clones should be classified as *heat* sensitive, as opposed to *cold* sensitive (Campbell, 1961). We shall, however, use the term temperature sensitive (ts) for the present paper.

C. Mutation Rate

Unlike auxotrophic mutants (Puck and Kao, 1967) and drug-resistant mutants (Harris, 1971; Mezger-Freed, 1971), we have been unable to determine the mutation frequency and dose response of a particular cell line with respect to a mutagen in inducing temperature-sensitive mutations for the following reasons: (1) variable reversion frequencies in different clones (10^{-2} to 10^{-5}), density dependence and increasing leakiness with every passage, (2) variable survival percentage of cloned cultures in different experiments, (3) variable morphology in the surviving fraction of cells (enlarged cytoplasm, loose texture), which does not strictly conform to the definition of a visible, healthy clone. Thompson *et al.*, (1971), however, measured a spontaneous mutation frequency of 6×10^{-3}, in which the mutants were temperature sensitive for growth.

D. Cloning

The various cloning techniques have been currently reviewed by Ham (1972). In our experiments we pick discrete clones directly from the experimental plates with Pasteur pipettes and purify them by end-point dilution in Falcon dilution plate (plastic) and growing cultures from single cells.

III. Methods

A. Cell Line

The stable cell line of heteroploid SV40-sensitive, African green monkey (*Cercopithecus aethiope*) kidney cells, BSC-1 (Meyer *et al.*, 1962), was used in these experiments. Stock cultures were maintained by planting about 10^6 trypsinized cells in 20 ml of L-15 medium (Leibovitz, 1963) containing 10% fetal calf serum (FCS). A homogeneous clone was isolated and cultured through three successive passages in L-15 medium supplemented with 10% FCS.

B. Treatment with Mutagens

About 200 cells were inoculated into each of a series of Falcon petri dishes (60 × 15 mm) in L-15 medium with 10% FCS. After 6 hours of incubation at 39.5°C, *N*-methyl-*N*′-nitro-*N*-nitrosoguanidine (obtained from Aldrich Chemical Company, Milwaukee, Wisconsin) was added, using a concentrated stock solution prepared in L-15 medium, to make a final concentration 0.50 μg/ml of mutagen in contact with the cells. Incubation was continued at 39.5° C for an additional 16 hours, which is a little less than a generation time. The medium was then removed from the culture plates, the cells were washed twice in L-15 medium (with 10% FCS), and fresh growth medium was added; incubation was continued for a further 24 hours.

At the end of this period, 5-bromo-2′-deoxyuridine (BUdR, obtained from Sigma Chemicals) was added in solution of prewarmed L-15 medium so as to make its concentration 4 to 5 × 10^{-6} *M*/ml in each plate. The cultures were again incubated at 39.5°C for 5–7 days before being exposed to visible light. After exposure to light, the medium containing BUdR was removed, fresh L-15 medium was added, plates were transferred to 33° for 12–15 days until clones of dividing cells were visible under the microscope. These clones were isolated as described above, cultured, and tested for efficiency of plating at the permissive (pm^+) and nonpermissive (pm^-) temperatures of 33°C and 39.5°C, respectively.

In other experiments, BUdR treatment was replaced with thymidine-^{3}H or uridine-^{3}H to select specifically for "mutant" clones that were defective for thymidine or uridine incorporation. Thompson *et al.* (1971) have described methods of using cytosine arabinoside as well. The combination of nitrosoguanidine and BUdR or radioactive precursors of nucleic acids was expected to select for "mutants" of the desired type.

Nitrosoguanidine induces a high frequency of mutation is bacterial systems (Loveless and Howarth, 1959; Zamenhof, 1966; Adelburg *et al.*, 1965), and the incorporation of BUdR in DNA (Djordjecive and Szybalski, 1960) renders bacteriophage (Stahl *et al.*, 1961) sensitive to visible or near-visible light. A shift in incubation temperature for selection and isolation of temperature-sensitive mutants of bacteriophage T4 has been successfully performed by Edgar and Lielausis (1964).

The principle of the technique described is to allow the normal (wild type) nonmutated and temperature-independent mutants to divide and grow at 39.5°C. The incorporation of bromodeoxyuridine (BUdR) in the DNA of growing cells rendered these cells hypersensitive to visible light. Transfer of cells to a lower temperature (33°), following preincubation at an elevated temperature (39.5°), is designed to select out the pm^+ temperature-sensitive cells, which are then allowed to divide and form clones at 33°C.

IV. Discussion

The temperature-sensitive mutants of mammalian cells can be used (1) for study of cellular regulatory mechanisms, (2) for study of host–virus interaction, and (3) for morphological analyses in cells synchronized by appropriate exposure to the limiting temperature. It has become apparent, however, that many cultivated cells with mutated phenotypes have high reversion rates, which not only hinders biochemical analyses but also raises questions as to the nature of the genetic changes that lead to the changed phenotype. Chromosomal and therefore gene dosage effects may, for example, be important.

High reversion frequencies are not unique to temperature-sensitive "mutation" in mammalian cells; they are almost equally common for mutants in microbial organisms (Campbell, 1961; Edgar and Lielausis, 1964; Naha, 1966). One additional complication in mammalian cell "mutants" is that the reversion frequency in these cases is dependent on cell density (Meiss and Basilico, 1972). We have used 1×10^5 cells per milliliter as a standard cell density for studying reversion frequency and also biochemical analysis. A second selection by testing the "mutants" through successive passages for efficiency of plating at the restricted temperature has been found to be helpful. It is quite possible that at least some of the "mutants" are non-Mendelian. In the absence of recombination and segregation of markers in mammalian cells, the best evidence must come from autoradiographic studies by use of suitable precursors (Naha, 1970, 1973).

References

Adelberg, E. A., Mandel, M., and Chien Ching Chen, G. (1965). *Biochem. Biophys. Res. Commun.* **18**, 788.

Campbell, A. (1961). *Virology* **14**, 22.

Djordjevic, B., and Szybalski, W. (1960). *J. Exp. Med.* **112**, 509.

Edgar, R. S., and Lielausis, I. (1964). *Genetics* **49**, 649.

Ham, R. G. (1972). *In* "Methods in Cell Physiology" (D. M. Prescott, ed.), Vol. 5, p. 37. Academy Press, New York.

Harris, M. (1971). *J. Cell. Physiol.* **78**, 117.

Leibovitz, A. (1963). *Amer. J. Hyg.* **78**, 173.

Loveless, A., and Howarth, S. (1959). *Nature* (*London*) **184**, 1780.

Meiss, H. K., and Basilico, C. (1972). *Nature* (*London*), *New Biol.* **239**, 66.

Meyer, M. H., Jr., Hopps, H. E., Rogers, N. G., Brooks, B. E., Bernheim, B. C., Jones, W. P., Nisalak, A., and Douglas, R. D. (1962). *J. Immunol.* **88**, 796.

Mezger-Freed, L. (1971). *J. Cell. Biol.* **51**, 742.

Naha, P. M. (1966). *Virology* **29**, 676.

Naha, P. M. (1969). *Nature* (*London*) **223**, 1380.

Naha, P. M. (1970). *Nature* (*London*) **228**, 166.

Naha, P. M. (1973). *Nature* (*London*), *New Biol.* **241**, 13.

Puck, T. T., and Kao, F. T. (1967). *Proc. Nat. Acad. Sci. U.S.* **58**, 1227.

Stahl, F. W., Craseman, J. M., Oknu, L., Fox, E., and Laird, C. (1961). *Virology* **13**, 98.

Thompson, L. H., Mankovitz, R., Baker, R. M., Till J. E., Siminovitch, L., and Whitmore G. F. (1970). *Proc. Nat. Acad. Sci. U.S.* **66**, 377.

Thompson, L. H., Mankovitz, R., Baker, R. M., Wright, J. A., Till, J. E., Simonvitch, L., and Whitmore, G. F. (1971). *J. Cell. Physiol.* **78**, 431.

Zamenhof, P. J. (1966). *Proc. Nat. Acad. Sci. U.S.* **55**, 50.

Chapter 4

Preparation and Use of Replicate Mammalian Cell Cultures

WILLIAM G. TAYLOR AND VIRGINIA J. EVANS

Tissue Culture Section, Laboratory of Biology,
National Cancer Insitute, Bethesda, Maryland

I. Introduction

A. Objective

Many present day investigators are employing tissue culture as a simple experimental model removed from the physiological constraints of the intact animal. Several developments have facilitated this multidisciplinary application of mammalian cell culture. These include the formulation of several inexpensive culture media (see Morton, 1970, for a compendium), the availability of contaminant-free, well characterized cell lines from several species (Stulberg *et al.*, 1970), the development of better, low cost materials including "clean" work areas (Coriell and McGarrity, 1968; Kreider, 1968; McGarrity and Coriell, 1971), and the refinements in techniques. One such technical development has been the fabrication of equipment for preparing replicate mammalian cell cultures (Evans *et al.*, 1951; Taylor *et al.*, 1972a).

Two requisite elements of any analytical or quantitative experimental model are *replicacy* and *randomness*. To assess accurately the effect of any agent in a biological assay, large numbers of statistically identical (or replicate) tissue cultures must be planted with a minimum physiological trauma to the cell population. These cultures are then separated randomly into control and experimental sets, each containing a number of culture flasks from which statistically valid conclusions can be drawn. The variables are introduced, and, if an intervening change(s) of culture medium is required, the change must be made with a minimum of cell trauma and loss. After the appropriate incubation period, the cell populations must be harvested and the results quantitated. Without such precautions, an *in vitro* model also will yield "anomalous data" best characterized as "biological variation"; with these precautions, the variation from experiment to experiment can be minimized.

The objective of this chapter is to outline in detail the methods and equipment used in this laboratory for quantitative tissue culture studies. A treatment of other problems encountered by the authors will be included. No attempt has been made to discuss the methods for preparing replicate microtest plate (Goldsby and Zipser, 1969; Suzuki *et al.*, 1971) or large-scale suspension cultures (Weirether *et al.*, 1968).

It must be remembered that the "perfect tissue culture system" is not as yet a *fait accompli*. The routine use of antibiotics and the seemingly obligate requirement of cells in culture for serum contribute to *in vitro* variation; presented elsewhere are practical considerations of the problems associated with adventitious contaminants (Fogh *et al.*, 1971; Stanbridge, 1971) and serum (Fedoroff *et al.*, 1971; Molander *et al.*, 1971; Boone *et al.*, 1971; Merril *et al.*, 1972).

B. Historical Evolution

Until 1951, no methods for preparation of replicate mammalian cell cultures were reported. Evans and co-workers then published the design of an all-glass reservoir and stirring assembly (Evans *et al.*, 1951). The apparatus delivered uniform samples of 0.5 or 1.0 ml from a stirred cell suspension contained in the cylindrical reservoir; the suspension was constantly aerated with 5% CO_2–air to maintain a physiological pH. The original apparatus, however, was large and fragile and expensive to fabricate, and the volume of cell sample dispensed was restricted to the size of calibrated delivery tip. Also, statistically uniform cell samples in good physiological condition were influenced by mechanical considerations, such as the design and pitch of the stirring blades and their alignment in the reservoir and their rotational speed.

These limitations were soon recognized (Katsuta *et al.*, 1954), and modified designs were introduced. Syverton and co-workers (Syverton *et al.*, 1954) retained the original cell suspension reservoir with its limitations, but eliminated the sidearms used to aerate the suspension and replaced the graduated delivery tip with an automatic syringe pipette. The pH was maintained by reducing the bicarbonate concentration of the medium. Others subsequently reported the use of a syringe pipette or similar devices (Wallis and Melnick, 1960; Martin, 1964; Malinin, 1966). The device described herein combines the simplicity and versatility of the syringe pipette and the gassing facility of the original reservoir and stirring assembly. The large numbers of replicate cultures which can be quickly prepared—up to 200 per hour—and the ease with which this device can be used make it a highly useful tool.

II. Replicate Culture Planting Device

A. Fabrication

A typical replicate culture planting device is shown in Fig. 1b. It can be made easily and with inexpensive laboratory glassware. The precise shape and size of the cap, the sample withdrawal tube, etc., will vary with the geometry of the cell and medium reservoir, but the principles inherent in the design remain. The following instructions correspond to those used in this laboratory.

First, a closure for the cell and medium reservoir is prepared. Beginning at the broad end, a cylindrical hole is bored through approximately two-thirds the depth of a No. 6 silicone stopper. For our purposes, this is done best with a No. 13 size cork bore. The silicone plug can be removed by gripping the stopper and cork bore obliquely and turning the stopper so that core is sliced by the cutting edge of the bore. The resulting cap has an internal diameter of 21–22 mm and securely fits the mouth of a flat-bottomed

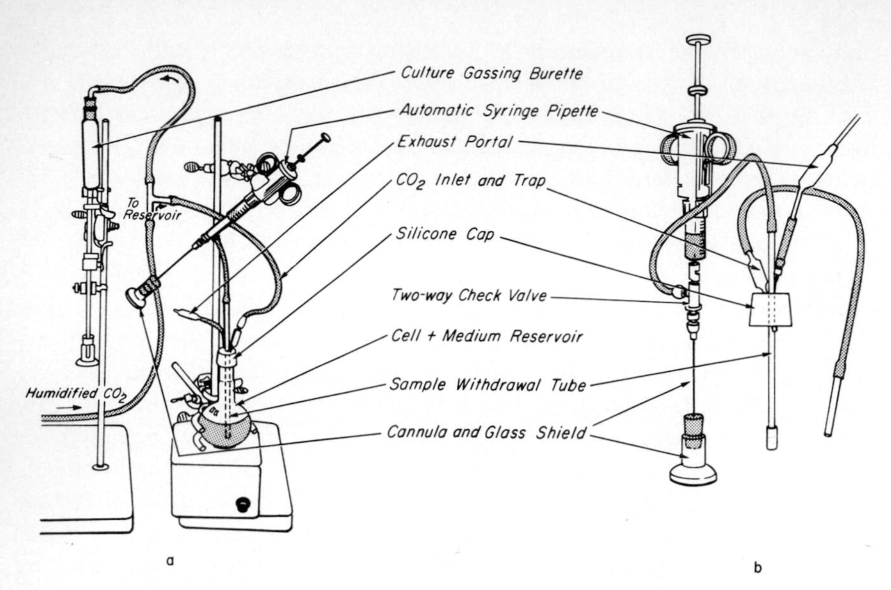

FIG. 1. The replicate culture planting device. (1a) The replicate planting device with other necessary equipment; the syringe pipette assembly should be positioned vertically if flasks are to be inoculated, and in the position shown for seeding petri dishes. (1b) A replicate culture planting device with a 5-ml automatic syringe pipette.

boiling flask (o.d. = 20 mm). Silicone rubber stoppers are far more pliable than black rubber stoppers, are nontoxic for cells *in vitro*, and withstand repeated sterilization.

A 7–8-inch length of glass tubing (5 mm o.d.) and a cotton-stoppered trap are inserted through the narrow end (top) of the cap. A length of latex tubing, precleaned by boiling in triple-distilled water, is attached to the cotton-stoppered trap and with the glass connector forms the CO_2 inlet and trap. The end of the glass tube—the sample withdrawal tube—which extends into the boiling flask, is shielded with 1–2 cm of prewashed silicone rubber tubing (Silatube, Labtician Products Company, Hollis, New York). This minimizes the change of breakage when the depth of the sample withdrawal tube is adjusted downward. The opposite end of the sample withdrawal tube is connected to the two-way check valve of either a 2-ml or 5-ml Cornwall syringe pipette (Becton, Dickinson & Company, Rutherford, New Jersey) by 12–15 inches of prewashed Silatube. Finally, a bell-shaped glass shield is used to protect the tip of the cannula, and the mouth of the culture flask being inoculated, from airborne contamination. The glass shield is attached to a silicone stopper affixed to the end of the cannula.

In earlier designs (Taylor *et al.*, 1972a), the cap was fabricated to fit the reservoir snugly, necessitating an exhaust portal for the flowing CO_2/air.

To overcome space limitations in the cap, a hypodermic needle was used to vent the reservoir with the aid of a cotton-stoppered trap. Repeated autoclaving corroded the needles, occluding the already small bore. Better venting of the flowing gas and a simpler design are achieved with the loose-fitting, larger-diameter cap.

B. Sterilization

Immediately before use, the replicate culture planting device is flushed with triple glass-distilled water. The withdrawal tube and silicone cap, the cannula and glass shield, and the glass connector attached to the gas inlet are wrapped separately with Patapar paper (A. J. Buck, Cockeysville, Maryland). The device is placed in an aluminum can, wrapped again with Patapar paper, and autoclaved (15 minutes, 121°C). In wrapping the unit for sterilization, pinching the silicone or latex tubing must be avoided. Pinching brings the inside walls into close proximity and, with the heat of sterilization, they adhere to each other; in time, the creased point tends to lose rigidity and become weakened and misshapen. These all impair proper flow of the cell suspension through the device. It is equally important that triple-distilled water be present in the syringe and barrel assembly during and after sterilization so as to avoid drying and adherence to the syringe barrel and valves to the surrounding assembly. In practice, the autoclave chamber should be vented to atmospheric pressure *slowly* (as with any liquid) and the replicate culture planting removed from the chamber immediately. Once cool, the device can be used or stored temporarily. Although the unit will remain functionally trouble free as long as the valve assembly and syringe are wet; in practice we limit the storage time to 5 days.

III. Preparation and Use of Replicate Mammalian Cell Cultures

A. Necessary Glassware and Equipment

The following is a list of equipment and glassware used to set up metabolic studies in this laboratory. Obviously, all glassware is sterilized prior to use.

1 Magnetic stirrer
2 Ring stands
1 Burette clamp
2 Extension clamps (maximum grip 2.5 inch)
1 Glass "T" with 15-inch latex tubing on each arm

5 Cotton-stoppered test tubes (15 × 150 mm)
2 Burettes, 25ml
2 Graduated pipettes, 50-ml
2 Cell and medium reservoirs with two 15 × 1.5-mm magnetic stirring bars each
Glass rods or a bacteriological loop for scraping stock cultures (if adherent to the culture flask)
Pipettes, 1-, 5-, and 10-ml
Stoppers or caps if glass vessels[1] are to be used; plastic flasks or dishes
1 Container for used culture medium and extra fluid to be discarded
Prewarmed culture medium

The planting medium and/or those media containing the experimental variables should be prepared 24 hours prior to use to allow time for preliminary sterility testing. Care should be taken to prepare each medium at the same pH, osmolality, etc.

B. The Cell Suspension

1. The Parent Culture

Factors influencing the preparation of a well dispersed cell suspension and the ultimate success of the biological assay will be discussed in Section IV, A–C. The instructions below pertain to the adherent or stationary phase cultures.

Several parent cultures are planted and grown in a medium of choice—preferably the one to be used in the bioassay. The cultures are incubated until a mid- to late-logarithmic phase population is obtained. For a rapidly proliferating cell line such as strain L or MK_2, this may require 4–6 days, with a final cell population of 4 to 5 × 10^6 cells/T-15. For a slow growing or more fastidious cell strain, a greater number of parent flasks should be prepared to obtain an adequate exponential phase population. For example, low passage level WI-38 (Hayflick and Moorhead, 1961) grow rapidly but produce fewer cells per square centimeter than, for example, strain L. An incubation period of 3–5 days results in a confluent monolayer in excellent physiological condition (90–92% viability), but in our hands this represents *no more than* 0.6 to 1.0 × 10^6 cells/T-15. Further incubation (5–7 days total) results in a slightly increased cell number with a decreased viability (80–85%) unless a strict schedule for replacing spent culture medium with fresh fluid is used (N. J. Lewis, personal communication).

[1]Better than average growth is obtained if the acid cleaned Pyrex T-15 flasks (Earle and Highhouse, 1954) are boiled in dilute, aqueous Calgolac (Calgon Corp., St Louis, Missouri), then washed thoroughly and sterilized.

Since the ultimate success of the bioassay is partly contingent upon the physiological state of the cells employed, factors to be considered in planning the experiment include the cell population doubling time, the culture medium used, the ultimate viable cell population achieved and the incubation period needed, and the usefulness of the complete replacement of the old culture medium with fresh fluid during the incubation period.

Prior to planting replicate cultures, it may be useful to replace the growth medium in the parent flasks with a "maintenance" medium free of the variable(s) under evaluation. This helps to reduce the biological "background" of the bioassay. For example, one of the ongoing research programs in this laboratory is the analysis of autoclavable serum substitutes, specifically bacteriological peptone (Taylor *et al.*, 1972b). In preparation for a typical study, the test cells are incubated 4 days, at which time the growth medium is replaced with unsupplemented culture medium; after an additional 18–24 hours of incubation, these cells can be used to evaluate fractionated serum substitutes with the knowledge that the negative control cultures (i.e., incubated in unsupplemented medium) will undergo only 2–3 population doublings during the subsequent 5-day test period. In the case of cells grown in serum-supplemented medium, growth of the parent cultures in 0–2% serum for the entire incubation period may be required to achieve a minimum proliferation in the negative control cultures. Obviously, this will vary for each cell line.

2. Harvesting the Parent Cell Population

After the appropriate incubation period, the cell population in the parent flasks is harvested and the cell suspension is prepared. Cell lines propagated in serum-supplemented medium are most frequently dissociated and removed from the parent flasks with 0.125–0.5% "trypsin." Depending upon the cell strain or line, this dissociation may require 10–30 minutes at 37°C or may occur immediately at room temperature. A secondary effect frequently observed is the adherence of trypsinized cells to each other resulting in dumps of 10–500 cells. These clusters can be disrupted by gently aspirating the suspension; a sieving pipette, as described in Section IV, may prove to be very useful. The use of Ca^{2+}- Mg^{2+}-free saline as a diluent for the trypsin (Zwilling, 1954) and the use of chelating agents such as EDTA (Versene) (Anderson, 1953) in conjunction with trypsin may mitigate against clumping. Other enzymes that have been used successfully for dissociating mammalian cells include elastase (Rinaldini, 1959; Phillips, 1972), collagenase (Kono, 1969; Lasfargues, 1957; Lasfargues and Moore, 1971), and Pronase (Wilson and Lau, 1963; Gwatkin and Thompson, 1964; Poste, 1971); a compilation of these and related procedures has been prepared recently (Kruse and Patterson, 1973).

Mechanical disaggregation methods can be used in the case of cell lines continuously subcultured in serum- or protein-free growth medium. Such lines include Pumper's RH line (Pumper *et al.*, 1965), Merchant's L-M line (Hsu and Merchant, 1961; Merchant and Hellman, 1962), or the many mouse, hamster, monkey, and human lines which grow in protein-free NCTC 135 (Evans *et al.*, 1964). Also, in those instances wherein enzymatic and/or chemical disaggregation is undesirable, mechanical methods can be employed. Cells are removed from the floor of the parent flask by: (a) disrupting the cell sheet with a piece of sterile, perforated Cellophane (Microbiological Associates, Bethesda, Maryland) mopped across the flask floor with a sterile, bent-tip glass rod, or (b) disrupting the monolayer with a platinum-iridium bacteriological loop. Sterile rubber policemen can also be used, though the rubber may introduce cytotoxic components. Clusters of cells are dispersed by aspirating the suspension.

Irrespective of whether they are grown in serum-supplemented or serum-free culture medium, the following rules apply. The parent cell suspension must be maintained in a physiological pH (pH 7.2–7.4) at all times. The parent cells should be diluted to the desired inoculum level and planted to replicate cultures as soon as possible.

3. Preparing the Test-Cell Suspension

In general a greater volume of cell suspension, 50–100 ml depending on the syringe size, must be prepared than is needed. This is due to the loss of some suspension when flushing the planting device (see below), and because some fluid will be lost in the syringe, tubing, and reservoir once the desired number of culture vessels are planted.

Two methods can be used to resuspend the parent cells in the variable-free planting medium. The pooled suspension can be centrifuged gently (110–140 g, 10–15 minutes, 25°C), and the packed cell mass resuspended in a small (5–10 ml) volume of planting medium. After performing a viable count on the concentrated suspension, the appropriate volumes of suspension and planting medium are mixed to yield the desired inoculum level per milliliter of planting medium. Alternatively, an arbitrary volume of culture fluid can be dispensed into the planting reservoir, and the suspensions from individual parent flasks added to this medium until the desired inoculum size is reached. The disadvantages of the first technique are that: (a) cell clumps which form or cells which attach to the sides of the centrifuge tube while the viable count is being performed reduce the actual cell number per milliliter of suspension, and (b) errors in dilution frequently occur. The disadvantage of the second method is that the inoculum size is a relatively small number (typically 4 to 5×10^4 cells/ml for nutritional studies in this laboratory) and the inoculum size is, therefore, best estimated by an electronic counter. Obviously, an electronic counter will not distinguish between dead and viable cells nor large bits of debris. The two methods used in conjunction will give relative accurate data.

Periodically it is necessary to "hold" a cell suspension for a brief period of time. WI-38 cells, for example, will remain in good physiological condition and will not attach to the side of the vessel (thus altering the cell number per milliliter of suspension) if the suspension is placed in a beaker of ice. Obviously, this should be kept to a minimum time span, of the order of 30 minutes.

Once the cell and medium reservoir has been inoculated with culture medium and cells at the desired inoculum level, the vessel—with two Teflon-coated 15 × 1.5 mm magnetic stirring bars—should be positioned eccentrically (for best stirring efficiency) on a magnetic stirrer. The suspension must be maintained at a physiological pH by gassing with 10% CO_2 and should be planted as quickly as possible.

4. Using the Replicate Culture Planting Device

Once the cell and medium reservoir is clamped in position (Fig. 1a), the following should be done:

a. Connect the base of the glass "T" to the source of CO_2 + air. Connect one arm to a culture gassing burette (Fig. 1a). Connect the tubing on the other arm to the gas inlet by means of the glass connector. Turn on CO_2 + air supply.

b. Using an extension clamp, secure the automatic syringe pipette 6–8 inches above the cell and medium reservoir. Carefully unwrap the sample withdrawal tube and cap. Aseptically place the sample withdrawal tube into the planting medium. Adjust the withdrawal tube so that the Silatube tip is 3–5 mm from the bottom of the reservoir. Adjust the control on the magnetic stirrer so that the cell suspension is gently but thoroughly mixed. Position the reservoir eccentrically to aid mixing.

c. Position the syringe pipette vertically with the cannula and shield *upward*. Aspirate the syringe to draw cells and medium into the barrel of the syringe (see Section IV, D in regard to priming the device). Remove *all* air bubbles from the barrel of this syringe. This is particularly important if serum-supplemented medium is being used; air bubbles will cause frothing of the medium and lead to planting error.

d. Turn the syringe pipette 180°; the cannula and shield should now point directly at the table top. Flame the glass shield until dry; allow it to cool. Aspirate the syringe until 15–20 ml have been collected in the sterile test tube; discard this volume of suspension. This flushes any remaining water from the syringe, water and residue from the cannula, and assures a representative cell population in each subsequent aliquot planted. Petri dishes are more easily seeded if the syringe pipette is clamped at a 45° angle and the cell suspension is introduced beneath the partly opened lid.

e. Position a culture vessel beneath the tip of the cannula and aspirate the syringe. Flush the residual air from the culture vessel with CO_2-air delivered with the culture gassing burette (PGC Scientifics, Rockville, Maryland and stopper the flask. Incubate all cultures.

f. Immediately after use, thoroughly rinse the entire replicate culture planting device with triple-distilled water. In this way neither cells nor medium will dry inside the device.

C. A Statistical Analysis

The ultimate utility of the replicate culture planting device lies in preparing statistically identical sister cultures while maintaining the cell suspension at a physiological pH. To determine whether this device met these qualifications, several parent cultures of NCTC 2071 (Evans *et al.*, 1964) were prepared. NCTC 2071 is a derivative of strain L-929 (Sanford *et al.*, 1949), which has been continuously subcultured in serum-free NCTC 109 or NCTC 135 since 1955. Monodispersed cell suspensions were prepared; aliquots of 0.5 and 1.0 ml were collected from a planting device consisting in part of a 2-ml Cornwall automatic syringe pipette, while 3.0-ml samples were obtained with a planting device fitted with a 5-ml syringe pipette. Sixty samples per test were collected for each designated volume. The pH of the odd-numbered samples was estimated immediately by comparison with a La Motte phenol red pH indicator set (A. H. Thomas, Philadelphia, Pennsylvania). The thirty remaining even-numbered samples were brought to a volume of 10 ml with Isoton (Coulter Electronics, Hialeah, Florida), and the cell number was determined in triplicate with a Coulter Model B counter. The average of these three counts was used for further calculations.

The results are summarized in Table I. The pH of the cell suspension remained physiologically constant throughout the 30-minute period required per test. A slight drop ($<$ 0.1 pH unit) was observed during the first minutes of continuous gassing, after which no further color changes were observed. The same operator evaluated this parameter for each test.

The original reservoir and stirring assembly (Evans *et al.*, 1951) was designed so that the experimental error of the planting procedure was 5%, 95% of the time. As seen in Table I, the percent deviation observed with the replicate culture planting device was in good agreement with the earlier model. When aliquots were collected with devices consisting in part of 2-ml and 5-ml automatic syringe pipettes, the greatest variation was observed with 0.5-ml samples delivered from the 2-ml Cornwall. Although not tested, it is reasonable to assume that a reduced variation would be

TABLE I

VARIATIONS IN CELL NUMBER AND IN pH FOR REPLICATE SAMPLES DISPENSED WITH THE REPLICATE CULTURE UNIT[a]

Automatic pipette size (ml)	Sample size (ml)	Observed pH	Sample number tested	Mean cell number per sample ($\bar{X}$)[b]	Standard deviation[c] (S)	Deviation ($S/\bar{X}$) (%)
2	0.5	7.4–7.5	28	29,968	2,371	7.9
		ND[d]	30	20,084	1,655	8.2
		ND	29	22,989	1,670	7.3
2	1.0	7.4–7.5	30	84,125	4,653	5.5
		7.3–7.4	30	89,563	4,871	5.4
		ND	29	97,300	5,627	5.8
5	3.0	7.3–7.4	28	253,737	17,104	6.7
		7.4	26	301,448	19,223	6.4

[a]Reproduced with permission of the Tissue Culture Association, Inc. (Taylor *et al.*, 1972a).

[b]$\bar{X}$ represents the sum of the average cell counts observed per total sample number tested, or $\Sigma x/n$.

[c]S represents the square root of the variance, or $(\Sigma(\bar{X} - X)^2/n - 1)^{1/2}$.

[d]ND signifies not determined.

observed if a 1-ml pipette were used; the sample volume would be 50% of the syringe's capacity rather than 25%. The best agreement for sample to sample was observed with a 1.0-ml aliquot delivered from a 2-ml syringe pipette (% deviation = 5.4–5.8), and the deviation from sample to sample increased only slightly (6.4–6.7%) when the 3.0-ml sample size delivered from a 5-ml Cornwall syringe was tested. Inherent in these data is the intrinsic error of the counting procedure itself.

Factors which influence sample variability include: (a) the degree to which the parent cells are monodispersed, (b) how well the suspension is mixed throughout the planting procedure (eccentric positioning of the reservoir enhances mixing), and (c) how quickly the cultures are planted. Mammalian cells will readily settle if not stirred constantly, and if an extended delay (more than a minute) is encountered between samples, the device should be flushed as described in 4D above to assure uniformity between the aliquot in the syringe and the cell suspension.

D. Randomization of Replicate Cultures

As stated in the introduction, replicacy *and* randomness are requisite features of an analytical or quantitative procedure. To minimize the inherent variability of any replicate planting procedure, the cultures planted

TABLE II

RANDOMIZATION TABLE FOR PREPARING SETS OF CONTROL AND EXPERIMENTAL CULTURES

2	36	5	33	18	43	31	52
23	1	16	32	42	53	11	30
56	21	20	4	38	7	35	47
29	25	15	17	40	41	14	9
45	50	24	34	12	54	28	55
37	3	46	44	26	8	27	51
49	39	48	22	10	19	6	13

should be randomly distributed into sets. In this way the planting sequence will not unduly influence any one variable or control set.

Culture flasks are inoculated as quickly as possible and incubated at 37.5° ±0.15°C. As soon as the cells begin to attach, the flasks are identified with prenumbered labels in the sequence planted, i.e., planting number. The cultures are distributed into experimental and control sets by means of the planting number and a random numbers table. An example chart for the randomization of fifty-six cultures is shown in Table II, although any random numbers table can be used.

E. Addition of Experimental Variables

Two methods can be used successfully for the addition of experimental variables. The selection is dictated by whether or not the test calls attach rapidly and whether they can withstand a change of culture medium soon after planting.

1. BY COMPLETE CHANGE OF CULTURE MEDIUM

This is the method used for the serum-substitute studies described above. Both Pumper's rabbit heart line and Merchant's L-M line attach readily; after 2–3 hr. incubation, greater than 99% of the cells have attached. As a consequence, the planting medium can be completely replaced with the appropriate control of experimental medium.

2. USING A FRACTION OF THE FINAL VOLUME

If the test-cell line attaches poorly or is a suspension culture, the entire seed inoculum can be introduced into the culture vessel in one-half to one-third the final medium volume. The variables (at 2–3 times the desired concentration, respectively) can be added in the remaining volume of medium. *N.B.*: Use of this method may lead to greater variability in the osmolality of the culture media; while cells in culture can tolerate wide variations in osmotic strength (Waymouth, 1970), care should be taken to assure a relatively constant osmolality within an experiment. Hence, both the variable free planting medium and the fluid containing the variables should be prepared at an osmolality equivalent to that of the routine culture medium.

IV. Problems, and Some Suggestions

Most investigators using tissue culture use serum or serum fractions as a component of the growth medium. It is generally agreed that serum functions, in part, as a "blotter" which efficiently detoxifies the *in vitro* milieu. Historically, this laboratory has sought to define the physiological requirements of cells *in vitro*, and thereby develop stringently controlled models for the study of "spontaneous" malignant conversion (transformation) of cultured cells (Gey, 1941; Earle and Nettleship, 1943). Specific emphasis has been placed on developing serum- and protein-free culture media. Owing to the absence of serum, cells continuously grown in chemically defined media usually are more susceptible to *in vitro* perturbations. The examples cited below can be extended validly to other cell lines, but the sensitivity of these serum-free lines to physicochemical or biological trauma probably exceeds that encountered with similar cell cultures propagated in serum-supplemented growth medium.

A. Minimum Inoculum Size

Physiological studies using mammalian cells are valueless if the test cells fail to survive and proliferate owing to an inadequate inoculum size. This property varies for each cell line and may be influenced by the presence and concentration of serum or protein supplement, or even by the characterized culture medium itself (see below). The minimum inoculum required can be ascertained by preparing a cell suspension at the approximate density used for routine culture transfers; several inocula sizes are prepared by dilution with culture medium, and the cell suspensions are each planted to 4–5 culture vessels. The lowest inoculum size (or highest dilution) at which all cultures *survive* and *proliferate* optimally can be considered the minimum inoculum level.

Fioramonti and co-workers (Fioramonti *et al.*, 1958) analyzed the minimum inoculum requirements of strain L-929 adapted to growth in chemically defined medium NCTC 109 (Evans *et al.*, 1956a,b) and designated

TABLE III

MINIMUM INOCULUM REQUIREMENTS OF NCTC 2071[a]

Seed inocula ($\times 10^{-5}$ cells /T-15)	Cell number $\times$ 10^{-5} per T-15 at		Population doubling time[c] (hours/population doubling)
	48 hours[b]	120 hours[b]	
3.83	7.45	15.95	65.6
2.44	4.90	9.70	73.1
1.38	1.16	0.77	—

[a]After 830 subcultures in chemically defined medium.

[b]Average cell population in four replicate cultures per inoculum size.

[c]Calculated from cell numbers observed on days 2 and 5. The calculated population doubling time is significantly longer than that determined earlier by cinemicrographic analysis (McQuilkin and Earle, 1962).

NCTC 2071. They found that subsequent survival and proliferation was unpredictable at inocula levels of 1.55 to 3.66 $\times$ 10^4 cells per T-15 (in 3 ml of NCTC 109). Proliferation was observed in all cultures with inocula greater than 3.66 $\times$ 10^4 cells/T-15, and that the maximum proliferation was observed in the range between 1.6 and 4.1 $\times$ 10^5 cells/T-15.

Recently we reevaluated the minimum inoculum requirements of NCTC 2071, now continuously subcultured in medium NCTC 135 (Evans *et al.*, 1964). The preliminary results are summarized in Table III. Under these conditions, the lowest inoculum survived the first 48 hours but subsequently failed to proliferate. The highest and intermediate inocula sizes were equally adequate for growth, as reflected by the very similar population doubling times. The minimum inoculum level, therefore lies between 1.38 and 2.44 $\times$ 10^5 cells/T-15, and further studies are required to determine its exact value.

B. The Culture Medium

One of the ongoing programs in this laboratory is the biochemical characterization of a heat-stable serum substitute, Bacto-peptone dialyzate (PD). A rabbit heart line (RH) (Pumper *et al.*, 1965) has been used to characterize further the biologically efficacious components of PD (Taylor *et al.*, 1972c). To establish that the efficacy of PD was unrelated to the culture medium used, derivatives of the RH line were grown for at least ten subcultures in five different PD-supplemented media. Table IV summarizes the difference observed when the *same cell line* is grown in different culture media with the same supplement. The surprising aspect was not that NCTC 135 or Dulbecco-Vogt's MEM supported the greatest degree of proliferation, nor that MEM required the largest seed inoculum while producing the smallest cell population. What was surprising, as revealed in repeated experimental

TABLE IV

INFLUENCE OF CULTURE MEDIUM ON THE GROWTH OF PUMPER'S RABBIT HEART CELLS

Culture medium	Reference	Efficacy of medium for Growth of stock cultures	Typical experimental growth data	
			Seed inoculum $\times 10^{-5}$ /T-15	Cell number $\times 10^{-5}$/T-15 after 5 days' incubation
NCTC 135	Evans *et al.* (1964)	Excellent	1.01	39.4
Dulbecco-Vogt's Modified MEM	(a) Vogt and Dulbecco (1963) (b) Morton (1970)	Excellent	0.91	30.8
MB 752/1	(a) Waymouth (1959) (b) Waymouth (1965)	Above average	1–2.5	No growth
199	Morgan *et al.* (1950)	Above average	1.32	34.82
MEM	Eagle (1959)	Fair	1.69	25.2

TABLE V

MINIMUM INOCULUM REQUIREMENTS OF PUMPER'S RABBIT HEART CELLS CONTINUOUSLY SUBCULTURED IN MB 752/1

Seed inocula ($\times 10^{-5}$/T-15)	Cell number $\times 10^{-5}$ /T-15 after 24 hours[a]	96 hours[a]	Population doubling time[b] (hours/population doubling)
4.5	3.97	17.75	33.3
3.0	2.93	1.72	NC[c]
1.5	NI[d]	NI	—

[a]Average of five replicate cultures per variable per day.
[b]Calculated from cell numbers observed on days 1 and 4.
[c]NC = not calculated.
[d]NI = no intact cells.

attempts, was that the RH derivative continuously subcultured in MB 752/1 rapidly became pycnotic and died when planted at the low inoculum level used for the other RH derivative, though high cell densities were achieved in stock cultures. An evaluation of the minimum inoculum size (Table V) supported the theory that a more fastidious cell population had been selected by continuous growth in MB 752/1. The calculated population doubling time observed at the highest inoculum level, 33.3 hours, is in good agreement with that for RH cells continuously subcultured in MEM-PD (Taylor *et al.*, 1972c). In an attempt to mitigate against physical trauma for low cell inocula, protective agents, such as 0.12% w/v Methocel (Bryant *et al.*, 1961; Nagel *et al.*, 1963; Bryant, 1969; Medsona and Merchant, 1971) or 0.5 mg of polyvinylpyrrolidone (PVP) per milliliter or culture medium (Katsuta *et al.*, 1959; Gwatkin and Haidri 1973), were incorporated into the planting medium, unsupplemented MB 752/1. In cultures inoculated with $\sim 1.5 \times 10^5$ RH cells/T-15, the cells attached to the floor of the flask, then died as before despite the protective agents. This suggested that, rather than a fragile subpopulation of cells, the medium per se was exerting a deleterious influence, which was exaggerated at a sparse cell density.

To test this explanation, replicate cultures of RH continuously subcultured in Dulbecco-Vogt's MEM were planted and randomized. After 2 hours of incubation to allow attachment, the planting medium was renewed and half the culture received PD-supplemented MB 752/1 while the others were fluid changed with PD-supplemented Dulbecco-Vogt's MEM. As seen in Table VI, the populations achieved in Dulbecco-Vogt's were quantitatively greater at the higher inocula, but the differences were not as remarkable as the 2.3-fold difference observed at the lowest inoculum level, 1.07×10^5 RH cells/T-15. This strongly suggested that the failure of RH cells to survive and proliferate when continuously subcultured and

TABLE VI

GROWTH OF PUMPER'S RABBIT HEART CELLS IN PEPTONE DIALYZATE SUPPLEMENTED DULBECCO-VOGT'S MEM AND MB 752/1

Seed inoculum ($\times 10^{-5}$ cells/T-15)	Cell number $\times$ 10^{-5}/T-15 after incubation in	
	Dulbecco-Vogt's MEM[a]	MB752/1[a]
4.5	61.6	51.8
3.0	44.9	37.0
1.07	24.7	10.9

[a] Average of cell populations in six replicate cultures, per variable.

planted at a low seed inoculum in MB 752/1 was a function of the medium itself. Waymouth found that the inability of MB 752/1 to support low L-cell inocula was due to a poor redox balance. The addition of a trace metal mixture stabilized the redox potential and permitted the successful initiation of strain L cell cultures at inocula levels less than half that required with MB 752/1 (C. Waymouth, personal communication).

In summary, if test cells do not survive the planting procedure, one reason may be the formulation or quality of the culture medium per se.

C. The Monodispersed Cell Suspension

Ideally, replicate mammalian cell cultures should be seeded with a monodispersed cell suspension. In the case of NCTC 2071 or RH cells, the mechanical disruption of the cell sheet results in a monodispersed suspension. Unlike these lines and many bacteria, however, most mammalian cell lines aggregate into multicellular clusters when the cell sheet is disrupted mechanically, enzymatically, or chemically. When planted in a culture vessel, these clusters attach and soon necrose because of the high cell number to surface area ratio. Thus, for quantitative studies, clumping must be kept to a minimum. Because of its known propensity for clumping, Hull's MK_2 line (Hull *et al.*, 1956) adapted to continuous growth in peptone-supplemented medium (Pumper *et al.*, 1974) was used to evaluate dispersing methods. Several sister T-15 cultures were prepared, and after 6 days' incubation at 37.5° ± 0.15°C (Fig. 2a) each were subcultured to two Falcon plastic T-25 flasks by different methods. The first culture was mechanically disrupted, and the cell population was gently aspirated in about 1 ml of culture medium. When visually well dispersed, the suspension was planted (Fig. 2b). Clumping to the degree seen in this culture would be unsatisfactory for replicate culture preparation. To reduce the number and size of the clumps, a sieving pipette fitted at the tip with a disk of No. 80 mesh platinum

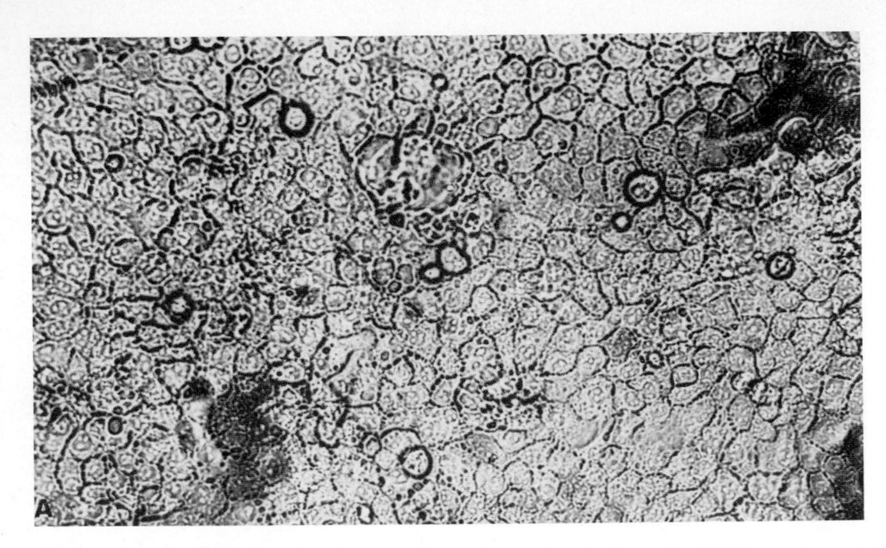

FIG. 2a. Six-day-old culture of MK_2 cells continuously subcultured in MEM supplemented only with 1% w/v Bacto-peptone and dextrose at a final concentration of 200 mg/100 ml (Pumper *et al.*, 1974). These cells are designated MK_2-BPD. ×41.

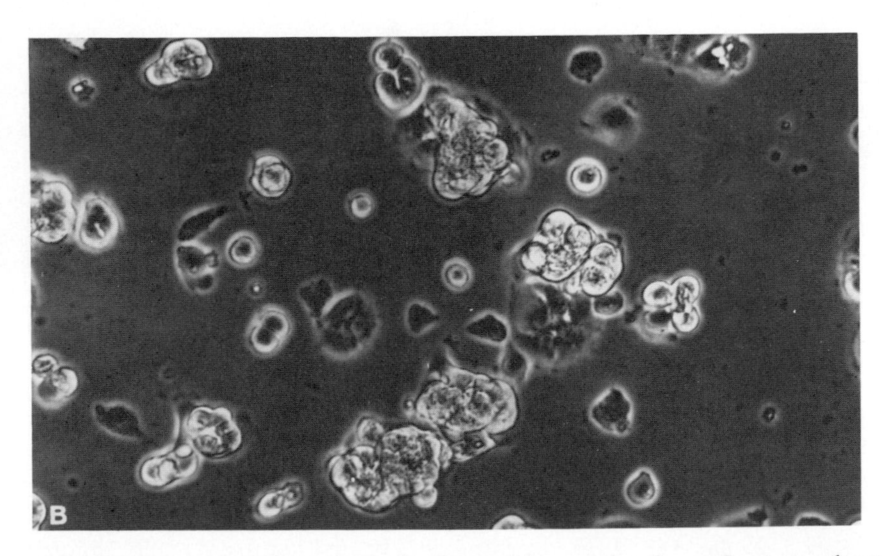

FIG. 2b. MK_2-BPD cells mechanically dispersed by mopping the confluent monolayer (Fig. 2a) with a sterile Cellophane strip, photographed approximately 2 hours after planting. The round refracted cells are in the process of attachment. Phase contrast photomicrography. ×41.

gauze (Kontes Glass Co., Vineland, New Jersey) was used. The cell suspension was aseptically transferred to a sterile test tube and gently aspirated 5–10 times by means of a sieving pipette. As can be seen in Fig. 2c, the number of dispersed cells is greatly enhanced and the remaining clumps are smaller than those in Fig. 2b and each contain approximately the same number of cells; for replicate culture purposes, this suspension should be aspirated 5–10 additional times before use. No significant loss of viability was observed in growth studies performed with cells dispersed in this way. A limitation of the sieving pipette is that very large clumps of cells may not be disrupted and will not pass through the gauze, but merely be held against it as the culture medium is pulled into the pipette. When one aspirates the fluid back into the test tube, these clumps will be washed back into the "monodispersed" suspension. Hence, one should discard the first 1–1.5 ml of suspension discharged from the sieving pipette to eliminate these clumps from the planting suspension.

A third sister flask was treated with 0.25% "trypsin" in Dulbecco's PBS. The cells dispersed well, but when planted in MEM supplemented with 10% v/v FBS, huge clusters of cells appeared (Fig. 2d). When, however, the fourth sister culture was treated with a trypsin-EDTA preparation (American Type Culture Collection formulation) a well dispersed cell population was obtained (Fig. 2e). While these cultures serve as examples, it must be

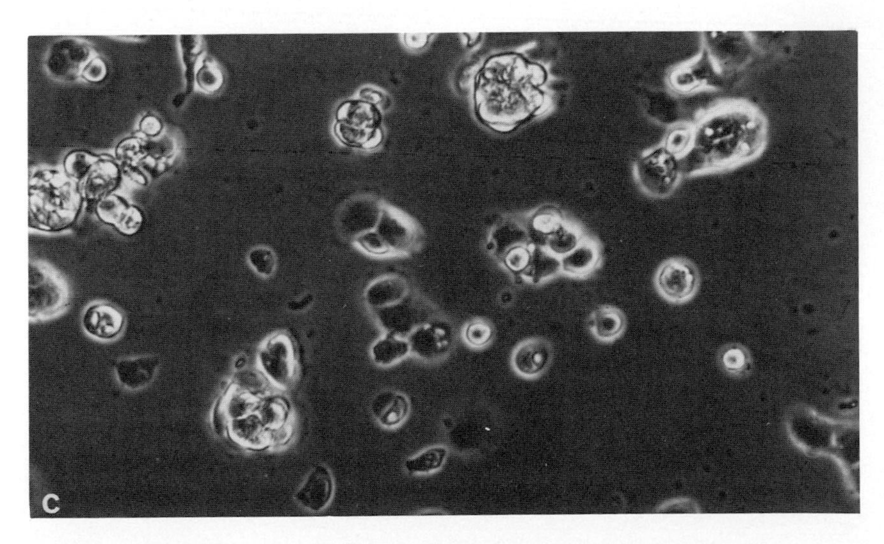

FIG 2c. MK_2-BPD cells mechanically disrupted by scraping with a Cellophane mop, then further dispersed with a sieving pipette. Phase contrast photomicrography, approximately 2 hours after planting. ×41.

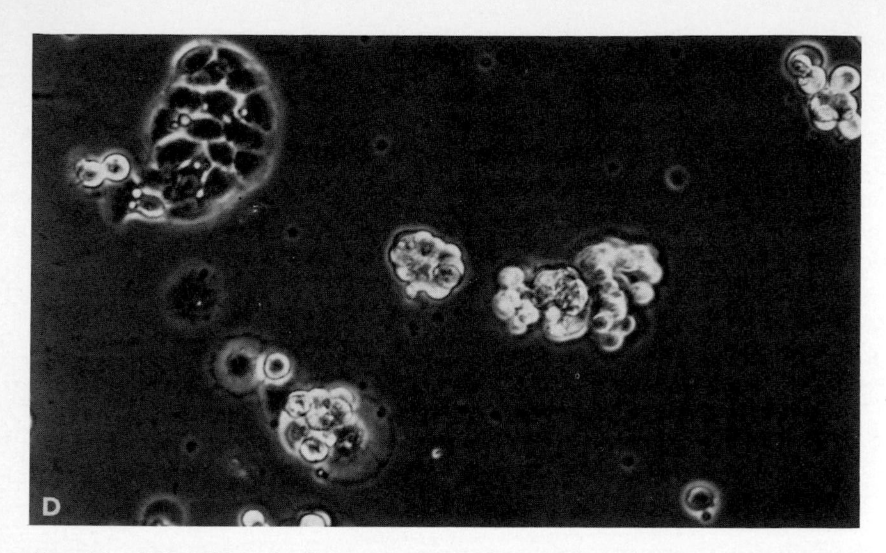

FIG. 2d. MK_2-BPD cells dispersed with 0.25% trypsin in Dulbecco's PBS. The cell suspension was aspirated until a well dispersed suspension was obtained, then planted in MEM + 10% FBS. Upon incubation, massive clumps of cells formed and remained suspended in the culture medium. Phase contrast photomicrography, approximately 2 hours after planting. ×41.

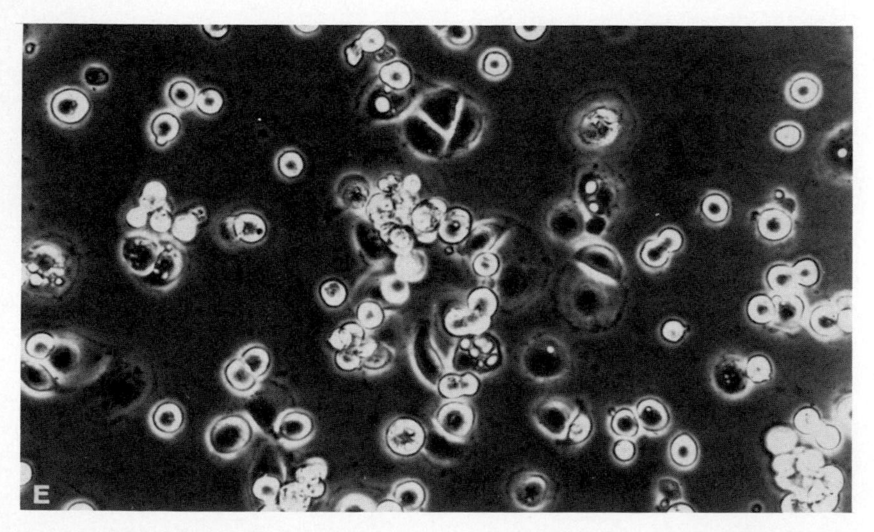

FIG. 2e. MK_2-BPD cell dispersed with a trypsin–EDTA–saline preparation (American Type Culture Collection formulation). Note the well dispersed population. Pilot studies to determine whether EDTA is cytotoxic or inhibits cell growth are advisable. Phase contrast photomicrography, approximately 2 hours after planting. ×41.

remembered that cells continuously subcultured in serum-free culture medium *cannot* be trypsinized because of the absence of a serum antitryptic factor in the culture medium. It should also be remembered that "tissue culture trypsin" is a desiccated mixture of gastric enzymes; unfortunately, crystalline trypsin dissociates explanted tissue poorly (Parker, 1961) and exhibits variable efficiency for confluent monolayers. Pine and co-workers analyzed several lots of "trypsin" for their ability to dissociate monkey kidney tissue and also for the presence of elastase, trypsin, chymotrypsin, ribonuclease, lipase, collagenase, and phosphatase (Pine *et al.*, 1969). They were unable to predict the biological efficacy of a "trypsin" preparation by the enzymes it contained or to specify which enzyme(s) promoted dissociation. While proteolytic enzymes including trypsin are able to release confluent monolayers from density-dependent growth inhibition (Burger, 1970; Sefton and Rubin, 1970; Pumper *et al.*, 1971), the precise mechanism is unknown. Germane to this may be the presence of physiologically significant (Thomas and Johnson, 1967; Higuchi, 1970) trace elements in "trypsin"; $Zn^{2}+$, $Fe^{2}+$, $Ca^{2}+$, and $Mg^{2}+$, among others, were found in two different "trypsin" preparations used successfully in this laboratory. In the same context, the use of chelating agents, such as EDTA, is thought to interfere with DNA synthesis (Lieberman and Ove, 1962; Rubin, 1972).

D. Priming the Device

In our hands, two problems have arisen which are artifacts of either the design or sterilization of the replicate culture planting device. Both occur at the outset of planting after the device is in place (Fig. 1a). The first is characterized by an inability to draw culture medium and cells up the withdrawal tube and into the syringe. The culture fluid may rise slightly in the withdrawal tube—only to fall back again, as if there were a leak in the system. This frequently is due to a sticking valve, and may be corrected by gently pinching the tubing immediately adjacent to the two-way check valve, depressing rapidly, then releasing the syringe barrel, then releasing the tubing. This usually will solve the problem, although the process may need to be repeated several times.

The second problem is characteristic of units with a 1- or 2-ml Cornwall syringe pipette. Since the volume displaced by the syringe barrel is relatively small, this type is best primed by siphoning the culture fluid and cells into the syringe assembly. Once the reservoir is positioned eccentrically, remove the syringe pipette from the extension clamp (see Section III, B,4,b) and position it below the level of the culture medium in the reservoir. Aspirate the syringe to draw the cell suspension into it, then resecure the syringe pipette with the slamp. Proceed as in Section III,B,4,c.

E. The Gas Mixture

In planting low numbers of cells in the fashion described above, certain of the parameters normally associated with clonal growth undoubtedly influence the survival and growth of the test cells. One of these factors is the oxygen content of the gas used to maintain a physiological pH in both the cell and medium reservoir and the culture flask. While earlier studies showed that the O_2 content of the gas mixture used did not prevent or consistently delay the neoplastic conversion of C3Hf/He mouse embryo cells *in vitro*, increasing the gas O_2 content to 40 or 60% inhibited cell proliferation (Sanford and Parshad, 1968). Subsequent studies on the karyotypic stability of these cells demonstrated that cultures continuously gassed intermittently with 10% CO_2/20% O_2N_2 underwent a more rapid shift from diploidy and had a higher frequency of abnormal chromosomes than sister cultures gassed with 10% CO_2/2% O_2/N_2 (Parshad and Sanford, 1971). Recent studies indicate that both rat and mouse cultures have an equal or better single cell plating efficiency when gassed with a low O_2 mixture (10% CO_2/1–3% O_2/N_2) than when an atmospheric O_2 content is employed (10% CO_2/20% O_2/N_2) (Richter *et al.*, 1972). Hence, the gas used to aerate the culture medium must be considered a component of the cellular milieu and may be influential when low inoculum sizes are used.

V. Quantitation of Results

Several methods are available for quantitating *in vitro* cellular responses to metabolites, drugs, microbes, etc. These include total and viable cell counts, isotopic labeling of cellular constituents which increase linearly with cell number during exponential phase growth, and total protein determination. These and other methods are detailed elsewhere (Merchant *et al.*, 1964). Described below are the procedures used for quantitative studies in this laboratory.

A. Harvesting the Cell Population

After an appropriate incubation period, the adherent cell population in each replicate culture flask is dispersed by trypsinization. The precise way in which this is done, however, will depend upon: (a) whether the culture medium contained serum, and (b) whether a microscopic (visual) or electronic counting procedure will be used.

If the culture medium used in the experiment contained serum, the cell sheet may need to be rinsed twice (3–5 seconds each time) with the trypsin preparation. In this way the residual serum will be removed and the cell dispersal facilitated. If the culture medium did not contain serum or a protein which is an antitrypsin factor, a small volume of serum must be used to (a) prevent cellular digestion, and (b) prevent clumping of the intact, dispersed cells. For example, in metabolic studies using NCTC 2071 (strain L) continuously subcultured in NCTC 135, we noticed that a large percentage of a confluent cell population was lost in the harvesting procedure. To ascertain whether this were due to cell digestion by trypsin, the cell population in several 5-day-old sister cultures were dispersed, each in a slightly different fashion. It was found that, for this cell line, if serum were added to the culture vessel *prior* to the "trypsin" preparation, a typical cell count of 2.75×10^6 cells/T-15 was obtained; if serum or Methocel (Dow Chemical Co., Midland, Michigan) were added after 15 minutes of incubation at $37.5° \pm 0.15°C$, greater than 90% of the cell population was digested by the enzyme mix. The serum, therefore, probably acts as a crude titrant for the "trypsin" preparation for this cell line. Equally important, a small volume of serum efficiently prevents clumping of the dispersed cells. Finally, the trypsinization time is a variable; in the case of strain L or L-M, we add 0.1 ml of serum/T-15, then 5 ml of 0.25% "trypsin" and incubate at $37.5° \pm 0.15°C$ for 10–15 minutes; for Pumper's RH cells, we use 5 ml 0.25% "trypsin" at room temperature for 10–15 seconds, at which time serum is added to prevent clumping. It is obvious that the best conditions must be determined for each cell line used; in practice we select the simplest set of conditions which dispense $> 99\%$ of the cell population with the least cell trauma.

B. Quantitative Procedures

1. Viable Cell Counts

Because of the inherent error in hemacytometer counts, we use a minimum volume (0.9–1.4 ml) of 0.25% trypsin per T-15 or T-25 with 0.1 ml of serum. Counting error can be minimized by resuspending a maximum number of cells in a minimum volume of trypsin, since this provides a greater number of cells per 0.5 ml. For the cell number enumeration, 0.5 ml of the suspension is mixed with 1.0 ml of saline containing 0.1% Trypan blue and counted by the standard method.

Due to the clumping problems described in Section IV, a combination of trypsin and EDTA has proved useful for MK_2 cells. A well dispersed suspension of WI-38 can be maintained without clumping in *ice cold* Hank's saline and serum (L. P. Rutzky, personal communication).

2. Coulter Counting

This is the method routinely used for nutritional studies in this laboratory. To each T-15 culture (containing 3 ml of culture medium), 5–6 ml of 0.25% trypsin is added together with 0.1 ml of serum in the appropriate sequence, and the cultures are incubated on a shaker table. Once > 99% of the population is monodispersed, 5–6 ml of 20% v/v formaldehyde solution in Isoton is added to prevent microbial contamination, inactivate the "trypsin," and stabilize the cell membrane (Pruden and Winstead, 1964).[2] The suspension is decanted to a 25-ml graduate, the flask is rinsed with 3–5 ml of Isoton, and the rinse fluid is used to bring the cell suspension to a final volume of 25 ml. The suspension is transferred to a 6-dram screw-cap vial for counting.

There are two points to be remembered regarding electronic counters. First, the average cell numbers observed by hemacytometer enumeration and electronic counting may vary by as much as 100%. We have observed this variation with fibroblastlike cells, and particularly in the case of primary or secondary cell strain. This disparity may be due to cellular geometry and/or electrical properties, or in human and mechanical error intrinsic in visual cell counts. For this reason, a concentrated cell suspension should be counted both ways to determine whether there are any significant differences, and if so a correction factor should be calculated. The procedure used to standardize our Coulter Counter appears in the Appendix. Though perhaps less accurate, we have found this easier than adjusting the 1/amplification and l/aperture current settings for each cell line. Second, some investigators have recommended the use of a formalized erythocyte preparation as standard reference for the proper function and calibration of the instrument (Torlontano and Tata, 1971). Appropriately sized latex particles can serve the same purpose. In practice, we interpret a count of < 50 from a freshly drawn Isoton sample as evidence of proper instrument function. Commercially available pollens are also available as standards, but we recently discontinued their use since they are allergins and constitute a potential health hazard for personnel. Hematology standards are commercially available but generally have a limited half life.

3. Total Protein

The Lowry method for protein determination as modified for tissue culture systems (Oyama and Eagle, 1956) is a convenient means of augmenting

[2]Since Isoton contains 0.1% sodium azide, a reduced formalin concentration may be desirable. Care should be exercised in adding this to the dispersed cell suspension as, in the absence of exogenous protein, the formaldehyde-Isoton mixture may cause clumping of the dispersed cells and preclude an accurate cell count.

cell count data. Obviously the thorough washing of the cell sheet with saline is imperative for cultures grown in protein-supplemented medium.

Appendix
Standardization of Hemacytometer and Coulter Counting Methods

1. *Cell suspension*. Select a typical, late-logarithmic phase culture (e.g. 5 days, confluent) and trypsinize (0.125%) as usual. Then add 20% v/v formaldehyde solution in Isoton mixture. Shake well. Decant to graduate and label as concentrated suspension.

2. *Hemacytometer count*. Mix 0.5 ml concentrated cell suspension with 1.0 ml saline–0.1% trypan blue (i.e., 1.5 ml final volume). To determine the cell number/0.5 ml of concentrate, use

$$\frac{\text{(total cell number per 9 chambers) 1500}}{0.9} = \text{cells/0.5 ml}$$

where the total cell number is the sum of the cell number in the 9 squares examined. Count a total of 36 hemacytometer chambers; average the four values obtained. Divide by 12.5. This is a *corrected* value with respect to the Coulter Counter (see below); from the correction chart, find *uncorrected* value.

3. *Coulter Counter*. Dilute 2.0 ml of concentrated cell suspension. Set the 1/amplification = 8, 1/aperture current = 2; lower threshold = 10, upper threshold = 100. Take 10 readings and record. Average. This figure represents the *uncorrected* cell number/0.5 ml.

4. To calculate correction factor for these settings, divide uncorrected hemacytometer count by uncorrected Coulter count.

Acknowledgments

The authors thank Drs. Marilyn J. Taylor and Robert W. Tucker for critical evaluation of the manuscript; Floyd M. Price, Robert A. Dworkin, and Avery Kerr for technical assistance; Miriam Hursey and Raymond Steinberg for assistance with the photomicrography; and Marian Gundy of the Statistical Drafting Unit, Medical Arts Section, for the line drawing. Finally, we gratefully acknowledge the cheerful assistance of Mrs. Nancy V. Carter during all phases of this manuscript's preparation.

References

Anderson, N. G. (1953). *Science* **117**, 627.

Boone, C. W., Mantel, N., Caruso, T. D., Jr., Kazam, E., and Stevenson, R. E. (1971). *In Vitro* **7**, 174.

Bryant, J. C. (1969). *Bioeng. Biotechnol.* **XI**, 155.

Bryant, J. C., Evans, V. J., Schilling, E. L., and Earle, W. R. (1961). *J. Nat. Cancer Inst.* **26**, 239.
Burger, M. M. (1970). *Nature (London)* **227**, 170.
Coriell, L. L., and McGarrity, G. J. (1968). *Appl. Microbiol.* **16**, 1895.
Eagle, H. (1959). *Science* **130**, 432.
Earle, W. R., and Highhouse, F. (1954). *J. Nat. Cancer Inst.* **14**, 841.
Earle, W. R., and Nettleship, A. (1943). *J. Nat. Cancer Inst.* **4**, 213.
Evans, V. J., Earle, W. R., Sanford, K. K., Shannon, J. E., and Waltz, H. K. (1951). *J. Nat. Cancer Inst.* **11**, 907.
Evans, V. J., Bryant, J. C., Fioramonti, M. C., McQuilkin, W. T., Sanford, K. K., and Earle, W. R. (1956a). *Cancer Res.* **16**, 77.
Evans, V. J., Bryant, J. C., McQuilkin, W. T., Firoamonti, M. C., Sanford, K. K., Westfall, B. B., and Earle, W. R. (1956b). *Cancer Res.* **16**, 87.
Evans, V. J., Bryant, J. C., Kerr, H. A., and Schilling, E. L. (1964). *Exp. Cell Res.* **36**, 439.
Fedoroff, S., Evans, V. J., Hopps, H. E., Sanford, K. K., and Boone, C. W. (1971). *In Vitro* **7**, 161.
Fioramonti, M. C., Evans, V. J., and Earle, W. R. (1958). *J. Nat. Cancer Inst.* **21**, 579.
Fogh, J., Holmgren, N. B., and Ludovici, P. P. (1971). *In Vitro* **7**, 26.
Gey, G. O. (1941). *Cancer Res.* **1**, 737.
Goldsby, R. A., and Zipser, E. (1969). *Exp. Cell Res.* **54**, 271.
Gwatkin, R. B. L., and Haidri, A. A. (1973). *Exp. Cell Res.* **76**, 1.
Gwatkin, R. B. L., and Thompson, J. L. (1964). *Nature (London)* **201**, 1242.
Hayflick, L., and Moorhead, P. S. (1961). *Exp. Cell Res.* **25**, 585.
Higuchi, K. (1970). *J. Cell. Physiol.* **75**, 65.
Hsu, T. C., and Merchant, D. J. (1961). *J. Nat. Cancer Inst.* **26**, 1075.
Hull, R. N., Cherry, W. R., and Johnson, I. S. (1956). *Anat. Rec.* **124**, 490.
Katsuta, H., Takaoka, T., Oishi, Y., Baba, F., and Chang, K. C. (1954). *Jap. J. Exp. Med.* **24**, 125.
Katsuta, H., Takaoka, T., Mitamura, K., Kawada, I., Kuwabara, H., and Kuwabara, S. (1959). *Jap. J. Exp. Med.* **29**, 191.
Kono, T. (1969). *Biochim. Biophys. Acta* **178**, 397.
Kreider, J. W. (1968). *Appl. Microbiol.* **16**, 1804.
Kruse, P. F., and Patterson, M. K., ed. (1973). "Tissue Culture: Methods and Applications." Academic Press, New York.
Lasfargues, E. Y. (1957). *Anat. Rec.* **127**, 117.
Lasfargues, E. Y., and Moore, D. H. (1971). *In Vitro* **7**, 21.
Lieberman, I., and Ove, P. (1962). *J. Biol. Chem.* **237**, 1634.
McGarrity, G. J., and Coriell, L. L. (1971). *In Vitro* **6**, 257.
McQuilkin, W. T., and Earle, W. R. (1962). *J. Nat. Cancer Inst.* **28**, 763.
Malinin, T. (1966). "Processing and Storage of Viable Human Tissues," Pub. Health Serv. Publ. No. 1442, p. 67. U.S. Gov. Printing Office, Washington, D.C.
Martin, G. M. (1964). *Nature* (*London*) **201**, 1338.
Medsona, E. L., and Merchant, D. J. (1971). *In Vitro* **7**, 46.
Merchant, D. J., and Hellman, K. B. (1962). *Proc. Soc. Exp. Biol. Med.* **110**, 194.
Merchant, D. J., Kahn, R. H., and Murphy, W. H. (1964). "Handbook of Cell and Organ Culture", pp. 155–174. Burgess, Minneapolis, Minnesota.
Merril, C. R., Friedman, T. B., Attallah, A., Geier, M. R., Krell, K., and Yarkin, R. (1972). *In Vitro* **8**, 91.
Molander, C. W., Kniazeff, A. J., Boone, C. W., Paley, A., and Imagawa, D. T. (1971). *In Vitro* **7**, 168.

Morgan, J. F., Morton, H. J., and Parker, R. C. (1950). *Proc. Soc. Exp. Biol. Med.* **73**, 1.
Morton, H. J. (1970). *In Vitro* **6**, 89.
Nagle, S. C., Jr., Tribble, H. R., Anderson, R. E., and Gray, N. D. (1963). *Proc. Soc. Exp. Biol. Med.* **112**, 340.
Oyama, V. I., and Eagle, H. (1956). *Proc. Soc. Exp. Biol. Med.* **91**, 305.
Parker, R. C. (1961). "Methods in Tissue Culture," p. 120. Harper (Hoeber), New York.
Parshad, R., and Sanford, K. K. (1971). *J. Nat. Cancer Inst.* **47**, 1033.
Phillips, H. (1972). *In Vitro* **8**, 101.
Pine, L., Taylor, G. C., Miller, D. M., Bradley, G., and Wetmore, H. R. (1969). *Cytobios* **2**, 197.
Poste, G. (1971). *Exp. Cell Res.* **65**, 359.
Pruden, E. L., and Winstead, M. E. (1964). *Amer. J. Med. Technol.* **30**, 1.
Pumper, R. W., Yamashiroya, H. M., and Molander, L. T. (1965). *Nature (London)* **207**, 662.
Pumper, R. W., Fagan, P., and Taylor, W. G. (1971). *In Vitro* **6**, 266.
Pumper, R. W., Taylor, W. G., Parshad, R., and Rutzky, L. P. (1974). In preparation.
Richter, A., Sanford, K. K., and Evans, V. J. (1972). *J. Nat. Cancer Inst.* **49**, 1705.
Rinaldini, L. M. (1959). *Exp. Cell Res.* **16**, 477.
Rubin, H. (1972). *Proc. Nat. Acad. Sci. U.S.* **69**, 712.
Sanford, K. K., and Parshad, R. (1968). *J. Nat. Cancer Inst.* **41**, 1389.
Sanford, K. K., Earle, W. R., and Likely, G. D. (1949). *J. Nat. Cancer Inst.* **9**, 229.
Sefton, B. M., and Rubin, H. (1970). *Nature (London)* **227**, 843.
Stanbridge, E. (1971). *Bacteriol. Rev.* **35**, 206.
Stulberg, C. S., Coriell, L. L., Kniazeff, A. J., and Shannon, J. E. (1970). *In Vitro* **5**, 1.
Suzuki, F., Kashimoto, M., and Horikawa, M. (1971). *Exp. Cell Res.* **68**, 476.
Syverton, J. T., Scherer, W. F., and Elwood, P. M. (1954). *J. Lab. Clin. Med.* **43**, 286.
Taylor, W. G., Price, F. M., Dworkin, R. A., and Evans, V. J. (1972a). *In Vitro* **7**, 295.
Taylor, W. G., Dworkin, R. A., Pumper, R. W., and Evans, V. J. (1972b). *Exp. Cell Res.* **74**, 275.
Taylor, W. G., Taylor, M. J., Lewis, N. J., and Pumper, R. W. (1972c). *Proc. Soc. Exp. Biol. Med.* **139**, 96.
Thomas, J. A., and Johnson, M. J. (1967). *J. Nat. Cancer Inst.* **39**, 337.
Torlontano, G., and Tata, A. (1971). *Acta Haematol.* **45**, 325.
Vogt. M., and Dulbecco, R. (1963). *Proc. Nat. Acad. Sci. U.S.* **49**, 171.
Wallis, C., and Melnick, J. L. (1960). *Tex. Rep. Biol. Med.* **18**, 670.
Waymouth, C. (1959). *J. Nat. Cancer Inst.* **22**, 1003.
Waymouth, C. (1965). *In* "Tissue Culture" (C. W. Ramakrishnan, ed.), p. 168. Junk, The Hague.
Waymouth, C. (1970). *In Vitro* **6**, 109.
Weirether, F. J., Walker, J. S., and Lincoln, R. E. (1968). *Applied Microbiol.* **16**, 841.
Wilson, B. W., and Lau, T. L. (1963). *Proc. Soc. Exp. Biol. Med.* **114**, 649.
Zwilling, E. (1954). *Science* **120**, 219.

Chapter 5

Methods for Obtaining Revertants of Transformed Cells[1]

A. VOGEL[2] AND ROBERT POLLACK

Cold Spring Harbor Laboratory, Cold Spring Harbor, New York

I. Introduction

Although *in vitro* cell culture techniques have facilitated the investigation of the mechanisms involved in viral carcinogenesis, we as yet know very little about the functions involved in maintaining the "normal" state of growth

[1]This investigation was supported by NIH Cancer Center Grant No. 1-PO1-CA13106-01.
[2]A. Vogel is supported by NIH training grant 5 TO GMO1668 from the National Institute of General Medical Science.

control, and consequently we know little about the mechanism(s) used by viruses to alter this regulation. There is no single path to understanding these functions.

Some investigators are searching for the existence of viral coded molecules (proteins and nucleic acids) in transformed cells with the hope of discovering a molecule required for either the induction or the maintenance of the transformed state.

Other investigators are looking at the properties of the transformed cells themselves, to understand how they differ from normal cells. Many differences have been established between normal and transformed cells, but no one yet knows if these differences are the primary result of the transformation process, or if they reflect changes secondary to the initial alternation. An excellent recent review by Sambrook summarizes much of this work (Sambrook, 1972).

Our way of investigating these alterations has been to isolate sublines of transformed cells which have reverted in one or many of the transformed properties.

Three systems for the isolation of transformed cells have been developed, based upon the ability of the transformed cells to grow in conditions where the normal cells cannot (Table I). These three assays measure (1) the maximum cell density attained by a line in excess serum, (2) the ability of a cell to establish an isolated colony suspended in agar or Methocel, and (3) the ability of a cell line to grow in limiting or depleted sera.

A. Density

Most normal cell lines grow in an oriented fashion and exhibit a density-dependent cessation of cell division. For example, untransformed 3T3 mouse cells stop growing at a saturation density of 5×10^4 cells/cm^2 whereas

TABLE I
SELECTIVE ASSAY FOR TRANSFORMATION

	System		
Assay	Normal	Transforming agent	Initial reference
Growth to high density	3T3 mouse	SV40	Todaro and Green (1964)
	Chick cells	RSV	Temin and Rubin (1958)
Growth suspended in gel	BHK hamster	polyoma	Macpherson and Montagnier (1964)
Growth in reduced serum	3T3 mouse	SV40	Smith *et al.* (1971)

BHK cells cease division at 25×10^4 cells/cm^2. Infection of 3T3 cells by SV40 or murine sarcoma virus results in dense transformed foci appearing on the contact-inhibited monolayer. These dense clones, when isolated from the monolayer and grown on plastic dishes, grow to higher saturation densities ($> 20 \times 10^4$ cells/cm^2) than the untransformed 3T3 cells. Transformants of BHK cannot be isolated in such a way because these normal cells grow to high densities.

B. Anchorage

Polyoma virus transformants of BHK can be isolated by an assay involving growth in semisolid medium. Normal cells (both BHK and 3T3) will grow only when attached to a plastic or glass surface and are therefore said to have a high anchorage requirement for growth. Infection by oncogenic viruses (containing either DNA or RNA) results in a decrease in this anchorage requirement. Transformed cells will grow into spherical colonies when suspended in a semisolid medium of agar or methocel. The transformed colonies picked in agar grow to high saturation density when cultured on plastic. When cultured on plastic tissue culture dishes, polyoma-transformed BHK cells grow in a randomly oriented manner and continue to divide until they eventually turn the medium acidic and die.

C. Serum

γ-Globulin-free calf serum incubated with confluent monolayers of Balb 3T3 loses the ability to support the growth of normal 3T3 cells. This depleted serum will allow SV40-transformed 3T3 cells to grow. Thus, serum transformants can be isolated by infecting 3T3 cells with SV40 and plating the infected cells in this agamma-depleted medium, because only transformed cells will grow (Smith *et al.*, 1971).

Transformed cells isolated in each of these systems display other properties which differentiate them from the parental normal cells (Table II). The most important difference is in tumorigenicity. In general, transformants form tumors when injected into an animal, while normal cells do not. Transformed cells exhibit new antigens. Some of these are presumably virus-coded (Black *et al.*, 1963), while other antigens are cell-coded, but induced by the virus (Hayri and Defendi, 1970). Alterations in the surfaces of cells occur upon transformation. The most completely studied such alteration is measured as an increased agglutinability by plant lectins such as concanavalin A and wheat germ agglutinin. Finally, transformed cells maintain a lower internal concentration of cyclic AMP than do normal cells.

TABLE II
NONSELECTIVE PROPERTIES OF TRANSFORMED LINES

Transformed property	System: Normal	System: Transforming agent	Reference
Increased tumorigenicity	3T3	SV40	Aaronson and Todaro (1968)
	Hamster	DMBA	Chen and Heidelberger (1969)
Increased agglutinability with lectins	3T3	SV40	Burger (1969)
	Mouse	MSV	Burger and Martin (1972)
	Hamster	Py	Inbar and Sachs (1969)
Decreased level of cAMP	3T3	SV40, Py	Otten *et al.* (1971); Sheppard (1972), Oey *et al.* (1973)
Increased sensitivity to drugs affecting cell morphology			
Colchicine	3T3	SV40	Vogel *et al.* (1973)
Cytochalasin B	3T3	SV40	Kelly and Sambrook (1973)
Growth on monolayer of normal cells	3T3	SV40	Pollack *et al.* (1968)
New antigens	3T3	SV40	Black *et al.* (1963)

II. Isolation of Revertants from Transformed Cell Lines

A. Definition of a Revertant

A revertant cell line lacks at least one of the properties of a transformed cell line from which it is descended. A revertant may be *selected* by a modification of the protocol used to isolate the transformed parent. Three different assays for selecting transformants are in current use (Table I); therefore three different selective assays for reversion will be described. Once selected, a revertant may be *described* as either transformed or not with regard to the other two transformation properties in Table I, and also with regard to other nonselective properties of transformed cell lines (Table II).

One obvious pathway to reversion is to cure the transformant of its virus. This approach was used successfully by Marin and Littlefield (1968). Using somatic cell hybridization and the appropriate selection system, they were able to induce the loss of the viral genome from polyoma-transformed BHK cells. Having lost the viral genome, these cell lines now appear normal. They no longer contain the polyoma tumor antigen, are unable to grow in agar medium, and grow in a similar manner to the parental BHK cells.

In most cases, however, reversion is not the result of a lost viral genome, since most revertant cell lines still contain the viral coded antigens associated

TABLE III

REVERTANT CELL LINES ISOLATED THROUGH NEGATIVE SELECTION

Revertant selected for	Parent transformant	Selective agent	Transformed properties				
			Density	Serum	Anchorage	Agglutinability	Tumorigenicity
Density	SV3T3	FUdR[a,b]	Normal[p]	Transformed	Normal	Normal[n]	NT
Density	SV3T3	BUdR-light[c]	Normal	Transformed	NT[o]	NT	NT
Density	SV3T3	Colchicine[d]	Normal	Normal	Normal	NT	NT
Density	PyBHK	FUdR[a]	Normal	NT	NT	NT	Reduced [a,m]
Density	Methylcholanthrene[l] transformed C3H mouse prostate cells[l]	FUdR	Normal	NT	NT	NT	Reduced[l]
Serum–1% calf serum	SV3T3	BUdR-light[c]	Normal	Normal	Normal	Normal	NT
Serum-agamma depleted calf serum	SV3T3	BUdR-light[c]	Normal	Normal	Transformed	Transformed	NT
Anchorage	PyBHK	BUdR-light[e]	Normal	Normal	Normal	NT	Transformed
Concanavalin A resistance	SV3T3	Concanavalin A[f,g]	Normal	Transformed	Normal	Normal	NT
Growth on glutaraldehyde monolayers	Py-hamster	Glutaraldehyde fixed cells[h]	Normal	NT	Normal	Normal	Reduced
	Methylcholanthrene-transformed C_3H mouse	Glutaralde hyde fixed cells[l]	Normal	NT	NT	NT	Reduced
No selection spontaneously arising revertants	MSV-3T3	—	Normal	NT	Normal	NT	NT
	MSV-NRK[j]	—	Normal	NT	NT	NT	NT
	RSV-NRK[j]	—	Normal	NT	NT	NT	NT
	RSV-BHK[k]	—	Normal	NT	Normal	NT	NT

[a] Pollack *et al.* (1968).
[b] Culp *et al.* (1971).
[c] Vogel and Pollack (1973).
[d] Vogel *et al.* (1973).
[e] Wyke (1971a,b).
[f] Ozanne and Sambrook (1970).
[g] Culp and Black (1972).
[h] Rabinowitz and Sachs (1968).
[i] Fischinger *et al.* (1972).
[j] Stephenson *et al.* (1972).
[k] Macpherson (1965).
[l] Mondal *et al.* (1971).
[m] Pollack and Teebor (1969).
[n] Pollack and Burger (1969).
[o] Not tested.
[p] Pollack and Vogel (1973).

with transformation (Culp *et al.*, 1971; Culp and Black, 1972; Ozanne with Sambrook, 1970; Pollack *et al.* (1968); Vogel *et al.*, 1973; Vogel and Pollack, 1973) as well as virus-coded RNA and DNA (Ozanne *et al.*, 1973; Shani *et al.*, 1972). The next most obvious way to isolate revertants is by observation. Cells with variant morphologies are present in cultures of transformed cells, and these variant cells have been isolated from the cultures and found to have lost transformed properties.

Finally, revertant cell lines have been isolated from both RNA and DNA virus transformed cell lines, and from populations of chemically transformed cells (Table III), through negative selection. These protocols will be described in detail below.

B. Negative Selection

This selection procedure involves placing a population of transformed cells in a situation which supports the growth of only transformed cells and then adding an agent which is toxic only to growing cells. The toxic drug and restrictive conditions are then removed, and conditions supporting the growth of cells with normal properties are restored to allow the recovery of surviving cells. Some of these cells escape the toxic effects of the drug because of their inability to grow and thereby have reverted to a more normal state.

For example, selection of revertants unable to grow beyond a monolayer in density involves plating transformed cells at high density and killing any cells capable of overgrowing the monolayer. Serum revertants are selected by placing the transformants in the restrictive serum and killing the cells capable of growth. Cells reverted in their ability to grow in semisolid medium are obtained by killing the cells in this medium, and recovering the survivors. Colonies grown up from cells surviving these selections can be tested directly for the reversion of the specific property or they can be recycled through the selection once or twice more to eliminate cells with transformed phenotype which have slipped through the killing process.

C. Agents Used in Negative Selection of Revertants

The choice of drug depends upon the type of selection desired. Selection for reversion in an ability to grow in certain conditions requires the use of a drug that kills growing cells. Selection for reversion in a surface property involves the use of compounds, such as concanavalin A, which are more toxic to transformed cells than to normal cells.

Among the substances specifically toxic growing cells are antimetabolites known to affect cells in different phases of the cell cycle. The most commonly used antimetabolites are the thymidine analogs bromodeoxyuridine (BUdR) and fluorodeoxyuridine (FUdR). Both these agents specifically affect cells synthesizing DNA. Although these drugs are both thymidine analogs, they kill by different mechanisms. BUdR is incorporated into newly synthesized DNA, and blue light will cause breaks in the DNA and lead to cell death (Kao and Puck, 1968). FUdR inhibits endogenous thymidine synthesis by inhibiting thymidylate synthetase, the enzyme that converts d-UMP to d-TMP (Rueckert and Mueller, 1960). The cells no longer synthesize thymidine and are killed. Thymidine can prevent the toxic effects of both FUdR and BUdR.

Dividing cells can also be killed by colchicine because of its ability to block cells in mitosis. This drug prevents the polymerization of microtubules (Borisey and Taylor, 1967) and thus prevents the completion of cytokinesis. Cells blokced in mitosis for long periods of time will die. In addition, colchicine is more toxic to growing SV40-transformed 3T3 cells than to growing 3T3 cells (Vogel *et al.*, 1973). This differential toxicity is not understood.

There are many other drugs that could probably be used as agents specifically toxic for growing cells. Cytosine arabinoside, 6-thioguanine, and aminopterin, are examples.

Toxicity of Selective Agents

Sparse and confluent cultures of the contact-inhibited mouse line 3T3 should be used to test a drug for specific toxicity to growing cells. At confluence, 3T3 cells are blocked in G_1 and do not synthesize DNA or enter mitosis (Nilhausen and Green, 1964). At sparse densities, 3T3 cells double in about a day. Treatment of confluent 3T3 cells with FUdR, BUdR, or colchicine does not affect the viability of the cells, but treatment of sparse 3T3 cultures with these agents will leave fewer than 1 cell in 10^4 alive. The following procedure can be used to evaluate the relative toxicity of drugs on growing versus nongrowing cells.

Seed 3T3 cells at varying cell densities in regular growth medium. Densities chosen should represent cell number which correspond to confluence, one-tenth confluence, and one-hundredth confluence (3T3 typically has a confluent saturation density of 5×10^4 cells/cm^2). One day after plating, count a plate at each density, and add the drug to a test plate at each density. A range of the times of exposure to the drug and of concentrations of the drug should be tested to give each cell a chance to go through the cycle. After exposure for at least a day, remove the medium containing the drug and add fresh medium for 2–3 hours. This incubation with fresh medium serves to remove any residual drug. Then remove this medium and trypsinize the remaining cells. Count and plate them at suitable dilutions to determine the number of cells capable of forming colonies. Plot the ratio of colony-forming survivors of the treated plates to colony formers on the control plates vs cell density at the time the drug was added. Drugs which are specific for growing cells kill the cells at non-confluent cell densities, but do not affect the confluent cells. Drugs toxic for nongrowing cells will show very few survivors at any cell density.

D. Selection of Density Revertants

The following selection procedures have been used to isolate variants of transformed cells which are reverted in the denisty, serum, or anchorage properties.

Transformed cells are seeded at a density of 5×10^4 cells/cm^2, and 24 hours later, the toxic agent is added. One can use per milliliter 25 μg of FUdR, 100 μg of BUdR, or 0.5 μg of colchicine. Selections with FUdR or BUdR should be done in the presence of a 10-fold excess of uridine to prevent the incorporation of any fluorouracil or bromouracil into RNA. The optimum time of exposure of these drugs is 2 days. Stock solutions of FUdR (1 mg/ml), BUdR (10 mg/ml), uridine (10 mg/ml), and colchicine (10 μg/ml) are dissolved in distilled water, sterilized by Millipore filtration, and stored frozen. After the 2 days in drug, the medium is removed, the cells are trypsinized and counted, and serial dilutions (with 10^{-1}, 10^{-2}, 10^{-3}, 10^{-4}, and 10^{-5} of the original population) are plated to determine the number of surviving colonies. Survivors will arise at a frequency of approximately 1 per 10^4 cells initially plated.

It is generally a good idea to put the cells through a second cycle of selection before colonies are picked and tested. After 7–10 days of growth, the 10^{-1} or 10^{-2} plates form the colony surviving experiment are trypsinized, pooled and replated at 5×10^4 cells/cm^2. The next day, the selective agent should be added and the procedure repeated as described above. A third selection may be done, but in general, two selections with FUdR, BUdR, or colchicine are sufficient. Colonies with revertant properties are recognized by their variant, more normal, flat morphology. These generally arise at a frequency of 1 per 10^5 cells initially plated. Variant colonies should be recloned and their saturation densities be determined. Revertant lines generally have saturation densities of 7 to 15×10^4 cells/cm^2.

E. Selection of Serum Revertants

Either 1% calf serum or 10% agamma-depleted calf serum can be used to differentiate 3T3 from SV40-transformed 3T3, since neither support the growth of 3T3. The 1% calf serum requires no preparation and supports the growth of transformed cells, but it reduces their doubling time and saturation density. The agamma-depleted serum has the advantage of not appreciably reducing the saturation density or doubling time of the transformants, but requires a good deal of preparation. Also, different batches of treated serum vary in their ability to differentiate between 3T3 and SV3T3. We have tried the heating method (Smith *et al.*, 1971) to deplete newborn agamma serum, but our preparations do not support the growth of SV3T3. The following depletion procedure developed by Smith and her colleagues, has worked for us (Smith *et al.*, 1971).

Allow Balb/3T3 to grow to a confluent monolayer in 100-mm Falcon tissue culture dishes in medium containing 10% calf serum. Balb/3T3 must be used because agamma serum depleted on Swiss/3T3 allows the growth of both Balb/ and Swiss/3T3. Remove the medium and add 10 ml of medium containing 20% newborn agamma calf serum (North American Biologicals Incorporated). Allow the plates to incubate for 4 days at 37°C, then remove the medium and centrifuge for 20 minutes at 5000 rpm at 4°C in a Sorvall GSA rotor to remove debris. Then filter through a Millipore apparatus and add an equal volume of serum-free medium to restore any small molecules that may have been depleted. This 10% agamma-depleted medium can be stored frozen. Balb/3T3 monolayers can be used 2 to 3 times to deplete serum. After removing one batch of depleted medium, add fresh 20% agamma serum and incubate for 4 days.

To select cells revertant in their ability to grow in restrictive sera, plate transformed cells at sparse cell density in either 10% agamma-depleted calf serum or 1% whole calf serum. These conditions will allow SV3T3, but not 3T3, to grow. Then add an agent to kill dividing cells, remove the drug and permit survivors to grow in 10% calf serum. One must be certain to start the selection at very sparse cell densities to avoid selecting cells which have ceased dividing because of a density-dependent shut-off of cell division. A specific protocol follows.

Seed the transformed cells at a density of 10^4 cells/cm^2 in 10% calf serum. The next day wash the cells in serum-free medium and add the restrictive serum. One day after adding the restrictive serum, add 200 μg of BUdR per milliliter and incubate for 4 days. At the end of 4 days remove the medium containing BUdR and add fresh 10% calf serum medium for 2 hours. Remove the medium and expose the cells to blue light or a short pulse of UV light. The time of exposure necessary should be determined by the investigator. A two-second UV pulse from a 30-W GE germicidal lamp 18 inches from the cells decreases the plating efficiency of untreated cells by 50%, but reduces the plating efficiency of cells exposed to BUdR to 10^{-3} to 10^{-4}. The cells should then be trypsinized, counted and plated out for colony survivors in 10% calf serum containing 10^{-5} *M* thymidine and 8×10^{-5} *M* uridine (to prevent the incorporation of any residual BUdR or bromouracil into nucleic acid).

It is advisable to put the cells through 2–3 cycles of selection before picking colonies. Trypsinize the treated plates, dilute them 15-fold, and replate them in 10% calf serum with uridine and thymidine. When the cells reach a density of 10^4 cells/cm^2, add the restrictive serum and, 1 day later, add the BUdR and repeat the procedure. After 2–3 cycles, surviving colonies should be picked, grown, and tested for their growth in the restrictive serum. Serum revertants generally show no growth in the restrictive serum, but sometimes lines can be isolated that can grow slightly in the restrictive serum, with a doubling time of 3–4 days.

F. Selection of Revertants in Agar

Either agar suspension or methocel suspension cultures can be used to select cells with a reduced anchorage dependence for growth. Methocel is the more advantageous medium because the cells are more easily recovered from it. Methocel is quite viscous and can be a problem to pipette, but wide-bore pipettes (Bellco) facilitate the dispensing of methocel.

1. Methocel Medium

The method of preparation is that of methocel medium, Vogel *et al.* (1973), based on Stoker (1968).

Make up a 2.5% methocel suspension by adding 2.6 gm of methocel (Dow Chemical Co. standard grade 4000 centipoises) to 25 ml of 80–90°C distilled water. Shake vigorously to disperse all clumps. Make up the rest of the volume with 4°C distilled water, shake vigorously, and stir on a magnetic stirrer for 3 hours at 4°C. Autoclave the methocel and then store at 4°C. The suspension will clarify at 4°C. Allow the methocel to remain at 4°C for at least 1 day before use.

Methocel medium is prepared by adding to the methocel an equal volume of twice-concentrated growth medium containing 20% calf serum and twice the concentration of antibiotics (final concentration of methocel = 1.3%). The methocel medium should be stirred on a magnetic stirrer at room temperature for a few hours before use.

Plates are prepared by adding 3 ml of agar medium (equal volumes of sterilized 1.8% Bacto Difco agar in water, and twice-concentrated growth medium containing 20% serum) to 60-mm tissue culture plates, tilting the plates to coat the sides, and allowing the agar to harden. The agar prevents the cells from attaching to the bottom or sides of the dish. Different amounts of cells (10^1 to 10^5 per milliter) are then suspended in the methocel medium, and 4 ml is added per agar plate. The plates are incubated for 3 weeks, 4 ml of fresh methocel medium being added every week. Visible colonies (0.3 mm or larger) should appear within 2 weeks. SV101 cells typically give a plating efficiency of 10–50% in methocel, while 3T3 has a plating efficiency of 0.001%.

Isolated colonies can be picked from the methocel with a 100-μl micropipette. The colony is placed in growth medium, vigorously suspended with a pipette, centrifuged, resuspended, and replated. Alternatively the picked colony can be dispersed in a small volume of trypsin–EDTA, centrifuged, and replated. It is also possible to harvest all the colonies on the plate by placing the plates of 4°C for 1 hour (methocel is less viscous at lower temperatures), removing the methocel medium with a pipette, washing the plates with 4 times the volume of 4°C serum free medium and centrifuging at 300 *g* at 4°C. The cells form a pellet at the bottom of the tube and the methocel can be poured off.

2. Selection of Revertants with an Increased Anchorage Requirement for Growth

The following procedure has been used by Wyke (1971a) to isolate variants of PyBHK no longer capable of growth in methocel.

Plate 10^5 transformed cells per milliter of methocel medium. Incubate the cultures at 37°C for 45 hours to allow cells that can grow in methocel to begin to do so. Add 5×10^{-6} *M* BUdR in 0.5 ml of serum-free medium. Continue the incubation for an additional 45 hours, and then harvest all the cells in the methocel as described above. After centrifugation, seed the cells into tissue culture dishes and incubate for 4–6 hours to allow the cells to attach. Then expose the plates to blue light to kill the cells that have incorporated BUdR into DNA. The duration of exposure to the light should be determined by the investigator. Add fresh growth medium to allow the surviving colonies to grow. The cells should be cycled through the selection procedure three times before colonies are picked and tested for their ability to grow in methocel.

G. Selection of Revertants with Altered Surface Properties

Plant lectins such as concanavalin (Con A) and wheat germ agglutinin (WGA) are used to demonstrate differences in surface properties between transformed and normal cell. (Burger, 1969; Inbar and Sachs, 1969). Transformed cells that have been removed from plates with EDTA are agglutinable whereas normal cells treated in a similar fashion are not. Trypsinization renders all cells equally highly agglutinable.

Spontaneously transformed cells, transformants induced by various oncogenic viruses, and chemically transformed cells all manifest this agglutinability. However, it appears that this difference in agglutination between normal and transformed cells does not arise because of the transformed cells binding more lectin than normal cells (Ozanne and Sambrook, 1971; Cline and Livingstone, 1971; Arndt-Jovin and Berg, 1971). A possible explanation for this difference in agglutinability may be that the lectin binding sites are clustered together on the surface of transformed cells (Nicholson, 1971). If confirmed by other groups, this observation may be very important.

In addition to agglutinating transformed cells, Con A will kill them. Con A does not affect normal 3T3 cells. Thus Con A can be used as a selective agent to isolate Con A-resistant sublines from transformed population. This procedure is not based upon the ability of a transformed cell to grow in conditions that will not allow normal cell growth, but uses a presumed difference in the surface structure as the selective condition. The surviving sublines resemble 3T3 in their resistance to subsequent Con A treatment and in their low agglutinability. They are revertant, because they grow to low saturation densities in 10% serum.

Selection of Con A-Resistant Cells

Plate SV40-transformed cells at a density of 4×10^5 cells per 100-mm dish in 10% calf serum. Twenty-four hours later, add 300 μg of Con A per milliliter. Make stock solutions of Con A (25 mg/ml) in 5 *M* NaCl, and make a suitable dilution of the Con A in calcium- and magnesium-free phosphate-buffered saline (PBS) just prior to adding the Con A to the plates. Leave the cells in Con A for 24 hours, then remove the medium and wash the plates 3 times in PBS to remove dead cells. Add fresh growth medium, and allow the surviving colonies to grow. Survivors arise at a frequency of 1 in 10^4 treated SV3T3 cells. Cycle these survivors through Con A selection another 1–2 times. After the final selection, plate the cells at suitable dilutions to yield colonies. These surviving colonies can then be picked and further characterized. (Ozanne and Sambrook, 1970.)

H. Isolation of Revertants on Monolayers of Fixed Cells

Working with cultures of Syrian hamster embryo cells and polyoma virus, Sachs and co-workers have devised a technique to isolate revertants using growth on monolayers of glutaraldehyde-fixed normal cells as the selective system. This procedure presumably selects for cells with altered surface properties, but there is no direct evidence for this. As opposed to other selection procedures where a small minority of revertant cells arise, this procedure produces revertants at a very high frequency. However, these revertants are not very stable since they back-revert to a transformed phenotype within a few dozen generations in culture.

PREPARATION OF GLUTARALDEHYDE-FIXED MONOLAYERS

Mix a cold 25.3% glutaraldehyde solution containing barium carbonate (to absorb oxidized derivatives) with activated charcoal, and filter through Whatman filter paper. This stock solution should be stored at 4°C in the dark and not kept longer than 3 weeks.

Prepare the monolayers by seeding 3×10^6 syrian hamster secondary fibroblasts onto 50-mm petri dishes and fixing with glutaraldehyde 16 hours later. The fixing solution is 1% glutaraldehyde in PBS (pH 7.3) which has been sterilized by filtration through a 0.45-μm filter. Wash the monolayers twice with PBS and add 5 ml of cold fixative. Store the plates at 4°C for 24 hours. Remove the glutaraldehyde and wash the cells with 10 ml of cold PBS. Follow with 12 washings with PBS at room temperature, two successive washings every hour.

To select revertants from polyoma-transformed Syrian hamster cells plate approximately 100 transformed cells on fixed monolayers and incubate for 10–14 days. To visualize the growing colonies, add a solution of 50 μg/ml neutral red to the plates for 20 minutes. The viable cells will take up the stain and appear red, and they can be picked with a 2-ml pipette. The majority of the cells capable of growth on these fixed monolayers have an epithelioid morphology, when compared to the fibroblastic morphology of the transformed parent. These variant colonies also grow to low saturation densities, have a decreased ability to form colonies in agar, are less tumorigenic and less agglutinable than the transformed parents.

It is not clear why such a selection works, but 80–97% of the colonies that grow up on these fixed normal cells have a revertant phenotype.

I. Isolation of Revertants of Cells Transformed by RNA Tumor Viruses

Cells transformed by RNA tumor viruses grow to high saturation densities, form dense colonies on monolayers of normal cells, and can grow in suspension in agar. Presumably, then, any of the procedures mentioned previously can be used to isolate density or anchorage revertants from RNA transformed cells.

Spontaneous morphological revertants of MSV-transformed 3T3 and RSV-transformed hamster cells have been reported (Macpherson, 1965; Nomura *et al.*, 1972). These revertants were not selected, but appeared spontaneously in cultures of the transformed cells. Such cells have a "flat" morphology, grow to low saturation densities, and are unable to grow in agar suspension cultures.

Stephenson *et al.* have isolated revertants of MSV-transformed NRK rat kidney cells and RSV-transformed NRK rat kidney cells after mutagenesis with BUdR. The cells were exposed to 200 μg/ml of BUdR for 20 hours. Single cells were then plated out in separate microtest titer wells (Falcon) to score the number of morphological revertants. One revertant colony was recovered among the 3×10^3 colonies examined. This procedure did not use BUdR as a killing agent, but rather depended on the mutagenic affects of BUdR to yield mutant cells defective in some transformed function. Nevertheless these BUdR-induced revertants grew to low saturation densities.

III. Properties of Revertant Cells

A. Selection versus Induction

By the use of a fluctuation analysis (Luria and Delbruck, 1943), it has been shown that the density revertant of the type selected with FUdR does exist in untreated transformed populations (Pollack, 1970). This observation is consistent with the fact that some serum transformants, isolated for their ability to grow in a gamma-depleted serum, also grow to low saturation densities in 10% calf serum (Scher and Nelson-Rees, 1971). These cells directly isolated in a transformation assay are identical to the density revertants isolated by negative selection from fully transformed SV3T3 with FUdR. Revertants can be isolated without prior treatment of transformed cultures with mutagens. Some revertant cells therefore are preexistent in populations of transformed cells, and the above selection procedures allow the isolation and cloning of these preexisting cells as revertant cell lines. However, not all revertants need have arisen spontaneously. For example, any type of revertant isolated with BUdR as the killing agent could have undergone a mutagenic event permitting survival.

Similarly, colchicine may act simply as a selective agent by killing growing cells in mitosis. However, colchicine can cause the formation of polyploid cells by interfering with cytokinesis. Many revertant lines show an increased number of chromosomes when compared with the transformed parent cell (Pollack *et al.*, 1970; Culp *et al.*, 1971; Vogel *et al.*, 1973; Vogel and Pollack, 1973; Wyke, 1971b; Nomura *et al.*, 1972). If this polyploidiation is related to reversion, then it is possible that colchicine acts as an inducer of the revertant phenotype.

B. Virus versus Cell

One of the major hopes of workers who isolate revertants is to correlate the revertant phenotype with a defect in the viral genome integrated in the transformed cell. Ideally one would like to isolate a revertant cell, rescue the virus by fusion with a permissive cell, and demonstrate that the rescued virus can no longer transform. Such a defective virus has not yet been found. The majority of revertants of SV3T3 cells isolated yield little or no virus upon fusion with permissive monkey cells. The virus that has been rescued can transform 3T3 cells as well as virus rescued from the transformed parent cells. Infectious virus cannot be rescued from polyoma transformed cells, so one does not know about the genotype of the virus. FUdR selection on 3T3 cells that were infected with mutagenized SV40 yielded temperature-sensitive transformants which grow to high saturation density at 32°C, but grow only to a confluent monolayer at 39°C (Renger and Basilico, 1972). The SV40 virus rescued from these cells is wild type at both temperatures with regard to growth and transforming ability. This result does not rule out the possibility of the existence in these cells of a defective virus causing the temperature-sensitive phenotype, but it does raise the possibility that these cells are temperature sensitive in a cell function involved in regulating saturation density.

Another way to investigate the question of defective viruses is to attempt to retransform the revertants with the same and other oncogenic viruses. None of the revertants of SV3T3 cells are retransformable by SV40 (Renger, 1972; Ozanne *et al.*, 1973; Vogel, 1973). That is, SV40 will not induce the growth of dense areas on density revertants and will not restore the ability to grow in 1% calf serum in the serum revertants. The temperature-sensitive SV3T3 cells, the FUdR-selected revertant, the colchicine-selected revertants, and the serum revertants all can be transformed by murine sarcoma virus (Renger, 1972; Vogel, 1973). However, the con A revertants are not retransformable by MSV(Ozanne, 1973).

The flat MSV revertants can be transformed by MSV. However, further characterization of the MSV in these revertants has not been possible because the cells do not yield virus after infection with murine leukemia virus (Nomura *et al.*, 1972). The revertants of RNA virus transformed cells isolated by Stephenson are not retransformable by the virus which originally infected them, but can be retransformed by other DNA or RNA oncogenic viruses (Stephenson *et al.*, 1973).

C. Gratuitous Reversion in Properties not Selected Against

Once cells reverted in any one of the transformed properties (density, serum, anchorage, lectin susceptibility) have been isolated, one may ask whether reversion in one property leads to the gratuitous reversion of the other transformed properties and if any other alterations in cellular physiology consistently accompany reversion (Table III). Density revertants of SV40-transformed 3T3 cells isolated with FUdR or BUdR are reverted in their ability to grow in semisolid medium, but maintain a transformed serum requirement. Colchicine-selected density revertants also show a decreased ability to grow in methocel and have also reverted to a 3T3-like serum requirement. Thus, selection of density revertants with different drugs leads to the isolation of revertants with different properties.

The anchorage revertants of Wyke have reverted in other transformed properties in that they do not grow in 0.5% calf serum and grow to low saturation density when cultured on plastic culture dishes. Their susceptibility to lectins has not been determined.

Revertants selected with Con A grow to low saturation densities, but maintain a transformed serum requirement. They do not form colonies in methocel (Ozanne, personal communication).

Selection of variants with a higher serum requirement for growth surprisingly also leads to reversion in saturation density. Thus selection with either restrictive serum yielded serum-sensitive lines with reduced saturation densition. However, selection in different restrictive sera lead to the isolation of revertants with different properties with regard to growth in methocel. Serum revertants isolated in 1% calf serum are unable to form colonies in methocel, while variants selected in agamma-depleted serum can grow in methocel. The reason for this difference is unknown, but the behavior in methocel is correlated with the cyclic AMP levels of these lines. The 1% calf serum, the serum revertant isolated in 1% calf serum has high cyclic AMP levels, when grown in sparse culture, while the agamma-depleted revertant culture has a low SV3T3-like cyclic AMP level (Oey *et al.*, 1973).

Cyclic AMP levels are generally higher in normal cells than in transformed cells (Otten *et al.*, 1971; Sheppard, 1972; Oey *et al.*, 1973). Intracellular concentrations of cAMP increase when cells are deprived of serum. Recent studies with the mouse fibroblast line 3T3, an SV40-transformed subline of 3T3, and six different revertant lines derived from this clone, show that a marked increase in cAMP occurs only when the serum concentration is reduced below the minimum necessary for growth of a given line. Conversely, density-dependent inhibition of growth is not accompanied by an increase in cyclic AMP in any line. The relation between reversion and cAMP levels might be better understood by developing a system which selects directly for an alteration in cAMP level (Oey *et al.*, 1973).

All revertants of SV40-transformed 3T3 cells show an increase in chromosome number and DNA content (Pollack *et al.*, 1970; Vogel *et al.*, 1973; Vogel and Pollack, 1973; Culp and Black, 1972; Ozanne *et al.*, 1973. The reason for this alteration in ploidy is not understood, but this relation raises the possibility of selecting directly for polyploid cells to test whether these large cells manifest any revertant properties.

IV. Conclusion

It is a bit surprising that revertants exist, since unarrested tumors almost always get progressively more abnormal with time, rarely if ever throwing off more normal variants. Nevertheless, revertants are found whenever they are selected for properly. As their properties are uncovered, we hope that they will continue to contribute to an understanding of the regulation of normal control of growth and its loss in oncogenesis.

REFERENCES

Aaronson, S. A., and Todaro, G. J. (1968). *Science* **162**, 1024.
Arndt-Jovin, D., and Berg, P. (1971). *J. Virol.* **8**, 716.
Black, P., Rowe, W., Turner, H., and Huebner, R. (1963). *Proc. Nat. Acad. Sci. U.S.* **50**, 1148.
Borisey, G., and Taylor, E. (1967). *J. Cell Biol.* **34**, 525.
Burger, M. M. (1969). *Proc. Nat. Acad. Sci. U.S.* **62**, 994.
Burger, M. M., and Martin, G. S. (1972). *Nature* (*London*), *New Biol.* **237**, 19.
Chen, T., and Heidelberger, C. (1969). *Int. J. Cancer* **4**, 166.
Cline, M., and Livingston, D. (1971). *Nature* (*London*), *New Biol.* **232**, 156.
Culp, L., and Black, P. (1972). *J. Virol* **9**, 611.
Culp, L., Grimes, W., and Black, P. (1971). *J. Cell Biol.* **50**, 682.
Fischinger, P., Nomura, S., Peebles, P., Haapala, D., and Bassin, R. (1972). *Science* **176**, 1033.
Hayri, P., and Defendi, V. (1970). *Virology* **36**, 317.
Inbar, M., and Sachs, L. (1969). *Proc. Nat. Acad. Sci. U.S.* **63**, 1418.
Kao, F. T., and Puck, T. (1967). *Proc. Nat. Acad. Sci. U.S.* **58**, 1222.
Kelly, F., and Sambrook, J. (1973). *Nature* (*London*) *New Biol.* **242**, 217.
Luria, S., and Delbruck, M. (1943). *Genetics* **28**, 491.
Macpherson, I. (1965). *Science* **148**, 1731.
Macpherson, I., and Montagnier, L. (1964). *Virology* **23**, 291.
Marin, G., and Littlefield, J. (1968). *J. Virol.* **2**, 69.
Mondal, S., Embleton, M., Marquardt, H., and Heidelberger, C. (1971). *Int. J. Cancer* **8**, 410.
Nicholson, G. (1971). *Nature* (*London*), *New Biol.* **233**, 244.
Nilhausen, K., and Green, H. (1964). *Exp. Cell Res.* **40**, 166.
Nomura, S., Fischinger, P., Muttern, C., Peebles, P., Bassin, R., and Friedman, G. (1972). *Virology* **50**, 51.
Oey, J., Vogel, A., and Pollack, R. (1973). *Proc. Nat. Acad. Sci. U.S.* (in press).
Otten, J., Johnson, G., and Pastan, I. (1971). *J. Biol. Chem.* **247**, 7082.
Ozanne, B. (1973). *J. Virol.* (1973) **12**, 79.
Ozanne, B., and Sambrook, J. (1970). *Lepetit Colloq.* **2**, 247.
Ozanne, B., and Sambrook, J. (1971). *Nature* (*London*), *New Biol.* **232**, 156.
Ozanne, B., Sharp, P. and Sambrook, J. (1973). *J. Virol.* (in press).
Ozanne, B., Vogel, A., Sharp, P., Keller, W., and Sambrook, J. (1973). *Lepetit Colloq.* **4**, 176.
Pollack, R. (1970). *In Vitro* **6**, 58.
Pollack, R., and Burger, M. (1969). *Proc. Nat. Acad. Sci. U.S.* **62**, 1074.
Pollack, R., and Teebor, G. (1969). *Cancer Res.* **29**, 1770.
Pollack, R., and Vogel, A. (1973). *J. Cell. Physiol.* **82**, 93.
Pollack, R., Green, H., and Todaro, G. (1968). *Proc. Nat. Acad. Sci. U.S.* **60**, 126.
Pollack, R., Wolman, S., and Vogel, A. (1970). *Nature* (*London*) **228**, 938.

Rabinowitz, Z., and Sachs, L. (1968). *Nature (London)* **220**, 1203.
Renger, H. (1972). *Nature (London), New Biol.* **240**, 19.
Renger, H., and Basilico, C. (1972). *Proc. Nat. Acad. Sci. U.S.* **69**, 109.
Rueckert, R., and Mueller, G. (1960). *Cancer Res.* **20**, 1584.
Sambrook, J. (1972). *Advan. Cancer Res.* **16**, 144.
Scher, C., and Nelson-Rees, W. A. (1971). *Nature (London) New Biol.* **233**, 263.
Shani, M., Rabinowitz, Z., and Sachs, L. (1972). *J. Virol.* **10**, 456.
Sheppard, J. (1972). *Nature (London), New Biol.* **236**, 14.
Smith, H., Scher, C., and Todaro, G. (1971). *Virology* **44**, 359.
Stephenson, J., Scolnick, E., and Aaronson, S. (1972). *Int. J. Cancer* **9**, 577.
Stephenson, J., Reynolds, R., and Aaronson, S. (1973). *J. Virol.* **11**, 218.
Stoker, M. (1968). *Nature (London)* **218**, 234.
Temin, H., and Rubin, H. (1958). *Virology* **6**, 669.
Todaro, G., and Green, H. (1964). *Virology* **23**, 117.
Vogel, A. (1973). In preparation.
Vogel, A., and Pollack, R. (1973). *J. Cell. Physiol.* (in press).
Vogel, A., Risser, R., and Pollack, R. (1973). *J. Cell. Physiol.* (in press).
Wyke, J. (1971a). *Exp. Cell. Res.* **66**, 203.
Wyke, J. (1971b). *Exp. Cell. Res.* **66**, 209.

Chapter 6

Monolayer Mass Culture on Disposable Plastic Spirals

N. G. MAROUDAS

Imperial Cancer Research Fund Laboratories,
Lincoln's Inn Fields,
London, England

I. Introduction

A method of growing "anchorage-dependent" cells on a spiral of plastic film has been described (House *et al.*, 1972), and practical results were compared with those from other new methods (Maroudas, 1973). Described herein is a convenient form of this apparatus, designed for general laboratory use.

In its original form, the plastic spiral was made from an autoclavable film—"Melinex" polyester (I. C. I. Ltd.)—and assembly of the apparatus was done on the premises. However, as the number of vessels in use increased, so did the housekeeping problems of washing, reassembling, and sterilization. At the other extreme, small users found difficulties in constructing the apparatus on their own. Accordingly, ICRF decided to invest in a mass-produced unit: a disposable, radiation-sterilized 1.6-liter, spiral vessel. The idea was that the apparatus designed for polystyrene and manufactured by a specialist in disposable tissue-culture equipment, should be a product that had a better growth surface and was uniform, of guaranteed sterility, and convenient. These units have been in use since October 1972, and the majority of users trained in the use of conventional roller bottles have found that the new apparatus adapts very easily to their previous techniques and gives five to ten times the yield for the same effort. The low capital cost of the spirals compared to conventional bulk culture machinery, and the fact that they do not have to be rolled continuously, makes it possible for various users to set up preparations of 1.6 to 16 liters each week without competing for equipment, and to change their scale of operations as convenient.

Disposable spirals are manufactured and sold in the United Kingdom by Sterilin Ltd., 12–14 Hill Rise, Richmond, Surrey, TW 10 6UD, and distributed in the United States by Cooke Engineering, 900 Flatters Lane, Alexandria, Virginia 22314.

II. Apparatus

The disposable mass culture vessel is illustrated in Fig. 1.

The spiral film is made of transparent, unplasticized polystyrene, surface-treated for tissue culture. The spacing between the layers of the spiral is 3 mm; it is maintained by the two end layers of crimped spacer tape, 1 cm wide, at each end. Film and spacer tapes are cemented lightly to each other, making a cartridge that is rigid enough to handle but easy to unwind. This provides a total surface area of approximately 8500 cm^2, with cells growing on both sides of the film. The cartridge fits into a cylindrical container of molded, unplasticized, transparent polystyrene, not surface-treated at present. The outer dimensions of the cylinder (25 $\times$ 10 cm) are designed to make it compatible with existing 4-inch bottle-rollers. It is molded in two halves, with a lap joint; at present, after the spiral cartridge has been inserted, the joint in the middle is closed with permanent cement. Three screw-cap apertures are provided on the top; the larger one for filling and the smaller two for aeration. One of these apertures is connected to a central tube which runs almost the full length of the vessel. Special screw-on air filters are provided for aeration. These are also provided sterile, but, being made of polypropylene and cottonwool, are autoclavable and can be re-used.

The flow pattern in aeration has been worked out with some care. The principle of recirculation is as follows. The air/CO_2 mixture enters through the central sparge tube, and a bubble forms at the bottom.

When this bubble rises, up the central core of the spiral, it drags liquid up with it. The train of bubbles rising through the concentric annulus thus

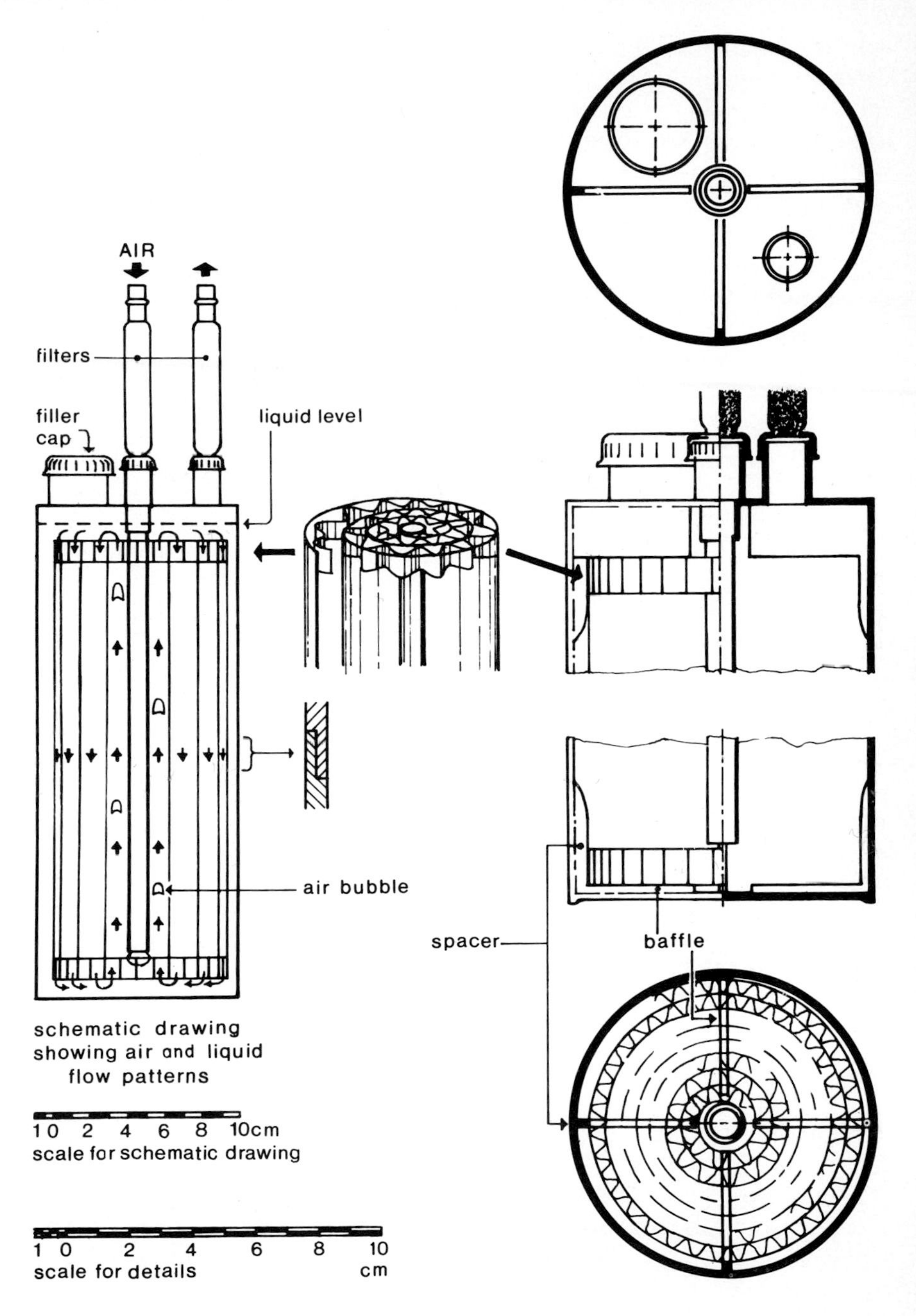

FIG. 1. Disposable mass culture vessel, ICRF spiral pattern. For details of construction and operation, see Section II.

acts as a hydraulic lift pump, which promotes recirculation through each layer of the spiral, in the pattern shown. The liquid flows up the central core, turns sideways at the top, flows down through the individual layers, and is again sucked inward toward the central lift zone, by means of the radial baffles at the bottom. To maintain the continuity of this flow pattern, it is important that the liquid level be 2–5 mm above the top of the cartridge.

When more vessels than one are being aerated simultaneously, there is the problem of adjusting the gas flow in each, without upsetting the others. This is very simple in principle: the pressure drop across the apparatus is less than 1 psi; hence, if the gas manifold is regulated to a supply pressure of, say 5–10 psi, a narrow constriction or fine needle valve must be placed upstream of each inlet filter, which exerts a resistance (i.e., pressure drop) of 4–8 psi at the maximum flow rate (300 cc/min). A more elaborate scheme entails an individual meter for each vessel (simple, cheap flow meters are supplied by G. A. Platon Ltd., Basingstoke, Hants. or by Glass Precision Engineering, Hemel Hempstead, Herts, U.K.); but, at the other extreme, screw clips and rubber tubing are quite adequate for a start. It is, however, a simple thing to have at least one flowmeter for checking the outlet gas flow rate at start-up. This flowmeter can then be removed for use on another vessel.

III. Method

The vessel is supplied sterile, with screw caps in place. Medium is added, followed by cells, to a final volume of 1.6 liters. We keep the medium in 1.4 liter lots, in Winchester bottles, and add serum, glutamine, and antifoam pro rata. The filler cap is replaced, and, by inverting the vessel several times, the contents are mixed and the cells are distributed evenly through the windings of the spiral (this is important). Alternatively, cells and medium can be premixed in the Winchester bottle.

The cells are then allowed to settle and spread evenly over both sides of the film, by rolling on a very slow machine, about 2 revolutions per hour. Normal roller speeds of 10 to 60 revolutions per hour tend to wash cells off the surface and produce aggregates. Slow rollers can be obtained from Luckham Ltd., Burgess Hill, Sussex. Conventional machines can either be equipped with slower motors, or run intermittently, with a timing cycle of approximately 15 seconds on and 4 minutes off. In the absence of slow rollers, the vessels can be laid on their sides in the warm room, and turned periodically, say one-third turn every 0.5 hour.

Cell attachment can be inspected periodically by placing the vessel horizontally on an inverted microscope as with roller bottles. A long-focus objective is desirable but not essential—the first layer starts close to the wall because of the variable pitch of the baffles (Fig. 1). The attachment and spreading of cells is usually complete in 3 hours, but overnight is sometimes more convenient (there is enough dissolved oxygen in the medium for the inoculum of cells to survive this period). Alternatively, the process can be speeded up by omitting serum until after attachment is complete, as this component of the medium retards initial adhesion of cells (although it is essential for the continued attachment of most cells). For uniform seeding, the roller must be level. Uneven distribution of cells is the most frequent cause of poor yields. It is almost always worth while to stain the first few spirals, then drying them and unwinding for inspection: the cell sheet should be uniform, top and bottom, inside and out, on both sides of the film.

After attachment of the cells, the vessel is removed from the rollers and placed upright for aeration. The two small caps are removed and replaced by the screw-on air filters. The outlet cap should be removed first, to avoid the danger of liquid siphoning up the central tube. The gas inlet tube should be connected firmly but with care (it can break away its welded cap) and inspected for leaks. Individual flowmeters facilitate testing for leaks, by comparison between the indicated flow of gas into each vessel, and the outlet flow rate. In the absence of antifoam, the rate of aeration is necessarily low; if more vigorous gassing is preferred, antifoam (e.g., Midland Silicone R. D. emulsion) allows gas rates up to 300 cc per minute. It is worth experimenting to find the optimum gas rate for each cell strain: too high a rate is wasteful of CO_2, requires possibly toxic concentrations of antifoam, causes splashing, and pressurizes the vessel; on the other hand, too little aeration can deprive the cells not only of oxygen, but of other nutrients, as well, through sluggish stirring of the medium. A pH stat can be used to operate a solenoid valve on the CO_2 line; with this device it is possible to maintain the required pH within ± 0.1 pH unit if required. However, for the comparative growth studies described below, a constant gas composition, with periodic manual adjustment, was found satisfactory.

Progress of the cells is inspected microscopically, by laying a microscope on its side, with the vessel remaining vertical to prevent liquid from wetting the filters. In this way the pattern of growth in the first few layers of the spiral can easily be followed. If the initial attachment of cells has been uniform, these layers will be representative of the whole spiral. A long-focus lens is here particularly useful, as a check on uniformity of growth throughout the layers. Cells which tend to detach from the film and to grow in suspension can also be grown with a high gas rate, as recirculation in the spiral is then powerful enough to keep them suspended. CHO cells give a high yield this way, as they grow first in monolayer, then in suspension (P. Davies, personal communication).

IV. Harvesting

For growth studies and for subculture, harvesting should be carried out as follows: (a) Remove medium, wash with approximately 400 ml of phosphate-buffered saline, and discard. (Suspension-prone cells, of course, must be centrifuged from the medium and washings, if these form an appreciable part of the yield.) (b) Add 200 ml of 0.05% trypsin/EDTA (Versene) solution. Gently invert the vessel from end to end and rotate slowly at the same time. Remove the cell suspension after a few minutes, and repeat with fresh trypsin/EDTA if necessary. Cell sheets that do not trypsinize well tend to get caught in the crimped tape.

Mechanical harvesting is also possible, by scraping or washing cells off the film, but tends to produce sheets or clumps of cells. However, for some biochemical studies this is no disadvantage. In this case the cartridge is removed, and passed between two rubber squeeze blades ("windscreen wipers"), hinged at one end. For convenience, the one blade can be fixed to the midline of a stainless steel tray, the top one hinged onto it at one end and fastened down with a catch at the other; the rolled-up cartridge is placed in the dish, its leading edge clamped between the wiper blades, the film is pulled through, and the scraped cells are collected in the dish.

At present, in order to remove the cartridge, it is necessary to knock off the bottom edge, with a pair of tin-snips. Later it is hoped to supply containers which come apart at the central lap joint (Fig. 1). Some cells, for instance CHO, can be harvested simply by vigorous agitation of the vessel.

V. Uses

The main use of the disposable spiral system so far has been to grow anchorage-dependent cells for laboratory investigation. Cell yields, from tests performed at the ICRF Laboratories, are given below. Cells were grown in Dulbecco/Eagle's medium with 10% calf serum, 10% CO_2/air and were harvested at confluence (3–4 days).

Cell type	Inoculum	Yield	Cells/ml
BHK	1×10^8	2×10^9	1.25×10^6
3T3	4×10^7	4×10^8	2.5×10^5
Whole mouse embryo primary	1×10^9	1×10^9	0.6×10^6

Other cells that have been grown successfully are Nil, CHO, and various transformed derivatives of BHK and 3T3. The latter tend to detach, and it is very much a matter of the individual cell strain that decides whether in this case it is going to grow in suspension and give a good yield, or not.

Virus production by this method has so far not been reported (although Dr. N. Mowatt and colleagues, at the Animal Virus Research Establishment, Pirbright, Surrey, have successfully grown both foot-and-mouth and swine-vesicular-disease virus on autoclavable "Melinex" spirals).

An interesting new use (Dr. Ruth Sager, personal communication) is for cloning, the idea being to make use of the large cell capacity to look for rare mutants (say, 1 in 10^7 to 10^9). Reconstruction experiments show that resistant colonies can be selectively grown, which become readily visible inside the spiral upon emptying the vessel and viewing it aginst a point source of light (stopped-down condenser). To isolate clones, the spiral cartridge is removed, hung on a horizontal rod, like a scroll, over a dish, and unrolled progressively, in front of a stopped-down lamp, to look for colonies. These are cut out from the sheet with a very sharply pointed, small pair of dissecting scissors, or they can be scraped off for subsequent passage.

The flexibility of the disposable spiral system, and its low capital cost, enable individual workers to make trials on their own particular applications, and to change at will. The present size seems to suit most workers, being a simple 10-fold scale-up from the roller bottle. However, should occasion warrant, it is estimated that a further 10-fold increase in scale, up to 15 liters, is feasible for a disposable design.

REFERENCES

House, W., Shearer, M., and Maroudas, N. G. (1972). *Exp. Cell Res.* **71**, 293.

Maroudas, N. G. (1973). *In* "Techniques in Biophysics and Cell Biology" (R. Pain and B. J. Smith, eds.), Vol. I. Wiley (Interscience), New York.

Chapter 7

A Method for Cloning Anchorage-Dependent Cells in Agarose

C. M. SCHMITT AND N. G. MAROUDAS

Imperial Cancer Research Fund Laboratories,
Lincoln's Inn Fields,
London, England

I. Introduction

The following method allows single cells to be isolated and cloned while retaining the conditioning effects of a high cell concentration (Maroudas and Schmitt, 1973). This method is an extension of the method of Metcalf, Bradley, and Robinson (1967), in which feeder cells were incorporated in an agar base layer, and bone marrow cells in an agar top layer: conditioning factors stimulated the growth of colonies of erythropoietic stem cells. This technique is restricted to cells which can grow in suspension in agar; however, most cells require anchorage to a solid substrate in order to grow (Stoker *et al.*, 1968). The normal cloning method for anchorage-dependent cells, using a feeder layer, is to irradiate the feeder cells, and seed the cells to be cloned on top, but in this case the two sorts of cells compete for substrate space.

The method here described incorporates cells and small pieces of solid substrate into an agarose gel. Some of the cells attach to the substrata and are able to grow. As in the method of Metcalf *et al.*, a feeder layer is incorporated into a bottom layer of gel, and conditioning effects are obtained, without interference between feeder and cloned cells occurring. The best substrate was found to be very thin glass "platelets" 100–200 μm long, giving a colony formation rate of 75 to 90% for BHK-21 and 3T3 cells in agar over a feeder layer (Maroudas, 1972, 1973).

Agarose was chosen instead of agar, because it does not contain the nonsetting sulfated, component of agar, which is growth-inhibitory to most cells. Agarose was selected in preference to methyl cellulose, because the latter forms merely a viscous liquid, whereas agarose sets to form an elastic solid, interlinked, hydrogen-bonded network, preventing movement and agglutination of cells, which would otherwise vitiate the technique as a cloning method (Dea *et al.*, 1972).

II. Procedure

In outline: Cells to be cloned are trypsinised to yield a single-cell suspension, and resuspended in an upper layer of 0.25% agarose, together with a number of glass platelets, over a preset base layer of 0.5% agarose, containing feeder cells. Two days later, platelets with only one cell attached are marked, and, after incubation, those with a confluent layer of cells are removed, and placed in tissue culture wells, for further growth in liquid medium.

Details of the procedure are as follows:

A. Preparation of Glass Platelets

Platelets are made from fine glass bubbles (glassblowers' "blowoff"), which are crushed in distilled water, using a silicone-rubber "policeman." The finer pieces are decanted, and sonicated with a Dawe sonicator probe for approximately 5 minutes. The resultant platelets are sieved through 300, 180, and 100 μm aperture, nickel, or stainless steel mesh, to give an average size product of 150 × 300 μm. The platelets are washed and stored in 70% ethanol. Before use in cell cultures, they are twice washed with serum-free medium and then resuspended in medium containing 10% calf serum.

B. Media Used

Dulbecco's modification of Eagle's medium (designated E4) is used, with calf or fetal-calf serum (F. B. S., Flow Laboratories).

Conditioned medium (c.m.) is made as follows: E4 medium with 10% calf serum is used to grow a subconfluent layer of BHK or 3T3 cells. This medium is then removed, centrifuged, and diluted with equal volume of fresh, serum-free E4. F.B.S. is added to a final concentration of 15%.

C. Preparation of Agarose Base Layers

The 0.5% agarose base layers are prepared by melting a 1% solution of agarose (B.D.H., or Marine Colloids, Rockland, Maine), and adding an equal volume of double-strength medium containing 20% calf serum. The mixture is kept at 44°C and cooled to 37°C immediately before use. Into 5-cm petri dishes, 2 ml of the agarose medium is poured, containing 2×10^5 whole mouse embryo cells; the gel is allowed to set at room temperature. Then a further 1 ml of agarose medium without cells is added to ensure complete separation of feeder cells from the top layer. After setting, the layers are rewarmed to 37°C.

D. Preparation of Agarose Top Layers

A single-cell suspension, of the cells to be cloned, is prepared, counted and kept at 37°C. Platelets in E4 medium are added to 0.5% agarose medium solution in a quantity sufficient so that, when finally poured on to the base layers, they are not less than 200 μm apart, to prevent overlap. The agarose medium mixture is cooled to 37°C, and the cells are added to give a cell concentration of 5×10^4 per dish. Enough E4 medium is added to give a final agarose concentration of 0.25%. The top layers are poured, and immediately thereafter, before they set, each tray of dishes is placed in a 37°C incubator for 1 hour, to allow cells to settle. The top layers are then set by cooling to room temperature for 10 minutes. All the platelets lie flat on top of the bottom layer, which makes them easily scanned by a microscope. After 2 days' incubation, a Leitz marker objective is used to mark the positions of platelets with only one attached cell. Under the above conditions, 5×10^4 cells per dish give an optimum number of platelets with a single cell attached.

E. Growth and Passage of Clones

The cells are allowed to grow to confluence on the marked platelets (5–7 days), and then transferred to 16 mm tissue culture wells (Linbro multi-dish disposo-trays FB-16-24TC). The transfers are effected by lifting individual platelets from the agarose with a spatula made of platinum wire (0.5 mm diameter), flattened at the tip. The platelets and cells are washed, and freed of any remaining traces of agarose and unattached suspended cells, in a 3-cm dish containing 2 ml of complete medium, before being pipetted by means of a finely drawn Pasteur pipette into the 16-mm wells. The platelets are incubated at 37°C as before, in 1 ml of conditioned medium containing 15% F.B.S.

Cells grow out from the platelets on the bottom of the wells, and at confluence can be trypsinized and passaged. Alternatively, the cells can be grown on to 12-mm glass coverslips, previously placed in the wells, and, when confluent, dispersed in 3-cm petri dishes by breaking the coverslips with sterile forceps: in this way the first two passages of the cells can be made without the use of trypsin. Thereafter the cells are passaged in the normal way: 0.05% trypsin in EDTA (Versene), with agitation of the dish until the cells are freed from the surface; then transferring to E4 medium containing 10% calf serum, centrifuging, and resuspending in fresh medium.

An alternative to the above method is found to give a higher yield of cells attached to platelets, but is a longer process, being carried out in two stages instead of one. The base layers are prepared as usual; after setting, sufficient platelets are added to each dish in 0.5 ml of E4 medium, taking care to distribute them as evenly as possible over the gel surface. The dishes are placed in a plastic desiccator containing silica gel. After gassing with 20% CO_2 in air, the desiccator is sealed and placed in a warm room at 37°C overnight to allow the free liquid on the surface to evaporate. After 24 hours, the cells to be cloned are added in 0.5 ml of the 0.25% agarose medium and allowed to set immediately at room temperature, before being incubated as usual at 37°C in a humidified CO_2 incubator. Thereafter, the method is the same as that described above.

III. Discussion

This technique has been used to clone both established cell lines (BHK 21 and Balb/C3T3 cells), and secondary cell cultures of baby mouse kidney and whole mouse embryo. Because BHK21 cells were found to form colonies in suspension when grown in agarose, agar (Difco Noble) was substituted, so that colony formation was confined to the platelets. The secondary cultures were 3-day confluent monolayers, obtained from freshly dissociated tissue.

The specific cloning efficiency, from anchored single cell to confluent platelet, was found to be high for all types of cells used: for example 66% for the WME secondary cells compared with 75% for the established 3T3 line. The high cloning efficiencies, in the suspended platelet method, is largely due to a self-conditioning effect of the high cell density (5×10^4) in the top layer. At this density it was found that BHK-21 cells did not require feeder cells in the base layer, in order to form colonies on platelets, and that secondary WME cells grew only marginally better with a feeder layer. The feeder cells could often be substituted by the addition of conditioned medium to both base and top layers, in a proportion of 1:3 parts of fresh medium, with equally satisfactory colony growth.

The success rate of growth from transferred confluent platelet to confluent well, which took from 12–14 days, varied from 36% for BMK and 72% for WME secondaries, to 88% for BHK and 96% for 3T3 cells. Subsequent passages from the wells to tissue dishes were completed without any loss of clones. Thus the crucial steps are the first and second: to grow up small, confluent cell sheets, and to grow them to confluence in the well. The ability to dispense with trypsinization in the sensitive first-transfer stage, we consider to be an important feature of the platelet method.

An advantage of the agarose cloning method is that, when it is poured, all the potential clones are inoculated simultaneously (unlike microwells), and after it has set, even clones which are only 200 μm apart will not interfere with each other. There is however the disadvantage that transferring the platelet to its well at present takes 3–6 minutes, but this could be expedited by the development of a special microtool. On the other hand, there is a marked saving of space, large numbers of clones occurring in one culture dish.

An advantage of the high cloning efficiency is that it gives a more representative sample of clones from the cell population chosen, as required in the study of rates of mutation and reversion.

Many types of cell can be used as feeders: one is not restricted, as in monolayer techniques, to feeder layers grown on a dish, and subsequently irradiated or poisoned. Another advantage of having the feeder cells dispersed in a separate agarose layer, rather than directly underneath the cloned cells as a monolayer, is that the former do not then interfere with the anchorage of the latter. It also provides a means of studying the effects, of one cell type on another, due to the secretion of diffusible factors into the medium.

The success of this method in cloning secondary cells indicates that the commonly held opinion, that primary or secondary cells have an inherently low cloning efficiency, is probably due to inadequacies in the standard methods so far employed.

References

Dea, I. C. M., McKinnon, A. A., and Rees, D. A. (1972). *J. Mol. Biol.* **68**, 153.
Maroudas, N. G. (1972). *Exp. Cell Res.* **74**, 337.
Maroudas, N. G. (1973). *Exp. Cell. Res.* **81**, 104.
Maroudas, N. G., and Schmitt, C. (1973). *J. Cell Differentiation* **2**, 171.
Metcalf, D., Bradley, T. R., and Robinson, W. A. (1967). *J. Cell. Physiol.* **69**, 93.
Stoker, M. G. P., O'Neill, C. H., Berryman, C., and Waxmann, V. (1968). *Int. J. Cancer* **3**, 683.

Chapter 8

Repetitive Synchronization of Human Lymphoblast Cultures with Excess Thymidine

H. RONALD ZIELKE AND JOHN W. LITTLEFIELD

Genetics Unit, Children's Service, Massachusetts General Hospital, and Department of Pediatrics, Harvard Medical School, Boston, Massachusetts

---

I. Introduction

Human lymphoblastoid, or hematopoietic, cell lines provide a source of biological material with advantages not found in human fibroblasts or aneuploid lines. Human lymphoblast lines usually remain diploid or near-diploid even after extended periods of growth, and some lines retain specialized functions such as immunoglobulin synthesis. The lines grow in suspension and do not become senescent, allowing production of large quantities of cells which can be readily sampled. Since Epstein-Barr virus (EBV) particles (Moore *et al.*, 1968) or genomes (zurHausen and Schulte-

Holthausen, 1970; Nonoyama and Pagano, 1972) are present in many lymphoblast lines, it is assumed that the indefinite life-span results from the presence of EBV. The origin and cultivation of hematopoietic cells has been recently reviewed by Lazarus and Foley (1972). We wish to describe here the considerations for obtaining synchronized human lymphoblast cultures.

Several recent reviews of methods for obtaining synchronized mammalian cells have been published (Stubblefield, 1968; Mitchison, 1971). Currently available techniques for the synchronization of mammalian cells generally depend on selection of a naturally synchronous population or imposition of a chemical block either by addition of an inhibitory substance or removal of an essential nutrient (Tobey and Ley, 1970), resulting in synchronized growth following release of the block. The mitotic selection technique of Terasima and Tolmach (1963) provides a synchronous population with the least physiological perturbation of the cells. However, the low yield of synchronous cells by this technique is a serious disadvantage for biochemical investigations. The total yield by mitotic selection can be increased by storing the mitotic cells in the cold to arrest their progress through the cell cycle until a sufficient number have accumulated (Tobey *et al.*, 1967), but even this mild treatment may distort the metabolism of the cells (Ehmann and Lett, 1972). Klevecz and associates (Klevecz, 1972; Klevecz *et al.*, 1972) have partially circumvented the problem of low yield by use of an automated system which yields at a particular time a series of samples each containing up to 1×10^7 cells staged various hours after mitotic selection. However, this system requires a large investment in equipment.

Synchronization of mammalian cells by imposition of a chemical block is followed by synchronized growth of the majority of the cell population, but may result in "unbalanced growth" (Rueckert and Mueller, 1960; Michison, 1971). However, the presence of "unbalanced growth" may be excluded if the parameter of interest varies in the same pattern through more than one cell cycle of synchronized growth.

Growth of human lymphoblast cells in suspension dictates the use of a metabolic block to achieve synchrony. Of several methods which we have tested, such as excess thymidine (dT), methotrexate, fluorodeoxyuridine, Colcemid, or isoleucine depletion, excess dT has proved the most successful. Excess dT arrests mammalian cells near the G_1/S boundary (Xeros, 1962) and also in S phase (Puck, 1964). Presumably excess dT blocks DNA synthesis by depletion of deoxycytidine derivatives, since addition of deoxycytidine (dC) reverses the dT inhibition (Xeros, 1962). Galavazi *et al.* (1966) have optimized the conditions for multiple dT blocks with heteroploid kidney cells. Doida and Okada (1967) obtained a greater degree of synchrony by following dT synchronization with a Colcemid block to arrest cells in mitosis.

Recent studies of human lymphoblast lines synchronized with excess dT have measured immunoglobulin synthesis (Buell and Fahey, 1969; Takahashi *et al.*, 1969) and sialic acid variation (Rosenberg and Einstein, 1972) during the cell cycle. However, the conditions necessary for optimal synchrony were not described in these studies. Therefore, before proceeding to investigate cell cycle-dependent changes in enzyme levels in human lymphoblasts, we identified the conditions resulting in maximal synchrony by varying the dT concentration, length of block, and time between blocks, for example, and have sought to explain why the fraction of cells dividing after the dT block is rarely more than 60% (range, 45 to 65%). Lack of cell viability after the dT treatment, induction of EBV, and the presence of nondividing cells in all cultures synchronized or exponentially growing, do not appear to produce the limited yield. The degree of synchrony is unchanged after repetitive synchronization or using freshly cloned lines.

Seven lymphoblast lines were tested for synchronization with excess dT. Three lines grown for some time synchronized well, while four recently established lines synchronized to a lesser extent.

II. Methods

A. Lymphoblast Cell Lines and Culture Maintenance

PGLC 33H, an established human lymphoblast line, was originated by Dr. P. R. Glade from a female patient with infectious mononucleosis. This line has been cloned three successive times in our laboratory (Sato *et al.*, 1972) and is designated L33-6-1-19. For simplicity it will be referred to in this chapter as line 19. LPM-10 is a once-cloned line obtained by Dr. George Moore from the peripheral blood of an apparently healthy male. CCRF-CEM originated from a 2.5-year-old female with acute lymphoblastic leukemia (Foley *et al.*, 1965). These three lines have been in culture for several years. Other lines were established in our laboratory from patients with infectious mononucleosis (MGH-5, MG-57) or from healthy donors (MGH-1, MGH-7) using phytohemagglutinin and cell-free lysate from established cell lines.

The predominant karyotype of line 19 presently is 47,XX,C+, with about 10% tetraploid cells. The predominant karyotype of LPM-10 is 47,XY,M+; the extra chromosome is a large metacentric similar in size to an A group chromosome. The karyotype of CCRF-CEM indicates loss of one or more C group chromosome combined with an extra D chromosome and an extra F chromosome. This is similar to the karyotype observed by Krishan *et al.* (1969). MGH-1, MGH-5, and MGH-7 have been in culture for less than

6 months and have a normal karyotype. MG-57 has been in culture for less than 2 years and is a mixture of 45,X/46,XY cells.

Lymphoblast cells were maintained at 37° under a 5% CO_2–95% air atmosphere in static cultures in a humidified incubator or shaken on a rotary shaker. Growth medium consisted of RPMI 1640 medium (Grand Island Biological Co.) supplemented with 15% (v/v) fetal calf serum (Gray Industries, Inc.) without antibiotics. During synchronization shaken cultures were continuously flushed with the above gas mixture. Monthly examinations for *Mycoplasma*, performed in the laboratory of Dr. L. Dienes at the Massachusetts General Hospital, were regularly negative during this study.

At designated times, samples were removed from the shaken culture by a timed peristaltic pump (Thilly, 1972) and stored in a refrigerated fraction collector. At 4°C the cell number was constant for over 12 hours. Three culture flasks could be monitored simultaneously by our apparatus. This automatic sampling system was a convenience, rather than a necessity, for the following work. Additions to the culture flasks were also made automatically at a preset time with a timed syringe pump. Cell number was determined with a Model Z_BI particle counter (Coulter Electronics, Inc.). Unless stated otherwise, all studies were performed on cultures of line L33-6-1-19.

B. Synchrony Procedure

1. Thymidine Concentration

Growth of most mammalian cell lines is inhibited by dT in the millimolar range. This is the concentration found to inhibit LPM-10, MGH-1, MGH-5, MGH-7, and MG-57 cultures. In contrast, 3×10^{-5} *M* dT was sufficient to inhibit growth of line 19. A similarly high sensitivity of the CCRF-CEM lymphoblast line to dT has been observed (Schachtschabel *et al.*, 1966). Since excess dT may result in abnormalities in the cell cycle (Rueckert and Mueller, 1960; Mitchison, 1971), the concentration of dT selected for synchronization was the minimum necessary to stop cell division for 24 hours. After 24 hours the cell number started to increase again even in the presence of dT, presumably because of its degradation.

2. Release of Block

The dT block was released by centrifugation of the cultures at 100 *g* for about 7 minutes at room temperature and resuspension in fresh medium lacking dT. Additional washing of cells was not necessary. As an alternative where indicated, the dT block was reversed by the addition of 1×10^{-6} *M* dC. Resuspension of cells in fresh medium with added dC neither increased the percentage of cells dividing nor shortened the time between release of the block and subsequent cell division when compared to resuspension in unsupplemented fresh medium. However, an interesting exception was observed with CCRF-CEM, since the time between release of the block and subsequent cell division was decreased if dC was added to the fresh medium. During synchronization the cell densities of all lines except CCRF-CEM were maintained between 2 and 5 × 10^5 cells per milliliter. The CCRF-CEM cell density was maintained between 5 and 15 × 10^5 cells per milliliter.

3. Initial Block

A dT block was applied to cultures of line 19 (3×10^{-5} *M*) and LPM-10 (3×10^{-3} *M*) for 4, 6, 8, 12, or 16 hours before release by centrifugation and resuspension in fresh medium. The results for line 19 are shown in Fig. 1. Four- or 6-hour blocks resulted in a small burst of synchronized growth; the longer blocks result in a sharper increase in cell number followed by a short plateau. Surprisingly little difference was observed between the improved synchrony in the remaining time intervals.

The minimum length of time necessary to accumulate cells at the G_1/S boundary with excess dT should equal the sum of (G_2 + M + G_1). This value is approximately 15 hours (see Section III,C). Although a greater number of cells may accumulate at the G_1/S boundary during longer blocks, prolonged arrest may be harmful to the cells that were initially near this phase of the cell cycle.

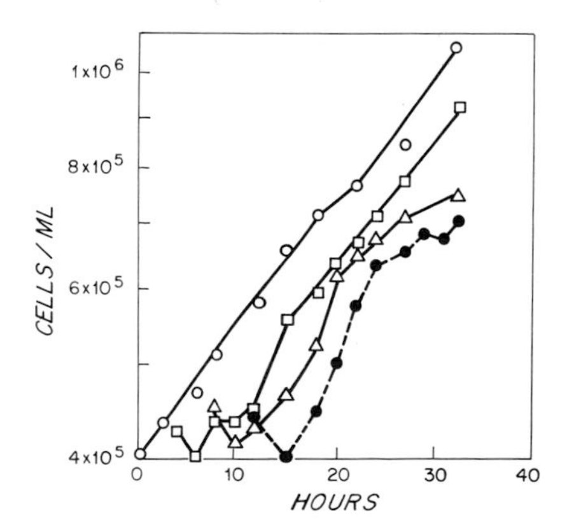

FIG. 1. Length of the initial thymidine (dT) block necessary to achieve synchronized growth. An exponentially growing culture of line 19 was blocked with 3×10^{-5} *M* dT for 4, 6, 8, 12, or 16 hours before release by centrifugation and resuspension in fresh medium. ○, no block; □, 6 hours; △, 8 hours; ●, 12 hours. The data for the 4- and 16-hour blocks are not shown.

4. Second and Subsequent Blocks

The main consideration in the timing of the second and subsequent blocks is that the interval between the release of one block and application of the next block be long enough to allow cells arrested at G_1/S to transverse S, but not so long that the cells can pass into the second S phase. Since S in an unsynchronized lymphoblast culture is about 10 hours in duration (see Section III,C), the calculated minimum time between release of the first block and application of the second block should be about 10 hours. If the second block were to be applied in the middle of the following G_1 phase, the time between release and subsequent block should be about 19 hours. These theoretical expectations do not take into consideration variability in generation time or alteration in phase duration due to the synchrony procedure.

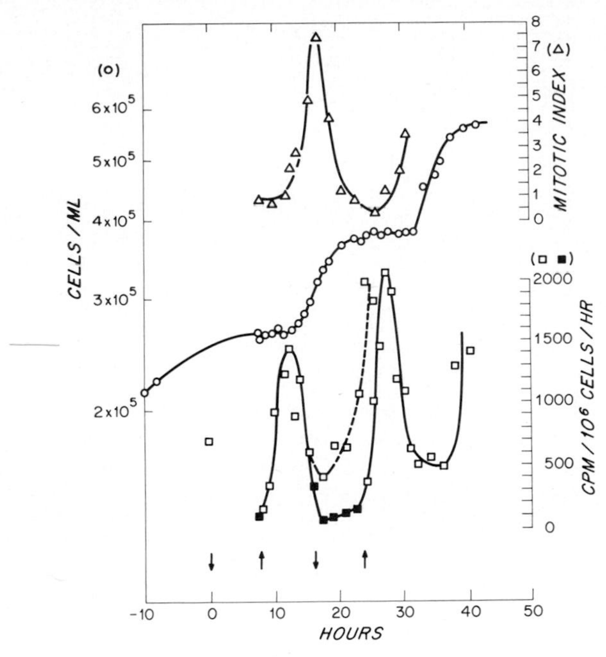

Fig. 2. Two-step synchronization of line 19. Thymidine (dT), 3×10^{-5} *M*, was added to the culture at time zero (arrow pointing down), and the block was released after 8 hours by centrifugation and resuspension in fresh medium (arrow pointing up). The culture was reblocked at 16 hours and released at 24 hours. The cell density is indicated by the open circles. The rate of DNA synthesis, determined by hypoxanthine-^{3}H (5 μCi/ml, specific activity = 1 Ci/mmole) incorporation into NaOH-resistant trichloroacetic acid-precipitated material during 1-hour pulses using 2-ml samples (Littlefield and Jacobs, 1965), is shown by the squares in the lower panel. Open squares represent incorporation of hypoxanthine-^{3}H in the absence of dT, and the closed squares in the presence of dT. The rate of DNA synthesis in a exponentially growing unsynchronized culture is shown by the open square at zero time. The mitotic index is shown in the upper panel by the open triangles.

To test these expectations, exponentially growing cells from line 19 were synchronized with two successive dT blocks (Fig. 2). The cell number began to increase 7 hours after the first release, plateaued shortly after application of the second block, and increased again after the second block was removed. The rate of DNA synthesis (as measured by hypoxanthine-^{3}H incorporation) increased during the first hour after release of the block and peaked at 3 to 4 hours after release of the block. With application of the second block, the rate of DNA synthesis decreased to the previous low level and remained there until the following release, when DNA synthesis again increased. The duration of S can thus be estimated to be 6 to 7 hours, compared to 10 hours or less in an unsynchronized population of human lymphoblasts (see Section III,C). A shortening of the S phase following dT synchronization has been observed previously (Bostock *et al.*, 1971).

If the second dT block was not applied, DNA synthesis did not fall to as low a rate, and began to increase sooner thereafter (Fig. 2). Similar timing was observed without application of the third dT block. During the dT block the rate of DNA synthesis in line 19 cells was reduced to 15% compared to zero time. This value can be reduced further if the dT concentration is raised to 5×10^{-5} *M*. Since DNA synthesis is rapidly inhibited after addition of excess dT, the second block should be applied when the rate of DNA synthesis reaches a minimum, viz., for line 19 cells, 9 to 10 hours after release of the first block.

The mitotic index peaked 9 hours after release of the first block (Fig. 2). Since the duration of S was 6 to 7 hours, the length of G_2 from these results was about 2 to 3 hours. This agrees with the estimate of approximately 3 hours for G_2 in an unsynchronized culture (see Section III,C).

The increase in cell number was studied during synchronized growth when the second dT block was applied to aliquots at various times after release of the first block (Fig. 3). If the second block was applied 2 or 5 hours after the initial release, no increase in cell number (or a limited increase) was observed. In contrast, the first burst of growth was of the expected size when the second block was applied 8 or 11 hours after the first release, and similar increase in cell number occurred following the second release. However, the second burst is dramatically decreased if the second block is applied 13 hours after the first release.

These results can be explained by the observed pattern of DNA synthesis during synchrony (Fig. 2). A block 2 hours after the first release occurs when the rate of DNA synthesis has just begun to rise, and the block at 5 hours occurs when DNA synthesis has passed the maximum, but probably before most cells have completed DNA synthesis. At 8 and 11 hours after the first release, replication of DNA has been completed and DNA synthesis is near a minimum. After 13 hours the cells are in the second S phase. Indeed the

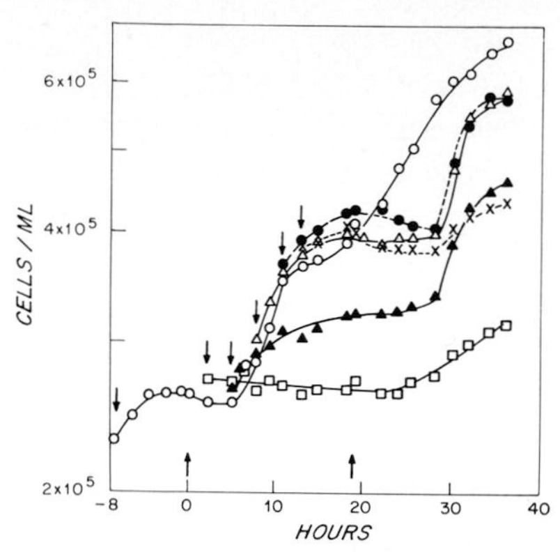

FIG. 3. Time of application of the second thymidine (dT) block. A culture of line 19 was blocked for 8 hours with 3×10^{-5} *M* dT. At zero time the block was removed and aliquots were reblocked at various times. At 18 hours all samples that had been reblocked were released. Arrows pointing down indicate times of application of dT blocks and arrows pointing up indicate times of release of the blocks. ○, No second block. Time between first release and application of second block: □, 2 hours; ▲, 5 hours; △, 8 hours; ●, 11 hours; X, 13 hours.

detrimental effect of a second block at this point suggests that cells blocked in S are injured and less likely to complete cell cycle transversal after release of the block. These results indicate that the second and any subsequent blocks must be applied when DNA synthesis has reached a minimum. With our cell lines this corresponds to the middle of the increase in cell number after release of the previous block.

5. Thymidine Synchronization Followed by Colcemid Treatment

With any method the degree of synchrony inevitably decays as the cells resume cell cycle transversal. Therefore after synchronization by excess dT the degree of synchrony is greatest during S and G_2, but less in G_1. Mitotic selection or inhibitors of mitotic spindle formation have been employed by others to obtain cells highly synchronized for G_1. Only the latter technique is applicable to cells in suspension, but our attempted synchronization of lymphoblasts by Colcemid alone has proved unsuccessful owing to cell death. Buell and Fahey (1969) compared human lymphoid cell lines synchronized with a double dT block to cells synchronized with a dT block followed by Colcemid; they observed that the patterns of immunoglobulin content during the cell cycle were similar in cells synchronized by the two methods.

Colcemid synchronization after an initial dT synchronization is shown in Fig. 4. The increase in cell number occurs between 1 and 2 hours after release of the Colcemid block and is initially more rapid than after a dT block. The percent of cells dividing is decreased slightly compared to dT

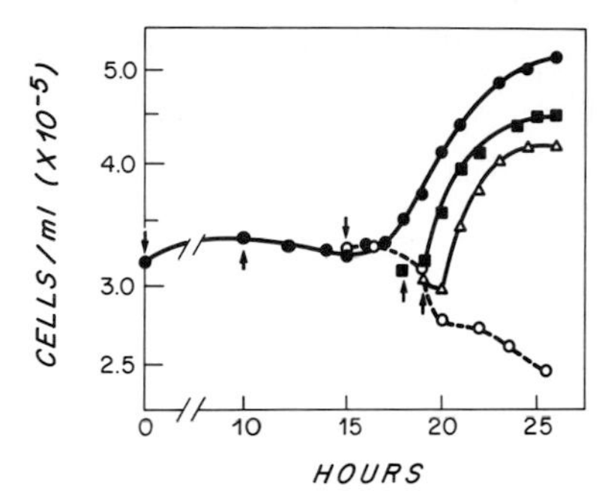

FIG. 4. Synchronization by thymidine (dT) followed by a Colcemid block. An exponentially growing culture of line 19 was blocked with 3×10^{-5} *M* dT at zero time and released by the addition of 1×10^{-6} *M* deoxycytidine at 10 hours. Aliquots were blocked with 0.02 μg of Colcemid per milliliter at 15 hours, and released at 18 or 19 hours by centrifugation and resuspension of cells in fresh, unsupplemented medium. Single synchronization by thymidine, ●; an additional Colcemid block, but not released, ○; Colcemid block released at 18 hours, ■; Colcemid block released at 19 hours, △.

alone. Furthermore, it appears that the Colcemid block should be applied as late as possible because if the block is maintained too long, cell death occurs, as measured by trypan blue staining and decrease in cell concentration. Thus it seems desirable to add Colcemid 1 or 2 hours before the anticipated rise in cell number after the dT synchronization, and release the Colcemid block about 1 hour after control dT synchronized cells have begun to divide; i.e., the total length of the Colcemid block should be 3 hours or less.

III. Characterization of the Synchronization

One measure of synchrony is the fraction of a population that divides after the dT block is removed. Ideally all the cells should divide, but during these studies the increase in cell number rarely exceeded 60%. Studies by others with dT synchronized human lymphoblast cultures have given similar results (Takahashi *et al.*, 1969; Rosenberg and Einstein, 1972).

A. Cell Viability during the Thymidine Block and Subsequent Synchrony

To test whether the limited increase in cell number after release was due to cell death during the dT block, cells were cloned in soft agarose (Sato *et al.*, 1972) at different times in the cell cycle (Table I) and for periods up to 24 hours of continuous exposure to dT (Table II). Viability as measured

TABLE I
VIABILITY DURING SYNCHRONY

Time after release (hr)	Predominant cell phase	Cloning efficiency[a]	Trypan Blue staining[b]
3	S	48	10.2
6	G_2	43	9.0
13	G_1	—	10.1

[a]The cloning was performed in a 0.24% agarose medium using a fibroblast feeder layer.

[a]Positive staining particles as small as one-fourth the size of whole cells were counted.

by cloning efficiency did not change. However, since it takes 2 to 3 weeks for a clone to become visible, delay in cell cycle transversal of up to several days could not be detected by this method. The other measure of viability, staining by the dye trypan blue, also indicated no killing of cells (Tables I and II). Unfortunately such techniques do not exclude transient effects on the growth of cells which might limit the increase in cell number after release of a block.

B. Epstein-Barr Virus

Replication of EBV during synchrony might produce cells incapable of division. To explore this possibility EBV antigen production was monitored during the cell cycle by the indirect immunofluorescence technique of Henle and Henle (1966). By means of this method, exponentially growing line 19 cells were found to contain no detectable cells expressing EBV antigen, and

TABLE II
VIABILITY DURING THYMIDINE (dT) BLOCK[a]

Time in dT (hr)	Cloning efficiency (%)	Trypan Blue staining (%)
0	49	9.5
8	52	9.9
16	43	9.5
24	41	8.9

[a]Procedures as in Table I.

the LPM-10 culture less than 1% of such cells. Line 19 cells were synchronized through two successive dT cycles and slides prepared during the 48 hours following the second release. No fluorescent cells were observed by the indirect immunofluorescent technique. These results indicated that mass induction of EBV did not occur, although variation on EBV content during the cell cycle in those few cells in LPM-10 cultures originally expressing EBV antigen is not excluded. Sairenji and Hinuma (1971) studied the variation of immunofluorescent cells in a culture of P3HR-1 lymphoblast cells synchronized with 5×10^{-3} *M* dT. At the time of release of the second dT block, 28% of the cells were immunofluorescence-positive. After 12 hours the immunofluorescence-positive cells increased to 34% and remained constant for an additional 12 hours. However, less than 30% of the cells divided after release of the dT block. The possibility of a toxic effect, rather than a cell cycle effect, was not excluded.

C. Generation Time of Unsynchronized, Exponentially Growing Cells

The observed 60% increase in cell number during synchronization could be explained if a fraction of the cells in any culture growing exponentially or synchronized were not dividing. Under such circumstances the average cell generation time would be shorter than the doubling time of the mass culture. The cell generation time of an unsynchronized, exponentially growing culture of line 19 was determined by the dT-^{3}H pulse method of Quastler

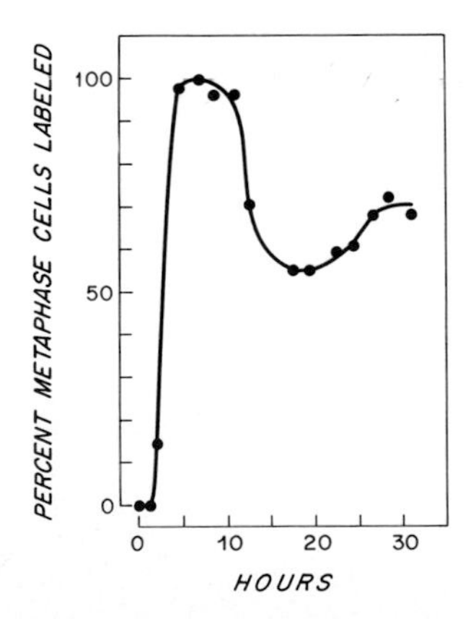

FIG. 5. Determination of the generation time of an exponentially growing, unsynchronized culture of line 19. The culture was pulsed for 15 minutes with thymidine (dT)-^{3}H (1 μCi/ml), centrifuged and resuspended in conditioned medium containing 1×10^{-5} *M* dT and 1×10^{-6} *M* deoxycytidine. Under these conditions the rate of the cell growth was not affected. The percentage of labeled metaphase cells was determined after autoradiography.

and Sherman (1959) to be about 22 ±1 hours (Fig. 5), while the doubling time of the mass culture was 24 hours. In a second study the generation time equaled the doubling time of the mass culture. It can be estimated from these data that approximately 90% of the cells in an unsynchronized population of lymphoblasts are dividing at any one time; in contrast to about 60% following the dT block.

The data in Fig. 5 indicate that G_2 is approximately 3 hours and S is 10 hours or less. Since 4.5% of the cells were in mitosis during exponential growth, M equals about 1.5 hours (Stanners and Till, 1960). By subtraction the remaining phase, G_1, equals 9 to 10 hours.

D. Repetitive Synchronization

An attempt was made to enrich cultures of line 19 and LPM-10 for cells responding favorably to synchronization. During a 3-week period, cultures were repetitively synchronized 4 times a week. The results obtained with line 19 are shown in Fig. 6. The percent increase in cell number after repetitive synchronization remained constant indicating that no enrichment occurred (Table III). This suggests again that a constant proportion of lymphoblasts fail to divide in each cycle under our conditions.

E. Synchronization of Freshly Cloned Lines

In another attempt to improve the synchrony, line 19 cells were cloned in agarose (Sato *et al.*, 1972), and 10 colonies were selected at random, with

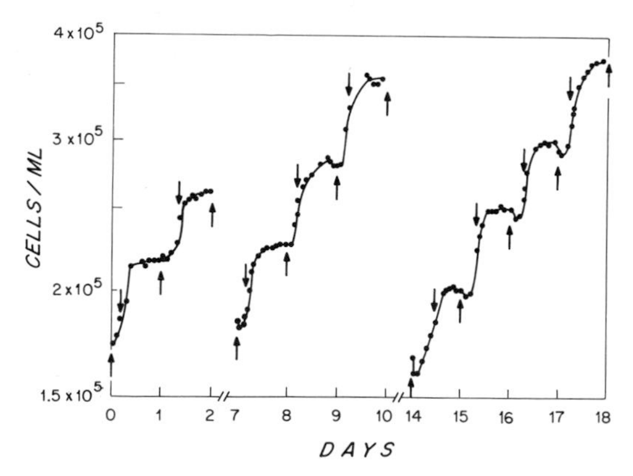

FIG. 6. Repetitive synchronization of line 19. The culture was blocked with $3 \times 10^{-5} M$ thymidine (dT) at times indicated by the arrows pointing down and released by centrifugation and resuspension in fresh medium at times indicated by arrows pointing up. A 24-hour growth cycle was imposed on the cells. During a 3-week period, cultures were repetitively synchronized 4 times a week. Only 9 out of 12 curves are shown, as the last synchrony burst per week was not followed.

TABLE III
INCREASE IN CELL NUMBER DURING SYNCHRONY

Day	Increase %
1	51
2	41
8	54
9	58
10	62
15	59
16	58
17	50
18	58

the expectation that such lines would be more uniform in generation time, cell size, and other parameters. However, no change was observed during subsequent synchronization attempts (data not shown).

F. Synchronization of Other Lymphoblast Lines

In addition to line 19, we have applied the dT synchronization technique to several other human lymphoblast lines (LPM-10, MGH-1, MGH-5, MGH-7, MG-57, and CCRF-CEM). Cell growth of the first five lines is inhibited by approximately 10^{-3} *M* dT while the CCRF-CEM culture, like line 19, is highly sensitive to dT inhibition. A culture of LPM-10 synchronized with 3×10^{-3} *M* dT showed essentially the same percent increase in cell number after release as obtained with line 19 (Fig. 7). Synchronization of CCRF-CEM, an acute lymphoblastic leukemia line, with 4×10^{-5} *M* dT resulted in a 50 to 70% increase in cell number after release (data not shown). During synchronization of four other lines by excess dT, fewer than one-third of the cells divided after release of the block. However, the specific conditions used may not have been optimal for these lines.

The lines that we have tested originated from apparently normal individuals and from patients with infectious mononucleosis or acute lymphoblastic leukemia. There was no relationship between the source of the cell line and the response to synchronization by excess dT. A notable difference, however, is the fact that three of the lines that did not respond favorably had been in culture less than 6 months and one less than 2 years (MG-57) while the lines that responded well, line 19, LPM-10, and CCRF-CEM have been in culture for several years. A correlation may exist between the degree of synchrony achieved by dT and the length of time a line has been in culture, as well as with the degree of diploidy of the lines.

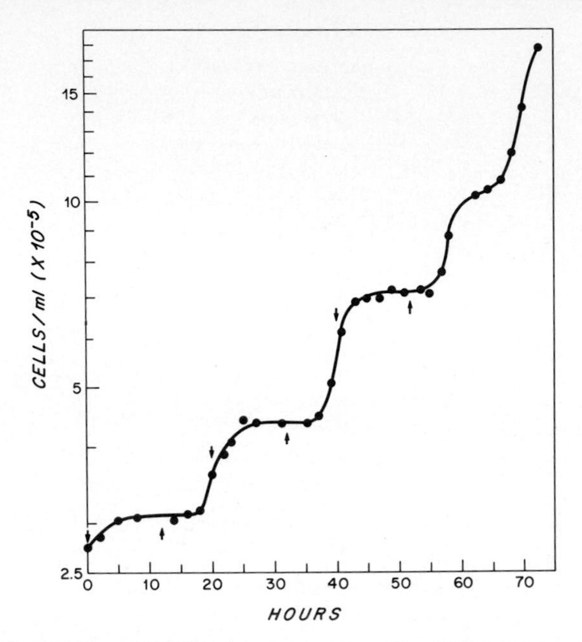

FIG. 7. Synchronization of LPM-10 using excess thymidine (dT). An exponentially growing culture of LPM-10 was blocked with 3×10^{-3} *M* dT at times indicated by arrows pointing down and released by centrifugation and resuspension in fresh medium at times indicated by arrows pointing up. At time of release the culture was diluted to maintain the cell density below 5×10^5 cell per milliliter.

IV. Concluding Remarks

Certain practical guidelines are provided by the results of these studies. The appropriate dT concentration must be determined for each lymphoblast line, since the sensitivity of lymphoblast lines to dT seems to be more variable than for fibroblast lines. Proper timing of the application and duration of the dT blocks is most important, especially for the second and any subsequent blocks, and should be applied when the rate of DNA synthesis is at a minimum. This necessitates determination of the pattern of DNA synthesis subsequent to the first release.

The participation of approximately 60% of the cells in the division after removal of the dT block remains unexplained. The nondividing cells do not appear to be killed, and may have been caused by this block to enter a temporary nongrowing state.

ACKNOWLEDGMENTS

During this study H. Ronald Zielke was supported by USPHS training grant HD-00362 and John W. Littlefield by a USPHS research grant CA-04670. Thanks are due to Mrs. Afina Broekman for the karyotyping and to Dr. George Miller for advice on the indirect immunofluorescence technique. Drs. P. R. Glade, George Moore, and Herb Lazarus kindly provided cultures of PGLC 33H, LPM, and CCRF-CEM, respectively. Special thanks are due to Dr. W. G. Thilly for information on the automatic sampling system prior to publication.

REFERENCES

Bostock, C. J., Prescott, D. M., and Kirkpatrick, J. B. (1971). *Exp. Cell Res.* **68**, 163.
Buell, D. N., and Fahey, J. L. (1969). *Science* **164**, 1524.
Doida, Y., and Okada, S. (1967). *Exp. Cell Res.* **48**, 540.
Ehmann, U. K., and Lett, J. T. (1972). *Exp. Cell Res.* **74**, 9.
Foley, G. E., Lazarus, H., Farber, S., Uzman, B. G., Boone, B. A., and McCarthy, R. E. (1965). *Cancer* **18**, 522.
Galavazi, G., Shenk, H., and Bootsma, D. (1966). *Exp. Cell Res.* **41**, 428.
Henle, G., and Henle, W. (1966). *J. Bacteriol.* **91**, 1248.
Klevecz, R. R. (1972). *Anal. Biochem.* **49**, 407.
Klevecz, R. R., Forrest, G. L., and Kapp, L. (1972). *J. Cell Biol.* **55**, 139a.
Krishan, A., Raychaudhuri, R., and Flowers, A. (1969). *J. Nat. Cancer Inst.* **43**, 1203.
Lazarus, H., and Foley, G. E. (1972). *In* "Growth, Nutrition and Metabolism of Cells in Culture" (G. Rothblat and V. J. Cristofalo, eds.), pp. 169–202. Academic Press, New York.
Littlefield, J. W., and Jacobs, P. S. (1965). *Biochim. Biophys. Acta* **108**, 652.
Mitchison, J. M. (1971). "The Biology of the Cell Cycle." Cambridge Univ. Press, London and New York.
Moore, G. E., Kitamura, H., and Gerner, R. E. (1968). *Proc. Amer. Ass. Cancer Res.* **9**, 52.
Nonoyama, M., and Pagano, J. S. (1972). *Nature (London), New Biol.* **238**, 169.
Puck, T. T. (1964). *Science* **144**, 565.
Quastler, H., and Sherman, F. G. (1959). *Exp. Cell Res.* **17**, 420.
Rosenberg, S. A., and Einstein, A. B., Jr. (1972). *J. Cell Biol.* **53**, 466.
Rueckert, R. R., and Mueller, G. C. (1960). *Cancer Res.* **20**, 1584.
Sairenji, T., and Hinuma, Y., (1971). *GANN Monogr.* **10**, 113.
Sato, K., Slesinski, R. S., and Littlefield, J. W. (1972). *Proc. Nat. Acad. Sci. U.S.* **69**, 1244.
Schachtschabel, D. O., Lazarus, H., Farber, S., and Foley, G. E. (1966). *Exp. Cell Res.* **43**, 512.
Stanners, C. P., and Till, J. E. (1960). *Biochim. Biophys. Acta* **37**, 406.
Stubblefield, E. (1968). *Methods Cell Physiol.* **3**, 25.
Takahashi, M., Yagi, Y., Moore, G. E., and Pressman, D. (1969). *J. Immunol.* **103**, 834.
Terasima, T., and Tolmach, L. J. (1963). *Exp. Cell Res.* **30**, 344.
Thilly, W. G. (1972). Ph.D. Thesis, Mass. Inst. of Technol., Cambridge, Massachusetts.
Tobey, R. A., and Ley, K. D. (1970). *J. Cell Biol.* **46**, 151.
Tobey, R. A., Anderson, E. C., and Petersen, D. F. (1967). *J. Cell. Physiol.* **70**, 63.
Xeros, N. (1962). *Nature (London)* **194**, 682.
zur Hausen, H., and Schulte-Holthausen, H. (1970). *Nature (London)* **227**, 245.

Chapter 9

Uses of Enucleated Cells

ROBERT D. GOLDMAN[1] AND ROBERT POLLACK

Department of Biological Sciences,
Carnegie Mellon University,
Mellon Institute,
Pittsburgh, Pennsylvania;
and Cold Spring Harbor Laboratory,
Cold Spring Harbor, New York

I. Introduction

Some of the most basic and interesting problems in eukaryotic cells are those involved in studying nuclear–cytoplasmic interactions. Cytologists and cell biologists have attempted to study the interactions and independent activities of these two major compartments of eukaryotes primarily by surgically removing the nucleus. The nucleus is then transplanted to other types of enucleated cytoplasm or the behavior and properties of the remaining enucleated cytoplasm are monitored (Chambers and Fell, 1931; Briggs and King, 1959; Kopac, 1959; Goldstein *et al.*, 1960a,b; Brachet, 1961; Goldstein, 1964; Goldstein and Eastwood, 1964; Marcus and Freiman,

[1] *Present address:* Cold Spring Harbor Laboratory, Cold Spring Harbor, New York.

1966). The microsurgical techniques utilized require a high degree of skill with micromanipulators and tremendous patience on the part of the investigators. In addition, the techniques have inherent limitations, such as the possibility of inflicting irreparable damage to the cells during surgery and the time factors involved in making relatively few observations.

Carter (1967, 1972) described the spontaneous enucleation of cultured L cells when they were treated with cytochalasin B (CB), a compound isolated from the mold *Helminthosporium dematioideum*. Unfortunately only a few percent of the cells in a culture were enucleated. Several other studies have also demonstrated that only a few cells in a population become enucleated in a variety of cell lines (Ladda and Estensen, 1970; Prescott *et al.*, 1972; Goldman, 1972). However, Poste and Reeve (1970) claimed that between 45% and 80% of cultures of murine macrophages and L929 cells treated with 30 μg of CB per milliliter for 8 hours, become enucleated. We have attempted to increase the yield of enucleates by treating several cell lines with concentrations of CB as high as 50 μg/ml without significant success.

TABLE I

CURRENT CONDITIONS FOR ENUCLEATING VARIOUS CELL LINES[a]

Cell line	Substrate (type of coverslip)	Speed of centrifugation (rpm)	Time (min) of centrifugation
BHK-21	Glass, collagen	6,500	50
	Plastic, collagen	11,000	15
PyBHK	Glass, collagen		
	Plastic, collagen	11,000	10
BSC-1	Glass	9,000	2 × 60[b]
	Plastic	16,500	2 × 15[b]
3T3	Glass, collagen	6,500	50
	Plastic, collagen	11,000	10
FLSV	Glass, collagen	7,000	60
	Plastic, collagen	11,000	15
SV3T3	Glass, collagen		
	Plastic, collagen	12,000	10
Py3T3	Glass, collagen		
	Plastic, collagen	12,000	10
CHO	Plastic	15,000	15

[a] It is to be emphasized that these are the present techniques we utilize for obtaining over 90% enucleation; however, we are constantly changing procedures in attempts to increase the percentage of enucleation and cell yields and to shorten times of centrifugation. All cell lines are enucleated at subconfluent densities.

[b] 2 × 60 or 2 × 15 means that the enucleation procedure is repeated following a recovery period in normal medium.

The enucleation phenomenon appears to be related to the response of cells to the drug at concentrations of 5–10 μg/ml. At these concentrations, the cells have a grossly altered morphology and the nucleus is raised up from the growth substrate onto a slender stalk of cytoplasm (Carter, 1972). In a few cases this stalk breaks and the nucleus is released from the cells, apparently surrounded by a thin sheet of cytoplasm and a plasma membrane. Most of the cytoplasm remains attached to the substrate (Carter, 1967, 1972; Prescott *et al.*, 1972; Goldman, 1972; Goldman *et al.*, 1973a). Recently, Prescott *et al.* (1972) have been able to greatly amplify the enucleation effect by growing the cells on round glass or plastic coverslips, placing them cell side down into the bottom of centrifuge tubes containing medium with 10 μg of CB per milliliter, and then centrifuging them at 37°C. In this way the nuclei can be forced out of the cells, while the cytoplasm remains behind on the coverslip. With this technique, Prescott *et al.* (1972) achieved up to 98% enucleated populations of CHO and L cells.

Utilizing slight modifications of the Prescott *et al.* (1972) procedure, we have been able to enucleate over 90% of the cells in populations of seven continuous cell lines (see Table I). We are now in the process of determining the uses of enucleates in a variety of problems involving nuclear–cytoplasmic interactions.

II. The Technology of Enucleation

As mentioned above, our technique is based on the Prescott *et al.* (1972) procedure. Cells are grown on round glass coverslips or on round plastic coverslips. The plastic coverslips are either cut from the bottom of Falcon tissue culture plates (Prescott, personal communication; Prescott *et al.*, 1972) with a cork borer or are punched out with a heated piece of stainless steel tubing. The coverslips are 12 mm in diameter for use with 15-ml Corex round-bottom centrifuge tubes or ~22 mm in diameter for use with 50-ml Sorvall plastic centrifuge tubes. In some cases, the coverslips are coated with rat tail collagen (Ehrmann and Gey, 1956), which makes the cells more adhesive to the glass or plastic substrate. The use of plastic or glass substrates coated with collagen minimizes the loss of cells during centrifugation [see Prescott *et al.* (1972) for a more extensive discussion of this problem].

Cells are grown on the coverslips to semiconfluence or confluence in normal growth medium. The coverslips are then placed into centrifuge tubes containing 10 μg of CB/ml so that the cell side is down. Within a few minutes the cells show a greatly altered morphology and the nucleus is pushed up onto a thin stalk of cytoplasm as described above. The tubes are placed in a

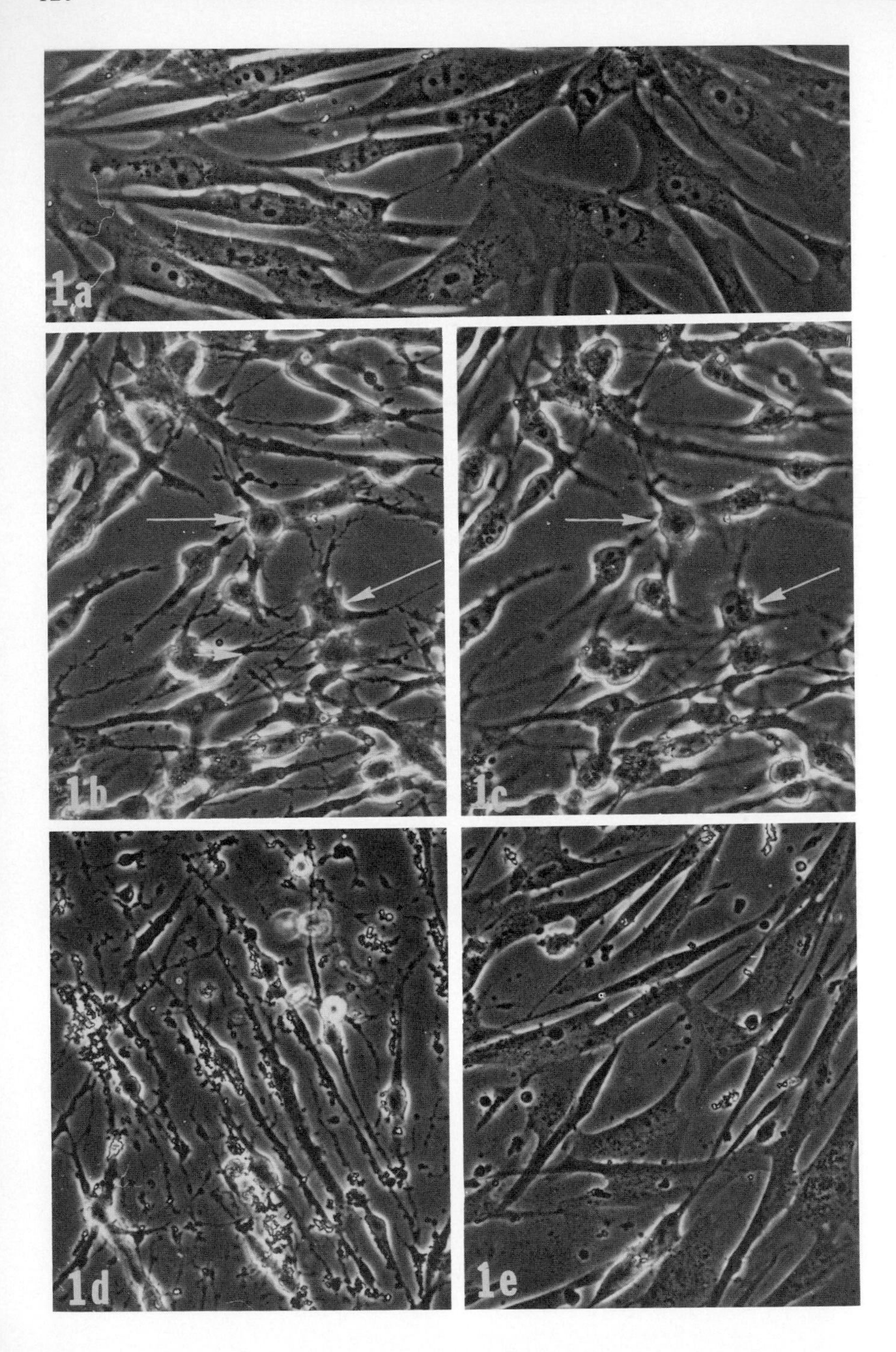
1a
1b
1c
1d
1e

Sorvall SS-34 head seated in a Sorvall RC-2 or RC-2B centrifuge which has been prewarmed to 37°C (see Prescott *et al.*, 1972). The tubes are spun at various speeds at 37°C for periods varying from 10 to 60 minutes. The conditions for centrifugation depend on the cell type, cell density, and substrate (Table I). After centrifugation the coverslips are removed from the tubes, and they are placed cell side up into 35-mm Falcon tissue culture dishes containing 2 ml of normal medium in order to recover from the effects of CB. Within 15–30 minutes the cells possess a normal morphology. The only obvious difference is that the nucleus is absent. Once coverslip cultures recover fully from the effects of the treatment they can be subjected to a second and even third cycle of enucleation. For example, in a two-cycle experiment with BSC-1 cells, 95% of the cells were enucleated in the first centrifugation, and approximately 95% of the cells that retained their nucleus in the first cycle lost them in the second cycle, leaving cultures that were more than 99% enucleated (Pollack and Goldman, 1973). Figures 1 and 2 illustrate the condition of cells prior to, during, and after enucleation of BHK-21 and BSC-1 cells.

The conditions used for enucleating various cell lines are included in Table I. By utilizing the procedures prescribed for each cell line, we are able routinely to obtain populations of cells that are between 90 and 99% enucleated.

III. Uses and Properties of Enucleated Cells

A. Trypsinization and Replating

After recovery of enucleates from the effects of CB we have been able to remove them from their substrates with 0.25% trypsin–EDTA solution (Grand Island Biological Co.). We carefully observe the cells during trypsinization, and as soon as they begin to round up and lose some of their contacts

FIG. 1. BHK-21 cells observed during the process of enucleation. (a) Subconfluent culture on a glass coverslip prior to the addition of CB. (b) Cells observed 10 minutes after the addition of CB. Note the morphology of the cells is greatly altered and that the nucleus is pushed up from the substrate and thus is out of focus. (c) The same field of cells with the nuclei in focus. Arrows in (b) and (c) point to the same cells. (d) Cells following centrifugation in CB. Note that the bulge toward the center of these cells, which contains the nucleus in (b) and (c), is missing or is reduced in most cases. (e) Cells 15 minutes after the removal of CB by washing with fresh normal medium. Most of the cells appear normal in shape, but lack nuclei. Phase contrast. (a) × 370; (b–e) × 320.

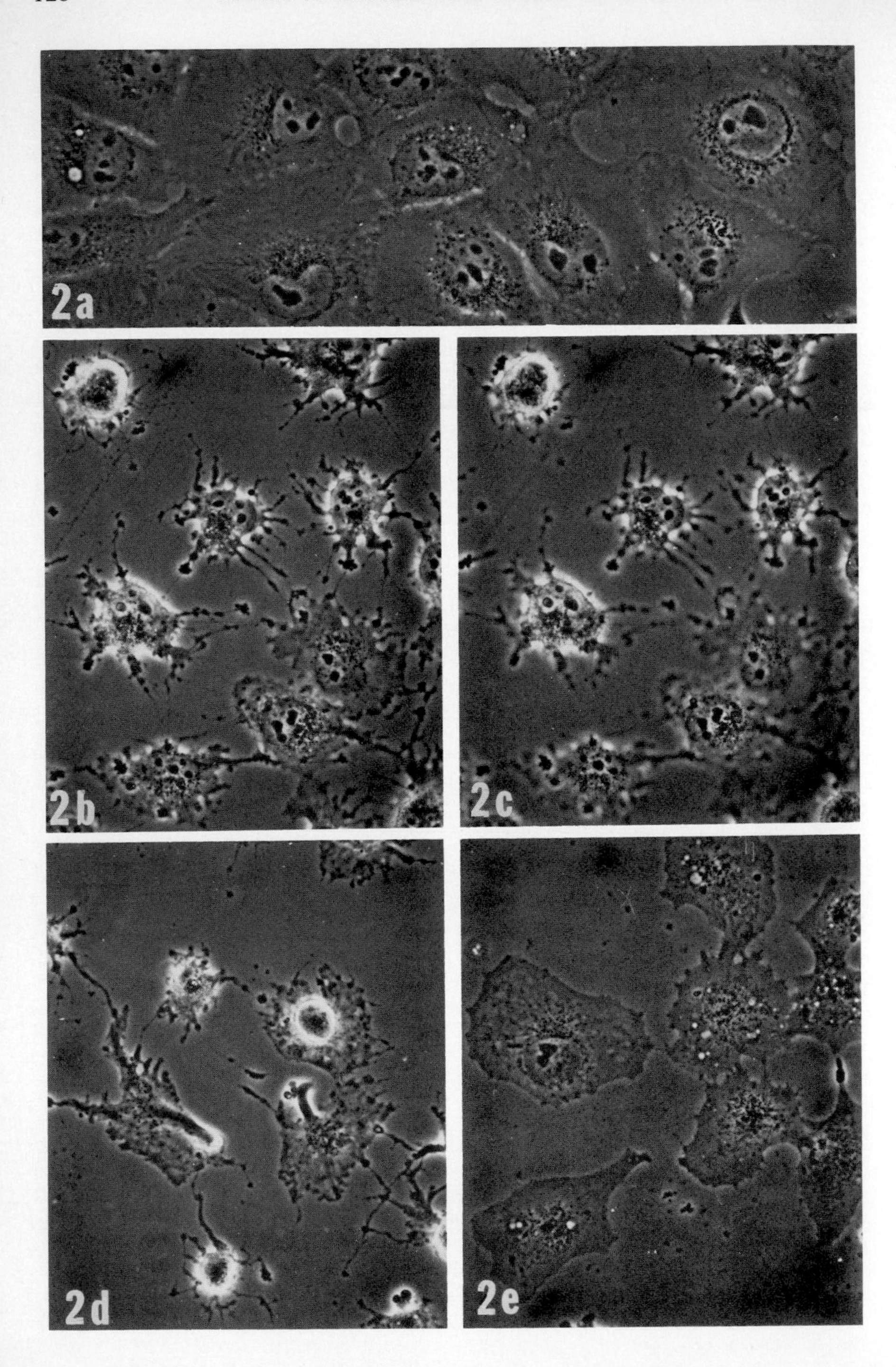
2a
2b
2c
2d
2e

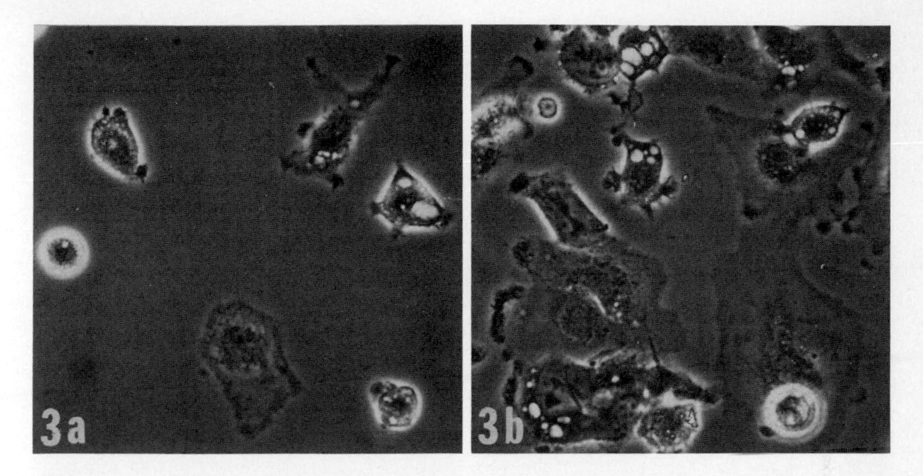

FIG. 3. (a) Sparse culture of BSC-1 enucleates which had been replated and observed 2 hours later. (b) Same time interval following replating in a crowded culture. Note that the enucleates are spread more in (b) than in (a). Phase contrast. ×480.

with the substrate, the trypsin-EDTA solution is removed and replaced with a few milliliters of normal medium. The medium is then rapidly pipetted in and out of the petri dish containing the coverslip, thereby removing the enucleates and suspending them in the normal medium.

The suspension of enucleates can be replated immediately into Falcon dishes or onto glass coverslips. In order to enhance reattachment and spreading of enucleates, they are concentrated into small drops of normal medium on the substrate. Within a few hours, the plate is flooded with medium. The effect of this crowding is to enhance the attachment and spreading of the enucleates (Fig. 3b). The enucleates are also capable of attachment and spreading at low cell densities; however, the process is slower (Fig. 3a).

Utilizing this procedure, we have demonstrated that rounded-up enucleated cells are capable of normal attachment, spreading, and shape formation (Goldman *et al.*, 1973a). For example, enucleated 3T3 cells attach and spread out over the substrate in a fashion which is identical to nucleated 3T3 cells (Fig. 4). The same is true for enucleated BSC-1 cells and BHK-21 cells

---

FIG. 2. BSC-1 cells observed during the process of enucleation. (a) Subconfluent culture on a glass coverslip prior to the addition of CB. (b) Cells observed 15 minutes after the addition of CB. (c) Same field of cells to demonstrate that following a change in focus up from the substrate the nucleus is still in focus. (d) Cells after centrifugation in CB. (e) Cells 30 minutes after removal of CB. Most cells appear normal in shape. Note cell with nucleus at left of field. Phase contrast. (a) ×370; (b–e), ×320.

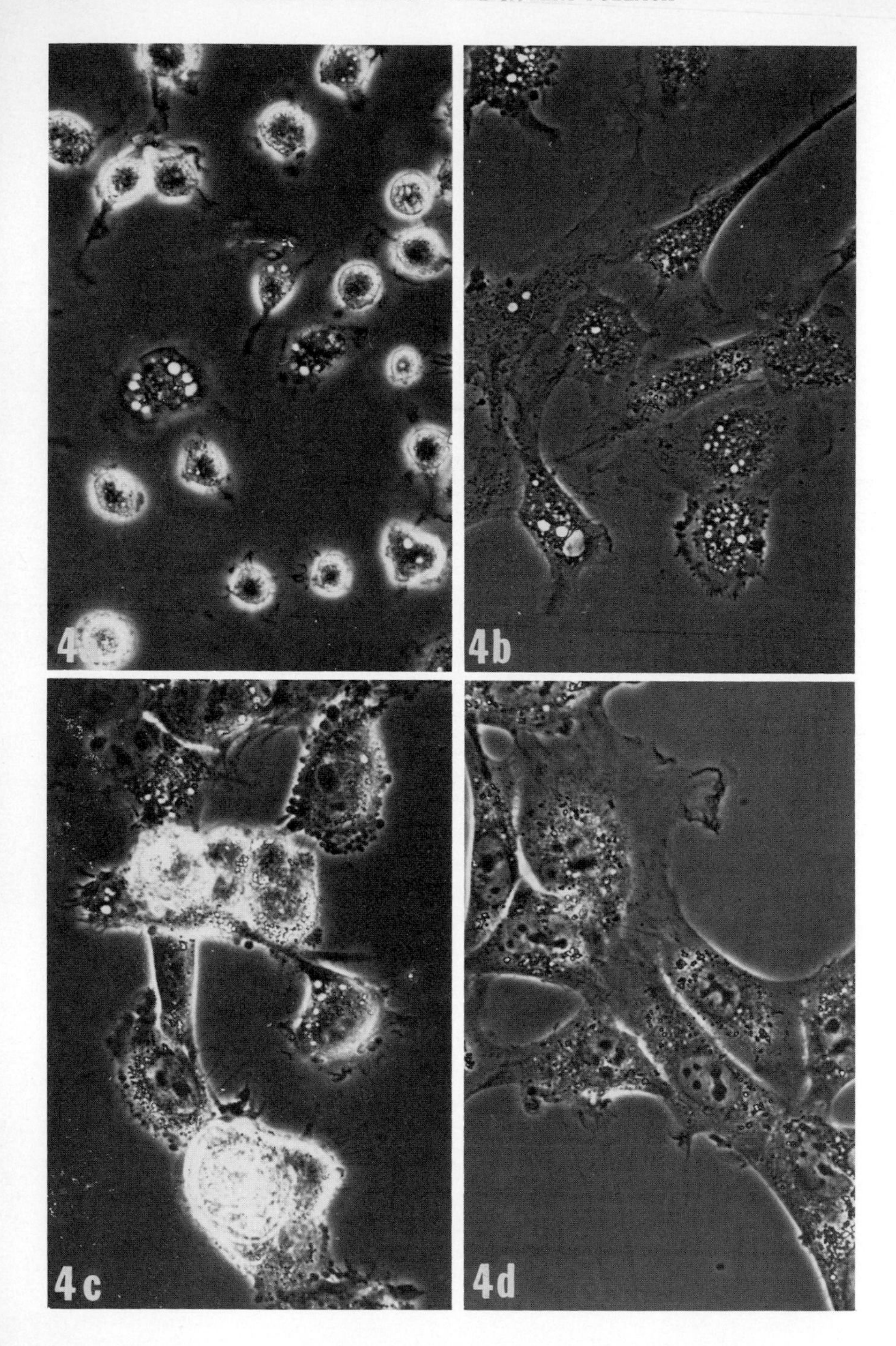
4a
4b
4c
4d

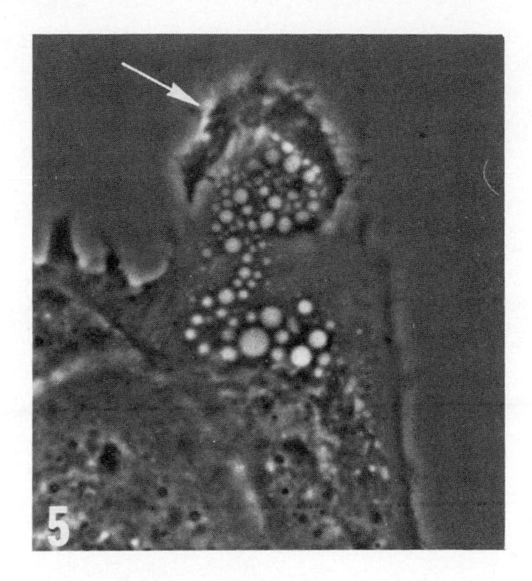

FIG. 5. Ruffling edge of a BSC-1 enucleate observed 2 hours after replating from a trypsinized suspension (arrow). Note phase bright pinocytotic vesicles forming beneath the ruffle. Phase contrast. × 1300.

(see Goldman *et al.*, 1973a). These observations demonstrate that cytoplasm contains the information necessary for cell attachment, spreading, and shape formation.

Other physiological phenomena are also seen in enucleates. They are capable of normal pinocytosis and membrane ruffling (Fig. 5). We have also demonstrated that contact inhibition of membrane ruffling occurs normally in enucleated BSC-1 cells and that cell locomotion may be seen in enucleated BHK-21 cells (Goldman *et al.*, 1973a).

B. Protein Synthesis in Enucleated Cells

With more than 10^6 enucleates easily obtainable in one cycle of centrifugation it has become possible to study the biochemistry as well as the physiology of cytoplasm. BSC-1 protein synthesis, as measured by the incorporation of leucine-^{3}H into TCA-precipitable material, continues for at

---

FIG. 4. Early and late stages of spreading in nucleated and enucleated 3T3 cells. (a) 3T3 enucleates observed 30 minutes after replating from a trypsinized suspension. Note rounded cell shapes and initiation of membrane ruffling at cell periphery. (b) Three hours after replating of 3T3 enucleates. Cell shape is well established and indistinguishable from nucleated 3T3 cells (see d). (c) Nucleated 3T3 cells observed 30 minutes after replating. (d) Nucleated 3T3 cells 3 hours after replating. The process of spreading from a spherical to a normal shape thus appears identical in both nucleated and enucleated 3T3 cells. Phase contrast. × 480.

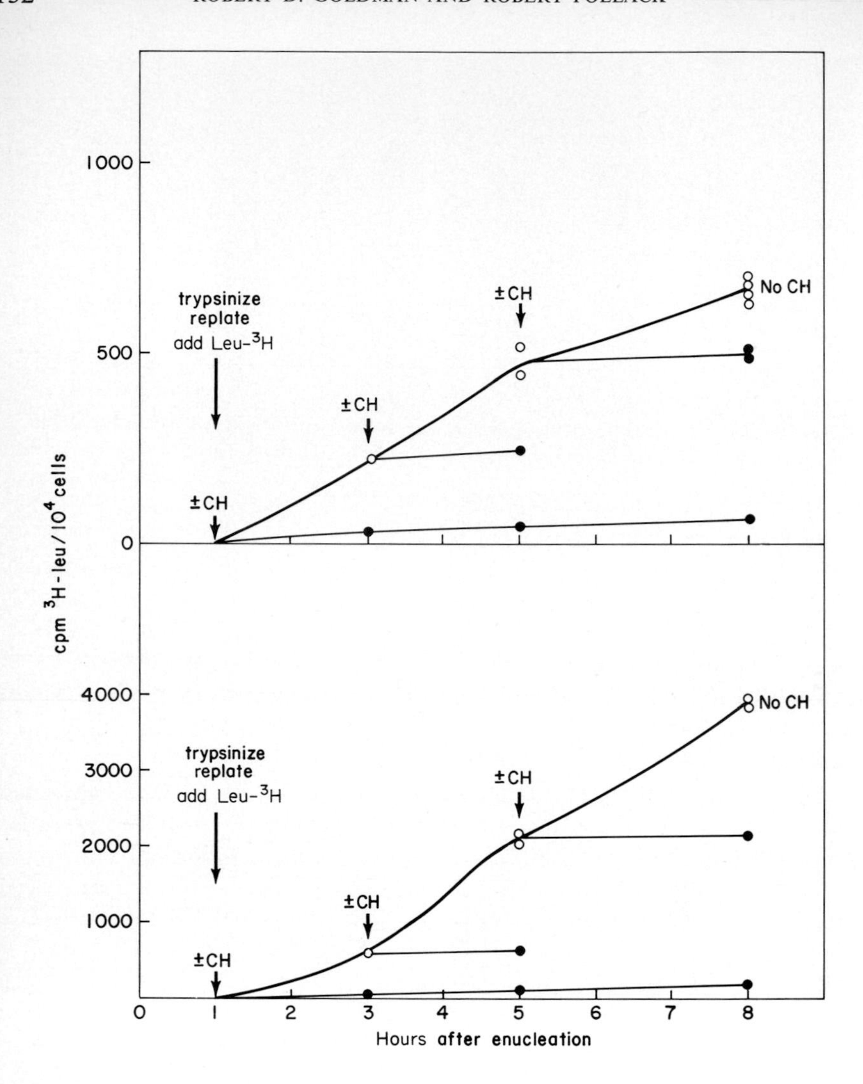

FIG. 6. BSC-1 cells (98.9%; top) were enucleated on plastic coverslips by centrifugation in 10 μg/ml CB at 37°C for 10 minutes at 12,000 rpm in the SS-34 head of a Sorvall centrifuge. Control cells (lower panel) were given CB but not spun. One hour after recovery from enucleation, cells on coverslips were trypsinized and replated onto many coverslips in medium containing 10% calf serum and 1 μCi/ml leucine-^{3}H (5 Ci/mmole). Cycloheximide (CH), 20 μg/ml, was added to some of these cultures at the times designated. Newly synthesized protein was assayed as incorporation of leucine-^{3}H into trichloroacetic acid-precipitable material. CH-sensitive protein synthesis occurred in the enucleates at one-fourth the rate per cell of control BSC-1 cells. Some coverslips from this experiment were prepared for autoradiography (Fig. 7).

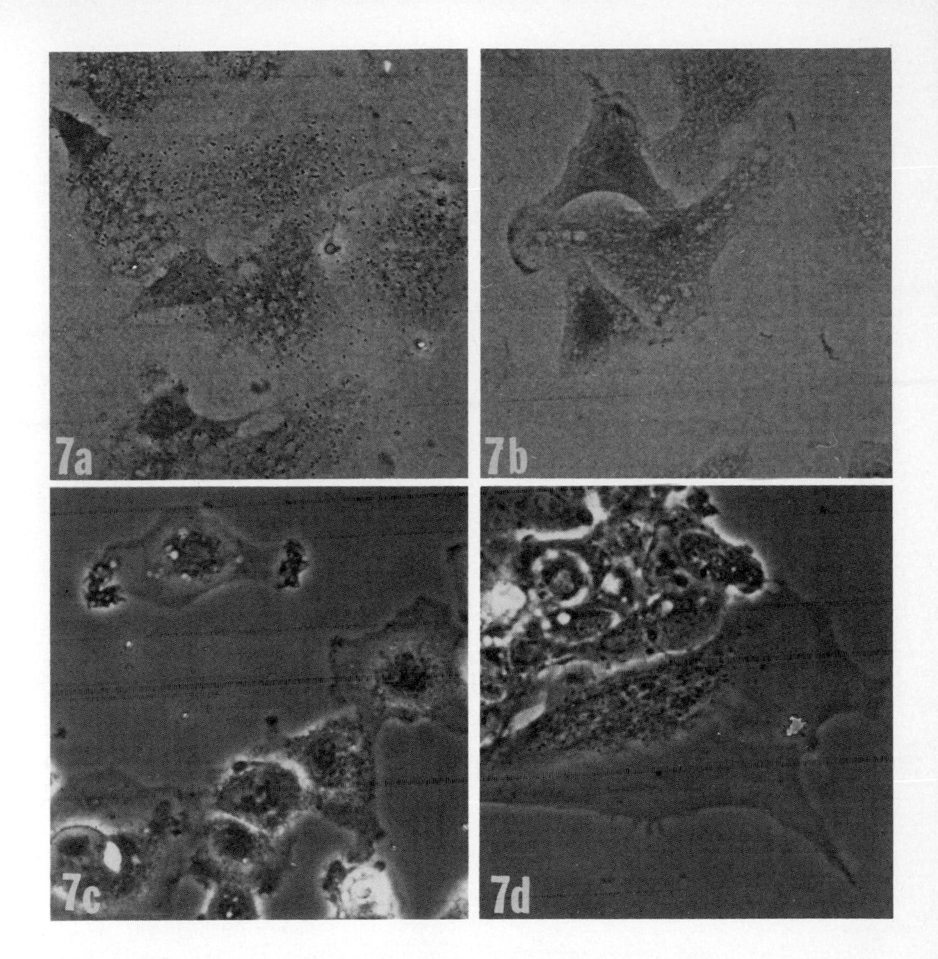

FIG. 7. (a and b). See legend to Fig. 6. Coverslips containing enucleated or control BSC-1 cells that had been replated in the presence of leucine $^{-3}$H were prepared for autoradiography at 7 hours after replating. The coverslip cultures were washed twice in PBS, fixed in ethanol–acetic acid (2:1), air-dried, and mounted cell side up on microscope slides. The slides were dipped in melted (40°C) Kodak NTB-2 emulsion, air-dried for 2 hours, stored at 4°C for 48 hours, developed in Dektol, and permitted to dry and harden overnight. The developed slides were then stained in hematoxylin (Harris), blued in $LiCl_2$, cleared through alcohol to xylene, mounted in Permount, and examined at 500 ×, phase contrast (Zeiss). Silver grains were seen only overlying the cytoplasm of nucleated control cells, and over entire enucleates (see a). Enucleates in 20 μg of cycloheximide per milliliter showed no silver grains over the cytoplasm (see b). Although protein synthesis occurred in the enucleates, it was not required for spreading in either the enucleates or in the nucleated control cells. (c and d). Enucleated BSC-1 cells at 1 and 4 hours after replating from a trypsinized suspension into medium containing 20 μg of cycloheximide per milliliter. Phase contrast. × 520.

least 8 hours after enucleation (Fig. 6). While the rate per cell is diminished by about 75% (Fig. 6), the enucleates are of course smaller than whole cells, and in fact the rate per milligram of protein is diminished by no more than 20% (P. Jeppeson, unpublished results).

In order to test for the necessity of protein synthesis in the attachment and spreading of enucleated BSC-1 and BHK-21 cells, we have suspended them in medium containing 20 μg of cycloheximide per milliliter (see Fig. 6) and replated them as described above. Under these conditions, cell attachment, spreading, and shape formation are normal for periods of observation to 7 hours after enucleation (Fig. 7).

Since microtubule and microfilament assembly seem to be involved directly in cell attachment, spreading, and shape formation in nucleated BHK-21 and BSC-1 cells (Goldman and Knipe, 1973; Goldman *et al.*, 1973b), it seems likely, based on the above observations, that pools of subunits to these cytoplasmic fibers are present in enucleated cytoplasm. These types of observations emphasize that the cytoplasm contains storage forms of molecules or subunits which may be assembled or disassembled during various types of physiological activity in the absence of nuclear information.

C. Ultrastructure of Enucleates

A preliminary electron microscopic survey of enucleated BHK-21 cells indicates that the ultrastructure of spread enucleates is normal. A normal distribution of mitochondria, Golgi, ribosomes, etc., is seen in enucleated BHK-21 cells. In addition there is a normal distribution of microfilaments, microtubules, and 100 Å filaments along the fibroblastic processes (Fig. 8). The only difference between these and nucleated cells is the lack of any obvious nuclear structure. Even the centrioles frequently appear in the center of the cell in the region which normally contains the nucleus (Fig. 9). We are in the process of doing a more detailed study of the ultrastructure of enucleated cells during the spreading process to determine whether or not the assembly and distribution of microtubules, microfilaments, and 100 Å filaments is normal (Goldman and Pollack, 1973).

D. Maintenance of the Normal and Transformed Cell State in Enucleates

We have initiated attempts to answer some basic questions about the behavior of normal vs transformed cells in the absence of nuclear genetic information by utilizing the trypsinization and replating procedure. Our initial experiments have involved the use of enucleated 3T3 and Py3T3 cells.

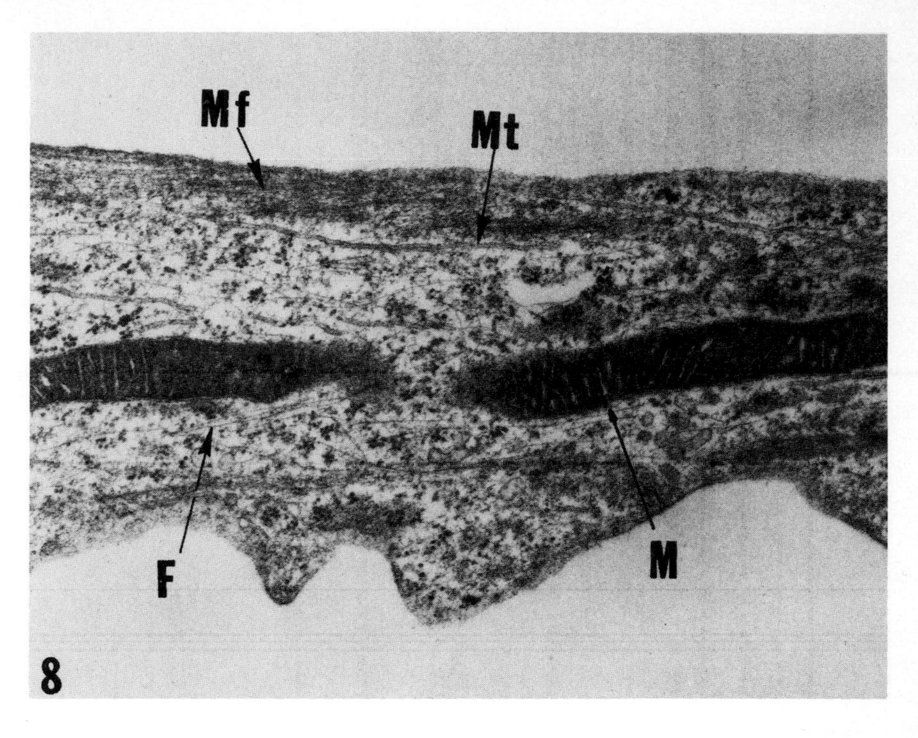

FIG. 8. Electron micrograph of a thin section of an Epon-embedded BHK-21 enucleate. The enucleates were fixed in 1% glutaraldehyde in phosphate buffer pH 7.2 and postfixed in 1% osmium tetroxide in the same buffer; after dehydration they were flat-embedded in Epon 812. Thin sections were taken parallel to the long axis of a fibroblastic process of this enucleated cell. Note filamentous mitochondria (M), longitudinally oriented microtubules (Mt), 100 Å filaments (F), and submembranous microfilaments (Mf). The ultrastructure is identical to that seen in similar fibroblastic processes of nucleated BHK-21 cells (see Goldman and Knipe, 1973). × 25,600.

When these enucleates are replated into drops and cells are observed at the periphery of the drops, it becomes obvious that the property of piling up or overlapping of cells is maintained by Py3T3 enucleates whereas 3T3 enucleates at the edge of the drop show relatively minimal overlaps (Fig. 10). Thus it appears that nuclei are not necessary for the maintenance of these properties of normal and transformed cells. These observations also indicate that the cytoplasm is capable of maintaining normal contact inhibition in enucleated 3T3 cells and a lack of contact inhibition in enucleated Py3T3 cells. From this type of experiment, it looks as though this aspect of the behavior of transformed cells (overlapping or lack of contact inhibition) does not

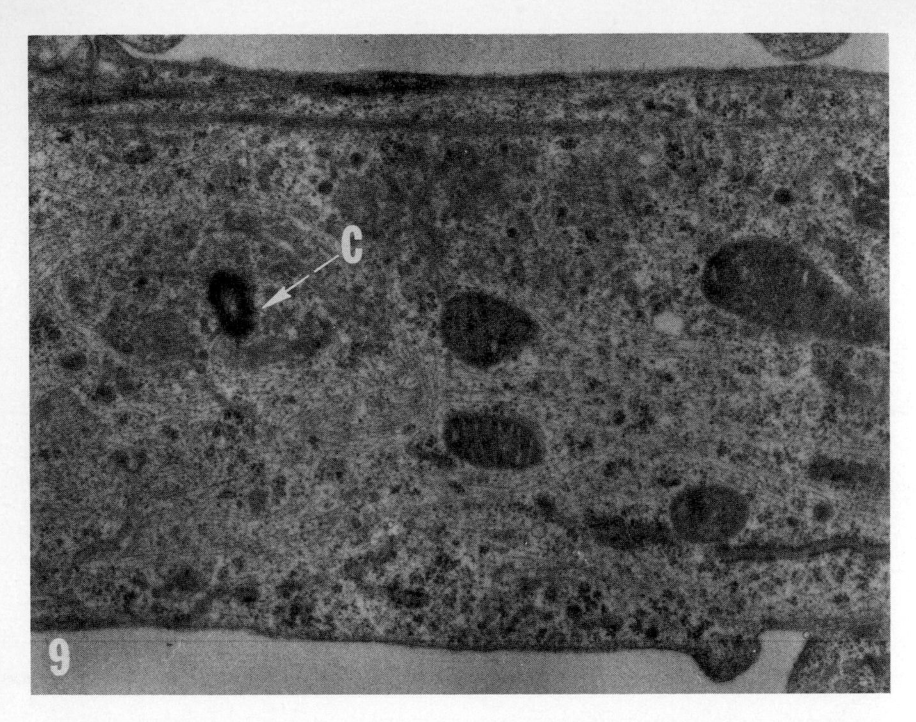

FIG. 9. Electron micrograph of the nuclear region of a BHK-21 enucleate prepared as in Fig. 8. Note the absence of any obvious nuclear structure and the presence of a centriole (C). ×17,600.

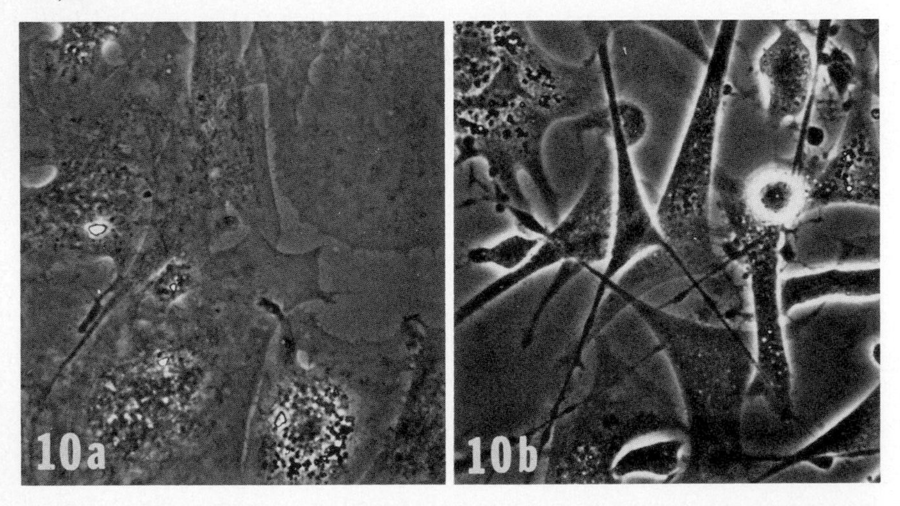

FIG. 10. 3T3 and Py3T3 (polyoma-transformed 3T3) enucleates observed at the edge of a drop after replating from a trypsinized suspension. Minimal overlaps of cytoplasmic processes are seen in 3T3 preparations (a). This is especially evident when compared to the many overlaps observed in Py3T3 preparations (b). Phase contrast. × 520.

require the presence of oncogenic virus genes which are located in the host cell nucleus, although other aspects of the transformed state probably do require a nucleus (Eckhart *et al.*, 1971).

E. Longevity of Enucleated Cells

Although the enucleates appear quite healthy for periods between 12 and 18 hours, they begin to look rather unhealthy after longer time intervals. We are in the process of attempting to prolong the viability of enucleates by crowding and the use of conditioned medium. Crowding the enucleated cells into small areas seems to increase the length of time during which the cells appear healthy with regard to such parameters as membrane ruffling, pinocytosis, and cell shape. For example, by crowding 3T3 enucleated cells some of them retain a fairly normal shape, even at 28 hours following enucleation and replating.

F. The Use of Enucleation in Virology

In early experiments with enucleated cell fragments prepared by microsurgery (Crocker *et al.*, 1964; Marcus and Freiman, 1966), synthesis of poliovirus capsid antigen in bits of human cell (HeLa) cytoplasm was detected by fluorescence microscopy. In addition, RNA synthesis in the presence of actinomycin—presumably poliovirus RNA replication (Franklin and Baltimore, 1962; Darnell, 1962)—was detected by autoradiography with uridine-^{3}H. Later, Prescott *et al.* (1971) enucleated a small minority of mouse L cells by CB treatment without centrifugation and were able to demonstrate by autoradiography with thymidine-^{3}H that vaccinia virus uncoats and establishes viral DNA synthesis in enucleated cytoplasm.

No newly synthesized virus could be recovered from enucleate fragments in these studies because cells containing nuclei were always present in large numbers during the infection. Recently, however, when we infected relatively pure enucleated BSC-1 monolayers with poliovirus, we were able to recover newly synthesized infective poliovirus (Pollack and Goldman, 1973). This work showed that monkey cell nuclei are not necessary for uncoating, translation, and replication of poliovirus RNA, and for the assembly of infectious virus.

The replication of more complex RNA viruses such as SV5 and influenza may also proceed in enucleated cytoplasms of susceptible cells, and those viruses are under examination. By the same token, enucleates should, in principal, serve as excellent material for the synthesis of protein directed by exogenous RNA, whether viral or cellular.

G. Other Possible Uses of Enucleated Cells

Sendai, a myxovirus, fuses cells together in the course of infection (Okada, 1962). Harris and Watkins (1965) showed that inactivating Sendai virus with UV prevented it from replicating, but did not prevent cell fusion. They then made dramatic use of this observation, fusing a human cell (HeLa) to a mouse cell (Ehrlich ascites tumor). The double parentage of these multinucleate cells was proved by autoradiography. Heterokaryons rarely divide, but even those heterokaryons that do not divide live long enough to permit observation of interactions between novel mixtures of nuclei and cytoplasms (Harris, 1965). In combination, Sendai virus and cytochalasin B treatment permits preparation of chimeric cells in which cytoplasm from one cell bears the nucleus of another. Ladda and Estensen (1970) fused nucleated avian erythrocytes to enucleated L cells and showed that the mammalian cytoplasm alone was sufficient to activate the avian erythrocyte nucleus. Poste and Reeve (1970) have recently extended this line of research.

Heterokaryons live long enough to permit virus growth. They were used by Koprowski *et al.* (1967) to recover SV40 from SV40-transformed human, hamster, and monkey cells. Fusion to primary monkey kidney cells revealed infectious SV40, which was detected by serial passage of extracts of the fused cells on permissive monkey kidney cultures. The exact contribution of the monkey cell to the liberation of SV40 from the transformed cell is not known. Recently SV40 has reportedly been recovered from enucleated monkey cells (CV-1) fused with Sendai to SV40-transformed mouse cells (Croce and Koprowski, 1973).

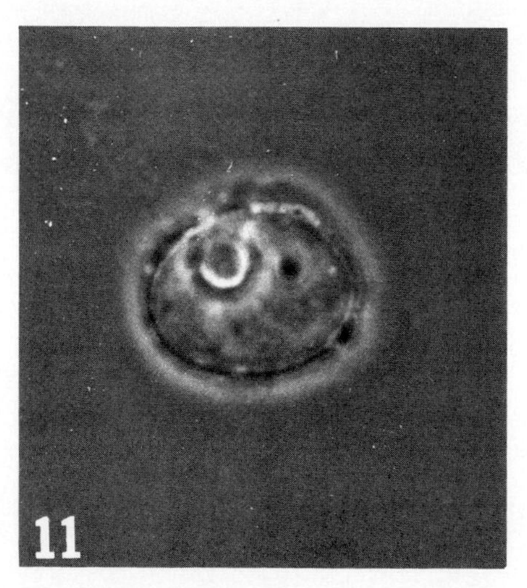

FIG. 11. A 3T3 nucleus obtained from the bottom of the centrifuge tube during the preparation of enucleated cells. Phase contrast. × 1520.

TABLE II

RESPONSE OF ENUCLEATES TO TRYPSINIZATION AND REPLATING[a]

Cell line	Substrate	Number of experiments	Cell density ($cells/cm^2$) Before centrifugation	Cell density ($cells/cm^2$) After centrifugation	Percent of original cell density after centrifugation	Percent enucleation	Cell density ($cells/cm^2$) after replating enucleated preparations	Percent enucleation after replating
BHK-21	Plastic + collagen	4	2.5×10^5	2.1×10^5	82	96	1.7×10^5	92
BSC-1	Glass	10	7.2×10^4	4.1×10^4	56	91[b]	2.9×10^4	80
PyBHK-21	Plastic + collagen	5	5.2×10^5	1.5×10^5	35	92	1.4×10^4	87

[a] Cell density was calculated by counting cells in 5 microscope fields; the counts were then averaged to give the above data. The total number counted in each experiment was between 150 and 500 cells. For calculating density of cells after replating, three coverslips were trypsinized following enucleation and were replated onto 1 coverslip of the same size. The total number of cells counted on this latter coverslip was divided by 3 to correct for the change in cell density and to attempt to obtain a more accurate estimate of the replating efficiency of enucleated cells. Percentage of enucleation was determined by counting the number of nucleates vs. enucleates.

[b] The percentage of BSC-1 enucleates can be increased as high as 99.8% by repeating the procedure for enucleation. Thus BSC-1 cells are enucleated once, allowed to recover in normal medium, and then enucleated a second time (see Table I).

It may also be possible to utilize the nuclei which are removed from cells after centrifugation. Since the extruded nucleus seems to be surrounded by plasma membrane, we will attempt to fuse these nuclei back into enucleated cytoplasm of different cells by utilizing Sendai virus (Okada, 1962; Harris and Watkins, 1965; Rao and Johnson, 1970). Preliminary observations of the nuclear fraction indicate that they possess a normal morphology shortly after removal from the cells (Fig. 11), but that within a short period of time they deteriorate, seemingly owing to swelling and subsequent lysis. By utilizing different concentrations of sucrose, we hope to maintain the nuclei for longer periods of time so that we can attempt fusing them into cells. In this way, for example, we hope to fuse BHK-21 fibroblast nuclei with BSC-1 epithelial cell cytoplasm, and subsequently determine whether the cytoplasmic or the nuclear information is expressed in the resulting cells with regard to such parameters as cell shape and locomotion.

Utilizing enucleated BHK-21 and BSC-1 cells, we are presently determining whether or not enucleates are capable of cell division. Our preliminary observations indicate that in some instances enucleates may divide into "daughter" enucleates. However, we have not yet followed individual cells very carefully. One of the major obstacles in this type of study is the relatively low number of cells going into mitosis in asynchronous populations. In order to increase the yield of cells about to enter cell division we intend to synchronize cells utilizing excess thymidine, which causes cells to become arrested either in S or to accumulate at the G_1-S boundary (Tobey *et al.*, 1967). After removal of the thymidine, the cells continue through S, into G_2, and finally there is a burst of mitosis. By synchronizing cells in this manner and then enucleating them in G_2, we should increase the potential number of enucleated cells going into cell division, and hopefully this will increase our chances of determining whether or not enucleated cells divide. In support of the possibility that cell division occurs, we have seen some structures in enucleated BHK-21 cells which are reminiscent of mitotic spindles without chromosomes (Fig. 12).

FIG. 12. An electron micrograph of a BHK-21 enucleate which contains a spindlelike configuration of microtubules (MT) converging on a centriole (C). Similar configurations have been seen only in normal dividing BHK-21 cells. The obvious difference is the lack of chromosome structure. × 11,200.

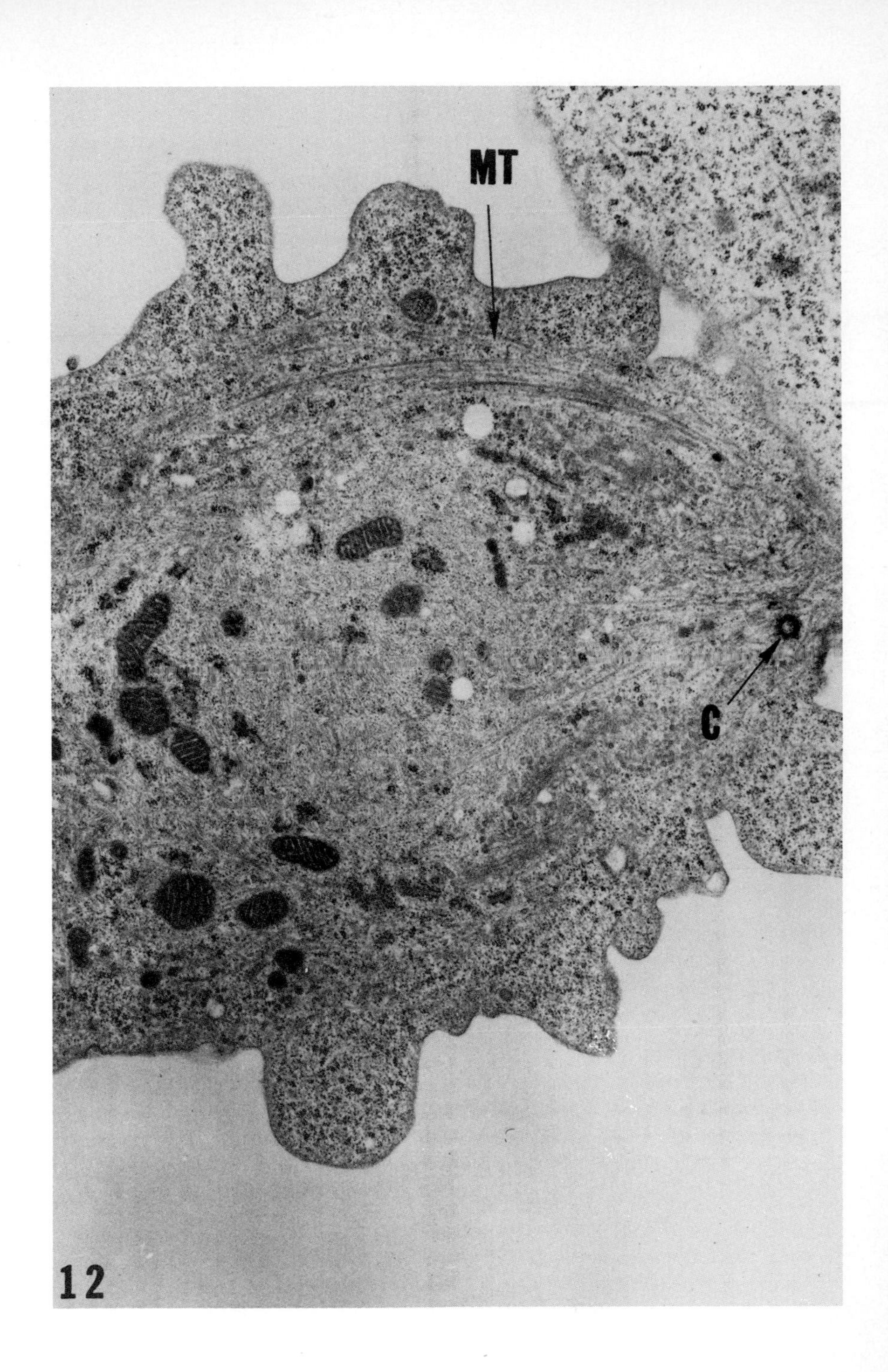
MT
C
12

IV. Summary

Enucleation with cytochalasin B does not result in irreversible damage to cultured cells, and the normality of these cytoplasms is rather surprising. For at least 12 hours, enucleates are viable by such disparate criteria as cell movement, virus replication, and protein synthesis. They should serve as an excellent system for a broad spectrum of studies on nuclear–cytoplasmic interactions in eukaryotic cells.

Acknowledgments

We thank Sue Arelt, Anne Bushnell, and Germaine Grant for their excellent assistance. This work was supported by grants from the National Institutes of Health (Cancer Center Grant No. 1-PO1-CA13106-01) the American Cancer Society (E-639) the National Science Foundation (GB-23185) and the Damon Runyon Fund (DRG-1083A). R.D.G. is the recipient of an N.I.H. Public Health Service Career Development Award (1-K4-GM-32,249-01).

References

Brachet, J. (1961). *In* "The Cell" (J. Brachet and A. E. Mirsky, eds.), Vol. 2, p. 771. Academic Press, New York.

Briggs, R., and King, T. J. (1959). *In* "The Cell" (J. Brachet and A. E. Mirsky, eds.), Vol. 1, p. 537. Academic Press, New York.

Carter, S. B. (1967). *Nature (London)* **213**, 261.

Carter, S. B. (1972). *Endeavour* **31**, 77.

Chambers, R., and Fell, H. B. (1931). *Proc. Roy. Soc. Ser. B* **109**, 380.

Croce, C., and Koprowski, H. (1973). *Virology* **51**, 227.

Crocker, T., Pfendt, R., and Spendlove, R. (1964). *Science* **145**, 40.

Darnell, J. (1962). *Cold Spring Harbor Symp. Quant. Biol.* **27**, 149.

Eckhart, W., Dulbecco, R., and Burger, M. (1971). *Proc. Nat. Acad. Sci. U.S.* **68**, 283.

Ehrmann, R. L., and Gey, G. (1956). *J. Nat. Cancer Inst.* **16**, 1375.

Franklin, R. M., and Baltimore, D. (1962). *Cold Spring Harbor Symp. Quant. Biol.* **27**, 175.

Goldman, R. D. (1972). *J. Cell Biol.* **52**, 246.

Goldman, R. D., and Knipe, D. (1973). *Cold Spring Harbor Symp. Quant. Biol.* **37**, 523.

Goldman, R. D., and Pollack, R. (1973). In preparation.

Goldman, R. D., Pollack, R., and Hopkins, N. (1973a). *Proc. Nat. Acad. Sci. U.S.* **70**, 750.

Goldman, R. D., Berg, G., Bushnell, A., Chang, C. M., Dickerman, L., Hopkins, N., Miller, M. L., Pollack, R., and Wang, E. (1973b). *In* "Locomotion of Tissue Cells," Ciba Found. Symp. 14 (New Series), pp. 83–107. Elsevier, Amsterdam.

Goldstein, L. (1964). *In* "Methods in Cell Physiology", (D. M. Prescott, ed.), Vol. 1, p. 97. Academic Press, New York.

Goldstein, L., and Eastwood, J. M. (1964). *In* "Methods in Cell Physiology" (D. M. Prescott, ed.), Vol. 1, p. 403. Academic Press, New York.

Goldstein, L., Cailleau, R., and Crocker, T. (1960a). *Exp. Cell Res.* **19**, 332.

Goldstein, L., Micon, J., and Crocker, T. (1960b). *Biochim. Biophys. Acta* **45**, 82.

Harris, H. (1965). *Nature (London)* **206**, 583.

Harris, H., and Watkins, J. (1965). *Nature (London)* **205**, 640.

Kopac, M. J. (1959). *In* "The Cell" (J. Brachet and A. E. Mirsky, eds.), Vol. 1, p. 161. Academic Press, New York.
Koprowski, H., Jensen, F., and Steplewski, Z. (1967). *Proc. Nat. Acad. Sci. U.S.* **58**, 127.
Ladda, R. L., and Estensen, R. D. (1970). *Proc. Nat. Acad. Sci. U.S.* **67**, 1528.
Marcus, P. I., and Freiman, M. E. (1966). *In* "Methods in Cell Physiology" (D. M. Prescott, ed.), Vol. 2, p. 93. Academic Press, New York.
Okada, Y. (1962). *Exp. Cell Res.* **26**, 98.
Pollack, R., and Goldman, R. D. (1973). *Science* **179**, 915.
Poste, G., and Reeve, P. (1970). *Nature* (*London*), *New Biol.* **229**, 123.
Prescott, D., Kates, J., and Kirkpatrick, J. (1971). *J. Mol. Biol.* **59**, 505.
Prescott, D., Myerson, D., and Wallace, J. (1972). *Exp. Cell Res.* **71**, 480.
Rao, P., and Johnson, R. T., (1970). *Nature* (*London*) **225**, 159.
Tobey, R. A., Anderson, E., and Petersen, D. (1967). *J. Cell Biol.* **35**, 53.

Chapter 10

Enucleation of Somatic Cells with Cytochalasin B[1]

CARLO M. CROCE, NATALE TOMASSINI, AND HILARY KOPROWSKI

Wistar Institute of Anatomy and Biology, Philadelphia, Pennsylvania; Division of Experimental Pathology, Childrens Hospital, Philadelphia, Pennsylvania

---

I. Introduction

Somatic cells can be enucleated by cytochalasin B, a metabolite produced by the fungus *Helminthosporium dematiodeum* (Carter, 1967). This drug was also found to inhibit cytokinesis and, therefore, to induce multinucleation of somatic cells (Carter, 1967).

In order to obtain extensive cell enucleation, Prescott *et al.* (1972) developed a method utilizing cytochalasin B treatment and centrifugation of the somatic cells. By means of such a method, enucleation was obtained in more than 95% of the cells. The percentage of enucleated somatic cells with cytochalasin B at a concentration of 10 μg/ml at 37°C varies according to the type of cell.

The high percentage of enucleation obtained by Prescott's method or by a modification of it, has made possible a number of different studies of

[1]This work was supported in part by the National Institutes of Health Grant CA 04534, CORE, Grant CA 10815, PHS Research Grant RR 05540 and the Commonwealth of Pennsylvania.

somatic cell cytoplasm with little, if any, damage to the cytoplasm. For example, fusion of enucleated African green monkey kidney cells with SV40-transformed mouse cells resulted in the rescue of the SV40 from the mouse cells (Croce and Koprowski, 1973). Furthermore, when enucleated BSC-1 monkey kidney cells were infected with poliovirus, this virus replicated and was recovered (Pollack and Goldman, 1973). Recently, Goldman *et al.* (1973) demonstrated that the information necessary for cell attachment, spreading, shape formation, pinocytosis, contact inhibition and cell locomotion was present in enucleated cytoplasm.

Two modifications of the original Prescott method for cell enucleation are described in this chapter.

II. Enucleation of Somatic Cells Attached to the Inner Surface of Ultracentrifuge Nitrocellulose Tubes

African green monkey kidney cells, CV-1 (4×10^6) (Jensen *et al.*, 1969) were grown directly on the inner surface of nitrocellulose tubes fitting either on SW 25.1 or SW 27 Beckman rotor. The nitrocellulose tubes had had prior treatment with concentrated sulfuric acid for 4 minutes, followed by thorough washing with 70% ethanol and distilled H_2O and sterilization by exposure to UV light.

The cells were seeded in the tubes in 5 ml of Basal Eagle's Medium (BME) containing 10% fetal calf serum (FCS), 2 m*M* glutamine, and antibiotics.

The tubes were stoppered with a rubber stopper and placed horizontally in a roller apparatus at 37°C for 48–72 hours to obtain uniform attachment of the cells to the inner surface of the tubes. The tubes were then removed from the roller apparatus and filled with BME with 10% FCS containing 10 μg/ml of cytochalasin B (from Imperial Chemical Industries). The cultures were left in this medium for 1–2 hours at 37°C. After this step, the medium was replaced by a 37°C prewarmed medium containing 2.5–5 μg of cytochalasin per milliliter. The tubes, placed in either SW 25.1 or SW 27 buckets sterilized by UV light were centrifuged at 23,000 rpm for 45 minutes at 25–30°C in a prewarmed L-2 Beckman ultracentrifuge. [It is also possible not to replace the medium containing 10 μg/ml of cytochalasin, and instead to centrifuge the cells at a lower speed (12,000–15,000 rpm) for 35–45 minutes.]

Examination of the cells attached to the surface of the tube after centrifugation revealed the presence of elongated thin structures, apparently without nuclei. The medium was removed, the pellet discarded and the tubes washed two or three times with BME with 10% FCS. The tubes were then refilled with BME with 10% FCS stoppered and incubated in a vertical position for 1–2 hours at 37°C. At this time, we observed that the elongated, thin structures seen immediately after centrifugation tended to regain the shape of the original cells, but without nuclei. These enucleated cells could be detached from the walls of the tubes with EDTA (Versene) in PBS (Ca^{2+} – Mg^{2+}-free) or with trypsin and EDTA. With this method it is possible to obtain large numbers of enucleated cells. The contamination with nucleated cells is always less than 3%.

III. Enucleation of Somatic Cells on Plastic Coverslips

African green monkey kidney CV-1 cells (4×10^6) were seeded in 100-mm glass petri dishes containing four round 25-mm plastic coverslips (Thermanos, Lux Scientific Corporation) each. After 24 hours, these cultures were treated with 10 μg/ml of cytochalasin B in BME with 10% FCS for 1–2 hours at 37°C. These coverslips containing monolayers of CV-1 cells were then placed in presterilized nitrocellulose tubes for either SW 25-1 or SW 27 rotor filled with 37°C prewarmed BME containing 10 μg/ml cytochalasin B. The round coverslips were placed horizontally at the bottom of the tubes with the cells facing the bottom. The tubes were then centrifuged at 12,000 to 18,000 rpm for 45 minutes at 30°C in a prewarmed L-2 ultra-centrifuge.

The coverslips were removed from the tubes, washed two or three times with BME and placed in petri dishes containing BME with 10% FCS. Within 1–2 hours the cells tended to regain their original shape and can be fixed and stained with Giemsa. The cells can be detached from the coverslips by the same technique used with the nitrocellulose tubes.

With this method, it is possible to obtain the enucleation of more than 95% of the cells. Studies of the incorporation of tritiated amino acids in cell proteins following enucleation show that more than 95% of the enucleated cytoplasms were able to synthesize proteins. In addition to CV-1 cells, many other cell types which adhere firmly to the surface of the centrifuge tubes or the coverslips can be similarly enucleated.

IV. Results and Discussion

The treatment of somatic cells with cytochalasin B and their centrifugation at a high speed result in the enucleation of more than 95% of the CV-1 cells with both the methods described. The same techniques can be applied to a variety of somatic cells from different species with analogous results.

The choice between the two methods depends on the purpose of the experiments. Method A is appropriate when a large number of enucleated cells are required, and Method B when only a limited number are required.

For approximately 24 hours after enucleation, the enucleated cells can be detached from the nitrocellulose tubes or from the plastic coverslips and

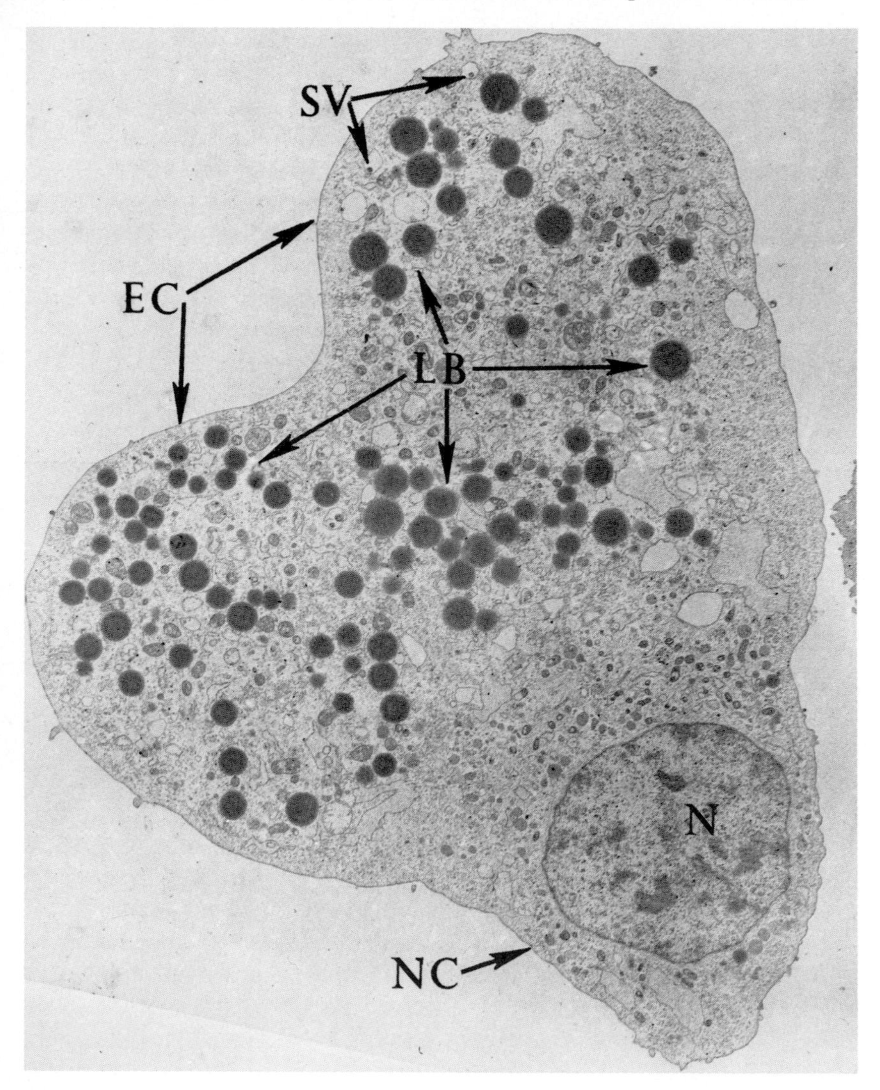

FIG. 1. Sendai virus-induced fusion between two monkey enucleated cells and one mouse nucleated cell. The enucleated monkey cells show typical lipid bodies (LB) which are not present in the mouse cell. EC, enucleated cells; NC, nucleated cell; SV, Sendai virus; LB, lipid bodies; N, nucleus × 4000.

TABLE I

RELEASE OF INFECTIOUS SV40 BY FUSED CELLS. EXPERIMENTS 1, 2, AND 3 IN THE PRESENCE OF 4000 HAU OF INACTIVATED SENDAI VIRUS[a]

Fusion experiment	Cells fused and ratio (in parentheses)	Dilution of the cells plated per petri dish[b]	Average number of infectious centers per petri dish
1	mKSBU100 + enucleated CV-1[c]	1:20	15.0
	(4.5×10^6 (4.5×10^6)	1:200	6.5
		1:2000	1.4
2	mKSBU100 + nucleated CV-1	1:20	63.0
	(4.5×10^6) (4.5×10^6)	1:200	11.0
		1:2000	1.4
3	mKSBU100 + nucleated CV-1	1:20	4.7
	(4.5×10^6) (1.8×10^5)	1:200	1.4
		1:2000	0.4
Control	mKSBU100 + nucleated CV-1[d]	1:20	0.7
	(4.5×10^6) (4.5×10^6)		

[a] From Croce and Koprowski (1973).
[b] Ten petri dishes containing CV-1 monolayers plated for each dilution of cells.
[c] Enucleated CV-1 cells 98.8%.
[d] Cells mixed in the absence of S endai virus.

fused in the presence of inactivated Sendai virus with nucleated cells and with other enucleated cells (Fig. 1).

When enucleated CV-1 cells were fused with SV40-transformed mouse cells, rescue of infectious SV40 virus resulted from such fused cells (Table I), indicating that a factor(s) which allows SV40 replication in the nucleus of the transformed mouse cells is present in the cytoplasm of the permissive CV-1 cells (Crose and Koprowski, 1973). Parallel studies of the fusion of enucleated cells in the presence of lysolecithin have shown the extreme sensitivity of the enucleated cytoplasm to the toxic effect of this compound.

References

Carter, S. (1967). *Nature (London)* **213**, 261.
Croce, C. M., and Koprowski, H. (1973). *Virology* **51**, 227.
Goldman, R. D., Pollack, R., and Hopkins, N. H. (1973). *Proc. Nat. Acad. Sci. U.S.* **70**, 750.
Jensen, L., Girardi, A., Gilden, R., and Koprowski, H. (1969). *Proc. Nat. Acad. Sci. U.S.* **52**, 53.
Pollack, R., and Goldman, R. (1973). *Science* **179**, 915.
Prescott, D., Myerson, D., and Wallace, J. (1972). *Exp. Cell Res.* **71**, 480.

Chapter 11

Isolation of Mammalian Heterochromatin and Euchromatin[1]

WALID G. YASMINEH AND JORGE J. YUNIS

Medical Genetics Division, Department of Laboratory Medicine and Pathology, University of Minnesota, Minneapolis, Minnesota

---

I. Introduction

The first realization of the existence of heterochromatic and euchromatic regions in the nuclei of eukaryotes came at the turn of this century in connection with the sex chromosomes. In 1904 Montgomery called "heterochromosomes" those chromosomes during the spermatogenesis of insects that "preserve to a great extent their compact form during the whole growth period of the spermatocytes, and during the rest stage of the spermatogonia, and retain during the whole period the deep staining characteristics which the other chromosomes exhibit only during the height of mitosis" (Montgomery, 1904). In 1928, Heitz observed in the nuclei of the liverwort (Pellia) chromosomes or chromosome regions that are condensed in interphase and prophase and do not decondense like the remainder of

[1] This work was supported in part by National Institutes of Health Grant No. HD-01961.

the chromosomes in telophase. He termed these regions "heterochromatin," and those that decondense in telophase "euchromatin" (Heitz, 1928). Subsequently, these darkly staining regions of the chromosomes were recognized in both germ and somatic cells of most eukaryotes (Heitz, 1935; Levan, 1946; Swanson, 1957), and various attributes of heterochromatin were described, including its preferential localization in certain areas of the chromosomes, such as the centromeres (Heitz, 1934; Levan, 1946), its late replication during the DNA synthetic period (Lima-de-Faria, 1959, 1969), and its dearth of genes and relative inability to synthesize RNA during interphase (Muller and Painter, 1932; Swanson, 1957). Furthermore, in 1961, with the advancement of the Lyon hypothesis (Ohno *et al.*, 1959; Lyon, 1961), explaining the inactivation and condensation of an X-chromosome as a means of dosage compensation in placental mammals, the idea that two main types of heterochromatin exist in mammals evolved: The first type is that resulting from inactivation of one of the two X-chromosomes in females early in embryogenesis. This type has been termed "facultative heterochromatin" and in mammals comprises about 2.5% of the diploid genome (Brown, 1966; Ohno, 1967). The second type was termed "constitutive heterochromatin" because it is found in homologous chromosomes or chromosome segments and remains condensed throughout the cell cycle (Brown, 1966). Again in mammals, constitutive heterochromatin represents about 15 to 20% of the genome, although some unique species, such as certain members of the family of rodents Microtinae, may possess smaller amounts (Schmid, 1967; Yunis and Yasmineh, 1971).

A special kind of condensed chromatin, often erroneously referred to as heterochromatin (Frenster *et al.*, 1963; Frenster, 1969), is the condensed inactive chromatin acquired during the maturation of certain cell types (e.g., thymocytes, lymphocytes, granulocytes, nucleated erythrocytes) (Littau *et al.*, 1965; Yasmineh and Yunis, 1970). This type of chromatin is largely inactive in RNA synthesis, occurs without intervening mitosis, and may engulf the bulk of the chromatin (Frenster *et al.*, 1963; Littau *et al.*, 1965; Dingman and Sporn, 1964).

Although in 1966 Spencer Brown recognized the two basic types of heterochromatin, there was a great deal of confusion with respect to the cytological expression, biochemical nature, and functional significance of constitutive heterochromatin. It was generally believed that constitutive heterochromatin represents a state of chromatin, rather than a specific entity, mainly owing to differences in its cytological expression among different tissues and early in embryogenesis (Brown, 1966; Lima-de-Faria, 1969). The first evidence to the contrary came from this laboratory in 1969 when constitutive heterochromatin was isolated from the brain and liver of male mice by the sonication of nuclei and fractionation of the chromatin into dense heterochromatin and light euchromatin by differential centrifugation (Yasmineh and Yunis, 1969, 1970). Density gradient centrifugation in CsCl and base composition of DNA from the heterochromatin and euchromatin fractions indicated that the former was composed primarily of satellite DNA while the latter reflected the properties of the bulk of the DNA. After these initial findings, work in this laboratory, as well as others, on a number of mammalian and nonmammalian species yielded essentially the same results (Yunis and Yasmineh, 1971). In all instances, the DNA of constitutive heterochromatin is enriched in one or more satellite DNAs, which differ in a number of characteristics from the DNA of euchromatin (buoyant density, base composition, separation of strands in alkaline CsCl gradients) and consist of short nucleotide sequences repeated in tandem a great number of times.

Considerations involving the origin and function of constitutive heterochromatin are outside the scope of this chapter. It should be stated, however, that current concepts in this regard involve structural roles of heterochromatin, such as the protection of vital regions of the genome from evolutionary changes, the initial alignment of homologous chromosomes during meiosis, the grouping during interphase of nonhomologous chromosomes or chromosome regions that are functionally related and the establishment of "fertility barriers" providing means for evolutionary diversity and speciation (Yunis and Yasmineh, 1971). These concepts are based mainly on (1) cytological and genetic evidence concerning heterochromatin and (2) investigation of the constituents of the DNA of heterochromatin by homology studies, sequence analysis, and localization within the genome by "*in situ*" hybridization (Yunis and Yasmineh, 1971, 1972; Walker, 1971; Bostock, 1971; Flamm, 1972; Sutton and McCallum, 1972).

The primary purpose of this paper is to relate the experience that we have acquired during the last few years in the isolation and characterization of heterochromatin and euchromatin from the nuclei of normal and transformed cells. It should be noted that most of the techniques to be described were intended primarily to investigate the DNA components of these chromatin fractions, although their use for the study of other components of chromatin (RNA, histone, acidic proteins) may be possible.

II. Isolation of Nuclei

One of the essential requirements for the isolation of relatively pure heterochromatin and euchromatin fractions from mammalian cells is the isolation of clean nuclei in which the architecture of the chromatin has not been significantly altered. As has already been reported by numerous investigators, such factors as the use of low temperatures, the medium and means of homogenization, pH, and the concentration of divalent cations are very important to this objective and may vary from tissue to tissue (Busch, 1967; Muramatsu, 1970; Roodyn, 1972). Furthermore, variations in technique are required in dealing with (1) normal cells where the cell membrane is easily disrupted and the cytoplasmic constituents may be voluminous and adhere loosely to the nuclear membrane, and (2) cells from old animals, tumor cells, or cells in culture, in which the cell membrane is difficult to disrupt and cytoplasmic constituents adhere tenaciously to the nuclear membrane (Dounce, 1963; Penman, 1969; Magliozzi *et al.*, 1971).

The use of low temperatures (2 to 4°C) is essential to minimize the action of autolytic enzymes. Care should also be taken, whenever excessive heat is generated, to dissipate the heat rapidly by jacketing devices (Busch, 1967). This occurs most frequently when cells are exposed to high shearing forces, as in their homogenization at high speeds in low clearance homogenizers or in highly viscous media, or when nuclei are disrupted by sonication, passage through a French pressure cell or other means, to isolate subnuclear components.

Speed in the isolation and processing of nuclei is also of essence, since in most cases conditions of pH, ionic strength, and divalent cation concentration have to be used that, unfortunately, are not unfavorable for the activation of autolytic enzymes. In this laboratory most of the techniques used require no more than 4 to 6 hours for the isolation of heterochromatin and euchromatin from normal tissues or tumor cells.

A. Isolation of Nuclei from Tissues

The techniques used in the isolation of nuclei from mammals, as well as other species, are divided into those utilizing aqueous or nonaqueous media, depending upon the nuclear component under investigation. Techniques involving nonaqueous media are usually time consuming, but are desirable when nuclear components, such as the soluble proteins, are to be retained in the nucleus. These techniques will not be discussed in this paper since they have not been used as a source of nuclei for chromatin fractionation in this laboratory and have been adequately evaluated by other investigators (Busch, 1967; Siebert, 1967; Roodyn, 1972). The techniques in which aqueous media are used differ with the type of tissue or cultured cells from which the nuclei are to be extracted. Sucrose solutions are usually most efficient for the isolation of nuclei from normal tissue cells, where the cytoplasmic constituents can be easily stripped off. In this case, numerous recipes exist that differ primarily in the concentration of sucrose and divalent cations (Busch, 1967; Muramatsu, 1970; Roodyn, 1972). In the one-step, old Chauveau method (Chauveau *et al.*, 1956), the tissue is homogenized directly in hypertonic sucrose (2.2 *M*) and centrifuged at 40,000 *g* for 1 hour. The recovery of nuclei is somewhat low (about 50%), although nuclei that are essentially free of cytoplasmic tags are obtained. In more recent modifications of this technique, various amounts of divalent cations in the form of Ca^{2+} and/or Mg^{2+} are added to prevent the chromatin from gelling and preserve its integrity. Since divalent cations also harden the nuclear membrane, a careful determination of the optimal concentrations for eash tissue has to be made, especially if the nuclei are to be disrupted subsequently for the isolation of various subnuclear components.

Nearly 100% recovery of nuclei may be obtained by homogenization of the tissue in isotonic sucrose solutions containing divalent cations. The nuclei, however, are contaminated to various degrees with cytoplasmic tags, depending upon the tissue used, the means of homogenization and the effort exerted in removing the tags by subsequent washing. Modifications in which the initial homogenization in isotonic sucrose is followed by a second homogenization of the crude nuclear pellet in heavy sucrose have also been described. Other modifications include the layering-over techniques in which an initial homogenate of the tissue is made in isotonic or hypertonic sucrose solutions containing Ca^{2+} and/or Mg^{2+} ions and then underlaid with more concentrated sucrose solutions. In the hypertonic sucrose modification, Maggio *et al.* (1963a) were able to isolate highly purified nuclei from guinea pig liver by homogenizing in a 0.88 *M* sucrose–1.5 m*M* $CaCl_2$ solution, underlaying the homogenate with 2.2 *M* sucrose–0.5 m*M* $CaCl_2$ and centrifuging at 53,000 *g* for 90 minutes. In the isotonic sucrose modification, Blobel and Potter (1967) homogenized rat liver in a solution containing 0.25 *M* sucrose, 5 m*M* $MgCl_2$, 25 m*M* KCl, 50 m*M* Tris-HCl, pH 7.5, adjusted the homogenate with heavy sucrose to 1.62 *M* and underlaid it with 2.2 *M* sucrose prior to centrifugation. The modification was designed to avoid the sedimentation of mitochondria at the interlayer between the homogenate and the heavy sucrose, that results in blocking the sedimentation of part of the nuclei and their loss.

The technique currently used in this laboratory for the isolation of nuclei from normal mammalian tissues (liver, brain, kidney) is a two-step procedure in which the tissue is initially homogenized in 0.32 *M* sucrose, usually containing 0.2 m*M* Ca^{2+} and 5 m*M* Mg^{2+} ions, and the crude nuclei pelleted by centrifugation, suspended in 2.2 *M* sucrose (containing 0.2 m*M* Ca^{2+} and 5 m*M* Mg^{2+}) and centrifuged. Previously, other concentrations of divalent cations have been successfully used, including 1.5 m*M* Ca^{2+}

for the isolation of nuclei from mouse brain and liver, and 0.5 m*M* Ca^{2+} plus 3 m*M* Mg^{2+}, for the isolation of nuclei from guinea pig and calf liver (Yasmineh and Yunis, 1970, 1971a; Yunis and Yasmineh, 1970).

Successful purification of nuclei by this technique has been achieved in tissues from various rodents including mouse, guinea pig, Syrian hamster, Chinese hamster, rat, and various Microtinae. For the isolation of nuclei from the tissues of larger mammals (such as human liver and placenta, and calf liver), minor modifications have been necessitated, mainly in the preparation of the tissue for homogenization. All steps are performed in the cold room at 4°C.

In the isolation of nuclei from liver, brain (cerebrum), or kidney of small mammals, the organ is excised and placed in powdered dry-ice, if it is not to be immediately processed, or otherwise in ice-cold physiological saline. The organs are thoroughly washed by stirring in physiological saline and then finely minced with scissors, washed several times with physiological saline, and homogenized with 3 to 5 volumes of 0.32 *M* sucrose–5 m*M* $MgCl_2$–0.2 m*M* $CaCl_2$ in a Potter-Elvehjem homogenizer (200 μm clearance; 50-ml capacity; Kontes Glass Co., Vineland, New Jersey). The Teflon pestle is mechanically operated with a 1/15 H.P. variable-speed motor (manufactured by Tri-R Instruments, Jamaica, New York). Nearly 100% disruption of the nuclei is normally achieved by 6 to 8 strokes of the pestle at 80 to 100% output of the motor. The resulting homogenate is filtered through 8 layers of gauze to remove fibrous connective tissue, diluted 4-fold with the same solution and centrifuged at 1000 *g* for 10 minutes. The supernatant fluid, containing the bulk of the cell membranes and cytoplasmic constituents, is carefully aspirated from atop the loose pellet of crude nuclei and discarded.

At this stage the nuclear pellet may contain, other than the cytoplasmic contaminants, a good number of erythrocytes and, in the case of brain tissue, an appreciable amount of lipid. The bulk of contaminants is effectively removed by suspending the nuclear pellet in 5 volumes of 2.2 *M* sucrose–5 m*M* $MgCl_2$–0.2 m*M* $CaCl_2$ with 2 or 3 mechanical strokes of the Potter homogenizer, after which the suspension is diluted 5-fold with the same solution of sucrose and centrifuged at 15,000 *g* for 1 hour in a Sorvall angle centrifuge. At the density of 2.2 *M* sucrose, the nuclei sediment at the bottom of the centrifuge tube while the contaminants float to the top as a leathery layer that can be easily removed by ringing with a wood applicator stick and decantating the supernatant fluid. At this point in the isolation procedure, the nuclei are equivalent in purity to those obtained by the Chauveau technique; only a few cytoplasmic tags can be seen upon low or high magnification under the phase contrast microscope and, in electron micrographs, the nuclei appear to be intact and are bounded by double-layered membranes. The nuclei can be kept frozen for a few days (at −70°C) at this stage with no adverse effect to the chromatin.

In the isolation of nuclei from tissues of larger mammals, the same procedure is used except that one more step is included in the preparation of the tissue for homogenization. For example, in the case of calf liver, the tissue is cut into approximately 0.5-inch cubes, washed several times with physiological saline and finely chopped for 30 seconds at low speed in a Waring Blendor in 3 to 5 volumes of the homogenizing medium. The mince of tissue is then homogenized without further dilution in a Potter-Elvehjem homogenizer, using 10 to 12 mechanical strokes, and the nuclei are extracted as above. For human nuclei, a readily available source of fresh tissue is the placenta. The procedure in this case involves washing the whole placenta thoroughly in physiological saline and stripping the cotyledons, containing the easily accessible nuclei, from the amnionic and chorionic layers of connective tissue with a sharp knife. The cotyledons are then washed again with physiological saline and chopped for approximately 15 seconds in a Waring Blendor prior to homogenization in the Potter-Elvehjem homogenizer.

It should be noted in the above procedures that no attempts have been made to separate the various cell types within each organ. The importance of working on single cell types in conducting biochemical analyses and metabolic studies on nuclei has been emphasized by other investigators (McBride and Peterson, 1970; Johnston and Mathias, 1972). This, however, has not been found to be essential for the isolation of heterochromatin and euchromatin for the purpose of studying their DNA components, although as indicated in Section III, B the purity of these chromatin fractions may be affected to some extent by the tissue and cell type used.

B. Isolation of Nuclei from Tumor and Cultured Cells

Isolation of nuclei from tissues of very old animals, whole tumors, transformed cells in culture and, frequently, normal cells in culture, is more difficult to achieve, and the methods utilized are, as a rule, harsher than those used for normal cells. As stated earlier, this is mainly due to difficulties in the disruption of the cell membrane and removal of the cytoplasmic constituents. These difficulties are compounded by the need of using divalent cations to preserve the integrity of the chromatin. Sucrose methods are not usually appropriate for the isolation of clean nuclei from cancer cells, although some success has been obtained in the case of certain tumor cells and in minimal deviation hepatomas (Muramatsu, 1970). In these cases, low clearance homogenizers (50 to 77μm) are used that would free the nuclei of cytoplasmic constituents with as little damage to the nuclear membrane as possible. Clean nuclei can be more easily obtained by using various modifications in which the crucial step(s) for removing the cytoplasmic constituents involves (1) swelling of the nuclei in hypotonic media; (2) the use of detergents, such as Triton X-100, NP-40, Tween 40, Tween 80, and sodium deoxycholate; (3) the use of high pH; and (4) the use of citric acid. For details concerning these methods, the reader is referred to the following articles: Dounce, 1963; Busch, 1967; Penman, 1969; Muramatsu, 1970; Magliozzi *et al.*, 1971; Roodyn, 1972. It is believed, however, that these treatments generally result in damage to the structure of the chromatin and more effort should be exerted in developing other milder procedures.

In this laboratory (de la Maza and Yunis, 1973), hypotonic media and citric acid media have been used for the isolation of nuclei from 3T3 mouse cell lines transformed with simian virus (SV40) [SV3T3-11A8 (Smith *et al.*, 1972), SV3T3-56 (Westphal and Dulbecco, 1968)] and Rous sarcoma virus (RSV) [RSV B77/3T3 (Varmus *et al.*, 1973)]. The method utilizing hypotonic treatment is a modification of the procedure described by Penman (1969): Normally nuclei are isolated from about 3×10^9 cells grown in 300 falcon flasks (250-ml capacity) or 25 roller bottles (0.5 gallon capacity). At the time of isolation, the original medium is poured off and the cells are scraped and collected in approximately 1 liter of RSB (reticulocyte standard buffer composed of 0.01 *M* NaCl, 0.01 *M* Tris, pH 7.4, and 1.5 m*M* $MgCl_2$). After centrifugation at 1000 *g* for 10 minutes, the cells are resuspended in 150 ml of a 1:3 dilution of RSB and allowed to swell for 5 to 10 minutes. The cells are monitored by phase contrast microscopy during the swelling period, since excessive swelling may result in cell burst and damage to the structure of chromatin. The cells are then transferred to a Potter-Elvehjem homogenizer of low clearance (77 to 100 μm; 50 ml capacity) and disrupted with 5 to 10 mechanical strokes at 80% output of the motor. This results in disruption of 95 to 100% of the cells. The homogenate is centrifuged at 1000 *g* for 5 minutes, and the pellet of crude nuclei and unbroken cells washed once with 150 ml of 1:3 RSB and centrifuged again. To remove the bulk of cytoplasmic tags and residual unbroken cells, the pellet is resuspended in 300 ml of 2.2 *M* sucrose–1.5 m*M* $MgCl_2$ by 2 or 3 mechanical strokes in a loose Potter-Elvehjem homogenizer (200 μm clearance) and the suspension is centrifuged at 15,000 *g* for 1 hour in a Sorvall angle centrifuge. The unbroken cells and cytoplasmic tags float to the top of the heavy sucrose solution, forming a thin pellet that is ringed and decanted together with the heavy sucrose. The remaining nuclear pellets at the bottom of the centrifuge tubes still contain some cytoplasmic tags but no unbroken cells, and the nuclear membranes are double-layered.

Nuclei from 3T3 mouse cells transformed with SV40 and RSV have sometimes been obtained in this laboratory by the citric acid procedure (Busch,

1967). Although the morphology of these nuclei is different from that normally seen in whole cells and the chromatin appears to be clumped, the nuclei are cleaner than those obtained by the hypotonic procedure and are more amenable to disruption and chromatin fractionation. In a careful study, Dounce (1963) showed that the pH or concentration of acid is a critical factor in the isolation of pure nuclei from abnormal cells. For each type of cell there apparently exists an optimal pH to free the nuclei from cytoplasmic constituents. In a typical isolation procedure, about 3×10^9 cells are collected in approximately 1 liter of RSB and pelleted by centrifugation at 1000 *g* for 10 minutes. The cells are resuspended in 150 ml of a solution containing 2.5% citric acid, pH 2.5, and 0.8 *M* sucrose, and homogenized by 5 to 10 mechanical strokes of a low clearance Potter-Elvehjem homogenizer (50 to 100 μm; 50 ml capacity) at 100% output of the motor. The homogenate is then centrifuged at 1,000 *g* for 10 minutes, the supernatant fluid is discarded, and the crude nuclear pellet is resuspended in 300 ml of 2.2 *M* sucrose–1.5 m*M* $CaCl_2$ with 2 to 3 strokes of a high clearance Elvehjem-Potter homogenizer (200 μm; 50 ml capacity) and centrifuged at 15,000 *g* in a Sorvall angle centrifuge.

C. Recovery of Nuclei and Integrity of Chromatin

Recovery of the nuclei is usually over 80% by the sucrose method described here for the isolation of nuclei from the normal tissues (liver, kidney) of smaller mammals. Slightly less recovery is obtained from transformed cells when using the hypotonic procedure (60 to 70%), and much less (40 to 50%) when using the citric acid procedure. In the case of tissues of bigger mammals, or tissues of high lipid content, such as brain, the recovery is usually low (40 to 60%). A good deal of this loss occurs during the centrifugation step in 2.2 *M* sucrose, where some 40% of the nuclei may be found in the top pellet. Exposing the suspension of nuclei to shear forces prior to centrifugation, by homogenizing with 2 or 3 strokes in a high clearance Potter-Elvehjem homogenizer, increases the recovery to about 70% in the case of brain from smaller mammals, but does not aid appreciably in the case of human placenta and calf liver.

Perhaps the single most important factor in the preservation of the integrity of chromatin is the kind and concentration of divalent cations used in isolating the nuclei. The divalent cations most frequently used are Mg^{2+} and Ca^{2+}, although others, including Zn^{2+}, Pb^{2+}, and Cd^{2+}, have also been investigated (Frenster, 1963; Dounce and Ickowicz, 1970; Magliozzi *et al.*, 1971). Inclusion of Mg^{2+} and/or Ca^{2+} in the initial homogenization of the cells of normal tissues is most crucial, as it appears that this is the step at which the chromatin is effectively stabilized. Omission of divalent cations in this step and its inclusion in subsequent steps may still result in irreversible damage to the structure of the chromatin (Maggio *et al.*, 1963a; Muramatsu, 1970).

Divalent cations also exert an effect in stabilizing or hardening both the nuclear and cellular membranes so that, depending upon the cell type, concentrations of Mg^{2+} and/or Ca^{2+} should be used that would (1) best preserve the integrity of the chromatin, (2) result in a maximum recovery of both cells and nuclei, and (3) afford optimal conditions for the disruption of cells and nuclei and the isolation of pure fractions of heterochromatin and euchromatin. Of the two divalent cations, Ca^{2+} exerts the stronger effect in these respects, so that it is frequently used to stabilize chromatin whenever disruption of the cellular and nuclear membranes can still be easily achieved. On the other hand, Mg^{2+} is sometimes used under the opposite conditions as, for example, in the case of tumor or transformed cells where the inclusion of very small amounts of Ca^{2+} may render the cells and nuclei impervious to disruption. As indicated in Section IIB, 3T3 mouse cells transformed with SV40 and RSV can be disrupted following swelling in very hypotonic media containing minimal amounts of Mg^{2+} (1:3 dilution of RSB).

In the citric acid procedure, it should be noted that the chromatin does not gel despite the omission of divalent cations in the initial homogenization step. This is presumably due to its stabilization by the low pH.

III. Isolation of Heterochromatin and Euchromatin

A. Preparation of Nuclei

Mechanical disruption of the nuclei for the isolation of chromatin fractions is best achieved following removal of the outer nuclear membrane. In this laboratory, a modification of the technique described by Frenster *et al.* (1960, 1963) is used. The method consists of repeated suspension of the nuclei (irrespective of the method used in their preparation to this point) by stirring in 0.01 *M* Tris buffer, pH 7.1, containing Ca^{2+} and/or Mg^{2+} ions, and centrifugation at 500 *g* for 5 minutes. About 100 ml of buffer are used per milliliter of packed nuclei. Aside from removing the outer nuclear membrane and cytoplasmic tags, this treatment also results in the extraction of nuclear ribosomes and soluble proteins. Various modifications in the treatment have to be used depending upon the tissue and the type and concentration of divalent cations used in the original homogenization medium and in the Tris buffer. In the treatment of normal nuclei isolated in the presence of Mg^{2+} alone (< 3 m*M*), the nuclei should be stirred gently in the Tris buffer and monitored carefully under the phase microscope following each suspension, since prolonged exposure of the nuclei (usually more than 3 suspensions) may result in dissolution of both nuclear membranes and gelation of the chromatin. This is not the case, however, when Ca^{2+} alone or Mg^{2+} and Ca^{2+} is used, nor is it the case when Mg^{2+} alone is used in the isolation of nuclei from transformed cells. In these latter cases, several suspensions with vigorous stirring are necessary to remove most of the outer nuclear membrane.

For the isolation of heterochromatin and euchromatin from the nuclei of most normal tissues, the ideal conditions of divalent cations in the Tris buffer have been found to be those used in the original homogenization medium (Section IIA). Low and high power phase contrast photographs of liver and brain nuclei from the field vole *Microtus agrestis* are shown in Fig. 1. In this species the bulk of the constitutive heterochromatin, representing about 17% of the total chromatin, resides in the sex chromosomes, which consequently appear as two giant masses in interphase nuclei (Lee and Yunis, 1971a,b). The nuclei were prepared by 4 suspensions in Tris buffer–0.2 m*M* Ca^{2+} – 5 m*M* Mg^{2+} with vigorous stirring for 30 seconds in a plastic tube by means of a Potter-Elvehjem pestle operated at 20% output of the Tri-R motor (clearance > 1 mm).

For cultured cells, the conditions of divalent actions depend upon the procedure used in the isolation of nuclei (Section II,B). Nuclei from transformed cells prepared by the hypotonic procedure are subjected to 4 or more suspensions of Tris buffer containing 1.5 m*M* Mg^{2+}. Nuclei prepared by the citric acid procedure are subjected to 4 or more suspensions of Tris buffer containing 1.5 m*M* Ca^{2+}. In both cases stirring should be quite vigorous and may be achieved by loose homogenization in a Potter-Elvehjem homogenizer (200 μm clearance; 3 to 5 strokes at 50% output of the motor). Although in the citric acid procedure the outer nuclear membrane and cytoplasmic tags appear to be removed prior to this treatment (Busch, 1967), it was found that the suspensions in Tris buffer render the nuclei more fragile and more amenable to disruption.

B. Disruption of Nuclei and Chromatin

Generally two methods, sonication and passage through a pressure cell, have been used for the disruption of nuclei and the breakdown of chromatin into fragments that can be subsequently separated on the basis of density by differential centrifugation. The intensity of disruption depends upon a number of factors including (1) whether the nuclei are derived from normal tissue cells, cultured cells or tumor cells, (2) the pattern of heterochromatin

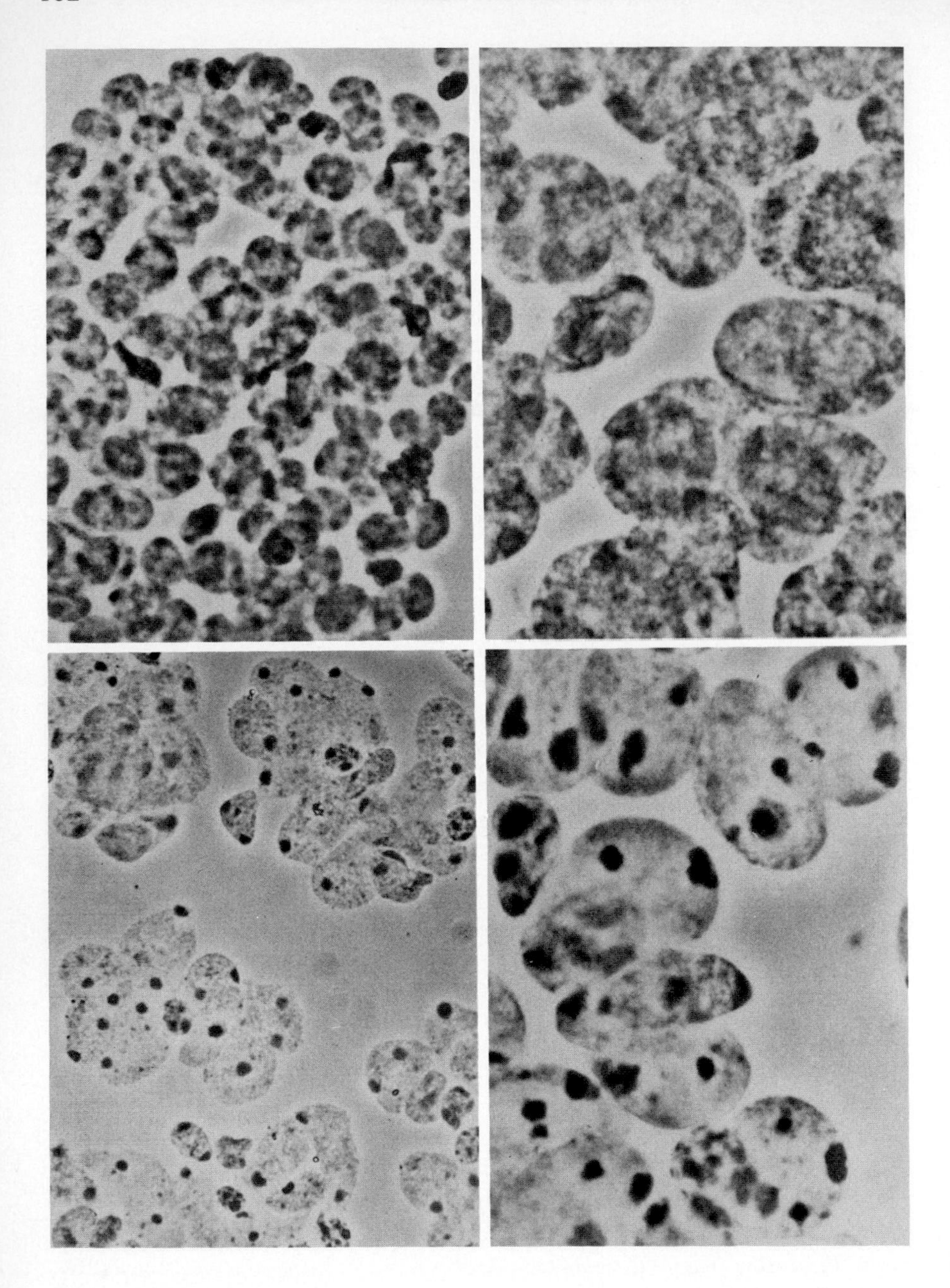

FIG. 1. Low and high magnification photographs of phase contrast nuclei from the liver (top left and right) and brain (bottom left and right) of *Microtus agrestis*. Note the diffuse giant masses in the liver nuclei, in contrast to the compact masses in brain nuclei.

within the interphase nuclei, and (3) the conditions used in the isolation of the nuclei, especially the concentration and type of divalent cations used in the initial homogenization medium.

For both methods, the nuclei that have been treated with Tris buffer are suspended in about 20 volumes of 0.25 *M* sucrose and stirred for 30 seconds in a plastic tube by means of a Potter-Elvehjem pestle (clearance >1 mm; 20% output of the motor). The optical density at 420 nm of the suspension is then adjusted to a value of 1 to 3 with 0.25 *M* sucrose, and the nuclei are allowed to swell with gentle stirring for about 20 minutes in the cold, after which they are passed through two layers of flannelette, to remove some fibrous material. The nuclei are monitored by phase contrast microscopy during the swelling period. Usually they are ready for disruption and chromatin fractionation when they assume spherical shapes of about twice their original size. At this time the interior of the nuclei shows a fine, homogeneous texture in which the masses of heterochromatin are sometimes difficult to outline.

1. Sonication of Nuclei

Sonication has frequently been used to isolate nucleoli from purified nuclei extracted in the presence of divalent cations (Frenster *et al.*, 1963; Maggio *et al.*, 1963b). Since heterochromatin is usually associated with nucleoli, it is not too surprising that the methods used in the isolation of nucleoli have been found to be also the most successful for the fractionation of chromatin (Yunis and Yasmineh, 1971, 1972). In this laboratory sonication is performed with a Branson sonifier that generates sonic waves at 20 kc/sec. The sonifier is normally operated at 7 to 11 A for various intervals, depending upon the method of preparation of nuclei. The sample is kept at 0 to 4°C, and overheating is avoided by using discontinuous sonic bursts of no longer than 15 seconds each. Aliquots, 15 to 18 ml, of the suspension of nuclei are sonicated at a time in 30 × 55 mm glass cups. In dealing with nuclei from normal tissue cells, the nuclei are normally sonicated for a 5-second burst at 7 to 11 A, after which the sonicate is examined for unbroken nuclei (or large fragments of nuclei). If unbroken nuclei (or large fragments of nuclei) are present, the suspension is sonicated for one or two more 5-second bursts, at which time essentially all the nuclei should be broken and no large fragments of nuclei are present. The presence of either of these in significant amounts (over 1% nuclei) after a total of 15 seconds of sonication is a good indication that the nuclei are unsuitable for the isolation of relatively pure fractions of heterochromatin and euchromatin. This occurs when the concentration and type of divalent cations used in the homogenization of the tissue and the extraction of the nuclei with Tris buffer are not optimal.

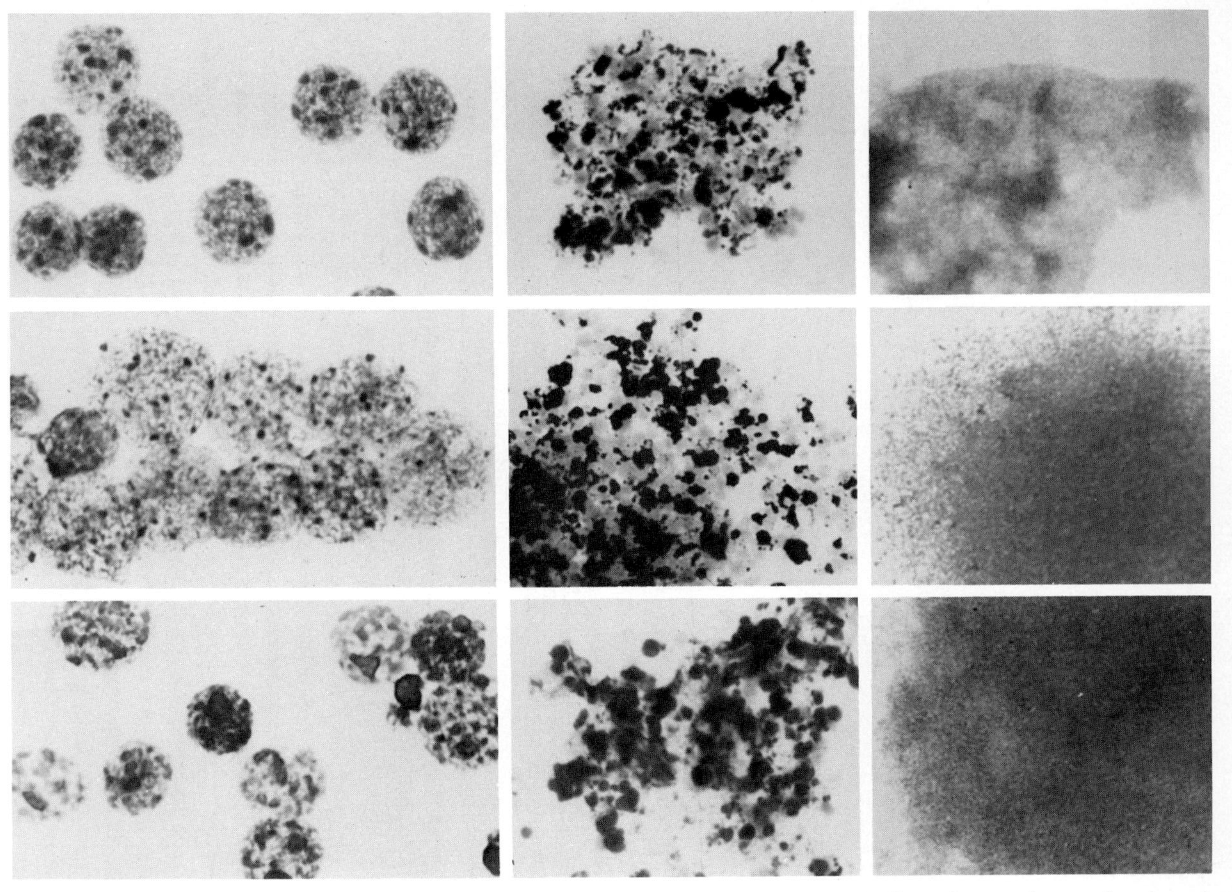

FIG. 2. Wright-stained preparations of nuclei, heterochromatin, and euchromation from the liver of mouse (top row), guinea pig (middle row), and calf (bottom row). Note the presence of nucleoli (lightly stained spheres) in the heterochromatin preparations.

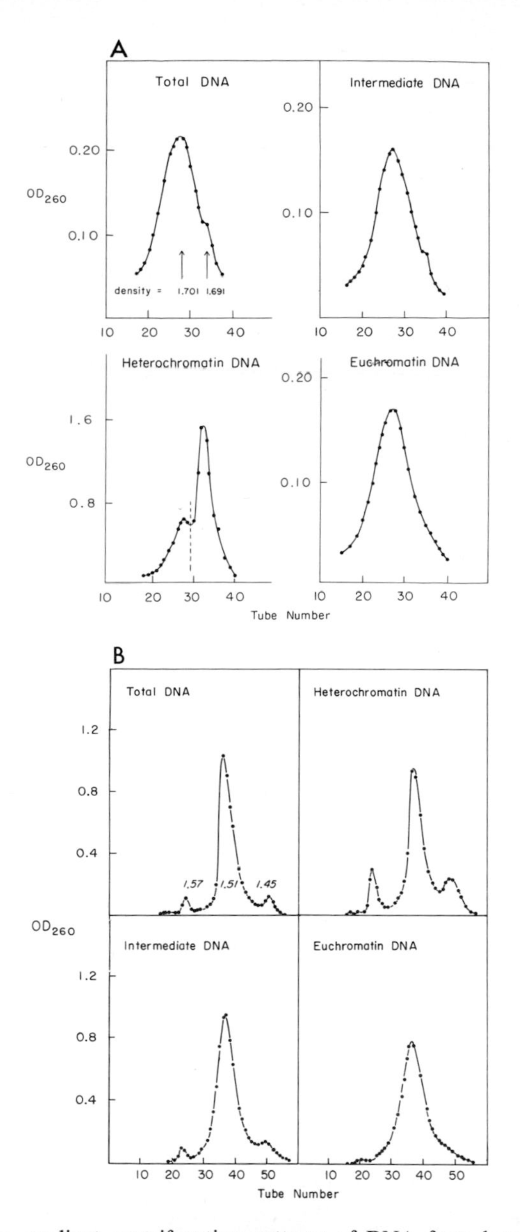

FIG. 3. Density gradient centrifugation patterns of DNA from heterochromatin (high density chromatin), intermediate chromatin (intermediate density chromatin), and euchromatin (low density chromatin) fractions of liver nuclei. (A) DNA patterns of mouse chromatin fractions in neutral CsCl (from Yasmineh and Yunis, 1970); (B) and (C) DNA patterns of guinea pig (from Yunis and Yasmineh, 1970; copyright 1970 by the American Association for the Advancement of Science) and calf chromatin fractions (from Yasmineh and Yunis, 1971a), respectively, in $Cs_2SO_4-Ag^+$. For (C) see p. 166.

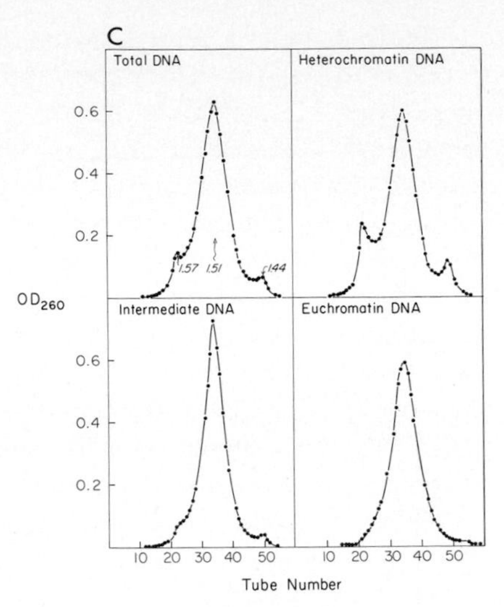

FIG. 3(C). For legend see p. 165.

Following sonication, the suspension of chromatin is passed through 2 layers of flannelette and subjected to a schedule of centrifugation designed to extract fractions of heterochromatin, intermediate chromatin, and euchromatin of different densities. The amount of chromatin that sediments at each speed differs with the tissue and species used, and to some extent with different sonicated preparations of nuclei from the same tissue. Because of this, several fractions of chromatin are isolated at various speeds to ensure that heterochromatin and euchromatin fractions, relatively free of contamination from each other, are obtained. This is most advisable when dealing with cells or tissues with which the investigator is unfamiliar, or when modifying the procedure with respect to the type and concentration of divalent cations. In these cases a schedule comparable to the following should be used: 500 *g* for 5 minutes; 1000, 2000, 4000, 6000, 12,000 *g* for 10 minutes each; 20,000 *g* for 1 hour and 78,000 *g* for 30 minutes. A last fraction of euchromatin of lowest density is isolated from the supernatant fluid, obtained following the last centrifugation, by making it 0.15 *M* with respect to NaCl and precipitation of the euchromatin with 2 volumes of cold ethanol. All chromatin fractions are monitored by fixation of small amounts in a solution of 1 part glacial acetic acid and 3 parts methanol and staining with Wright stain. In most cases the very small amounts of chromatin obtained in the 500 *g* fractions are contaminated with nuclei and are usually discarded following microscopic examination of the stained preparation.

The distribution of heterochromatin and euchromatin within the sedimented chromatin fractions varies with the species used. This is probably related to the degree of compactness of the masses of heterochromatin within the interphase nuclei, as well as their size and shape. Wright-stained preparations of nuclei and heterochromatin and euchromatin fractions from the liver of mouse, guinea pig, and calf are shown in Fig. 2. The preparations of nuclei and heterochromatin show condensed masses or chromocenters whereas the euchromatin preparations show none. It should be noted that chromocenters as seen in the preparations of whole nuclei and heterochromatin usually represent aggregates of chromocenters, each originating from more than one chromosome (Yunis and Yasmineh, 1971, 1972). Because of this, the distribution of heterochromatin within various fractions of chromatin, obtained by differential centrifugation, is difficult to predict solely on the basis of the appearance of heterochromatin in interphase nuclei.

A better indication of the distribution of heterochromatin within the various chromatin fractions involves the extraction of DNA from each fraction and determination of its satellite DNA content. This is shown in Fig. 3 in sonicates of liver nuclei from the mouse, guinea pig, and calf, where the chromatin has been fractionated into 3 fractions: a high density fraction, a fraction of intermediate density, and a fraction of low density obtained by alcohol precipitation (Yasmineh and Yunis, 1970, 1971a; Yunis and Yasmineh, 1970). DNA from each fraction was analyzed by density gradient centrifugation in CsCl or Cs_2SO_4-Ag^+.

In all 3 species, satellite DNA comprises about 10% of the DNA of the genome. In the mouse, this DNA appears as a minor component sedimenting at the light side of the main peak while in the calf and guinea pig it is composed mainly of two minor components sedimenting on either side of the main peak. As seen in Fig. 3, A–C, the high density fraction that is rich in heterochromatin is greatly enriched in satellite DNA in the case of the mouse (about 7-fold enrichment as determined from the areas under the curves), but is only moderately enriched in the case of the guinea pig (3 to 4-fold) and the calf (2 to 3-fold). The fractions of intermediate density show satellite DNA contents more or less comparable to those of total DNA, while the alcohol precipitates, which are rich in euchromatin, show little or none. Correspondingly, these variations in satellite DNA content are reflected in the amount of DNA represented by each fraction, as shown in Table I. For example in the mouse, the high density fraction represents only 10% of the total DNA; this, however, includes about 70% of the satellite DNA of the genome, on the basis of the 7-fold enrichment of this fraction in satellite DNA. This should be contrasted with the corresponding

TABLE I

PERCENT DISTRIBUTION OF CHROMATIN AMONG DIFFERENT DENSITY FRACTIONS OBTAINED FROM SONICATES OF MOUSE, GUINEA PIG, AND CALF LIVER NUCLEI BY DIFFERENTIAL CENTRIFUGATION

Source	Heterochromatin	Intermediate	Euchromatin
Mouse	10	8	82
Guinea pig	25	30	45
Calf	36	36	28

fractions from the guinea pig and calf that represent much larger portions of the total DNA (25 and 36%, respectively) and show only 3 to 4- and 2 to 3-fold enrichment in satellite DNA, respectively. At the other extreme, the low density fraction of chromatin, represented by the alcohol precipitate, comprises the bulk of the DNA of the mouse (82%) and only 45 and 28% of the DNA of the guinea pig and calf, respectively.

The tissue to be used is also of primary importance in attempting to

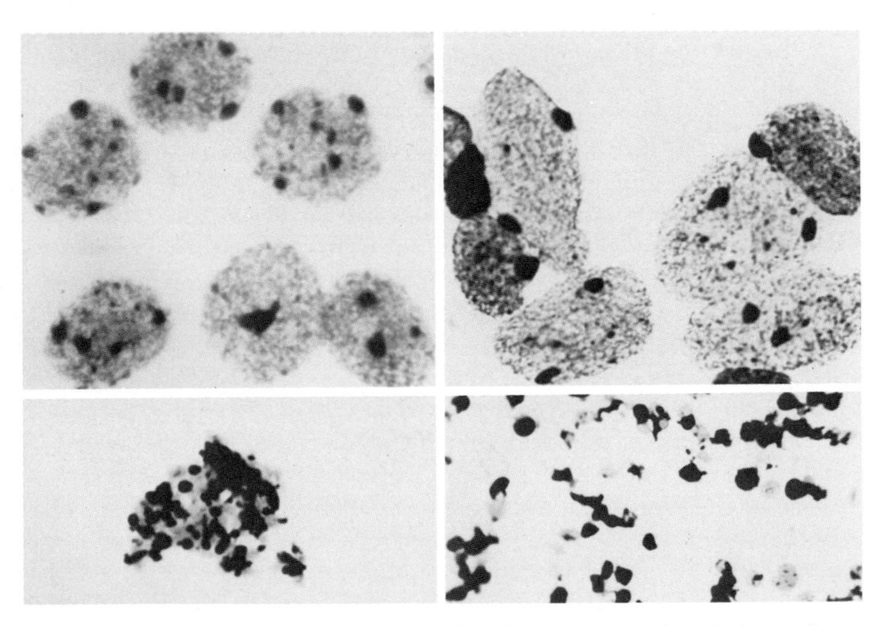

FIG. 4. Wright-stained preparations of nuclei and heterochromatin from the brain of mouse (top and bottom left) and *Microtus agrestis* (top and bottom right). The heterochromatin preparation from mouse brain was obtained by sonication of whole nuclei and sedimentation at 3500 *g*. The heterochromatin preparation from *M. agrestis* brain was obtained by passage of the nuclei through a French pressure cell at 3500 psi and sedimentation at 2000 *g* (see Table II).

isolate heterochromatin and euchromatin from mammalian cells. Ideally, the best sources of nuclei for this purpose are tissues in which constitutive heterochromatin is highly condensed and is consequently less apt to become contaminated with euchromatin upon disruption of the chromatin (Yunis and Yasmineh, 1971, 1972). This is true in the case of brain from most mammals and is illustrated in Fig. 4 for brain nuclei from the mouse and the field vole, *Microtus agrestis*. Mouse brain nuclei show masses that are distinct and more compact than those of liver shown in Fig. 2. In *M. agrestis*, the masses are also compact but much larger. Preparations of heterochromatin obtained from brain nuclei of both species are also shown. In *M. agrestis*, recent observations from this laboratory on the appearance of heterochromatin within almost every cell type have suggested that these tissue variations may be due to the fact that the giant masses are composed of long heteropycnotic fibers that assume various shapes and degrees of compactness in different tissues (Lee and Yunis, 1971a,b). As seen in Fig. 5, the masses are highly compacted and almost spherical in brain nuclei, somewhat diffuse with partly extended fibers in liver nuclei, almost completely extended in nuclei from fibroblasts, and in the form of compact rods in endothelial cells.

Another advantage of brain tissue lies in the fact that the DNA of chromatin fractions obtained by the sonication of brain nuclei (or their passage through a French pressure cell) is relatively undergraded as compared to the DNA of chromatin fractions from liver nuclei. This is presumably due in part to the compactness of the heterochromatin in brain nuclei and in part to the relative paucity of brain nuclei in nucleases. The use of relatively undergraded heterochromatin and euchromatin fractions is of importance for the study of chromatin-associated components, other than DNA. It should be noted, however, that one serious disadvantage of brain tissue is its low yield of nuclei due to its small size and high content of white matter. We have estimated, for example, that one liver of M. agrestis yields approximately ten times as much DNA as one brain (cerebrum).

The distribution of heterochromatin and satellite DNA in the various

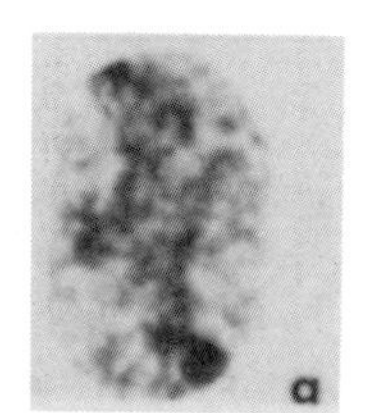

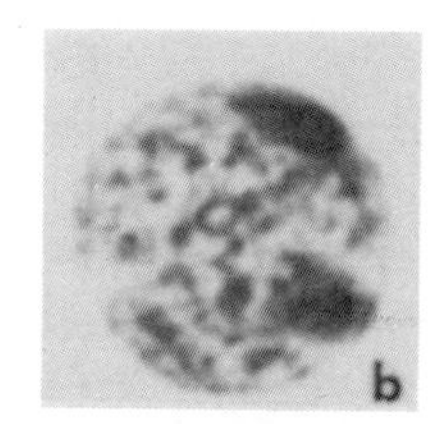

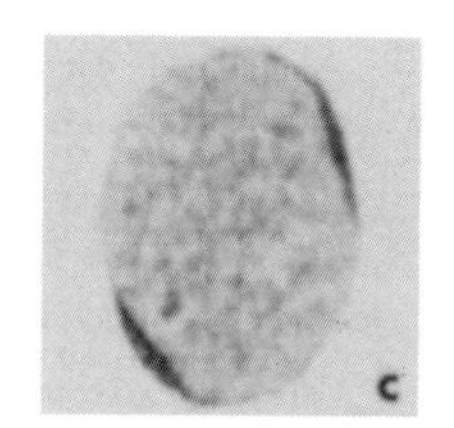

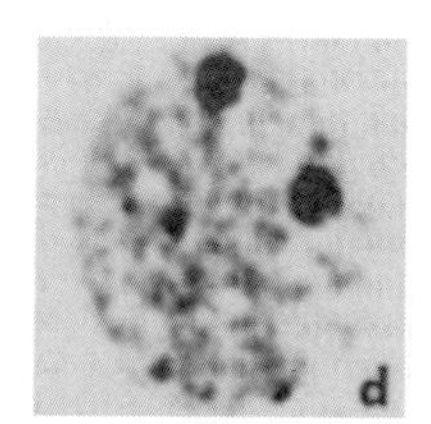

FIG. 5. Representative patterns of the giant chromosomes of *Microtus agrestis* exhibited by various interphase nuclei. (a) Fibroblast, (b) hepatocyte, (c) endothelial cell, (d) neuron. From Lee and Yunis (1971a) and Yasmineh and Yunis (1971a).

chromatin fractions obtained from nuclear sonicates may also be affected by the type and concentration of divalent cations used in the original homogenization of the tissue and/or during the suspension of the nuclei in Tris buffer. This is probably due to the effect that these exert in stabilizing the chromatin and in hardening the nuclear membrane, leading to variations in the size of the chromatin fractions obtained upon sonication. These variations, however, simply shift the schedule of centrifugation of the sonicates and do not appear to affect the extent of enrichment in heterochromatin and satellite DNA that is normally obtained from a specific tissue. As an example, two different lots of mouse liver nuclei, prepared by the same procedure, may require 10 and 15 seconds of total sonication time for disruption of all the nuclei. Upon subjecting both sonicates to the same schedule of centrifugation, the chromatin fractions in the preparation that required less sonication time would be obtained at relatively lower speeds, although the distribution of heterochromatin and satellite DNA in the fractions as a whole would remain unchanged. This does not apply, however, when the conditions of divalent cations are varied to a point that periods of sonication greatly exceeding 15 seconds are required for disruption of all the nuclei.

At the present time, the reasons for the observed tissue and species variations in the patterns of heterochromatin are not known. As shown in *M. agrestis*, extensive variations in the compactness and shape of heterochromatin are demonstrable in different tissues of the same species. One aspect of this tissue variation is observed in cultured fibroblasts, where the pattern of heterochromatin is generally extended (see *M. agrestis* fibroblast in Fig. 5) and, consequently, the heterochromatin is quite difficult to isolate in a pure form. Since these variations cannot be attributed to differences in DNA, it is probable that other factors, including differences in the divalent cation concentration of the nuclear milieu and the organization of the nucleoprotein complex, are of importance. These factors may also be involved in determining the physical characteristics of heterochromatin among tissues from different species, although in these cases the DNA of heterochromatin may also be a factor, since in most instances it includes satellite DNAs of different biochemical and biophysical characteristics (Yunis and Yasmineh, 1971, 1972).

The conditions of sonication when dealing with nuclei isolated from tumors or cells in culture appear to be different from those used for nuclei from normal tissue. In this laboratory, experience was derived mainly from nuclei isolated from 3T3 mouse cells transformed with SV40 (SV3T3–11A8 and SV3T3–56) and RSV (B77/3T3), and with an established cell line from *M. agrestis*. (de la Maza and Yunis, 1973). Preparations of nuclei from these cells are more resistant to sonication. Furthermore, they frequently contain a small proportion of nuclei that is much more resistant to sonication than the bulk of the nuclei. This is probably due in part to the persistence in these nuclei of cytoplasmic tags which hinder the extraction of the outer nuclear membrane. As a result, the nuclei are initially sonicated for approximately 10 seconds at 11 A, to disrupt about 90% of the nuclei, followed by sedimentation of the unbroken nuclei and large nuclear fragments at 300*g* for 10 minutes and careful aspiration of the supernatant fluid. At this stage, examination of the supernatant fluid by phase contrast microscopy should reveal no nuclei, although nuclear fragments may be seen. The supernatant fluid is sonicated for 10 more seconds, to achieve optimal separation of heterochromatin from euchromatin, and filtered through 2 layers of flannelette. The following fractions are then obtained as follows: a very small fraction, which is discarded, containing some fibrous material and some nucleoli and broken pieces of nuclei, obtained by centrifugation at 300*g* for 10 minutes; a dense fraction enriched in heterochromatin, usually comprising 20 to 30% of the chromatin, by centrifugation at 3000 *g* for 20 minutes; a fraction of intermediate density (10 to 20% of the chromatin) by centrifugation at 15,000 *g* for 20 minutes, and a light fraction (55 to 65% of the chromatin) obtained by precipitation with alcohol.

Differences in the time required for the sonication of nuclei with age of the animal have also been noted in this laboratory. Normally, nuclei from the very young adults are much more readily sonicated and yield better preparations of heterochromatin and euchromatin.

2. Disruption of Nuclei in a French Pressure Cell

Quite recently, relatively pure preparations of heterochromatin and euchromatin were obtained in this laboratory from nuclei of normal tissues by compression and decompression in a French pressure cell. The technique was suggested by its success in the hands of several investigators in isolating clean nucleoli from whole cells or nuclei (Liau *et al.*, 1968; Higashinakagawa *et al.*, 1971).

Nuclei from normal mammalian tissues (liver, brain) are isolated by the usual procedure (Section II,A), after which the outer nuclear membrane is removed by several suspensions in Tris buffer–0.2 m*M* Ca^{2+}–5 m*M* Mg^{2+}, and the nuclei are then made to swell in 0.25 *M* sucrose (Sections IIIA and B). When maximum swelling is attained, the suspension of nuclei is placed in a cold (2 to 4°C) 1-inch French pressure cell, and a pressure of 3000 to 5000 psi is applied by means of a mechanically operated press (American Instrument Co., Silver Spring, Maryland). The suspension is then released from the cell at this pressure by carefully opening the needle valve and allowing it to flow smoothly. Microscopic examination of the suspension at this time should show very few unbroken nuclei or nuclear fragments. The suspension is passed one more time through the cell at the same pressure, to ensure complete disruption of the nuclei for the isolation of chromatin fractions at varying densities.

The pressure used in disrupting nuclei is quite critical for the isolation of clean heterochromatin and euchromatin fractions, since the use of lower than optimal pressures would result in contamination of the heterochromatin fractions with euchromatin and fragments of whole nuclei, while the use of pressures higher than optimal would result in loss of heterochromatin to the euchromatin fractions and excessive degradation of the DNA. As with sonication, the optimal conditions for nuclear disruption may vary with the tissue and species used, and, therefore, it is advisable when working for the first time with nuclei from a certain tissue to use low pressures initially, with careful monitoring of the suspension, and increase the pressure gradually in subsequent passages, if required.

Using nuclei from mouse liver and brain, this technique has consistently yielded heterochromatin and euchromatin fractions comparable in purity to those obtained by sonication. In the case of the guinea pig, where sonication of liver nuclei yields only 3 to 4 fold enrichment of the heterochromatin in satellite DNA (Fig. 3), better enrichment (5 to 6 fold) was achieved from both liver and brain nuclei. No experiments have as yet been conducted with calf nuclei.

The technique was also recently used to isolate heterochromatin from liver and brain nuclei of *M. agrestis* (Yasmineh and Yunis, 1973). As stated earlier, the bulk of the constitutive heterochromatin of this species is in the form of two giant masses that show differences in compactness and size and shape, depending upon the tissue. These variations affect the purity in which the masses can be isolated. As in the nuclei from the respective tissues (Fig. 5), the masses in preparations of brain heterochromatin are usually large and compact, while those in liver preparations are smaller and more diffuse.

An estimate of the purity of giant masses prepared from brain and liver nuclei may be obtained by comparing their satellite and repetitive DNA content. Although satellite DNA is difficult to demonstrate in *M. agrestis* by the usual technique of density gradient centrifugation in CsCl, the presence of both satellite DNA and DNA of fast-intermediate repetitiveness has been recently established by reassociation of sheared-denatured DNA at Cot values of less than 1 and fractionation on hydroxyapatite (Yasmineh and Yunis, 1971b, 1973). The CsCl patterns of 3 repetitive DNA fractions isolated in this manner from total nuclear DNA are shown in Fig. 6. Figure 6A is the pattern of the most highly repetitive fraction that has been reassociated at a Cot of 4×10^{-3} and represents 5.5% of the total nuclear DNA. It contains two satellite DNAs, high in GC and AT, at a density of 1.716 and 1.700 gm/cm^3, respectively. Figure 6B is the pattern

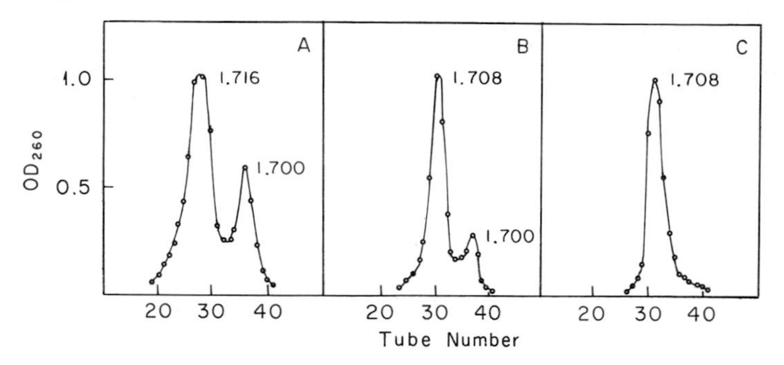

FIG. 6. Neutral CsCl density gradient centrifugation patterns of three repetitive DNA fractions isolated from the nuclear DNA of *Microtus agrestis*. The fractions were obtained by the reassociation of sheared-denatured DNA at Cot values of 4×10^{-3} (pattern A), 9×10^{-2} (pattern B), and 9×10^{-1} (pattern C), as described by Yasmineh and Yunis (1971b, 1973).

of a less repetitive fraction obtained by further reassociation at a Cot of 9×10^{-2} of the unreassociated DNA remaining after removal of the first repetitive fraction; it represents 3.8% of the total DNA and is made up of a minor peak of the high AT satellite DNA, at 1.700 gm/cm^3, and a major component at 1.708 gm/cm^3. The latter has a base composition comparable to that of total nuclear DNA. Figure 6C is the pattern of a still less repetitive fraction obtained by further reassociation of the unreassociated DNA at a Cot of 9×10^{-1}. This fraction represents 9.4% of the total DNA and is composed entirely of the 1.708 component. Estimates of the percentage of DNA in the 1.716, 1.700, and 1.708 peaks from the areas under the curves yield values of 4, 3, and 12% of the nuclear DNA, respectively.

Giant masses from both brain and liver nuclei, prepared by passage through a French pressure cell, show enrichment in all 3 repetitive components, although the enrichment in brain preparations, where the masses are compact, is much higher. A comparison of the results of experiments on brain and liver nuclei, in which the nuclei were passed through a French pressure cell at 3500 psi and six chromatin fractions obtained by differential centrifugation, is shown in Table II. The following observations can be made: Using the same schedule of centrifugation, the distribution of chromatin in the various fractions was strikingly different. For example, the high density fractions I and II from brain included 15.1 and 13.1% of the chromatin, whereas in liver they included 1.5 and 1.7%. At the other extreme, the low density fractions V and VI included 21.8 and 30.6% of the chromatin in brain, as compared to 22.3 and 57.7% in liver.

The extent of enrichment of the chromatin fractions in DNA that reassociates at a Cot of less than 1 is also different. Although for both tissues the early dense fractions are enriched in this DNA as compared to the lower density fractions, the enrichment is much more striking in brain

TABLE II

EXTENT OF DNA SEQUENCE REPETITION OF CHROMATIN FRACTIONS FROM *Microtus agrestis* OBTAINED BY PASSAGE OF BRAIN AND LIVER NUCLEI THROUGH A FRENCH PRESSURE CELL AND DIFFERENTIAL CENTRIFUGATION

Fraction	Gravitational force (×g)	Time (min)	Percent of total chromatin	Fraction Cot < 1[a]
Brain				
I	2,000	10	15.1	0.56
II	3,000	20	13.1	0.26
III	6,000	10	7.1	0.22
IV	12,000	10	12.4	0.19
V	19,000	30	21.8	0.17
VI	Alcohol precipitate	—	30.6	0.13
Liver				
I	2,000	10	1.5	0.33
II	3,000	20	1.7	0.31
III	6,000	10	8.0	0.25
IV	12,000	10	8.9	0.20
V	19,000	30	22.3	0.20
VI	Alcohol precipitate	—	57.7	0.18

[a]Yasmineh and Yunis (1973).

where, for example, 0.56 of the DNA of fraction I reassociates at a Cot of less than 1, as compared to 0.33 for fraction I of liver. However, because of differences between the two tissues in the distribution of chromatin among the various fractions, the bulk of the DNA that reassociates at a Cot of less than 1 is found in the dense fractions in the case of brain and in the lighter fractions in the case of liver. All these differences are compatible with the cytological picture of the heterochromatic giant masses in the two tissues.

When the DNA of Cot less than 1 from the various fractions is analyzed by density gradient centrifugation in neutral CsCl, it is found to be composed primarily of a component sedimenting at 1.708 gm/cm^3, as shown in Fig. 7 for this DNA from fractions I, III, and VI of brain (see Table II). This is the same nonsatellite DNA component that represents the bulk of the DNA of Cot less than 1 when nuclear DNA is fractionated according to repetitiveness (Fig. 6). It reassociates at a Cot between 10^{-2} and 1, and is intermediate in repetitiveness between the satellite DNAs and repeated sequences of Cot values higher than 1. Figure 7 also shows the presence of the two satellites DNAs in fraction I, at densities 1.716 and 1.700 gm/cm^3. Collectively they constitute about 30% of the DNA of Cot less than 1. Since

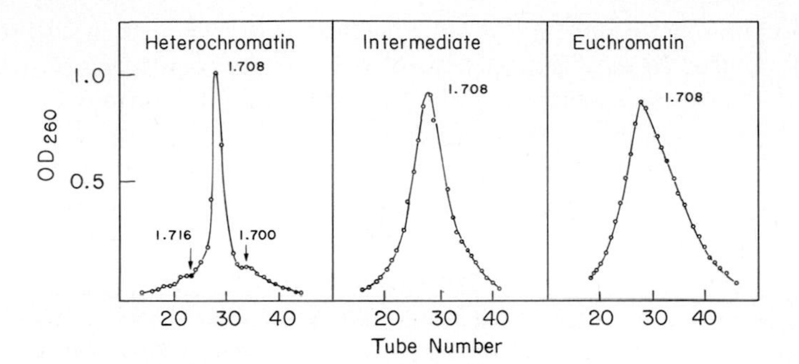

FIG. 7. Neutral CsCl density gradient centrifugation patterns of DNA of Cot less than 1 isolated from the heterochromatin, intermediate, and euchromatin fractions of brain nuclei from *Microtus agrestis*. The fractions correspond to fractions I, III, and VI, respectively, shown in Table II.

satellite DNA represents also about 30% of the nuclear DNA that reassociates at a Cot of less than 1 (Fig. 6), this indicates that fraction I is enriched in satellite DNA to an extent similar to that of the 1.708 component. Whether this enrichment in satellite DNA is due to preferential localization of satellite DNA in the giant chromosomes has not as yet been determined. However, on the basis of the presence of satellite DNA mainly in centromeric regions in other mammals, and the demonstration of the presence of centromeric heterochromatin in *M. agrestis* by "*in situ*" reassociation (Yunis *et al*., 1971), it is more likely that the enrichment of fraction I in satellite DNA is due to enrichment in both the giant masses of the sex chromosomes and the smaller masses of centromeric heterochromatin.

The above experiments with chromatin fractions derived from the disruption of nuclei in a French pressure cell indicate that this technique of chromatin fractionation is quite suitable for the isolation of heterochromatin and euchromatin from normal tissues. In some instances, as with guinea pig liver, it may be superior to the technique of sonication. We have as yet not examined its efficiency in isolating heterochromatin and euchromatin from tumor cells, transformed cells, or cells in culture.

IV. Conclusion

We have attempted in this chapter to relate the experience we have acquired during the last few years in the isolation of heterochromatin and euchromatin from mammalian cells. The state of the art is still young and a great deal more work needs to be done, especially in the isolation of these components in the pure form from tumor cells and cells in culture. The technique used in the fractionation of mammalian chromatin from normal tissues is highly reproducible and with some modification should be applicable to nonmammalian species. Heterochromatin fractions isolated by this technique are enriched in satellite DNA and provide excellent sources for the detection of satellite DNA and its isolation and study. They also provide a unique opportunity for the investigation of other DNA components that may reside in heterochromatin, such as fast-intermediate repetitive DNA, tandem gene duplicates and even single copy DNA. Furthermore, since these chromatin fractions have not been exposed to conditions that would seriously impair chromatin structure, they may be suitable for the investigation of constituents, other than DNA, including basic and acidic proteins and "chromatin-associated" RNA.

Currently, work in this laboratory is involved with the subfractionation of mammalian heterochromatin. In *M. agrestis*, for example, separation of constitutive heterochromatin into perinucleolar heterochromatin, centromeric heterochromatin, and the heterochromatin of the giant chromosomes, may be achieved by the use of discontinuous sucrose gradients. Of special interest is the characterization of the DNA of the giant chromosomes, which is of fast-intermediate repetitiveness and of base composition comparable to that of the bulk of the DNA. The intriguing possibility that this DNA represents mutated satellite DNA sequences and may have counterparts in other mammals, in the form of intercalary heterochromatin, is also being investigated.

References

Blobel, G., and Potter, V. R. (1967). *Science* **154**, 1662.

Bostock, C. J. (1971). *In* "Advances in Cell Biology" (D. M. Prescott, L. Goldstein, and E. McConkey, eds.), Vol. II, p. 153. Appleton, New York.

Brown, S. W. (1966). *Science* **151**, 417.

Busch, H. (1967). *In* "Nucleic Acids" (L. Grossman and K. Moldave, eds.), Methods in Enzymology, Vol. 12A, p. 421. Academic Press, New York.

Chauveau, J., Moulé, Y., and Rouiller, C. (1956). *Exp. Cell Res.* **11**, 317.

de la Maza, L., and Yunis, J. J. (1973). Personal communication.

Dingman, C. W., and Sporn, M. B. (1964). *J. Biol. Chem.* **239**, 3483.

Dounce, A. L. (1963). *Exp. Cell Res., Suppl.* **9**, 126.

Dounce, A. L., and Ichowicz, R. (1970). *Arch. Biochem. Biophys.* **137**, 143.

Flamm, W. G. (1972). *Int. Rev. Cytol.* **32**, 1.

Frenster, J. H. (1963). *Exp. Cell Res., Suppl.* **9**, 235.

Frenster, J. H. (1969). *In* "Handbook of Molecular Cytology" (A. Lima-de-Faria, ed.), p. 251. Amer. Elsevier, New York.

Frenster, J. H., Allfrey, V. G., and Mirsky, A. E. (1960). *Proc. Nat. Acad. Sci. U.S.* **46**, 432.

Frenster, J. H., Allfrey, V. G., and Mirsky, A. E. (1963). *Proc. Nat. Acad. Sci. U.S.* **50**, 1026.

Heitz, E. (1928). *Jahrb. Wiss. Bot.* **69**, 762.
Heitz, E. (1934). *Biol. Zentralbl.* **54**, 588.
Heitz, E. (1935). *Z. Indukt. Abstamm.-Vererbungsl.* **70**, **402.**
Higashinakagawa, T., Muramatsu, M., and Sugano, H. (1971). *Exp. Cell Res.* **71**, 65.
Johnston, I. R., and Mathias, A. P. (1972). *In* "Subcellular Components" (G. D. Birnie, ed.), 2nd Ed., p. 53. Plenum, New York.
Levan, A. (1946). *Hereditas* **32**, 449.
Lee, J. C., and Yunis, J. J. (1971a). *Chromosoma* **32**, 237.
Lee, J. C., and Yunis, J. J. (1971b). *Chromosoma* **35**, 117.
Liau, M. C., Craig, N. C., and Perry, R. P. (1968). *Biochim. Biophys. Acta* **169**, 196.
Lima-de-Faria, A. (1959). *J. Biophys. Biochem. Cytol.* **6**, 457.
Lima-de-Faria, A. (1969). *In* "Handbook of Molecular Cytology" (A. Lima-de-Faria, ed.), p. 227. Amer. Elsevier, New York.
Littau, V. C., Burdick, C. J., Allfrey, V. G., and Mirsky, A. E. (1965). *Proc. Nat. Acad. Sci. U.S.* **54**, 1204.
Lyon, M. F. (1961). *Nature* (*London*) **190**, 372.
McBride, O. W., and Peterson, E. A. (1970). *J. Cell Biol.* **47**, 132.
Maggio, R., Siekevitz, P., and Palade, G. E. (1963a). *J. Cell Biol.* **18**, 267.
Maggio, R., Siekevitz, P., and Palade, G. E. (1963b). *J. Cell Biol.* **18**, 293.
Maggliozzi, J., Puro, D., Lin, C., Ortman, R., and Dounce, A. L. (1971). *Exp. Cell Res.* **67**, 111.
Montgomery, T. H. (1904). *Biol. Bull.* **6**, 137.
Muller, H. J., and Painter, T. S. (1932). *Z. Indukt. Abstamm. Vererbungsl.* **62**, 316.
Muramatsu, M. (1970). *Methods Cell Physiol.* **4**, 195.
Ohno, S. (1967). "Sex Chromosomes and Sex Linked Genes." Springer-Verlag, Berlin and New York.
Ohno, S., Kaplan, W. D., and Kinosita, R. (1959). *Exp. Cell Res.* **18**, 415.
Penman, S. (1969). *In* "Fundamental Techniques in Virology" (K. Habel and N. P. Salzman, eds.), Vol. 1, p. 35. Academic Press, New York.
Roodyn, D. B. (1972). *In* "Subcellular Components" (G. D. Birnie, ed.), 2nd Ed., p. 15. Plenum, New York.
Schmid, W. (1967). *Arch. Julius Klaus-Stift. Vererbungsforsch.* (*Sozialanthropol. Rassenhyg.*) **42**, 1.
Siebert, G. (1967). *Methods Cancer Res.* **2**, 287.
Smith, H. S., Gelb, L. D., and Martin, M. A. (1972). *Proc. Nat. Acad. Sci. U.S.* **69**, 152.
Sutton, W. D., and McCallum, M. (1972). *J. Mol. Biol.* **71**, 633.
Swanson, C. P. (1957). "Cytology and Cytogenetics." Prentice-Hall. Englewood Cliffs, New Jersey.
Varmus, H. E., Vogt, P. K., and Bishop, J. M. (1973). *J. Mol. Biol.* **74**, 613.
Walker, P. M. B. (1971). *Prog. Biophys. Mol. Biol.* **23**, 145.
Westphal, H., and Dulbecco, R. (1968). *Proc. Nat. Acad. Sci. U.S.* **59**, 1158.
Yasmineh, W. G., and Yunis, J. J. (1969). *Biochem. Biophys. Res. Commun* **35**, 779.
Yasmineh, W. G., and Yunis, J. J. (1970). *Exp. Cell Res.* **59**, 69.
Yasmineh, W. G., and Yunis, J. J. (1971a). *Exp. Cell Res.* **64**, 41.
Yasmineh, W. G., and Yunis, J. J. (1971b). *Biochem. Biophys. Res. Commun.* **43**, 580.
Yasmineh, W. G., and Yunis, J. J. (1973). *Exp. Cell Res.* (in press).
Yunis, J. J., and Yasmineh, W. G. (1970). *Science* **168**, 263.
Yunis, J. J., and Yasmineh, W. G. (1971). *Science* **174**, 1200.
Yunis, J. J., and Yasmineh, W. G. (1972). *Advan. Cell Mol. Biol.* **2**, 1.
Yunis, J. J., Roldan, L., Yasmineh, W. G., and Lee, J. C. (1971). *Nature* (*London*) **231**, 532.

Chapter 12

Measurements of Mammalian Cellular DNA and Its Localization in Chromosomes[1]

L. L. DEAVEN AND D. F. PETERSEN

Cellular and Molecular Radiobiology Group,
Los Alamos Scientific Laboratory,
University of California
Los Alamos, New Mexico

I. Introduction

In spite of considerable scientific attention, evolution of cell populations *in vitro* remains a poorly understood phenomenon. Of particular interest, because of parallel behavior in carcinogenesis, are the chromosomal changes that accompany cellular evolution and the genetic consequences of transition from the euploid to the aneuploid or heteroploid state. Recent develop-

[1]This work is being performed under the auspices of the U.S. Atomic Energy Commission.

ments in methodology promise to shed new light on mechanisms involved in these transitions and are the subject of this chapter.

Kraemer *et al.* (1971, 1972) reported that, despite considerable variability in chromosome number and a significantly increased DNA content, many heteroploid mammalian cell lines are no more variable in cellular DNA content than euploid cells from which they were derived. This observation, which leads to a more optimistic view of cancer therapy than notions of multiple stem lines (Makino, 1957), has been documented extensively in a recent review (Kraemer *et al.*, 1972), and only the salient features are summarized here to provide the framework for a completely different line of evidence which also argues for more genetic homogeneity of tumor lines than heretofore suspected.

The original ideas of tumor cell heterogeneity were derived from investigations of DNA content and chromosome number variability in tumors and in cultured material derived from tumors (Atkin and Richards, 1956; Chu and Giles, 1958; Hsu and Klatt, 1958). These data, laboriously collected by measuring one cell at a time and based largely on comparisons of chromosome counts and Feulgen microphotometric measurements, lacked the inherent precision to attack the problem with the necessary statistical rigor. However, with the advent of techniques (Van Dilla *et al.*, 1969) permitting rapid, individual measurements of relative cellular DNA content (flow microfluorometry, abbreviated FMF), a survey of numerous euploid lines and their heteroploid derivatives became practical. From these surveys (Kraemer *et al.*, 1971, 1972), several conclusions have emerged which, taken together, convincingly challenge the notion of multiple stem lines in malignant transformation.

Regardless of source, euploid mammalian cells contain very close to 7.0×10^{-9} mg of DNA as a 2C value (Mandel *et al.*, 1950). Our previous studies confirm these estimates and indicate that flow microfluorometric measurements and Schmidt-Thannhauser determinations are in reasonable agreement (Kraemer *et al.*, 1972). Heteroploid populations contain more DNA than their respective euploid parents, reflecting generally the attendant increase in chromosome number accompanying transformation. Observations have been limited to readily available heteroploid lines (Kraemer *et al.*, 1972) and a number of mouse lines (3T3, 3T6, SV3T3, and polyoma transformants) examined in collaboration with Dr. Robert Pollack. Although the collection of lines on which measurements have been made is now large (in excess of 30 lines and strains), no exceptions have emerged which seriously challenge the conclusion that excess DNA accumulates in heteroploid cells in approximate multiples of the haploid set. That is, all cell lines appear to cluster around values of 1.0, 1.5, or 2.0 times the 2C DNA content. For example, in cells classified as pseudodiploid or low heteroploid, such as the

Chinese hamster line (CHO), L-5178Y (murine lymphoma), and Syrian hamster line BHK21, the DNA content differs from euploid parent cells by less than 5%. On the other hand, two HeLa lines contain 1.5 times as much DNA as WI-38, and another HeLa line and PK-15 (mixoploid porcine) cells contain twice the DNA found in their respective diploid parents. Except for some fused 3T3 lines, no instances have been observed in which the DNA contents of heteroploid cells exceed that of the diploid parent by more than a factor of two (Pollack and Petersen, unpublished). Furthermore, biological intervention leading to perturbation of genome content yields DNA distributions within a population which are entirely consistent with prediction. For instance, synchronization by mitotic selection (Petersen *et al.*, 1968) yields cell populations which reflect exclusively G_2 + M DNA content. After incubation for a brief interval to allow for completion of division, they exhibit exclusively a G_1 DNA distribution.

When introduced into growing cultures and subsequently removed, low doses of Colcemid induce nondisjunction (Kato and Yosida, 1970). After removal of Colcemid, cell populations in G_1 exhibit a marked broadening of DNA content per cell which is nearly symmetrical. Thus, in populations selected immediately and in which all cells are still viable, cells that have both gained and lost chromosomes can be detected. Blockade with higher concentrations of Colcemid, although not lethal to hamster cells, causes metaphase arrest, which is temporary in some cells and permanent in others. Cells that escape blockade reenter G_1 without dividing and eventually replicate a double amount of DNA. A number of cells in the population are capable of several cycles of DNA synthesis without cytokinesis, resulting in a mixed population with respect to DNA content. For example, analysis after several days of culture in Colcemid reveals cells with 2C, 4 C, 8C, 16C, and 32C values, but no intermediate DNA contents can be detected (Stubblefield, 1964; Kraemer *et al.*, 1972).

We conclude from these observations that replication of the genome, which appears to be faithful under these conditions, is equally faithfully reflected by the analytical method. Further examples can be obtained by subjecting exponentially growing cultures to thymidine blockade, exposure to ionizing radiation, etc. In each case, the predicted DNA content previously established by independent techniques has been found, and all manipulations employing the analytical method and known biological effects have resulted in acquisition of data which have remarkable internal consistency. However, independent methods for assessing the distribution of genome in cells which either have been subjected to some chemical intervention or have arisen from other defects resulting in aneuploidy would be desirable for establishing the validity of results summarized in the preceding brief review. An attractive alternative to direct measurement of total DNA content by analytical methods would be a method for identifying segments of the genome by relying on karyological features of metaphase chromosomes. It is now clear that euploid cells contain a constant amount of DNA (James, 1965) and that the amount of DNA contained in specific euploid chromosomes is also constant (Rudkin *et al.*, 1964). Analytical observations of the DNA content of Chinese hamster cells also exhibit constancy in terms of total DNA content (Kraemer *et al.*, 1971, 1972), and we have observed that, although rearranged extensively, banding sequences from euploid Chinese hamster cells are identifiable to a large extent in the aneuploid line CHO. Moreover, there appear to be pressures acting to maintain chromosomes of approximately the same size in aneuploid cells as originally existed in the euploid parent (Deaven and Petersen, 1973). It is attractive to speculate that a relationship between chromosome band patterns and DNA content exists and that morphologically identifiable band sequences represent specific regions of the genome. Hence, in surviving cells analysis of preservation of specific regions may eventually provide not only the ground rules for transition from euploidy to aneuploidy but also a better understanding of the paradox of chromosomal variability and DNA constancy.

Because of the widespread use of the Chinese hamster cell (line CHO) and because of our general lack of information concerning the nature of aneuploidy and its potential impact on interpretation of experimental observations made on the line, an extensive karyologic examination of line CHO was undertaken (Deaven and Petersen, 1973) to extend previous preliminary studies (Kao and Puck, 1969, 1970). In this chapter, line CHO is presented as a prototype system from which the methodology can be extended to other cell lines. The results are interesting in their own right but are presented here primarily to demonstrate the utility of combining several relatively new techniques to problems in tumor cell biology.

The idea of unique staining patterns as a means of identifying unequivocally specific chromosomes stems from work of Caspersson and his associates (1967) which clearly demonstrated differential staining properties of plant and mammalian chromosomes after treatment with quinacrine mustard. These studies were extended until it became possible to identify each human chromosome by its unique quinacrine fluorescence (Q-banding)[2] pattern (Caspersson *et al.*, 1970). Similar differentiation of chromosomal segments

[2] In this manuscript we use the nomenclature proposed at the International Congress of Human Genetics, held in Paris in 1971 (Hamerton *et al.*, 1972).

1. Q-bands—fluorescent bands revealed by quinacrine mustard or related fluorochromes.
2. C-bands—heavily stained regions revealed by alkaline treatment and SSC incubation, usually in the centromeric regions.
3. G-bands—chromosomal cross bands revealed by a variety of techniques using Giemsa stain.

was reported by Pardue and Gall (1970) and by Arrighi and Hsu (1971) using NaOH denaturation and Giemsa stain. They introduced the molecular interpretation that certain sequences in the DNA of metaphase chromosomes were more easily denatured and renatured than others and that typical Giemsa staining patterns were elicited by careful treatment of preparations to expose selectively renatured segments. More recent studies (Kato and Yosida, 1972; Deaven and Petersen, 1973) have questioned the chemical basis for production of typical patterns. However, as this methodology has been extended, resolution has been improved to the point where it is now possible to identify individual chromosomes with a high degree of certainty and, in the case of aneuploid cells, to identify both the origin and destination of genetic material translocated, rearranged and, to a lesser extent, deleted coincident with transformation of the cell line.

II. Banding Techniques

A. C-Bands

Localization of heterochromatic regions (C-bands) in mammalian chromosomes, first achieved by Pardue and Gall (1970) during the course of investigating hybridization of mouse satellite DNA and extended by Hsu and his associates (Arrighi *et al*., 1970; Arrighi and Hsu, 1971), led to the idea of specific association of heterochromatin with centromeric regions and to the contention that repetitious regions associated with centromeres somehow reflect a central role in organization of interphase chromatin.

Numerous methods, all based on alkaline hydrolysis and extended periods of reannealing at 60°C, have been explored in this laboratory. All these methods are variants of a procedure described by Arrighi and Hsu (1971) and require high quality chromosome preparations. The method which has afforded most consistent results involves a 3-minute treatment with saturated $Ba(OH)_2$, followed by a 17-hour incubation in 2 or 6X SSC at 60°C. Slides were subsequently stained for 10 minutes with 5 ml of stock Giemsa (Harleco) diluted with 50 ml of 0.01 M PO_4 buffer adjusted to pH 6.8. Results of the procedure, shown in Fig. 1, reveal that most of the heterochromatic material in mammalian chromosomes is indeed localized in centromeric regions or associated with the sex chromosomes. However, significant amounts of interstitial heterochromatin were also demonstrable, and the amount could be altered from several bands to none at all by varying the duration of alkaline treatment. For example, chromosome number 1 of the Chinese hamster has little centromeric heterochromatin, but has several

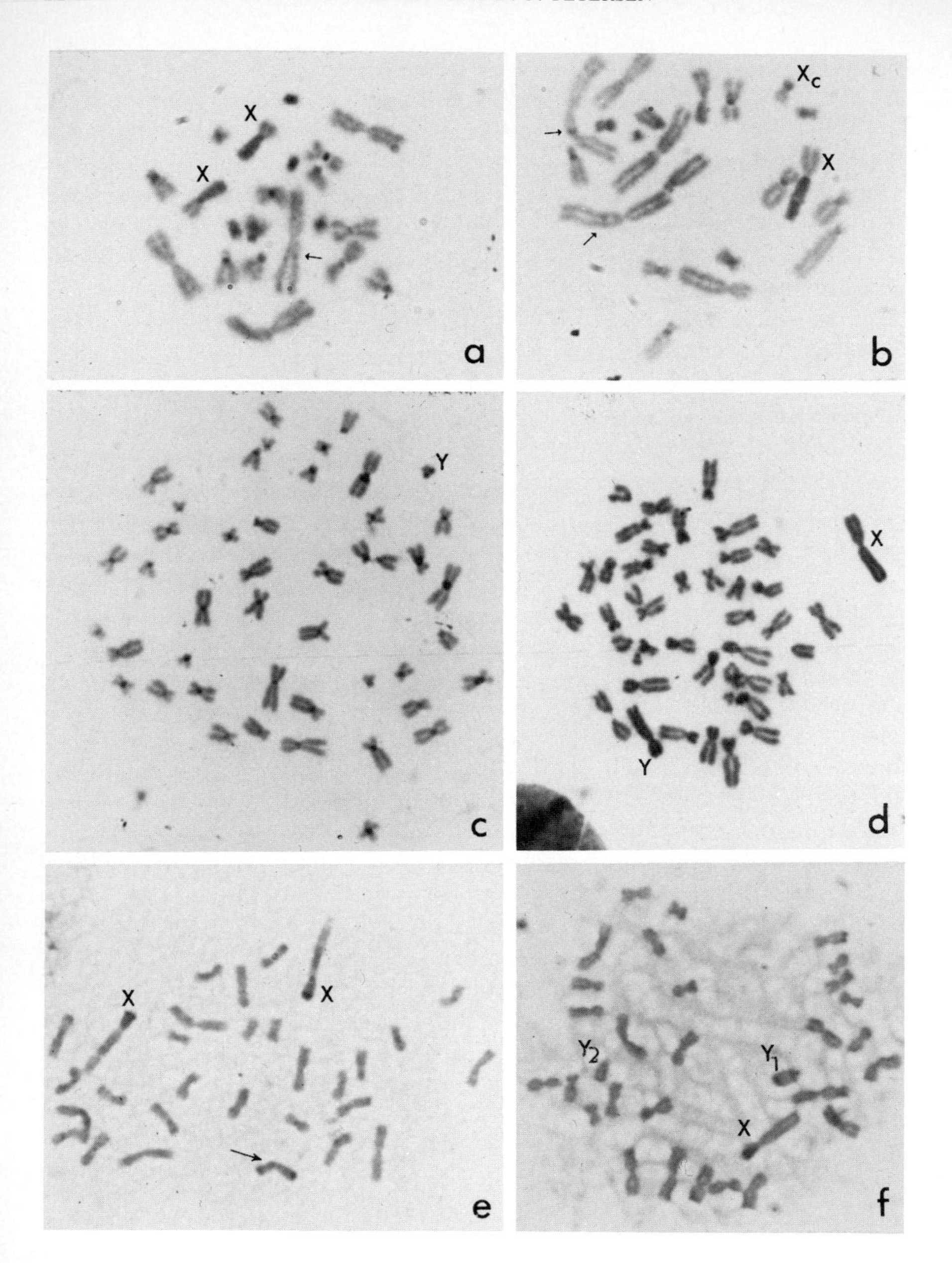
X
X
a
Xc
X
b
Y
c
X
Y
d
X
X
e
Y2
Y1
X
f

interstitial bands (identified by arrows in Fig. 1a and b). The most likely explanation for these bands, as well as those previously observed by Hsu and Arrighi (1971), is that they are remnants of the banding patterns induced by the alkaline–saline–Giemsa procedure (Schnedl, 1971). The critical dependence of C-banding patterns on treatment conditions severely limits routine application of the technique, and C-bands alone are not sufficiently distinctive to permit unequivocal identification of individual chromosomes. Nevertheless, the patterns do provide corroborative evidence for classifications based primarily on G-band analysis (Cooper and Hsu, 1972; Deaven and Petersen, 1973). In our hands, C-banding has been viewed not as a primary chromosome identification method, but, rather, as a means for rapidly and conveniently identifying sex chromosomes with a reliability equal to that experienced with quinacrine mustard fluorescence banding (Caspersson *et al.*, 1970). As an illustration, the absence of a major part of one of the X chromosomes of CHO can be readily deduced by inspection of Fig. 1b. In each of the other examples shown, the outstanding feature is the easily identified sex chromatin. The Persian gazelle is of particular interest because of the X, Y_1, Y_2 chromosomes in the male and the unpaired autosome in the female. A complete description of this sex differentiation system will be described elsewhere.

Treatment of chromosomes with either NaOH or $Ba(OH)_2$ results in rather severe degradation of chromosomal morphology and removes an appreciable amount of DNA. Autoradiographs of thymidine-^3H labeled chromosomes and nuclei before and after C-banding are shown in Fig. 2. The high background and loss of specificity following the C-band procedure suggest that the labeled DNA is extracted from the cells and reattached to the surface of the slide. Comings *et al.* (1972) have shown that treatment with 0.07 N NaOH for 30–180 seconds removes 16–81% of chromosomal DNA. For these reasons, it is important that chromosomes destined for more than one type of banding analysis (C- and G-banding or C- and Q-

FIG. 1. C-Bands of several mammalian species. (a) Euploid female Chinese hamster (*Cricetulus griseus*); the arrows indicate interstitial heterochromatin (see text). (b) Chinese hamster ovary cell (aneuploid line). The light long arms and dark short arms of one of the small metacentric chromosomes (X_c) suggest that it is the centromere of the missing X chromosome. (c) Euploid male human. The Y chromosome is indicated; the X cannot be distinguished from other C group chromosomes. (d) Euploid male Syrian hamster (*Mesocricetus auratus*). The X and Y chromosomes are heterochromatic as well as many autosomal arms. (e) Euploid female Persian gazelle (*Gazella subgutturosa*). The X chromosomes have a small amount of heterochromatin. Note the unpaired (on the basis of C-bands) metacentric chromosome with terminal heterochromatin (arrow). (f) Euploid male Persian gazelle (*Gazella subgutturosa*). The X, Y_1, and Y_2 chromosomes have small amounts of heterochromatin.

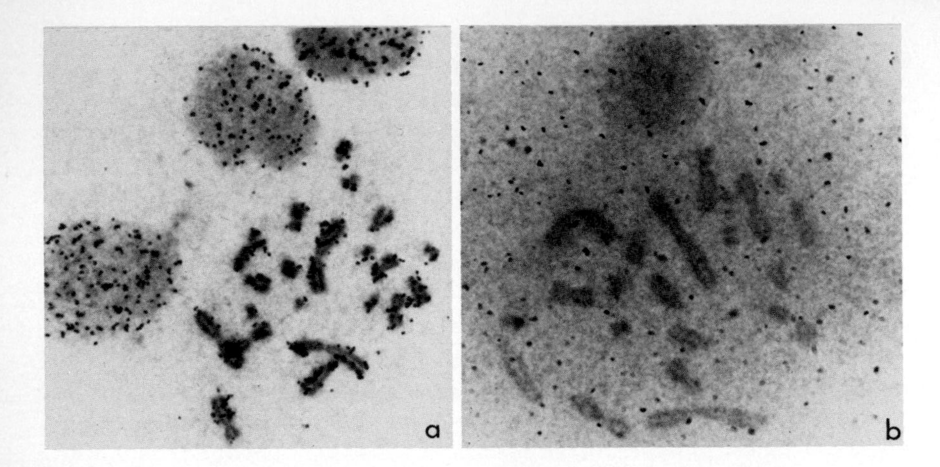

FIG. 2. Loss of DNA following C-band treatment. CHO cells were synchronized by mitotic selection, labeled with thymidine-^{3}H (0.1 μCi/ml) for one S period, and collected at the next mitosis. (a) Autoradiograph of chromosomes and nuclei without treatment. (b) Autoradiograph of chromosomes and nuclei after C-band treatment.

banding) should be treated with the relatively mild G- or Q-band methods first.

B. G-Bands

By far the most spectacular and informative patterns have been obtained by the Giemsa or G-banding techniques which we and others have segregated from hybridization-oriented procedures on the basis of putative mechanisms for band development. The early contention of Drets and Shaw (1971) and of Schnedl (1971) that the critical pattern development step was the result of DNA denaturation and renaturation does not appear to be consistent with more recent observations of conditions that produce similar but much higher resolution of banding patterns. These more recent procedures involve much milder conditions and are more consistent with notions which have developed concerning organization of chromosomal proteins. Ord and Stocken (1966) demonstrated that the highest proportion of disulfide bonds occurs in isolated, dense chromatin. In interphase chromatin, histone f3, which contains cysteine residues, occurs mainly in the reduced monomeric form. However, in metaphase chromatin, it is polymerized or complexed with acid-soluble nonhistone proteins by interpolypeptide disulfide bonds (Sadgopal and Bonner, 1970). Marushige and Bonner (1971) more recently have presented evidence for differences in histone binding to template-active and template-inactive DNA. These observations suggest an alternative view: namely, that high resolution G-banding is the result of peptide and/or disulfide bond cleavage of chromosomal proteins and has little to do with DNA denaturation.

In principle, treatment of chromosomes with proteolytic enzymes represents perhaps the least complicated banding technique, and early studies (Dutrillaux *et al.*, 1971; Seabright, 1971; Wang and Federoff, 1972) have indicated the appearance of banding which could be revealed either by Leishman or Giemsa staining. However, in our hands, the conditions for trypsinization were extremely sensitive and led to considerable variability in pattern from cell to cell within the same preparation.

The method outlined here has great latitude and reproducibility and appears to work with comparable results in several laboratories where it has been tried. It is a modification of the Seabright (1971) digestion technique (Deaven and Petersen, 1973) which involves trypsinization at 0°C. If hydrolysis is carried out for 3 minutes, bands appear reproducibly on all chromosomes of a slide preparation. Figure 3 is a typical preparation obtained by this technique from mitotically synchronized material (Petersen *et al.*, 1968). The exact procedure is as follows:

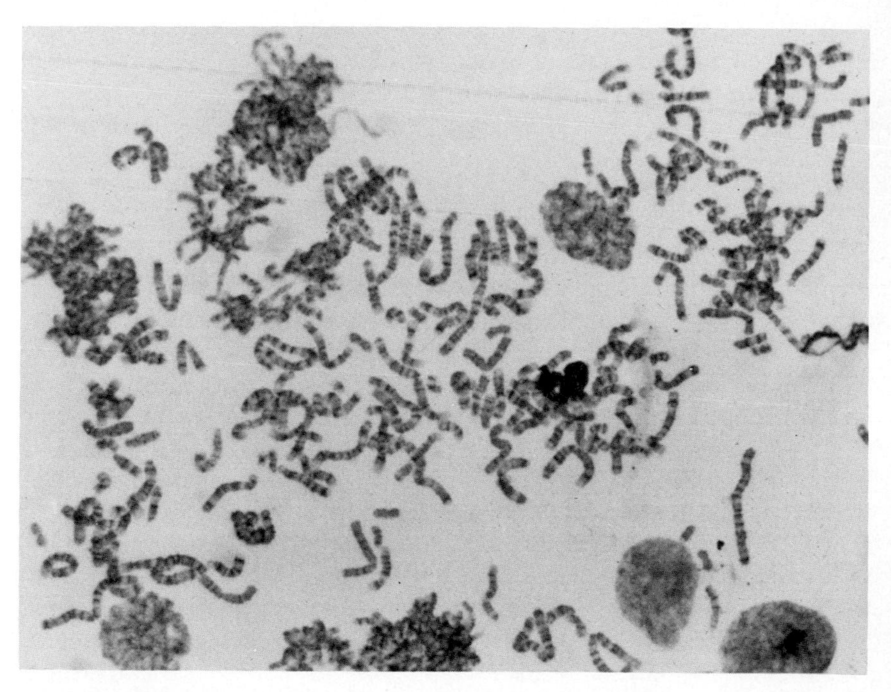

FIG. 3. Consistency of G-banding in isolated Chinese hamster chromosomes following low-temperature trypsin treatment.

1. Slide Preparation

Cell monolayers are blocked with Colcemid (0.06 μg/ml) for 2 hours; at this time, they are harvested by trypsinization and pelleted by slow-speed centrifugation. The pellet is thoroughly resuspended in 1 ml of medium with a Pasteur pipette. Distilled water (3 ml) is added to this suspension, and the cells are allowed to swell in the hypotonic solution for 7 minutes at room temperature. The first fixation is accomplished by adding 1 ml of freshly prepared 3:1 methanol–acetic acid to the hypotonic solution. This step avoids cell clumping upon subsequent centrifugation. The cells are then gently pelleted and fixed by repeated dispersion at 10-minute intervals in three changes of 3:1 methanol–acetic acid. After the third fixation, the cells are resuspended in approximately 1 ml of fixative solution (more or less, depending on size of the pellet). Several drops of this final suspension are applied to detergent-washed slides which were stored in distilled H_2O at 4°C. The backs of the slides are dried by wiping them with a tissue, but a film of H_2O is left on the upper surface to facilitate spreading of the cells. More usable spreads are found routinely on slides washed with detergent and water than on those precleaned in alcohol or acetone. The same procedure is used for lymphocyte cultures with two exceptions. The Colcemid dose should be reduced to 0.03 μg/ml to prevent excessive chromosomal condensation, and a more drastic hypotonic treatment must be used to swell the cells. We have had excellent results by using either 1 part medium to 3 parts distilled H_2O for 20 minutes at 37°C or hypotonic KCl (0.075 *M*) for 20 minutes at 37°C.

2. Banding Procedure

The air-dried slides are then washed twice in 95% EtOH and once in normal saline and are immersed in the trypsin solution held in an ice-water bath at 0°C. After a 3-minute trypsin hydrolysis at 0°C, the slides are rapidly washed twice in 70%, 95%, and absolute ethanol to stop the trypsin digestion. At this point, the slides may be air-dried for storage or washed in normal saline and stained immediately. The trypsin solution consists of 8 g NaCl, 0.2 g KCl, 0.2 g KH_2PO_4, 1.5 g Na_2HPO_4, 0.1 g $CaCl_2 \cdot 2H_2O$, 0.1 g $MgCl_2 \cdot 6 H_2O$, and 2.5 g trypsin (Difco 1:250) made up to 1 liter with distilled water. We have found that a 3-minute hydrolysis time is optimal for a variety of mammalian cell lines but, because the trypsin is crude and may vary slightly from one batch to another, we have arranged a series of pictures (Fig. 4) which show the appearance of undertreated and overtreated chromosomes. Chromosomes treated for 1 minute (Fig. 4a) remain practically unchanged; at 2 minutes (Fig. 4b) the bands are apparent but not as clear and distinct as is possible. At 3 minutes (Fig. 4c) the bands are well differentiated, and at 4 minutes Fig. 4d) they begin to appear fuzzy. This fuzzy appearance is more

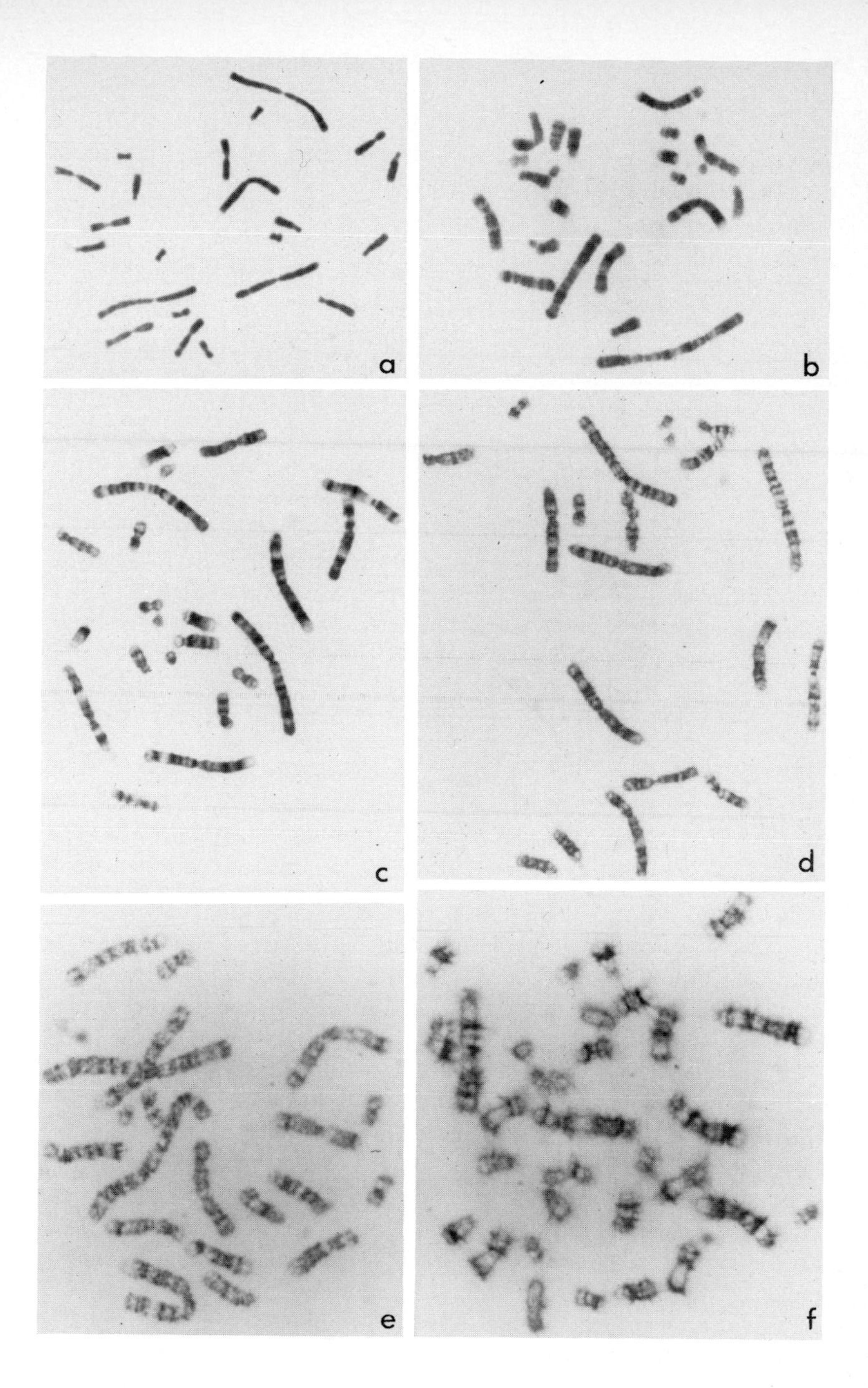

FIG. 4. The influence of hydrolysis time on quality of G-band patterns in the chromosomes of Chinese hamster line V79 (see text).

apparent at 5 minutes (Fig. 4e), and at 10 minutes (Fig. 4f) much of the dark staining material has sloughed away. The staining solution for G-bands is the same as for C-bands (5 ml of stock Giemsa diluted with 50 ml 0.01 M PO_4 buffer, pH 6.8). Chromosomes showing C-bands are stained for 10 minutes, but the staining time for G-bands must be reduced to 2 or 3 minutes. Prolonged staining times for G-banded material may result in fuzzy band patterns, as shown in Fig. 4d. The slides are then rinsed in 0.01 M PO_4 buffer (pH 6.8) and air-dried to be mounted with DePex (George T. Gurr, Ltd.) or Permount. This technique has evolved as the most reproducible from a large series of experiments employing both populations synchronized by mitotic selection and populations accumulated from cultures blocked with low doses of Colcemid (0.03–0.06 μg/ml) for periods not exceeding 2 hours. In general, higher Colcemid concentrations or longer blockade times yield preparations where condensation is excessive and banding patterns are too fused to be informative (Fig. 5). In terms of reproducibility, two conclusions can be drawn. First, prometaphase and early metaphase chromosomes possess the most useful potential banding patterns, and, second, although several other methods used in this laboratory result in similar banding patterns (Fig. 6), cold trypsin develops the bands with higher quality and more consistently than any of the other techniques.

III. Flow Microfluorometric Analysis

The ability to examine the DNA content of individual cells rapidly and conveniently permits contemplation and execution of experiments which otherwise would not be technically feasible. The technique for staining and measuring the DNA content of cells in a flowing stream has been reported by Van Dilla *et al.* (1969), and application to problems of biological interest has been a major preoccupation of this laboratory for the past 18 months (Kraemer *et al.*, 1972). In addition to questions of genome content reflecting acquisition of DNA during malignant transformation, data exhibiting the distribution of DNA in a cycling population immediately suggest the possibility of rapid and convenient life-cycle analysis. The only constraint is the availability of sufficient material to perform simultaneously both the experiment at hand and the life-cycle analysis. Figure 7a shows a typical distribution of DNA contents in exponentially growing populations of Chinese hamster cells where all cell-cycle phases are represented. The spectrum consists of two peaks representing G_1 and $G_2 + M$ with an intervening population representing the various degrees of completion of DNA replication by cells in S. A computer program which fits two normal distribution curves to the G_1 and $G_2 + M$ populations and a second-order

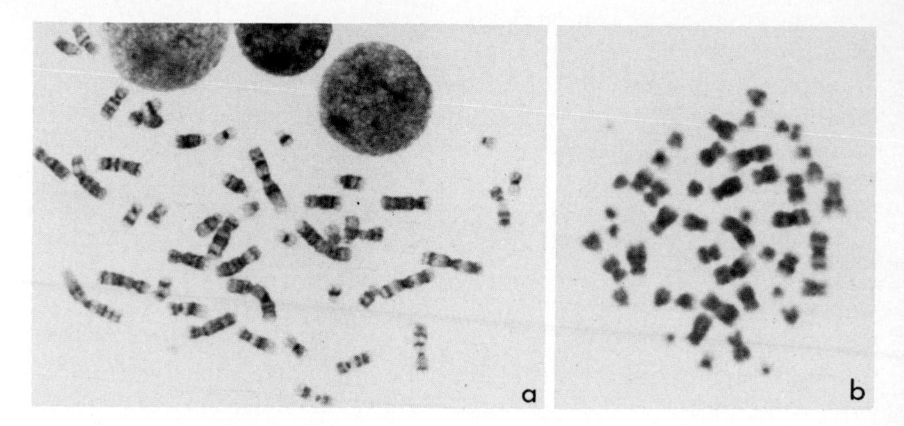

FIG. 5. Effect of prolonged or high Colcemid treatment on resolution of the G-band patterns of human lymphocytes. (a) Lymphocyte blocked in Colcemid (0.03 μg/ml) for 1 hour. (b) Lymphocyte blocked in Colcemid (0.03 μg/ml) for 3 hours.

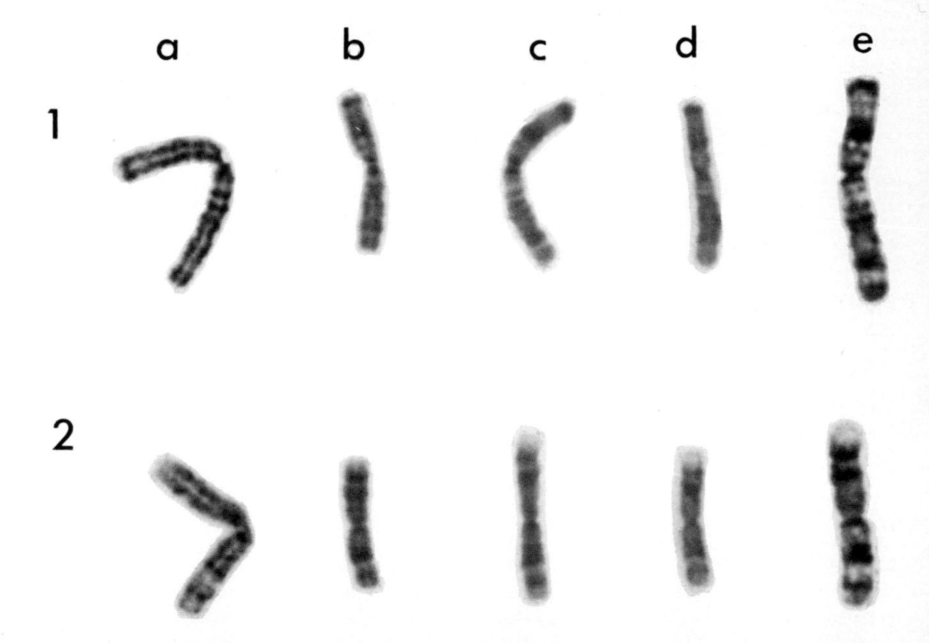

FIG. 6. Chromosomes 1 and 2 from Chinese hamster cells banded with five different methods: (a) Alkaline-saline-Giemsa procedure (Schnedl, 1971). (b) Acid-saline-Giemsa procedure (Sumner *et al.*, 1971). (c) Protein splitting solution (modified from Smithies, 1959) known to unfold proteins and to reduce their disulfide bonds. Formula: 48 gm of urea, 20 ml of borate buffer (0.5 *M* H_3BO_3, 0.2 *M* NaOH), 10 ml of 1% dithiothreitol, and 60 ml of distilled H_2O, pH 9.3. The slides were treated for 3–5 seconds, washed twice in 70%, 95%, and absolute EtOH, air-dried, and stained with Giemsa. (d) $KMnO_4$ procedure (Utakoji, 1972). (e) Cold trypsin procedure.

polynomial to the S population yields an estimate of the fraction of the population in each of the major cycle phases (Fig. 7b) (Crissman *et al.*, (1973).

Since the initial description (Van Dilla *et al.*, 1969), both the instrumentation and methods for sample preparation have undergone extensive modification. The current version of the flow microfluorometer has been described (Holm. and Cram, 1973), and an extensive discussion of sample preparation and staining techniques will be published by Crissman *et al.*, (1973). A brief description of the procedure is outlined here primarily to emphasize potential sources of error, but the reader is urged to consult the more detailed accounts (Crissman *et al.*, 1973; Holm and Cram, 1973) for construction, operation, and sample perparation procedures. For the present work, cells were isolated from suspension culture or from monolayers by trypsinization under conditions which provide very uniform, monodisperse populations.

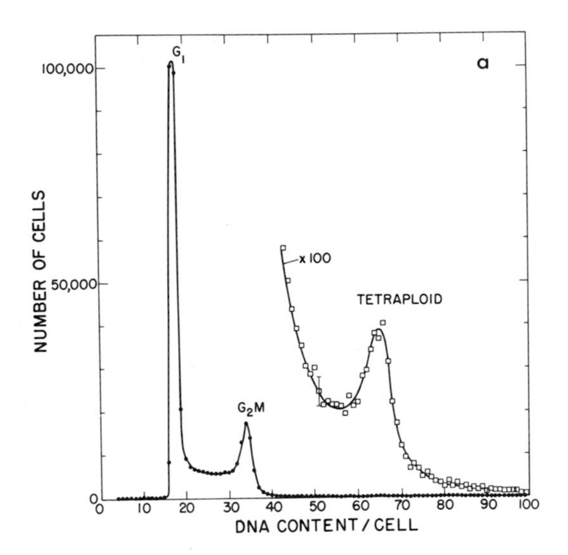

FIG. 7. Flow microfluorometric analysis of exponentially growing Chinese hamster cells (CHO). CHO cells were dispersed, fixed, and stained with acriflavine. (a) Distribution of population showing distinct G_1 and G_2 + M peaks with intervening S population. (b) Normal distributions for the G_1 and G_2 + M peaks and the second-order polynomial derived by computer fit of the input data. The program provides estimates of fraction of the population in each of the cycle phases in a small fraction of the time required for conventional life-cycle analysis and yields comparable results. ●, Experimental data points; —— computer fit; – – –, computer fit, S distribution.

	Area (%)	CV (%)	Mode
G_1 peak	59.8	3.3	36.2
S distribution	31.9	—	—
G_2M peak	8.3	3.2	70.6

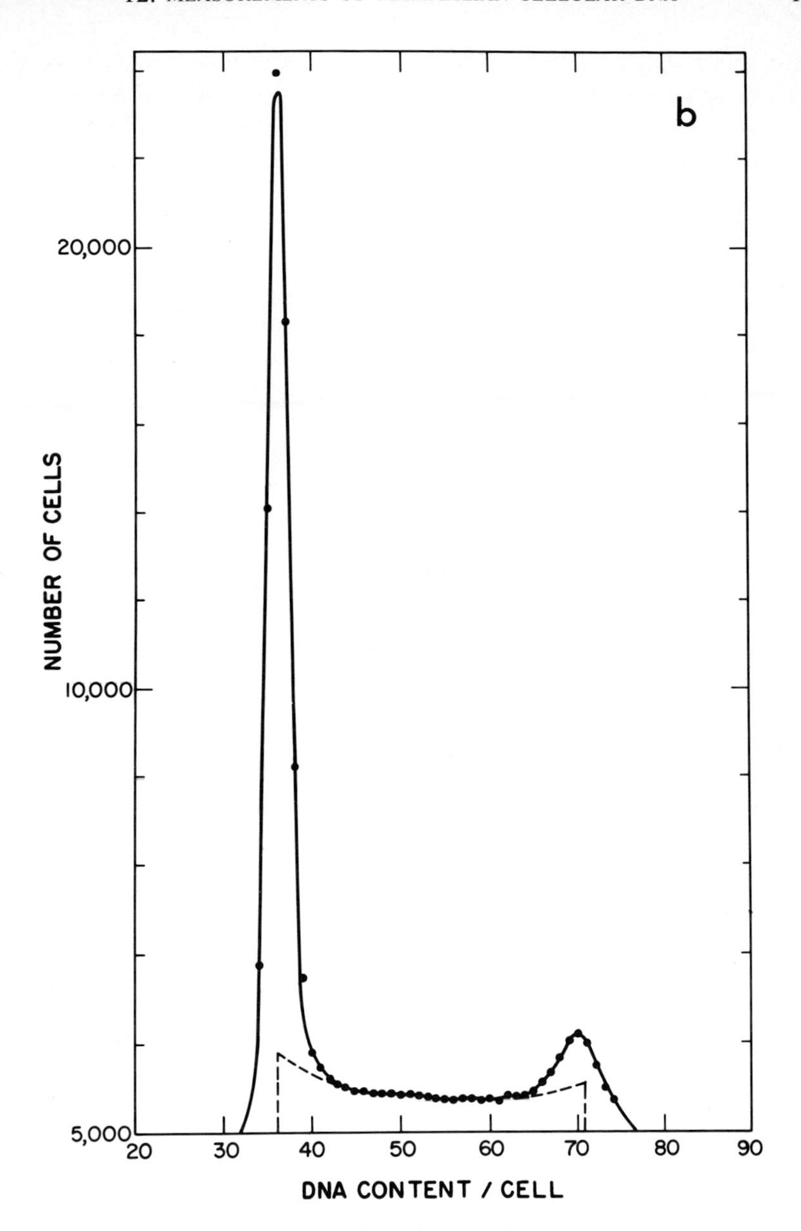

A. Cell Fixation Protocol

The procedure (Kraemer *et al.*, 1971, 1972) is applicable without modification to a large number of cell lines, and, although shortcuts can be taken with many lines, the standard procedure is less time-consuming than a systematic search for minimum protocols for individual lines. For example, in many cases, DNase can be omitted, but, because it is essential to obtain monodisperse suspensions from several lines, it is routinely used.

1. Solutions. Stock solution I (gm/liter): Glucose, 1.1; NaCl, 8.0; KCl, 0.4; $Na_2HPO_4 \cdot 12\ H_2O$, 0.39; and KH_2PO_4, 0.5. Stock solution II (gm/liter); $MgSO_4 \cdot 7\ H_2O$, 1.54; $CaCl_2 \cdot 2\ H_2O$, 0.16. Phenol red, saturated solution, approximately 1 gm/liter in saline; Puck's saline G, 900 ml of solution I plus 100 ml of solution II plus 1 ml of phenol red; Puck's saline GM, 1000 ml of solution I plus 1 ml of phenol red. Saline G and GM can be made conveniently in 5- to 10-liter volumes, sterilized by filtration, and stored at 4°C (yeasts and molds will stain and give spurious FMF signals).

2. Trypsin stock. Worthington 3 times crystalline, lyophilized, sterile 10 mg/ml in 0.001 *N* HCl (stable at 0°C).

3. Dispersing solution. Saline GM, 5 m*M* EDTA, 0.1 mg/ml trypsin.

4. Neutralization solution. Saline G containing 0.01 mg/ml DNase (Worthington), 0.2 mg/ml soybean trypsin inhibitor (Worthington), and 1 mg/ml bovine serum albumin (source not critical).

5. Fixing solution. Formalin (40%), 20 ml, q.s. 100 ml of saline G. Readjust to pH 7.0 (salmon pink) and store at 4°C.

Best results are obtained from sufficient cells to make a visible pellet in a 40-ml centrifuge cone (approximately 2 to 5×10^6 cells) from which a number of replicate measurements can then be made. Suspension cells are isolated from growth medium by low-speed centrifugation and from monolayers by carefully decanting and gently rinsing 3 times with cold saline GM. Dispersing solution (10 ml) is added, and cells are incubated for 20 minutes at 37°C. Detachment occurs much sooner, but prolonged treatment is useful in removing significant amounts of debris. The trypsinized suspension is drawn several times through a serological pipette, and 10 ml of the neutralization solution are added. The cell suspension is sedimented at low speed in a 40-ml centrifuge cone, and the neutralized solution is aspirated from the pellet. The pellet is then resuspended in 10 ml of saline G and drawn several more times through the pipette, and 10 ml of fix are added. Cells are fixed overnight at 4°C.

B. Staining Protocol

Fixed cells are pelleted at low speed, and the fixing solution is aspirated and discarded. Excess formalin is removed by suspension and recentrifugation in 2 washes with 15 ml of distilled water. Hydrolysis is accomplished by resuspending the pellet in 1 ml of the second H_2O wash, adding 6 ml of 4 *N* HCl, and incubating for 20 minutes at room temperature.

The staining solution consists of 60 ml of distilled water, 300 mg of potassium metabisulfite, and 18 mg of acriflavine or flavophosphine N. After combining the above, 6 ml of 0.5 *N* HCl are added, and the solution is filtered through a Millipore filter (~5 μm).

Hydrolyzed cells are pelleted at low speed, washed once with water, and repelleted. The wash is removed by aspiration except for approximately 0.5 ml, the cells are resuspended, and 7 ml of stain are added. After 20 minutes, the stained cells are pelleted, the stain is removed, and the pellet is resuspended in 10 ml of acid alcohol (1 ml concentrated HCl added to 99 ml of 70% ethanol). The cells are washed for 5 minutes in each of 3 changes of acid alcohol to remove nonspecific cytoplasmic fluorescence and are resuspended from the final pellet in distilled water. The water suspension of cells is then measured in the flow microfluorometer.

C. Sources of Error

This procedure yields coefficients of variation (CV in percent = standard deviation × 100/mean) of DNA contents of individual cells of comparable biochemical age which range from 3 to 5%. The data obtained are sufficiently precise to permit conclusions concerning DNA content and consequently cycle position for individual cells of a population, and the sources of error, both instrumental and biological, are not large enough to allow alternative explanations for the primary observation [namely, that the amount of DNA in the individual cells of a population is strikingly uniform (Kraemer *et al.*, 1971)]. However, under different conditions, the procedure can result in cells with considerably more fluorescence which is largely cytoplasmic and tends to increase the CV and to smear the apparent structure of the DNA distribution within a population.

In their original description of the acriflavine-Schiff procedure, Culling and Vassar (1961) cautioned against nonspecific cytoplasmic fluorescence and recommended acid alcohol for removal of nospecific stain. The original Feulgen procedure employing basic fuchsin traditionally used sulfurous acid for decoloration of nonspecific stain but, in the case of fluorescent Feulgen reagents such as acriflavine and flavophosphine N, the bisulfite–dye complex remains fluorescent and leads to a high nonspecific background signal. Acriflavine-Feulgen stained HeLa cells shown in Fig. 8 serve as an example of the potential error resulting from failure to remove nonspecific cytoplasmic stain. Thus, in any fluorescent staining procedure, microscopic examination for cytoplasmic staining is an essential feature, and conditions must be established for removal of nonspecific fluorescence.

The second major contributor to signal broadening is a slow change in staining intensity with time which results in a 5–10% decrement in signal intensity with attendant broadening over a period of 24 hours. This source of error can be avoided if fluorescence measurements are made within 1 or

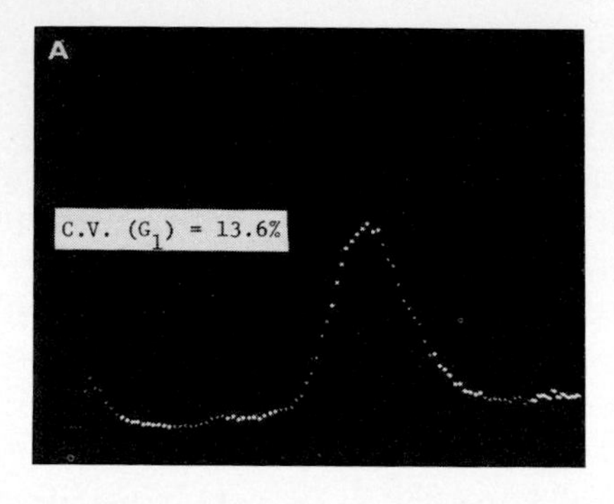

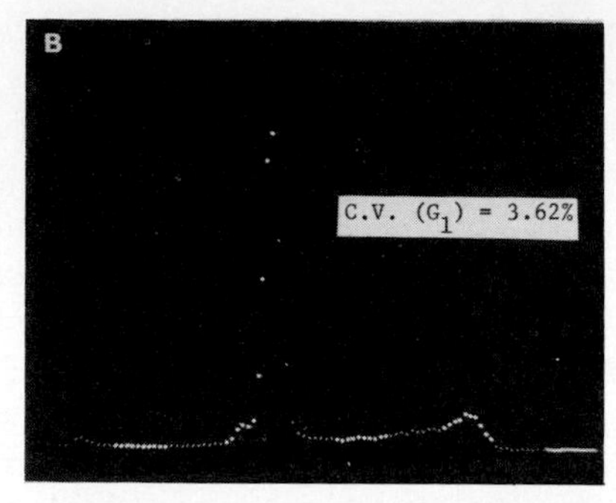

FIG. 8. Nonspecific cytoplasmic fluorescence of HeLa cells. The FMF spectra were obtained from identical HeLa cell preparations. (a) The acriflavine-Feulgen stained population containing significant nonspecific cytoplasmic fluorescence confirmed by fluorescence microscopy. Note the signal broadening and upward shift of the modal channel of the G_1 peak. (b) The same HeLa population after washing with acid–alcohol to remove nonspecific cytoplasmic fluorescence. The comparison shows the desirability of routine examination by fluorescence microscopy to ensure uniformity and specificity of the staining procedure.

2 hours after staining. These changes, which are best interpreted as increased variability of stain intensity from cell to cell, are poorly understood and must be avoided.

A third source of signal broadening appears to stem from prolonged fixation. The signal obtained from material stained several months after fixation is uniformly broader than that obtained from freshly fixed material, and the existing protocol calls for overnight fixation, staining, and measurement within 2 hours.

IV. Applications of Methods to Cells in Culture

Attempts to investigate differences between euploid and heteroploid cells inevitably lead to questions of changes in genome content, biochemical regulation, cell-cycle time, and response to external influences (for example, ionizing radiations or chemotherapeutic agents). The methodology outlined in this chapter offers unique advantages in approaching these questions, as illustrated by the following examples.

A. Colcemid-Induced Nondisjunction

Cox and Puck (1969) had as their major objective the production of cell lines either monosomic or trisomic for specific chromosomes and argued that monosomy was essential for effective use of mutagenic agents. They treated "wild type" CHO cells with low doses of Colcemid, which permitted cell division with a considerable number of nondisjunctive errors. Isolated clones from Colcemid-treated populations exposed to mutagens were predicted to be useful but were found unnecessary for isolation of mutant populations reported in subsequent studies (Kao and Puck, 1971, 1972). A similar study was reported by Kato and Yosida (1970) which demonstrated that, after reversal of Colcemid arrest, the degree of nondisjunction was dependent upon the time in Colcemid and that resulting cells with either high or low chromosome numbers displayed a marked decrease in viability. Quasi-stable aneusomies invariably involved the small chromosomes 9, 10, or 11; all others died or underwent remodeling within 3 to 4 mitoses (Kato and Yosida, 1971). However, their karyologic demonstration in hamster cells of a nondisjunctive defect of increasing severity with increasing exposure to Colcemid suggested a useful method for validating the fluorescence estimates of DNA content. From cells synchronized by Colcemid reversal (Stubblefield *et al.*, 1967) and stained with acriflavine for FMF determination of DNA content, the results shown in Fig. 9 were obtained. It is clear from these data that the microfluorometric estimate of genome content not only accurately reflected Colcemid-induced nondisjunction, but did so with a precision suggesting the gain or loss of single chromosomes. It is noteworthy that broadening of the curve was symmetrical, indicating that both deprived and overendowed cells existed in the population in equal numbers at the first division and that nondisjunction occurred randomly within the karyotype. However, over several succeeding generations, the cultures tended to revert to their original stable state with the dominant chromosome number 21. Thus, death resulting from loss of genome is not immediate, but defective distribution of genome generally leads to growth disadvantage and/or lethality.

B. Analysis of the CHO Genome

Kao and Puck (1969) estimated that line CHO contained 3% less DNA than euploid hamster chromosomes on the basis of arm-length measurements. An independent measurement employing a fluorescent Feulgen staining procedure yielded the data shown in Table I, which indicate a 4% decrement in genome (Deaven and Petersen, 1973). Thus, the two estimates are in excellent agreement and indicate that most of the DNA present in the euploid parent is present in the aneuploid derivative. Figure 10 shows the difference between line CHO and LA-CHE ♀ (Chinese hamster euploid female cell strain). The G_1 mode has shifted down in CHO by an amount equivalent to 4% of the genome. In arranging the CHO karyotype, Kao and Puck (1969) recognized a number of abnormalities which they characterized by assigning the 21 chromosomes to two groups. Twelve

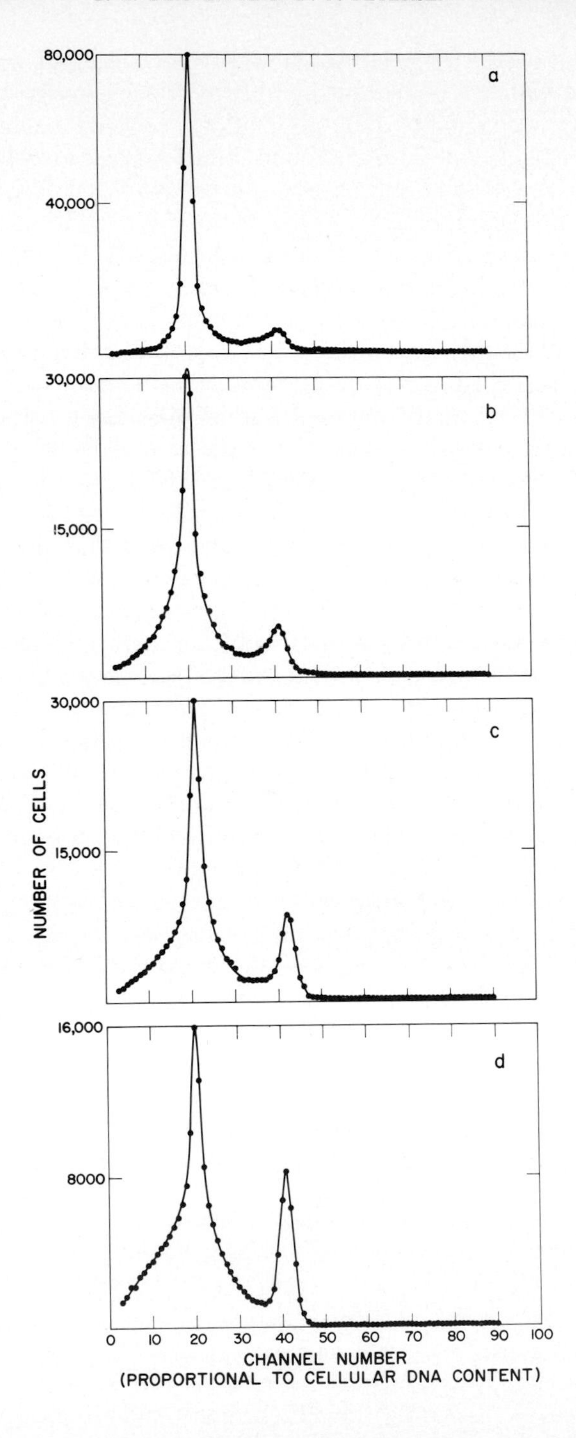
80,000
40,000
a
30,000
15,000
b
30,000
15,000
c
16,000
8000
d
NUMBER OF CELLS
0 10 20 30 40 50 60 70 80 90 100
CHANNEL NUMBER
(PROPORTIONAL TO CELLULAR DNA CONTENT)

TABLE I

CHROMOSOME NUMBER AND RELATIVE DNA CONTENT PER CELL FOR CHINESE HAMSTER LINE CHO AND STRAIN LA–CHE ♀ CULTURES

Culture	Chromosome number: number of cells					Microfluorometric data	
						G_1 peak[a]	G_1 mode normalized[b]
CHO	19	20	21	22	44	21.1	0.96
	1	6	38	4	1		
LA-CHE ♀	19	20	21	22	44	21.9	1.00
			1	46	3		

[a] G_1 peak is the modal channel of DNA content for G_1 cells in relative units.

[b] LA-CHE♀ (euploid) cells were considered to have a G_1 mode of 1.00. G_1 peak channel modes for CHO cells were normalized to this value.

were considered to be normal, and 9 which bore little or no resemblance to normal chromosomes were assigned to a "Z" group.

A karyotype analyzed by the G-band technique, shown in Fig. 11, indicates that there are no homologous pairs, that only 8 of the CHO chromosomes are normal, but that most of the genome can be accounted for in a series of rearrangments. In each case where sufficient banding detail to identify a chromosomal sequence is retained, both the origin and destination of the translocated material can be specified. In some instances a single band is involved, and particularly in the smaller chromosomes the identify of the translocations is ambiguous. A complete, comparative analysis of the karyotypes of CHO and LA-CHE ♀ has been reported (Deaven and Petersen, 1973), and only the major features of rearrangement of CHO are described here. Of particular interest are a reciprocal translocation between Z-3 and Z-7, the pericentric inversion of Z-4, and the absence of all but the centromere of X_2. Taken together with analytical evidence from flow microfluorometry, these data which show the presence of most of the original euploid genome, albeit in an extensively rearranged form, lead first to the conclusion that most of the hamster genome is retained in transition from euploidy to aneuploidy and, second, examination of the G-banded chromosomes indicates that 8 of the 21 chromosomes

FIG. 9. Progressive Colcemid-induced nondisjunction in Chinese hamster cells (line CHO). Exponential cultures were blocked with Colcemid; mitotic cells were selectively dislodged, reversed in fresh medium for 4 hours, collected, and stained with the acriflavine-Feulgen technique. The first three distributions (a–c) represent mitotic cells collected from the same monolayers at 2, 4, and 6 hours after addition of Colcemid. The lower distribution (d) is from a different monolayer blocked for 6 hours continuously, shaken, and reversed.

of CHO are normal, although none of them exist as homologous pairs. However, in several rearrangements the complete chromosome complement is retained and, for example in Z-9, a small translocation of ambiguous origin is the only change. With relatively minor changes, a haploid set of chromosomes appears in CHO and leads to the conclusion that severe constraints are placed on alterations which can occur within the CHO genome compatible with extended survival. Though extensively rearranged, several sublines of CHO exhibit banding patterns with remarkable similarity and confirm and extend notions originally introduced by Zakharov *et al.* (1964) regarding the necessity for preservation of chromosomal material to maintain viability. If these notions prove to be correct, the

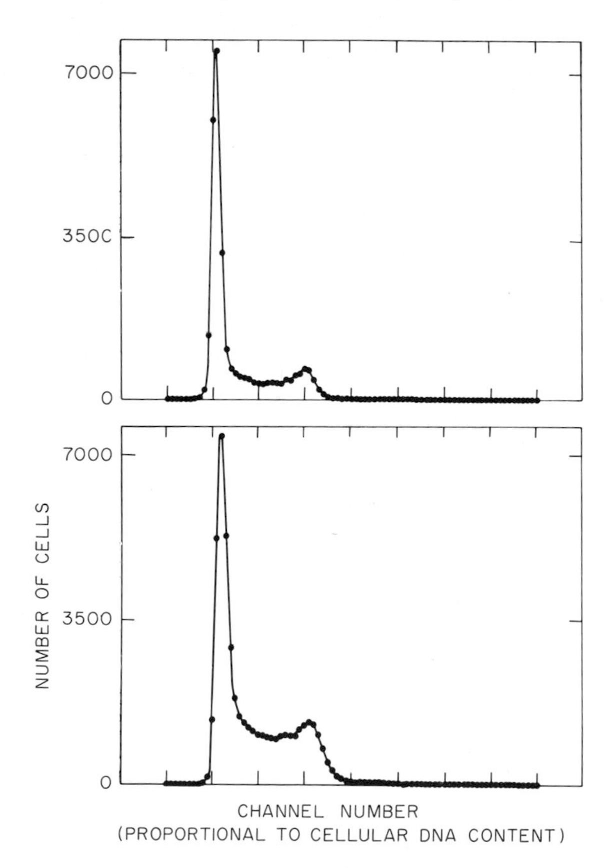

FIG. 10. FMF comparison of the relative DNA contents of CHO (upper figure) and LA-CHE ♀ (lower figure). CHO and LA-CHE ♀ cultures were exponentially grown and stained with acriflavine–Feulgen. Comparison of the normal distributions of the G_1 peaks indicates that CHO cells contain 4% less DNA than euploid Chinese hamster fibroblast LA-CHE ♀ (see text).

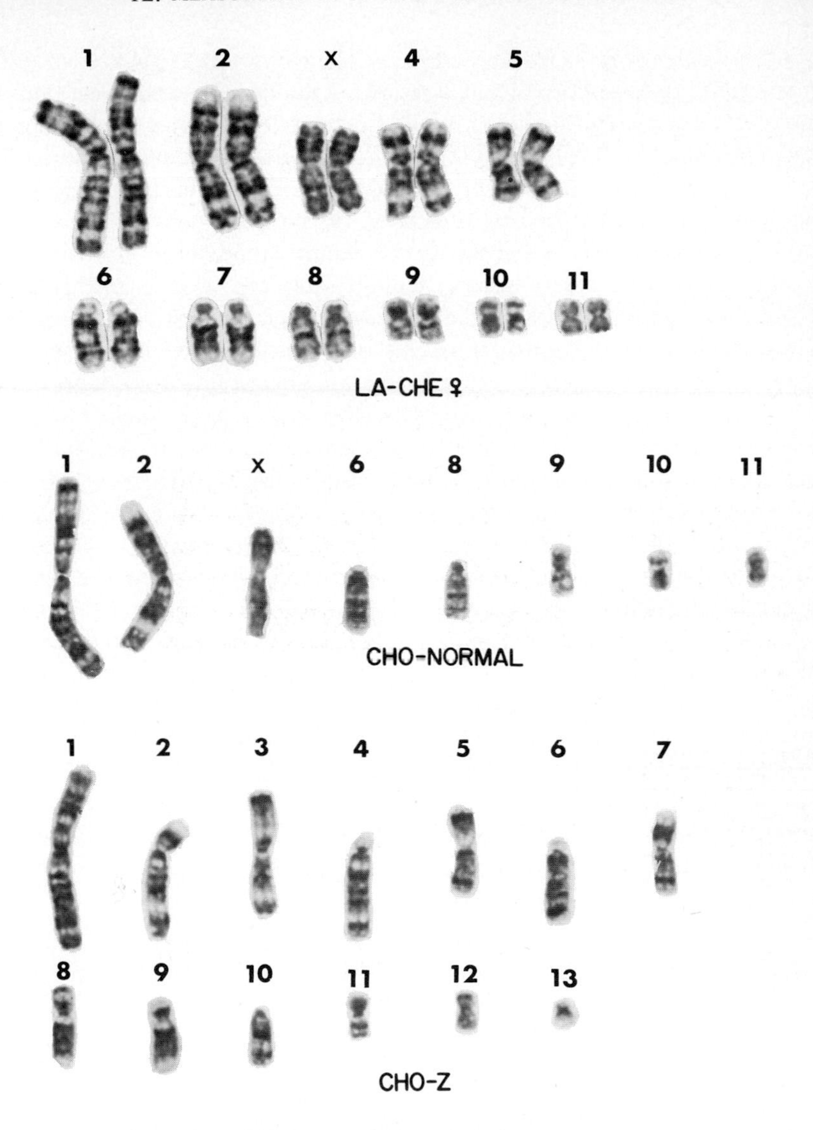

FIG. 11. LA-CHE ♀ and CHO karyotypes arranged by G-band comparison. Karyotypes demonstrate the unambiguous pairing of LA-CHE ♀ chromosomes and the absence of homologous pairs in CHO.

most likely implication is that aneuploid cells preserve genetically essential information in their rearranged chromosomes. However, an alternative idea contends that CHO cells are genetically stable despite their precarious position as operational haploids (i.e., extensively altered chromosomes are genetically inactive). Although no evidence except the experimental observation of remarkable stability of the karyotype exists, it is attractive to speculate that functional genes reside in the normal haploid set (Deaven and Petersen, 1972). This idea would account for the spectacular success Kao and Puck (1970, 1971, 1972) and others (Thomson and Stanners, 1972) have enjoyed in deriving nutritional and temperature mutants. The hypothesis is subject to experimental testing and predicts that CHO should be exquisitely sensitive to Colcemid nondisjunction, and more detailed studies of the type reported by Kato and Yosida (1971) should reveal that both deprived and endowed daughters rapidly disappear from a Colcemid-treated population. Second, in mutant or transformed populations, many though perhaps not all phenotypic revertants should have increased chromosome numbers. Rabinowitz and Sachs (1970) and Pollack *et al.* (1970) have reported revertants in Syrian hamster and mouse lines in which reacquisition of contact inhibition occurred exclusively in cells with increased numbers of chromosomes. However, Scher and Nelson-Rees (1971) contend that increased ploidy is not a prerequisite for phenotypic reversion in "flat" cells which continue to synthesize DNA at confluence. Finally, to establish such a notion convincingly, studies similar to those outlined here must be extended to include high heteroploid lines to determine the extent of rearrangement and number of homologies as well as their karyotypic stability. We know already that chromosome number is more variable than DNA content (Kraemer *et al.*, 1971, 1972), and the data for HeLa cells shown in Fig. 12 clearly show that cloned and wild-type HeLa cells contain identical amounts of DNA. If we now compare banding patterns, we may be able to explain the apparent paradox of chro-

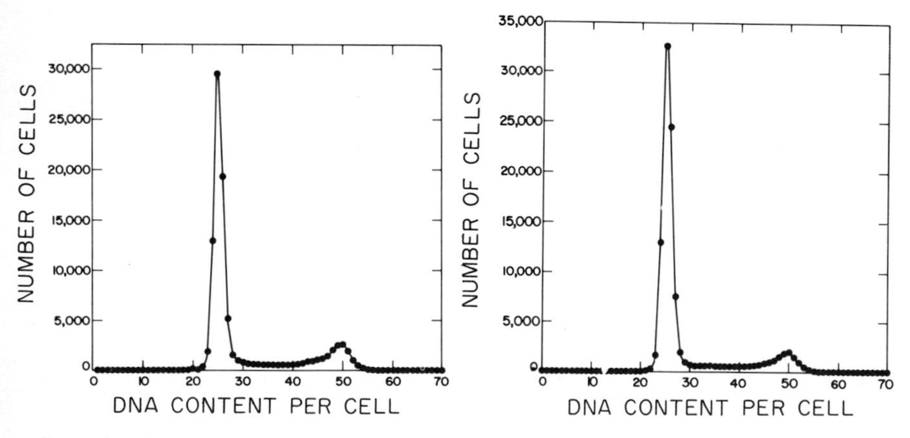

FIG. 12. DNA distribution of (a) wild-type HeLA cells and (b) a derived clone. There is identical DNA distribution in a "wild type" and a cloned population of HeLa cells. At the time the acriflavine-stained cells were measured, chromosome number in the cloned cells was almost constant (66–68), while that of the parent culture was typically more variable.

mosome variability despite DNA constancy and to examine critically the observation that malignant cells appear to acquire genome in haploid sets rather then a chromosome at a time by nondisjunctive mechanisms.

V. Summary

Two independent methods have been described in detail which permit investigation of the DNA content of mammalian cells. High-speed microfluorometric measurements allow collection of larger numbers of individual cellular DNA content measurements. The technique permits rapid, convenient cell-cycle analysis, quantitative estimates of the DNA content of euploid and aneuploid cells, and has already demonstrated a homogeneity of DNA content inconsistent with the observed variability of chromosome number in heteroploid lines. C-banding and G-banding techniques have been described which demonstrate extensive preservation of chromosomal segments in aneuploid chromosomes. Extended cell survival appears to be the result of strong selective pressures for genome preservation, and approximately haploid sets of normal chromosomes may be a minimum requirement for survival.

Acknowledgments

The authors wish to express their thanks to Mrs. P. C. Sanders and Mrs. J. Grilly for their excellent technical and photographic assistance.

References

Arrighi, F. E., and Hsu, T. C. (1971). *Cytogenetics* **10**, 81–86.
Arrighi, F. E., Hsu, T. C., Saunders, P., and Saunders, G. F. (1970). *Chromosoma* **32**, 224–236.
Atkin, H. B., and Richards, B. M. (1956). *Brit. J. Cancer* **10**, 769–786.
Caspersson, T., Farber, S., Foley, G. E., Kudynowski, J., Modest, F. E., Simonsson, E., Wagh, U., and Zech, L. (1967). *Exp. Cell Res.* **49**, 219–222.
Caspersson, T., Zech, L., and Johansson, C. (1970). *Exp. Cell Res.* **62**, 490–492.
Chu, E. H. Y., and Giles, N. H. (1958). *J. Nat. Cancer Inst.* **20**, 383–401.
Comings, D. E., Avelino, E., Okada, T. A., and Wyandt, H. E. (1972). *J. Cell Biol.* **55**, 48a.
Cooper, J. E., and Hsu, T. C. (1972). *Cytogenetics* **11**, 295–304.
Cox, D. M., and Puck, T. T. (1969). *Cytogenetics* **8**, 158–169.
Crissman, H. A., Van Dilla, M. A., and Petersen, D. F. (1973). *Exp. Cell Res.* (submitted for publication).
Culling, C., and Vassar, P. (1961). *Arch. Pathol.* **71**, 88/76–92/80.
Deaven, L. L., and Petersen, D. F. (1972). *J. Cell Biol.* **55**, 57a.
Deaven, L. L., and Petersen, D. F. (1973). *Chromosoma* 129–144.
Drets, M. E., and Shaw, M. W. (1971). *Proc. Nat. Acad. Sci. U.S.* **68**, 2073–2077.
Dutrillaux, B., DeGrouchy, J., Finaz, C., and Lejeune, J. (1971). *C. R. Acad. Sci.* **273**, 587–588.

Hamerton, J. L., Jacobs, P. A., and Klinger, H. P. (1972). *Cytogenetics* **11**, 313–362.
Holm, D. M., and Cram, L. S. (1973). *Exp. Cell Res.* **80**, 105–110.
Hsu, T. C., and Arrighi, F. E. (1971). *Chromosoma* **34**, 243–253.
Hsu, T. C., and Klatt, O. (1958). *J. Nat. Cancer Inst.* **21**, 437–473.
James, J. (1965). *Cytogenetics* **4**, 19–27.
Kao, F. T., and Puck, T. T. (1969). *J. Cell. Physiol.* **74**, 245–257.
Kao, F. T., and Puck, T. T. (1970). *Nature* (*London*) **228**, 329–332.
Kao, F. T., and Puck, T. T. (1971). *J. Cell. Physiol.* **78**, 139–144.
Kao, F. T., and Puck, T. T. (1972). *J. Cell. Physiol.* **80**, 41–49.
Kato, H., and Yosida, T. H. (1970). *Exp. Cell Res.* **60**, 459–464.
Kato, H., and Yosida, T. H. (1971). *Cytogenetics* **10**, 392–403.
Kato, H., and Yosida, T. H. (1972). *Chromosoma* **36**, 272–280.
Kraemer, P. M., Petersen, D. F., and Van Dilla, M. A. (1971). *Science* **174**, 714–717.
Kraemer, P. M., Deaven, L. L., Crissman, H. A., and Van Dilla, M. A. (1972). *In* "Advances in Cell and Molecular Biology" (E. J. DuPraw, ed.), Vol. 2, pp. 47–108. Academic Press, New York.
Makino, S. (1957). *Int. Rev. Cytol.* **6**, 26–84.
Mandel, P., Metais, P., and Cuny, S. (1950). *C. R. Acad. Sci.* **231**, 1172–1174.
Marushige, K., and Bonner, J. (1971). *Proc. Nat. Acad. Sci. U.S.* **68**, 2941–2944.
Ord, M. G., and Stocken, L. A. (1966). *Biochem. J.* **98**, 888–897.
Pardue, M. L., and Gall, J. G. (1970). *Science* **168**, 1356–1358.
Petersen, D. F., Anderson, E. C. and Tobey, R. A. (1968). *In* "Methods in Cell Physiology" (D. M. Prescott, ed.), Vol. 3, pp. 347–370. Academic Press, New York.
Pollack, R., Wolman, S., and Vogel, A. (1970). *Nature* (*London*) **228**, 938–970.
Rabinowitz, Z., and Sachs, L. (1970). *Nature* (*London*) **225**, 136–139.
Rudkin, G. T., Hungerford, D. A., and Nowell, P. C. (1964). *Science* **144**, 1229–1232.
Sadgopal, A., and Bonner, J. (1970). *Biochim. Biophys. Acta* **207**, 227–239.
Scher, C. D., and Nelson-Rees, W. A. (1971). *Nature* (*London*), *New Biol.* **233**, 263–265.
Schnedl, W. (1971). *Chromosoma* **34**, 448–454.
Seabright, M. (1971). *Lancet* **ii**, 971–972.
Smithies, O. (1959). *Biochem. J.* **71**, 585–587.
Stubblefield, E. (1964). *In* "Cytogenetics of Cells in Culture" (R. J. C. Harris, ed.), Symp. Int. Soc. Cell Biol., Vol. 3, pp. 223–248. Academic Press, New York.
Stubblefield, E., Klevecz, R., and Deaven, L. (1967). *J. Cell. Physiol.* **69**, 345–353.
Sumner, A. T., Evans, H. J., and Buckland, R. A. (1971). *Nature* (*London*), *New Biol.* **232**, 31–32.
Thompson, L. H., and Stanners, C. P. (1972). *J. Cell Biol.* **55**, 518.
Utakoji, T. (1972). *Nature* (*London*) **239**, 168–169.
Van Dilla, M. A., Trujillo, T. T., Mullaney, P. F., and Coulter, J. R. (1969). *Science* **163**, 1213–1214.
Wang, H. C., and Fedoroff, S. (1972). *Nature* (*London*), *New Biol.* **235**, 53.
Zakharov, A. F., Kakpakova, E. S., Egolina, N. A., and Pogosiany, H. E. (1964). *J. Nat. Cancer Inst.* **33**, 935–956.

Chapter 13

Large-Scale Isolation of Nuclear Membranes from Bovine Liver

RONALD BEREZNEY

Department of Pharmacology and Experimental Therapeutics. The Johns Hopkins University School of Medicine, Baltimore, Maryland

I. Introduction

The nuclear envelope (membrane) occupies a precise location in the eukaryotic cell representing the boundary between nucleus and cytoplasm. Figure 1 schematically illustrates some important organizational properties

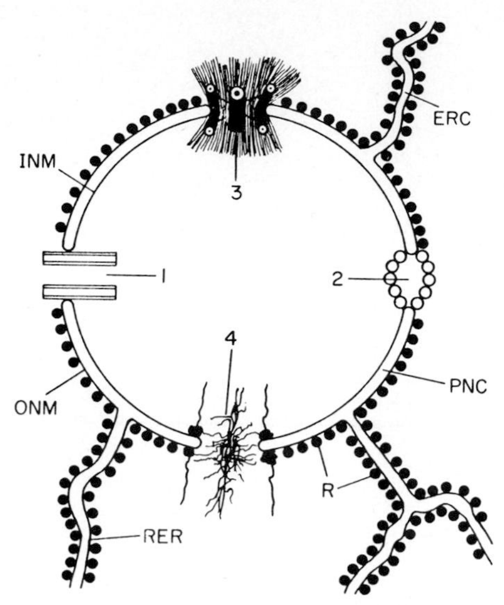

FIG. 1. Schematic representation of nuclear pore complexes, nuclear envelope, and connections with endoplasmic reticulum. The nuclear envelope is shown in cross section. ONM, outer nuclear membrane; INM, inner nuclear membrane; R, ribosome; RER, rough endoplasmic reticulum; ERC, endoplasmic reticulum cisternae; PNC, perinuclear cavity. The four models of the nuclear pore complexes are: (1) Cylinder model (Afzelius, 1955; Wischnitzer, 1958). (2) Globular model (Gall, 1954). (3) Recent cylinder model (Abelson and Smith, 1970); in this modification of the cylinder model the walls of the nuclear pore complex are tapered, octagonal, and of a filamentous nature; the cylinders are shown to be tubules by deviating from the "cross-section appearance" and giving them a three-dimensional aspect. (4) Recent globular model; a cross section of the filamentous version of Franke's model (Franke, 1970).

of the nuclear envelope. Current conceptions of the envelope emphasize that it is actually a mosaic of membranous and nonmembranous portions (Abelson and Smith, 1970).

The nonmembranous elements[1] are called nuclear pore complexes (Watson, 1959); they are believed to be organelles specialized for nucleocytoplasmic transport processes (Franke, 1970; Abelson and Smith, 1970). A number of models for the organization of the nuclear pore complex have appeared in the last 30 years (e.g., Gall, 1954, 1967; Afzelius, 1955; Wischnitzer, 1958; Yoo and Bayley, 1967; Vivier, 1967; Abelson and Smith, 1970; Franke, 1970; Franke and Scheer, 1970; Roberts and Northcote, 1970; LaCour and Wells, 1972; Engelhardt and Pusa, 1972), and this is still an active area of controversy as demonstrated by the fundamental differences of some current models (see, for example, pore models 3 and 4 of Fig. 1).

[1]The designation of nuclear pore complexes as nonmembranous elements is based solely on morphology and awaits biochemical verification.

The membranous portion of the envelope consists of an outer membrane with ribosomelike particles associated with the side facing the cytoplasm, and an inner membrane which lacks attached particles. The two membranes appear to fuse at the pore complexes, so that the outer and inner membranes most likely represent a continuous membrane cisterna. The observation that the outer membrane is connected with the rough endoplasmic reticulum led Watson (1955) to propose that the nuclear envelope is a form of endoplasmic reticulum cisterna specialized for nucleocytoplasmic transport via the pore complexes.

Procedures developed in the past four years for the mass isolation of nuclear membranes (Zbarsky *et al.*, 1969; Kashnig and Kasper, 1969; Berezney *et al.*, 1969, 1970a, 1972; Franke *et al.*, 1970; Zentgraf *et al.*, 1971; Kay *et al.*, 1971; Agutter, 1972; Price *et al.*, 1972; Moore and Wilson, 1972; Monneron *et al.*, 1972) has led to direct biochemical analyses of the isolated membranes and its comparison with endoplasmic reticulum membranes.

The procedure described here for the isolation of nuclear membranes from bovine liver is the largest preparation reported to date, making it extremely useful for nuclear membrane fractionation studies. The size of the preparation can be scaled up or down according to needs and facilities available. It is described here at a level within the means of most cellular biochemistry laboratories.

II. Large-Scale Isolation of Nuclei from Bovine Liver

A. Equipment

Model PR-2 International Centrifuge equipped with 4-liter capacity swinging bucket rotor

8 One-liter plastic bottles to fit the swinging-bucket rotor (A. H. Thomas Co.)

2 Beckman Spinco Ultracentrifuges Model L-2

2 Spinco type 21 rotors and 40 tubes and caps

Spinco type SW 25.2 rotor and tubes

Sorvall RC2-B centrifuge

Sorvall SS-34 rotor tubes

2 10-Liter plastic buckets preferably graduated and with a spout

1 Stainless steel Bucket (12 liters)

20-Liter cylindrical drum with spout

Cheesecloth

Stainless steel food strainer approximately 21 cm in diameter

Waring Blendor (single speed of 21,000 rpm with 1 liter capacity jar)
Powerstat continuous voltage regulator
Carving knives

B. Solutions

1. 20 liters of 0.25 *M* sucrose TKM (0.05 *M* Tris-HCl, pH 7.5, 0.025 *M* KCl, 0.005 *M* $MgCl_2$)
2. 4 liters of 2.3 *M* sucrose TKM
3. Preparation

a. 2.5 liters of 10-fold concentrated TKM (10 × TKM): Tris base, 151.4 gm; KCl, 46.5 gm; $MgCl_2$, 25.5 gm. Bring to pH 7.5 with HCl

b. 6 liters of 2.3 *M* sucrose TKM: Weigh out 6 times 783.3 gm of sucrose into the stainless steel bucket and add 600 ml of 10 × TKM followed by distilled water to the 6-liter mark. Heat vigorously under bunsen burners to get into solution with frequent stirring. Cool to room temperature and add distilled water to the 6 liter mark.

c. 20 liters of 0.25 *M* sucrose TKM: 10 × TKM, 1.783 liters; 2.3 *M* sucrose TKM, 2.174 liters; distilled water, 16.043 liters. Stir well

C. Procedure

1. Processing of the Liver for Homogenization

A beef liver weighing 10–15 pounds is obtained within an hour after removal from the animal. Excess blood is removed by washing with 0.25 *M* sucrose TKM. The surrounding membrane is then peeled from the surface of the liver. Slices are carved starting from the surface inward, avoiding large areas of connective tissue as much as possible, especially that associated with vessels of the hepatic portal system. The liver slices are cut into cubes of approximately [illegible] cm^3 and placed into ice-cold 0.25 *M* sucrose TKM. After collecting approximately 1600 ml, the cubes are transferred to a strainer and washed of excess blood by pouring through 0.25 *M* sucrose TKM while simultaneously stirring the liver.

2. Homogenization

The speed of homogenization with the Waring Blendor is carefully controlled with a voltage regulator such as a Powerstat. The actual speed is determined by calibration with a tachometer. One volume of liver cubes and three volumes of 0.25 *M* sucrose TKM are placed in the 1-liter jar of the blender. Effective cell disruption without destruction of cell nuclei is achieved at a speed of 12,000 rpm for 30 seconds. Occasionally, however, it is necessary to increase the speed up to 13,500 rpm or the time up to 60 seconds. It is therefore essential to always monitor the homogenization with light microscopy. Of the homogenate, 8.4 liters are passed through a stainless steel food strainer, followed by filtration through 4 and 8 layers of cheese cloth.

3. Low Speed Centrifugation

Crude nuclei pellets are isolated by differential centrifugation using the 4-liter capacity swinging-bucket rotor of the International PR-2 Centrifuge at 800 *g* for 20 minutes. The pellets are washed by resuspending in 0.25 *M* sucrose TKM to a final volume one-half of the initial volume, and centrifuged at 800 *g* for 10 minutes in the 4-liter capacity rotor.

4. High Speed Centrifugation

Nuclei are purified by centrifuging through a dense sucrose solution. The crude nuclear pellets are resuspended in 2.3 *M* sucrose TKM by vigorously shaking, and filtered through 8 layers of cheese cloth. To a final volume of 3600 ml, 2.3 *M* sucrose TKM is added; the suspension is centrifuged in the Spinco type 21 rotor for 75 minutes at 42,000 *g*. The use of two rotors speeds up this step.

The nuclei sediment to the bottom and often along one side of the centrifuge tube, while whole cells, mitochondria, endoplasmic reticulum, and other cellular membranes form a contamination band on the top of the solution and along the side directly opposite the nuclear pellet.

Nuclei are collected by inverting the tubes and removing visible contamination with a spatula and by swabbing with cotton gauze. The surface of the nuclear pellets is then rinsed several times with 0.25 *M* sucrose TKM, and the pellets are resuspended in the same solution and centrifuged at 800 *g* for 10 minutes in the Sorvall SS-34 rotor.

5. Final Purification through Dense Sucrose

Although the nuclear fraction was generally devoid of contamination as judged by morphological and biochemical parameters, the occasional presence of measurable contamination led to the development of an additional purification step which ensured reproducible purity of the isolated nuclei. This involves a discontinuous sucrose gradient as previously described by Blobel and Potter (1966) for rat liver nuclei directly purified from the intial homogenate. Two volumes of 2.3 *M* sucrose TKM are added to 1 volume of nuclear suspension (approximately 5 mg of protein per milliliter), and 40 ml are added to SW 25.2 centrifuge tubes. After underlayering with 15 ml of 2.3 *M* sucrose TKM with a 12-gauge needle, the gradients are centrifuged at 75,000 *g* for 60 minutes. The pellet sedimenting through 2.3 *M* sucrose TKM is washed once in 0.25 *M* sucrose TKM and centrifuged at 800 *g* for 10 minutes in the Sorvall SS-34 rotor.

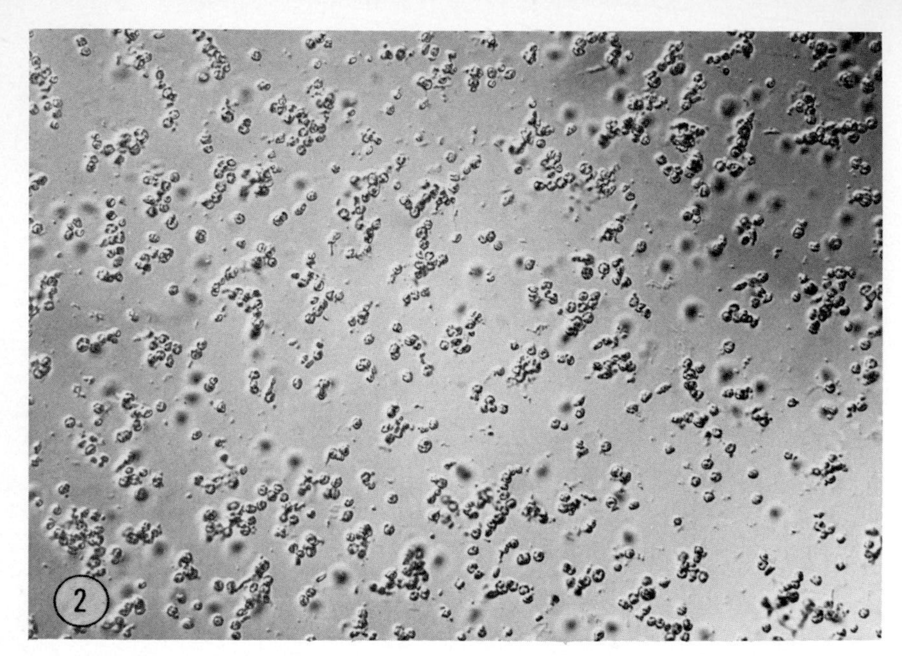

FIG. 2. Isolated bovine liver nuclei in 0.25 *M* sucrose TKM. Nomarski interference microscopy. × 80.

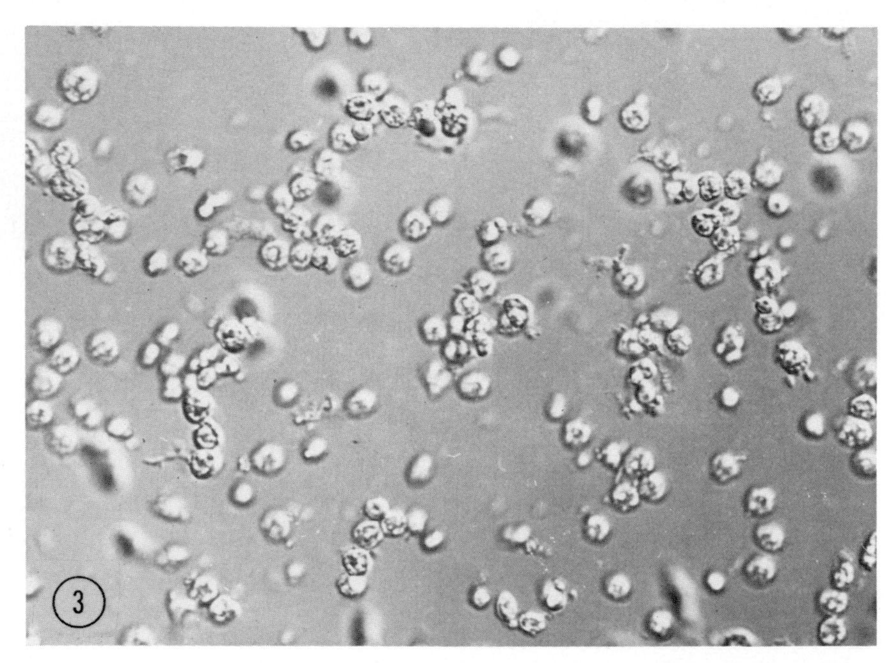

FIG. 3. Isolated bovine liver nuclei in 0.25 *M* sucrose TKM. Nomarski interference microcopy. × 512.

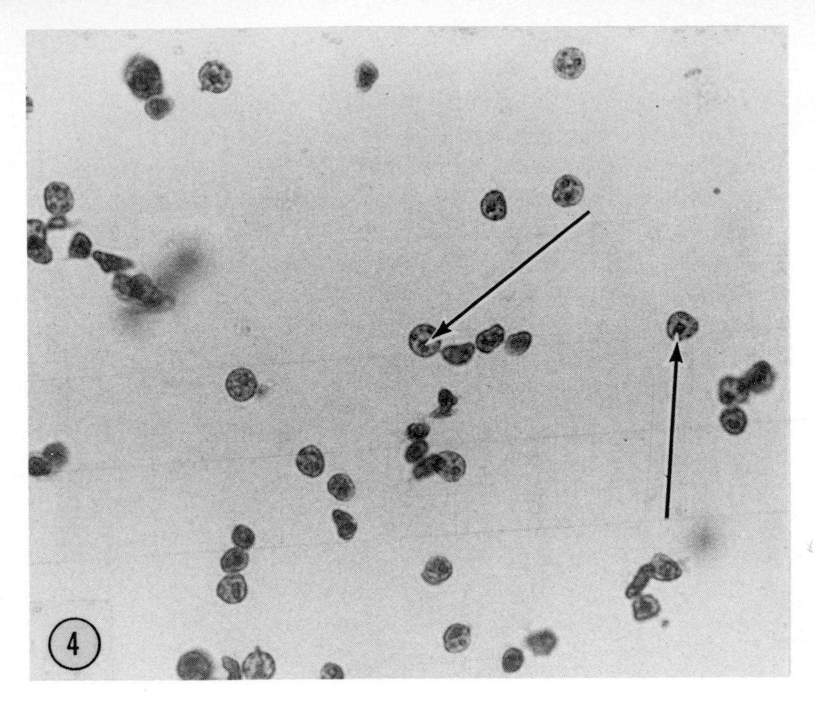

FIG. 4. Isolated bovine liver nuclei in 0.25 *M* sucrose TKM. Note prominent nucleoli within the nuclei denoted by arrows. Light microscopy; stained with methylene blue. × 512.

D. Structural Analysis

The nuclear preparation was analyzed by light, Nomarski interference, and electron microscopy. Figure 2 shows a large field of nuclei using Nomarski interference microscopy. No whole cell contamination is visible and the nuclei appear intact. At higher magnifications (Figs. 3–5) individual nuclei

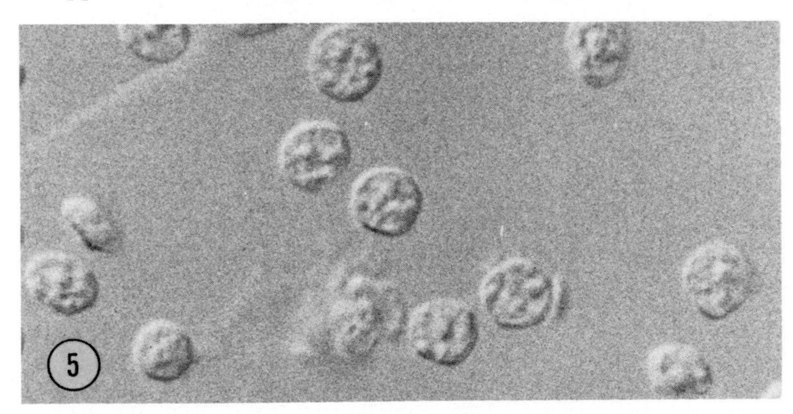

FIG. 5. Isolated bovine liver nuclei in 0.25 *M* sucrose TKM. Note the high degree of intactness and granularity of the nuclei. Nomarski interference microscopy. × 1024.

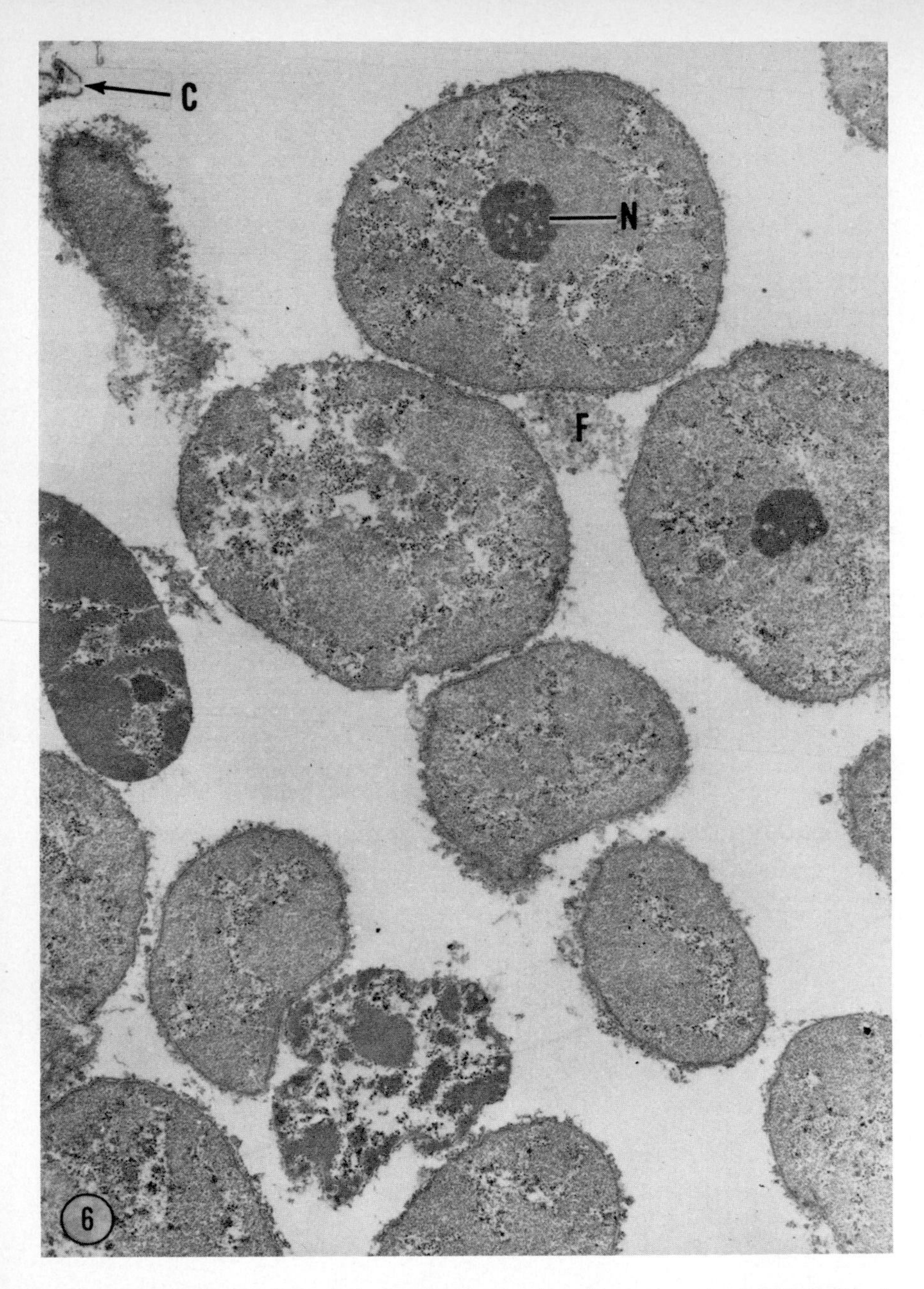

FIG. 6. Thin section through the nuclear preparation. A representative field is displayed. The nuclei are apparently intact. Only occasional fragments of definite cytoplasmic contamination are recognized (C). Fibrous material (F) is also present, generally in association with the nuclear surface. Higher magnification (see Fig. 7) demonstrates that most of this, at least, is probably chromatin derived from the nucleus. Membranes can be seen surrounding the nuclei, and nucleoli (N) are present. $\times$ 3000.

are seen in more detail with both light and Nomarski interference microscopy. The nuclei are of a granular nature (Fig. 5) and have prominent nucleoli (Fig. 4).

More definitive estimations of possible cytoplasmic contamination as well as a comprehensive analysis of nuclear ultrastructure were obtained using electron microscopy. Figure 6 is an electron micrograph of a thin section through the nuclear preparation. The relative intactness of the nuclei compares favorably with previous reports of purified nuclei prepared on a smaller scale (cf. Maggio *et al.*, 1963; Monneron *et al.*, 1972). Higher magnification clearly reveals that both outer and inner nuclear membranes are present. The outer nuclear membrane, however, is broken at various points (see Fig. 7, inset). The chromatin has a fine fibrous appearance, and the nucleoli are well defined. Fibrous and granular elements are often present in channels which lead up to the nuclear envelope (cf. Smetena *et al.*, 1963; Busch, 1967; Monneron and Bernard, 1969: Franke and Falk, 1970).

E. Biochemical Analysis

A quantitative estimation of mitochondrial and endoplasmic reticulum contamination was obtained using the marker enzyme approach. A chemical marker, coenzyme Q (ubiquinone), was also used to evaluate inner mitochondrial membrane contamination.

The results presented in Table I demonstrate very low levels of contamination for inner and outer mitochondrial membranes and endoplasmic reticulum in the isolated nuclear preparation.

III. Isolation of Nuclear Membranes

A. Equipment

Beckman Spinco Model L-2 Ultracentrifuge
Type SW 25.2 rotor
Type 40 or 50 rotor

B. Solutions

0.25 *M* sucrose TKM
0.25 *M* sucrose–0.5 *M* $MgCl_2$ TKM
2.2 *M* sucrose–0.5 *M* $MgCl_2$ TKM
1.6 *M* sucrose–0.5 *M* $MgCl_2$ TKM

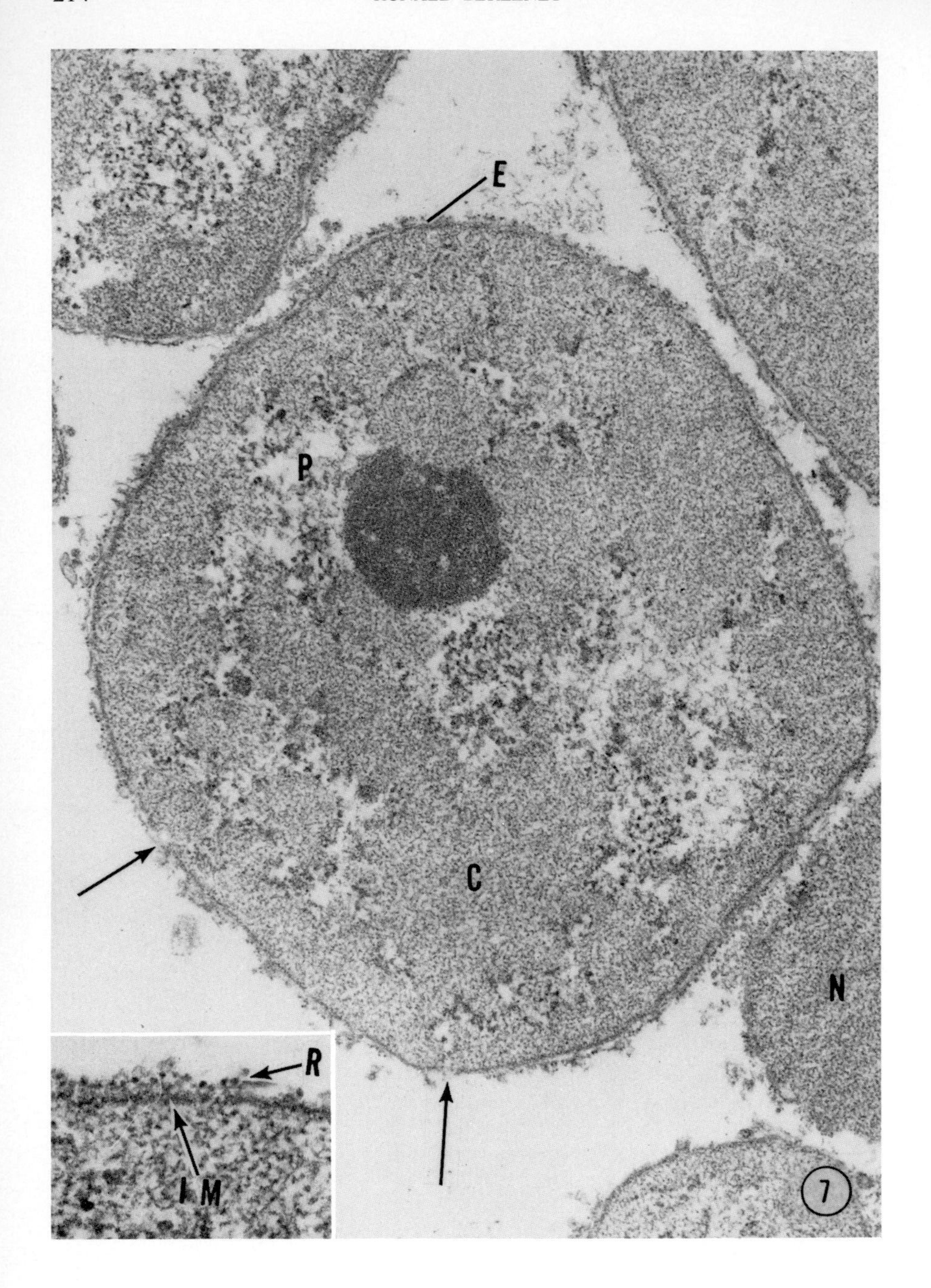
E
P
C
N
R
IM
7

Electrophoretically purified DNase I (Worthington). Stock solution of 1 mg/ml in distilled water. Store at −20°C.

C. Procedure

Nuclear membranes are isolated from nuclei by treatment with DNase I followed by short-term extraction with a high salt buffer [0.25 *M* sucrose–0.5 *M* $MgCl_2$ TKM (Berezney *et al.*, 1970a)]. A recent improvement of this method utilizes high salt sucrose gradients for further purifying nuclear membranes from nucleoli (Berezney *et al.*, 1972).

1. DNase Digestion

The methods previously reported involved incubation with DNase I for 10–14 hours at 2°C (Berezney *et al.*, 1969, 1970a, 1972). The low temperature enables a mild digestion resulting in the slow release of DNA from the nucleus. It was felt that such a situation would best maintain the integrity of the nuclear envelopes. However, it is often desirable to isolate nuclear membranes as rapidly as possible, and so a short-term digestion with DNase has been developed and will be reported here.

The isolated nuclei, suspended in 0.25 *M* sucrose TKM (2 mg nuclear protein per milliliter) are digested with DNase I (4 μg/ml) at 25°C for 1 hour. The digestion is terminated by placing the suspension on ice and centrifuging at 3000 *g* for 10 minutes in the Sorvall SS-34 rotor. The DNase treated nuclei are then washed two times by resuspending in 0.25 *M* sucrose TKM and centrifuging at 3000 *g* for 10 minutes in the SS-34 rotor.

2. High Salt Extraction and Purification on High Salt Sucrose Gradients

The washed DNase-treated pellets are resuspended in 0.25 *M* sucrose–0.5 *M* $MgCl_2$ TKM with a loose-fitting Teflon pestle homogenizer or a Dounce homogenizer type A. Suspension in 25-ml amounts is layered on dis-

---

FIG. 7. Thin section through the nuclear preparation. Details of a single nucleus are shown. The chromatin (C) forms a fibrous network. Channels within the nucleus (P) containing granular-fibrillar material are quite extensive and may represent the ribonucleoprotein network first described by Smetena *et al.* (1963) and Busch (1967). Note the apparent connection of these passageways with the nuclear envelope (arrows). Both membranes of the nuclear envelope (E) are visible but appear to be disrupted at many points. At higher magnification it is seen (see inset) that the disruption is limited to the outer membrane, which has ribosomelike particles (R) associated with it. The inner membrane remains intact (IM). In the lower right side is a portion of a nucleus with both the outer and inner nuclear membranes removed (N). The fibrous material seen closely associated with the outside surface of many of the nuclei (see Fig. 6) may therefore be derived from chromatin of such membraneless nuclei. × 9000; inset: × 50,000.

TABLE I

EVALUATION OF CONTAMINATION IN BOVINE LIVER NUCLEI[a]

Assay	Mitochondrial activity	Endoplasmic reticulum activity	Nuclear activity	% Mitochondrial or endoplasmic reticulum protein in nuclei
Mitochondrial markers (inner membrane)				
Succinoxidase (μmoles O_2/min/mg protein)	0.18 (12)	—	0.002 (12)	1.2
Succinate–cytochrome *c* reductase (μmoles cyt *c*/min/mg protein)	0.28 (12)	—	0.004 (12)	1.4
Succinate–PMS reductase (μmoles DCPIP/min/mg protein)	0.14 (12)	—	0.002 (12)	1.5
Coenzyme Q (nmoles/mg protein)	2.0 (2)	—	0.01 (2)	0.5
(outer membrane)				
Monoamine oxidase (ΔOD 250 nm/10 min /mg protein)	0.58 (6)	—	0.00 (3)	0
Endoplasmic reticulum marker				
NADPH-cytochrome *c* reductase (μmoles cyt *c*/min/mg protein)	—	0.080 (12)	0.0009 (12)	1.1

[a]Values in parentheses represent the number of determinations; taken in part from Berezney *et al.* (1970c).

continuous gradients consisting of 15 ml of 2.2 *M* sucrose–0.5 *M* $MgCl_2$ TKM and 15 ml of 1.6 *M* sucrose–0.5 *M* $MgCl_2$ TKM in SW 25.2 centrifuge tubes. Material from up to 500 mg of initial nuclear protein can be placed on a set of three gradient tubes. The gradients are centrifuged at 75,000 *g* for 1 hour in the SW 25.2 rotor.

A small pellet which sediments through the 2.2 *M* sucrose layer contains mainly nucleoli and other nonmembranous nuclear components (Berezney *et al.*, 1972). Material at the interface between 1.6 *M* and 2.2 *M* sucrose consists of nuclear membranes contaminated with nucleoli fragments

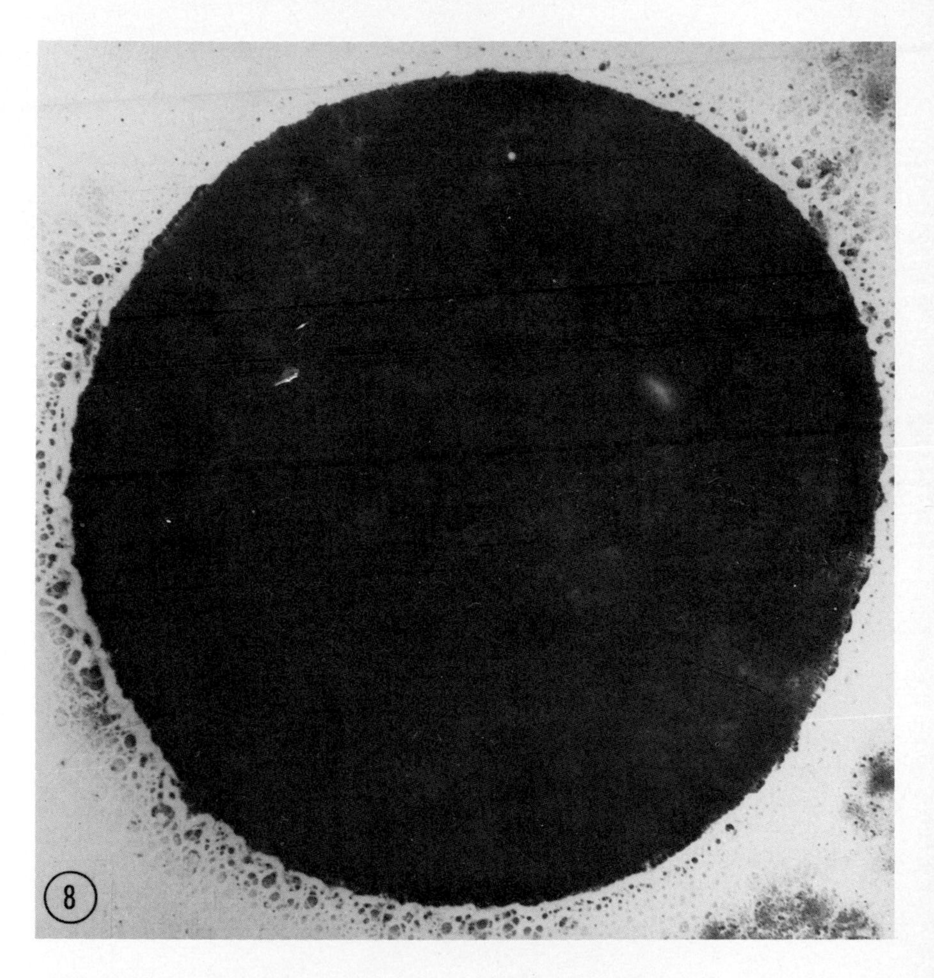

FIG. 8. Isolated nucleus in 0.25 *M* sucrose TKM. Negatively stained with PTA [2% phosphotungstic acid (pH 7.0)]. Although the nucleus is seen in surface view flattened out on an electron microscopic grid, very little of the nuclear envelope structure is visible. × 9000.

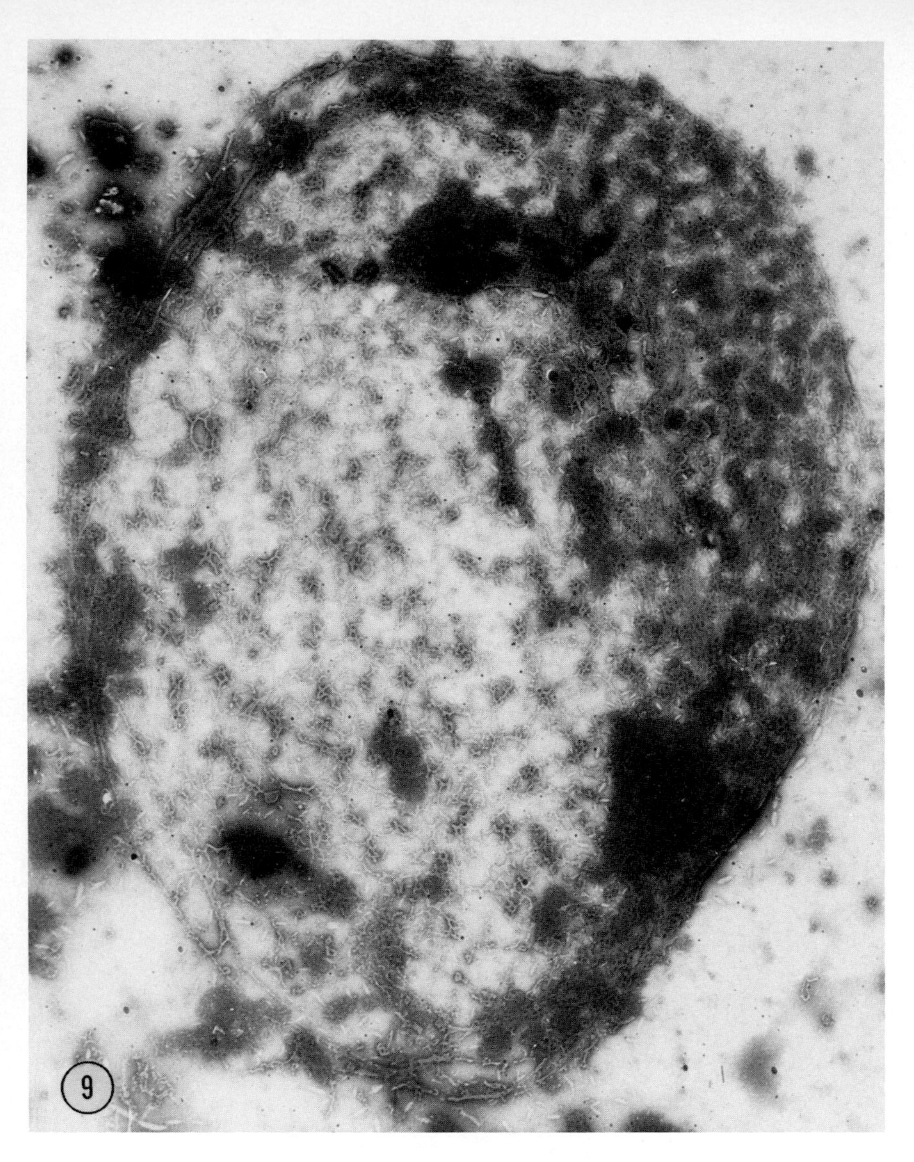

FIG. 9. Isolated nuclear membranes negatively stained with PTA. A nuclear ghost is demonstrated. The inner contents of the nucleus have been removed, revealing the nuclear envelope (cf Fig. 8). Although the ghost is highly fragmented, it maintains its integrity in size and shape (cf Figs. 7 and 8). The general speckled or pitted appearance is due to the nuclear pore complexes, which are sites of high concentration of PTA and appear as "black dots." × 8000.

FIG. 10. Isolated nuclear membranes negatively stained with PTA. The nuclear pore complexes (P) are clearly seen and identify the membrane fragments as nuclear membranes. Only very few membrane fragments appear devoid of the pore complexes. × 30,000.

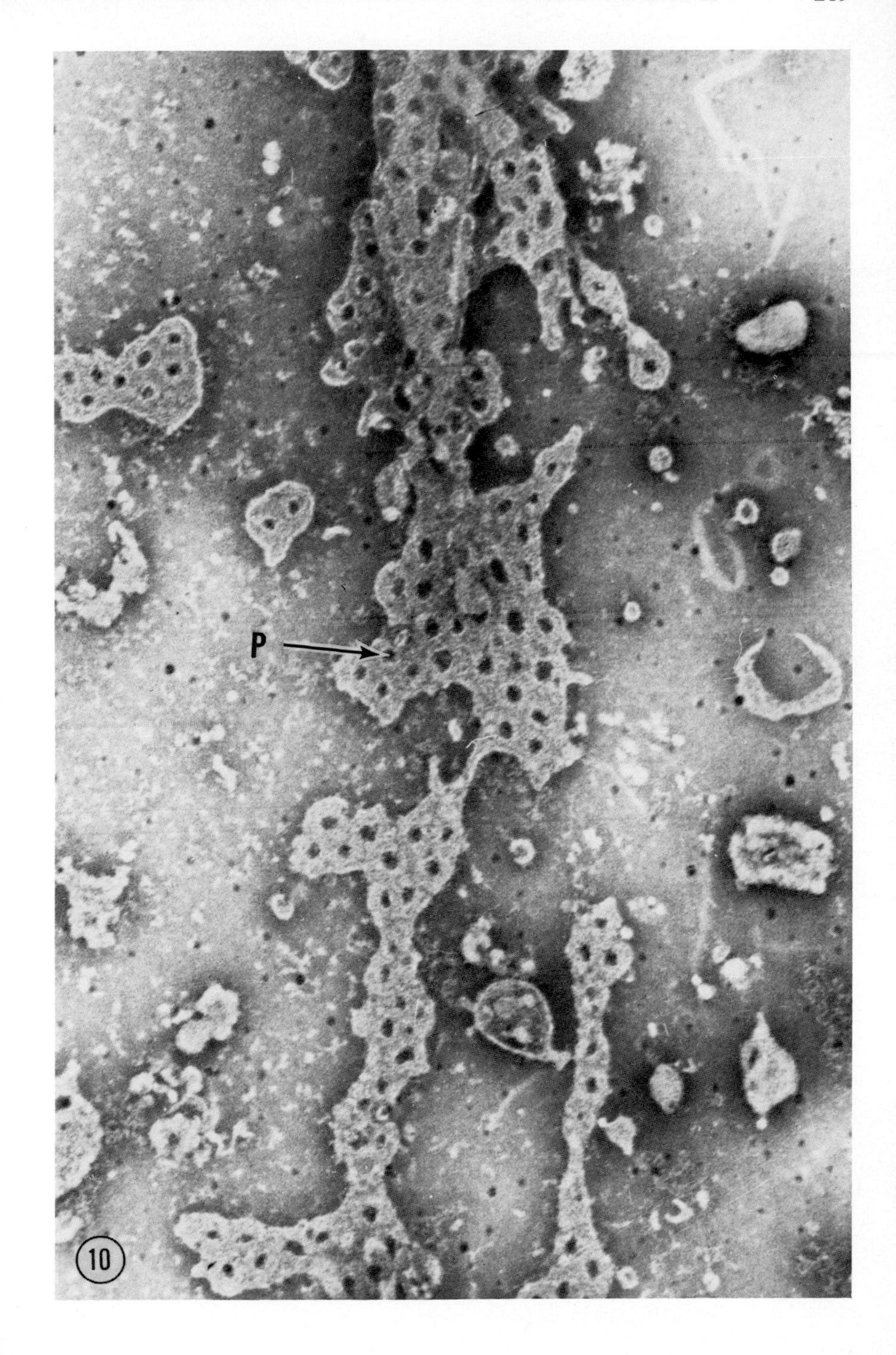
P
10

(Berezney *et al.*, 1972). Nuclear membranes are collected at the interface between 0.25 *M* sucrose and 1.6 *M* sucrose with a syringe and needle, and washed once with 0.25 *M* sucrose–0.5 *M* $MgCl_2$ TKM followed by 0.25 *M* sucrose TKM. In each case the fraction is resuspended with a Dounce A homogenizer and centrifuged at 100,000 *g* for 1 hour in the Spinco type 40 or 50 rotor.

D. Structural Analysis

Nuclear pore complexes are an excellent ultrastructural marker for identifying nuclear membranes, if the membranes are isolated in an intact enough state to demonstrate pore complexes. When visualized in surface view by negative staining, the pore complexes show up particularly well as annular or ringlike structures with a diameter of approximately 1000 Å. The width of the annular "wall" is about 200–250 Å, and appears to consist of a series of eight concentrically arranged subunits (Franke, 1966). In the center of the pore complex is often found a central granule of varying size and shape. Additional fine structure has also been observed (e.g., Yoo and Bayley, 1967; Franke, 1970; Abelson and Smith, 1970; LaCour and Wells, 1972).

Electron microscopic study of the nuclear membranes preparation indicated that it was extremely heterogenous with respect to size ranging from intact nuclear ghosts to small membrane vesicles and fragments.

Figure 8 demonstrates an isolated nucleus negatively stained with PTA (2% phosphotungstic acid, pH 7.0). Because of the thickness of the nucleus, very little of the surface structure is discernible. A nuclear ghost structure in the membrane preparation is seen in Fig. 9 negatively stained with PTA. Although the nuclear envelope is highly fragmented, it still maintains the basic outline of a nucleus. The nuclear pore complexes correspond to the intensely absorbing PTA areas or the "black spots" which give to the structure a speckled appearance. At higher magnificantion (Fig. 10) the pore complexes are more easily identified. The presence of annuli as well as substructure within the pore complexes (Fig. 11) indicates that the DNase-$MgCl_2$ treatment did not result in the complete extraction of nuclear pore material. Examination of nuclear membranes at high magnifications also substantiates the absence of inner mitochondrial membranes which are

FIG. 11. Isolated nuclear membranes negatively stained with PTA. At very high magnification it is demonstrated that the nuclear pore complexes are definitely not "empty" but contain substructures. The annuli (A) are preserved to varying extents, and particles (arrows) are seen inside the annuli. × 260,000. Taken from Berezney *et al.* (1972).

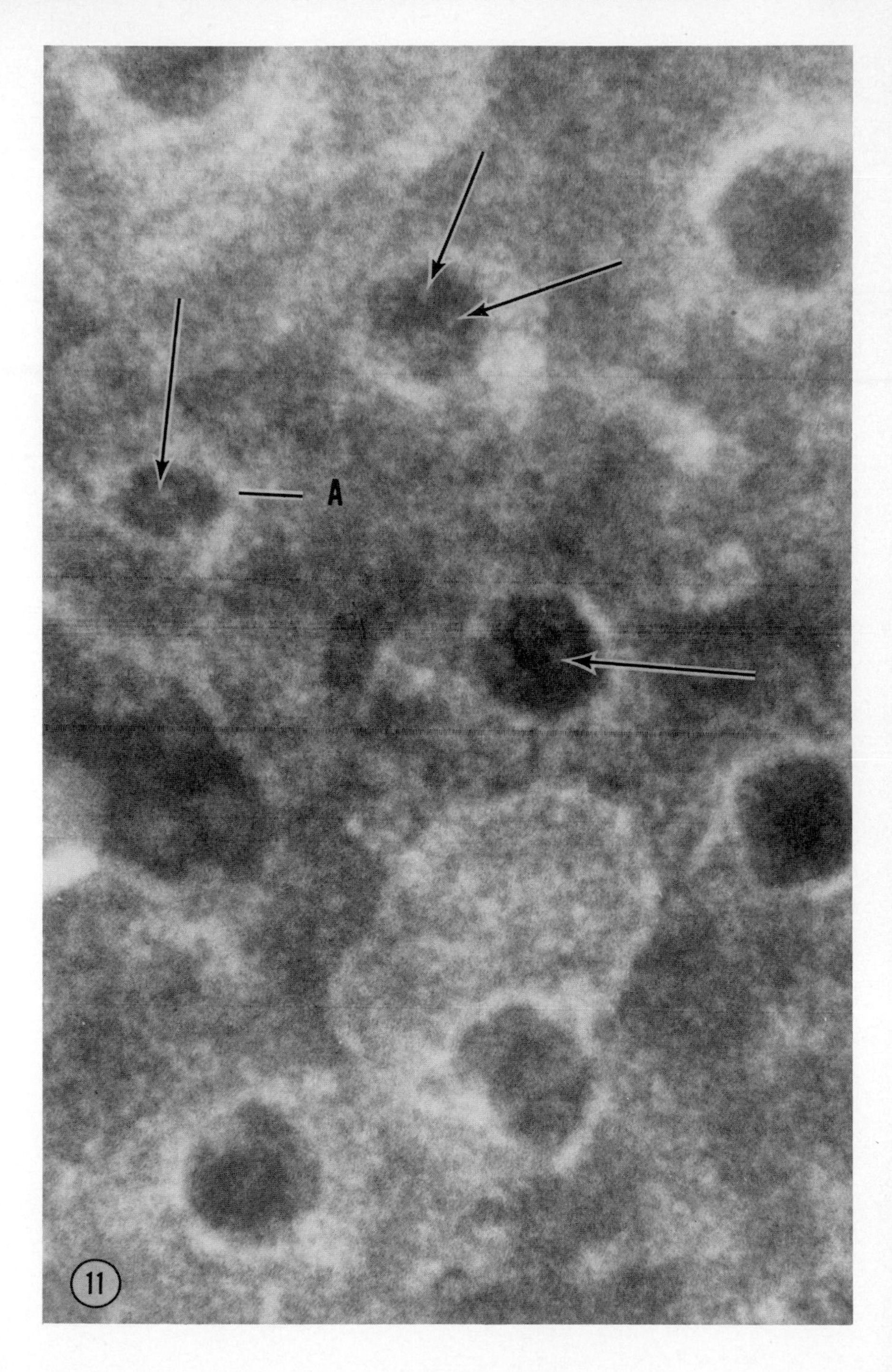
A
11

characterized by 90 Å diameter particles projecting from the membrane surface (Fernández-Morán, 1962; Parsons, 1963).

Nuclear pore complexes were also identified with thin sectioning in both transverse and tangential views. Figure 12 is a transverse section through a portion of a nuclear ghost in the nuclear membrane preparation. The outer nuclear membrane is readily identified by the ribosomelike particles attached to its surface. Material is seen within the confines of the pore complex. Note that the outer and inner membranes actually appear to join together at the

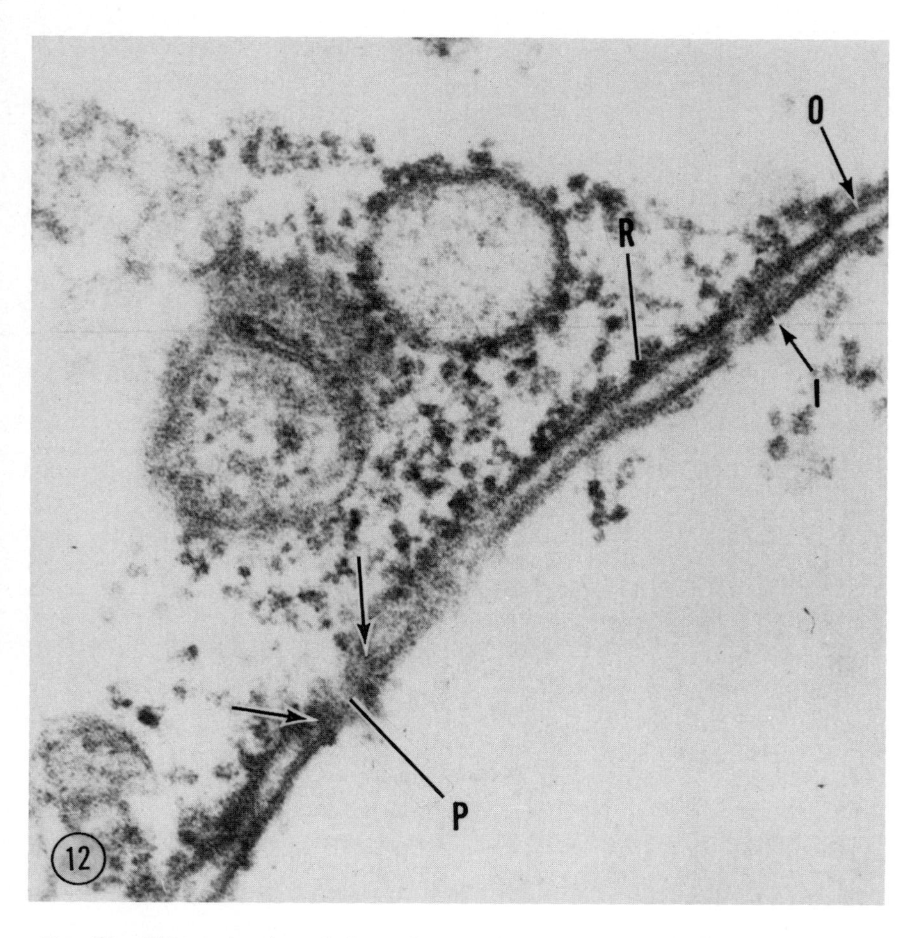

FIG. 12. Thin section through the nuclear membrane preparation. This micrograph shows a portion of a nuclear ghost sectioned transversely through membranes. Note that the outer (O) and inner (I) nuclear membranes appear to join together (see arrows) at the nuclear pore complex (P). Ribosomelike particles (R) are seen on the surface of the outer nuclear membrane. × 132,000. Taken from Berezney *et al.* (1970a).

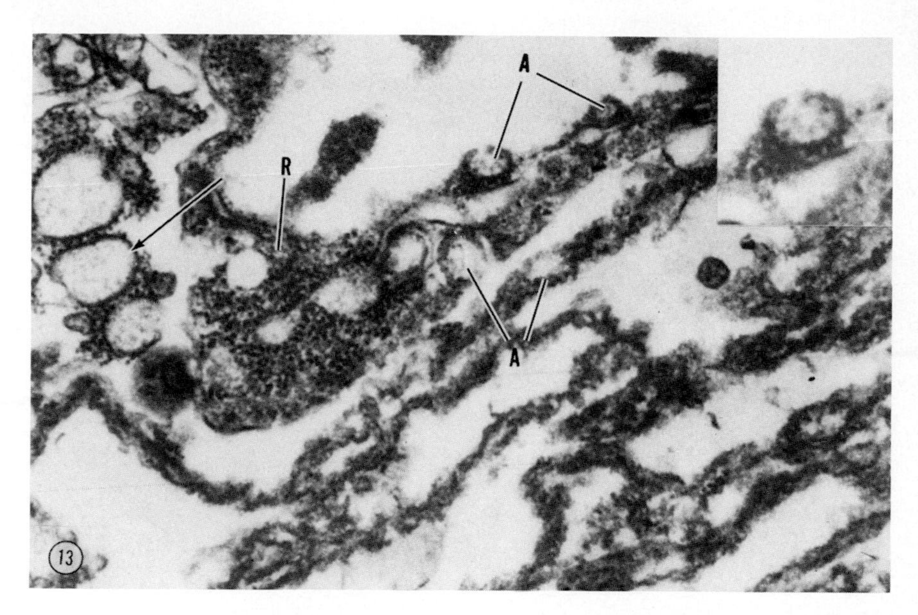

FIG. 13. Thin section through the nuclear membrane preparation. Nuclear membranes and the associated pore complexes are sectioned tangentially. The annular structures (A) of the pore complexes are seen and often appear to project from the surface of the membrane (see inset). Ribosomelike particles (R) are seen on some membrane surfaces, presumably the outer nuclear membrane. In the same field are seen membranes sectioned transversely (extreme left side, see arrow) × 66,300; inset: × 120,900.

pore complex, thus forming a continuous membrane cisternae between pore complexes. The two membranes are also similar, as judged by general morphology and thickness (see especially upper right of Fig. 12).

Tangential sectioning through the membranes is seen in Fig. 13. The annular structures corresponding to nuclear pore complexes are easily recognized. Ribosomelike particles in apparently polysomal formations are visible on the surface of some of the membranes (presumably the outer nuclear membrane). Some of the pore complexes (see Fig. 13 inset) appear to actually project from the surface of the membrane. This is consistent with the observations of Afzelius (1955), Wischnitzer (1958), and Abelson and Smith (1970).

E. Biochemical Analysis

Perhaps the best single criterion for demonstrating the isolation of nuclear membranes is by its unique and interesting ultrastructure. Biochemical analyses can strongly complement the ultrastructural analyses, however, by providing quantitative data on chemical and enzymatic composition.

TABLE II

EVALUATION OF MITOCHONDRIAL AND ENDOPLASMIC RETICULUM CONTAMINATION IN NUCLEAR MEMBRANES[a]

Assay	Mitochondrial activity	Endoplasmic reticulum activity	Nuclear membrane activity	% Mitochondrial or endoplasmic reticulum protein in nuclear membranes
Mitochondrial markers (inner membrane)				
Succinoxidase (μmoles O_2/min/mg protein)	0.10	—	0.0032	3.2
Succinate–cytochrome *c* reductase (μmoles cyt *c*/min/mg protein	0.12	—	0.0034	2.8
Succinate–PMS reductase (μmoles DCPIP/min/mg protein)	0.10	—	0.0037	3.7
Coenzyme Q (nmoles CoQ/mg protein)	2.06	—	<0.05	<2.4
Acid nonextractable, trypsin releasable flavin (nmoles flavin/mg protein)	0.27	—	<0.01	<3.7
Endoplasmic reticulum markers				
NADPH-cytochrome *c* reductase (μmoles cytochrome *c* reduced/min/mg protein)	—	0.080	0.0052	6.5
Cytochrome P-450 (nmoles cytochrome/mg protein)	—	1.24	0.055	4.4

[a]Taken in part from Berezney *et al.* (1972).

1. Chemical Composition of Nuclei and Nuclear Membranes

All cellular membranes contain considerable amounts of lipid. The nuclear membrane preparation should therefore be highly concentrated in lipid with respect to the starting nuclear fraction. This is confirmed in Table III, where the phospholipid content of isolated nuclear membranes is 7.8-fold higher than the nuclei and the total phospholipid recovered in 42%. In contrast to the localization of nuclear phospholipid, most of the DNA is not directly associated with the membranes, but is located within the interior of the nucleus as chromatin. Only 0.2% of the total DNA was recovered in the membrane fraction. The DNA which survived both DNase and high salt extraction might represent a DNA species which is actually a part of the membrane structure or alternatively a part of the pore complexes. Twelve percent recovery of total nuclear RNA in the membranes suggests that the nuclear envelope is a site of concentration of RNA in the nucleus (cf. Scheer, 1972). The actual RNA content is probably higher due to the extraction of at least some of the ribosomal RNA by the high salt treatment (Franke *et al.*, 1970; Monneron *et al.*, 1972).

Finally the 6% recovery of nuclear protein is consistent with the absence of nonmembrane nuclear proteins.

2. Quantitative Evaluation of Contamination in Isolated Nuclear Membranes

An analysis similar to that for isolated nuclei involving both enzymatic and chemical markers demonstrates very low levels of contaminating mitochondrial and endoplasmic reticulum membranes in the isolated membranes (Table II).

IV. Biochemical Applications

A. Biochemical Comparison of Nuclear and Endoplasmic Reticulum Membranes

Until recently comparison of nuclear and endoplasmic reticulum membranes was limited to two ultrastructural properties: nuclear membranes have pore complexes that are absent in endoplasmic reticulum, and both membrane systems have associated ribosomes. The large-scale preparation provided a practical means for a rapid and comprehensive comparison of the biochemical properties of nuclear and endoplasmic reticulum membranes. These results are summarized in Table IV.

TABLE III

CHEMICAL COMPOSITION OF NUCLEI AND NUCLEAR MEMBRANES[a]

	Protein	Phospholipid	DNA	RNA
Nuclei[b]	68.4	2.9	26	2.7
Nuclear membranes[b]	70.4	22.7	1.1	5.8
Ratio nuclear membrane: nuclei	1.03	7.83	0.04	2.14
% Recovery in nuclear membranes	6	42	0.2	12

[a]Taken in part from Berezney *et al.* (1970a) and Berezney *et al.* (1972).
[b]Given as % of total wet weight.

The biochemical parameters fall into three categories: those found only in nuclear membranes, those found only in endoplasmic reticulum, and those found in both membranes. The differences between nuclear and endoplasmic reticulum membranes may result from membrane differentiation and specialization from a common membrane system reflected in the many similar properties.

TABLE IV

QUALITATIVE COMPARISON OF ISOLATED NUCLEAR MEMBRANES AND ENDOPLASMIC RETICULUM[a]

	Endoplasmic reticulum	Nuclear membranes
Only in nuclear membranes		
Pore complex	—[b]	+
DNA	—	+
Cholesterol	—	+
Cytochrome oxidase	—	+
NADH oxidase	—	+
Only in endoplasmic reticulum		
Cytochrome P-450	+	—
NADPH–cytochrome *c* reductase	+	—
Present in both		
Ribosomes	+	+
RNA	+	+
Phospholipids pattern	+	+
Cytochrome b_5	+	+
b_5 reductase flavoprotein	+	+
Glucose-6-phosphatase	+	+
Mg^{2+}-ATPase	+	+

[a]Based in part on the findings of Berezney *et al.* (1972), Berezney and Crane (1972), and Keenan *et al.* (1972).
[b]—, Absent or present in trace amounts.

B. Elucidation of an Electron Transport System in Isolated Nuclear Membranes

Study of electron transport components in isolated nuclear membranes has established it to be the site of nuclear electron transport (Zbarsky *et al.*, 1969; Kuzmina and Zbarsky, 1969; Berezney *et al.*, 1969, 1970b, 1972; Berezney and Crane, 1971). A more thorough investigation of the NADH oxidase electron transport system has revealed that it is the result of the coupling of two distinct enzyme systems (Berezney, 1972; Berezney and Crane, 1972): an endoplasmic reticulum-like cytochrome b_5 reductase system consisting of the catalytic components NADH-cytochrome b_5 reductase and cytochrome b_5, and cytochrome *c* oxidase as the terminal oxidase. The precise role of this electron transport system has yet to be elucidated. The fact that the first part of the chain is apparently identical to the endoplasmic reticulum and outer mitochondrial membrane NADH oxidase systems is suggestive of a biogenetic relationship among these three electron transport systems (cf. Berezney *et al.*, 1970c). The further similarities of the nuclear membrane and mitochondrial terminal oxidases suggests that these four electron transport systems (nuclear membranes, endoplasmic reticulum, outer and inner mitochondrial membranes) may have had a common origin in the evolution of the eukaryotic cell (Berezney, 1972).

Further studies comparing nuclear membrane and mitochondrial cytochrome *c* oxidases are in progress and will be aided greatly by the large-scale membrane preparations.

C. Further Fractionation of Nuclear Membranes

The isolation of nuclear membranes on a large scale enables extensive fractionation studies of this membrane system. Such questions as the comparison of the outer and inner nuclear membranes, the nature of the nuclear pore complexes and their role in nucleocytoplasmic transport, and the characterization of specific nuclear membrane proteins and nucleic acids can now be approached by direct biochemical analysis.

ACKNOWLEDGMENTS

Deep appreciation is expressed to Dr. Frederick L. Crane for support and valuable guidance in these studies. I thank, Ms. Linda K. Macaulay for collaboration, especially with respect to electron microscopy, and Dr. Thomas W. Keenan on lipid analysis. The electron microscopy of isolated nuclei by Ms. Barbara Roschke is gratefully acknowledged. I am grateful to Drs. Donald S. Coffey and Paul Talalay for encouragement and support.

References

Abelson, H. T., and Smith, G. H. (1970). *J. Ultrastruct. Res.* **30**, 558–588.
Afzelius, B. A. (1955). *Exp. Cell Res.* **8**, 147–158.
Agutter, P. S. (1972). *Biochim. Biophys. Acta* **255**, 397–401.
Berezney, R. (1972). *J. Cell Biol.* **55**, 18a.
Berezney, R., and Crane, F. L. (1971). *Biochem. Biophys. Res. Commun.* **43**, 1017–1023.
Berezney, R., and Crane, F. L. (1972). *J. Biol. Chem.* **247**, 5562–5568.
Berezney, R., Funk, L. K., and Crane, F. L. (1969). *J. Cell Biol.* **43**, 12a.
Berezney, R., Funk, L. K., and Crane, F. L. (1970a). *Biochim. Biophys. Acta* **203**, 531–546.
Berezney, R., Funk, L. K., and Crane, F. L. (1970b). *Biochem. Biophys. Res. Comun* **38**, 93–98.
Berezney, R., Funk, L. K., and Crane, F. L. (1970c). *Biochim. Biophys. Acta* **223**, 61–70.
Berezney, R., Macaulay, L. K., and Crane, F. L. (1972). *J. Biol. Chem.* **247**, 5549–5561.
Blobel, G., and Potter, V. R. (1966). *Science* **154**, 1662–1665.
Busch, H. (1967). *In* "Nucleic Acids" (L. Grossman and K. Moldave, eds.), Methods in Enzymology, Vol 12, Part A, pp. 448–468, Academic Press, New York.
Engelhardt, P., and Pusa, K. (1972). *Nature (London)* **240**, 163–168.
Fernández-Morán, H. (1962). *Circulation* **26**, 1039–1065.
Franke, W. W. (1966). *J. Cell Biol.* **31**, 619–623.
Franke, W. W. (1970). *Z. Zellforsch. Mikrosk. Anat.* **105**, 405–429.
Franke, W. W., and Falk, H. (1970). *Histochemie* **24**, 266–278.
Franke, W. W., and Scheer, U. (1970). *J. Ultrastruct. Res.* **30**, 288–316.
Franke, W. W., Deumling, B., Ermen, B., Jarasch, E.-D., Kleinig, H. (1970). *J. Cell Biol.* **46**, 379–395.
Gall, J. G., (1954). *Exp. Cell Res.* **7**, 197–200.
Gall, J. G. (1967). *J. Cell Biol.* **32**, 391–399.
Kashnig, D. M., and Kasper, C. B. (1969). *J. Biol. Chem.* **244**, 3786–3792.
Kay, R. R., Haines, M. E., and Johnston, I. R. (1971). *FEBS (Fed. Eur. Biochem. Soc.), Lett.* **16**, 233–236.
Keenan, T. W., Berezney, R., and Crane, F. L. (1972). *Lipids* **7**, 212–215.
Kuzmina, S. N., and Zbarsky, I. B. (1969). *FEBS (Fed. Eur. Biochem. Soc.), Lett.* **5**, 34–36.
LaCour, L. F., and Wells, B. (1972). *Z. Zellforsch. Mikrosk. Anat.* **123**, 178–194.
Maggio, R., Siekevitz, P., and Palade, G. E. (1963). *J. Cell Biol.* **18**, 267–292.
Monneron, A., and Bernard, W. (1969). *J. Ultrastruct. Res.* **27**, 266–288.
Monneron, A., Blobel, G., and Palade, G. E. (1972). *J. Cell. Biol.* **55**, 104–125.
Moore, R. J., and Wilson, J. D. (1972). *J. Biol. Chem.* **247**, 958–967.
Parsons, D. F. (1963). *Science* **140**, 985–987.
Price, M. R., Harris, J. R., and Baldwin, R. W. (1972). *J. Ultrastruct. Res.* **40**, 178–196.
Roberts, K. and Northcote, D. H. (1970). *Nature (London)* **228**, 385–386.
Scheer, U. (1972). *Z. Zellforsch. Mikrosk. Anat.* **127**, 127–148.
Smetena, K., Steele, W. J., and Busch, H. (1963). *Exp. Cell Res.* **31**, 198–201.
Vivier, E. (1967). *J. Microsc, (Paris)* **6**, 371–390.
Watson, M. L. (1955). *J. Biophys. Biochem. Cytol.* **1**, 257–270.
Watson, M. L. (1959). *J. Cell Biol.* **6**, 147–155.
Wischnitzer, S. (1958). *J. Ultrastruct. Res.* **1**, 201–222.
Yoo, B. Y., and Bayley, S. T. (1967). *J. Ultrastruct. Res.* **18**, 651–660.
Zbarsky, I. B., Perevoschikova, K. A., Delektorskaya, L. N., and Delektorsky, V. V. (1969). *Nature (London)* **221**, 257–259.
Zentgraf, H., Deumling, B., Jarasch, E.-D., and Franke, W. W. (1971). *J. Biol. Chem.* **246**, 2986–2995.

Chapter 14

A Simplified Method for the Detection of Mycoplasma

ELLIOT M. LEVINE

The Wistar Institute of Anatomy and Biology, Philadelphia, Pennsylvania

---

I. Detection of Mycoplasma Contamination: Introduction and Background

The increasing recognition that mycoplasma infection of animal cell cultures has widespread effects has emphasized the need for rapid, facile detection methods for these microorganisms, which are frequent contaminants of serially propagated cells and have also been detected in primary explants and chick embryos (Hayflick and Stanbridge, 1967; Stanbridge, 1971). Since the initial description of mycoplasma as cell culture contaminants, a wide variety of alterations in characteristics of infected cultures has been reported, including cytopathic changes which mimic viral infections, deleterious changes stemming from arginine depletion, alterations in nucleic acid metabolism, effects on drug metabolism, chromosome breakage, and permanent changes in culture morphology (for specific references, see Levine, 1972). On the other hand, many laboratories have reported no visible effects either on morphology or growth (see Levine, 1972). Although some antibiotics (but not penicillin and streptomycin) afford a measure of protection against mycoplasma infection, antibiotic resistance frequently develops (Hayflick and Chanock, 1965; Newnham and Chu, 1965; Macpherson, 1966; Hayflick and Stanbridge, 1967; Perlman *et al.*, 1967; Levine *et al.*, 1968; Sabin, 1968; Stanbridge, 1971). In the absence of a completely effective preventative or cure, and because primary sources of contamination have not been identified, it becomes necessary to monitor cell cultures for mycoplasma and discard contaminated cultures to prevent cross infection (Hayflick and Chanock, 1965; Macpherson, 1966; Levine *et al.*, 1968; Sabin, 1968). Ideally, cultures being used for experimental purposes should be monitored on the day of an experiment, and results should be available that same day.

Most available detection methods are neither simple nor rapid, and, in some cases, specialized equipment or skills are required. Visualization by electron microscopy, or light microscopy of orcein or fluorescent stained material is often time consuming and not always conclusive (Fogh and Fogh, 1964; Hayflick and Chanock, 1965; Macpherson, 1966; Shedden and Cole, 1966; Zucker-Franklin *et al.*, 1966; Hummler and Armstrong, 1967; Wolanski and Maramorosch, 1970). Cultivation of mycoplasma on agar plates in Hayflick's medium or on feeder layers of animal cells will detect most mycoplasma species (Hayflick, 1965; House and Waddell, 1967; Zgorniak-Nowosielska *et al.*, 1967). However, both these methods as well as "plaque" type assay employing mouse lymphoma cells usually require incubation time of 4–14 days (Kraemer *et al.*, 1963; Kraemer, 1964a,b). They also suffer from the drawback that some strains of mycoplasma found in contaminated cell cultures cannot be cultivated in artificial media (see Table III), or will not lyse lymphoma cells (Kraemer, 1964b; Markov *et al.*, 1969; Todaro *et al.*, 1971; Schneider *et al.*, 1973a,b).

In the past, two enzymatic methods have been proposed for detection of mycoplasma in infected cells: determination of arginine deiminase and "thymidine phosphorylase" (Schimke and Barile, 1963a,b; Horoszewicz and Grace, 1964; Schmike, 1967). Although these do not require extended incubation times, false negative results have been obtained in several instances (Horoszewicz and Grace, 1964; Kraemer, 1964b; Barile *et al.*, 1966; Macpherson, 1966; House and Waddell, 1967). In particular *Mycoplasma hyorhinis*, a common contaminant in our laboratory, does not exhibit significant deiminase activity (Barile *et al.*, 1966).

Adherent mycoplasma on the cell periphery may be demonstrated by autoradiography of contaminated cell cultures labeled with thymidine-^{3}H or uridine-^{3}H (Nardone *et al.*, 1965; Levine *et al.*, 1967, 1968). As in any autoradiographic procedure, there is a significant time delay before assay results are known, as well as the usual ambiguities of interpretation. In addition, the sensitivity of these procedures compared with other detection methods remains to be evaluated, and experiments reported in this manuscript indicate that certain mycoplasma strains are impermeable to some nucleic acid precursors.

Characteristic sedimentation properties and electrophoretic mobilities of various labeled mycoplasma components differing from those of the host cell may also be used to detect mycoplasma-contaminated cultures. A particularly sensitive method based on equilibrium centrifugation of ammonium sulfate-precipitated mycoplasma present in infected culture fluid has been developed (Todaro *et al.*, 1971). Also, mycoplasma RNA and DNA can be separated from host nucleic acids either by sedimentation analysis or agar gel electrophoresis, as was observed several years ago, and subsequently confirmed (Levine *et al.*, 1967, 1968; Markov *et al.*, 1969; Harley *et al.*, 1970; Schneider *et al.*, 1973a). All these procedures, however, involve rather lengthy times for both separation and analysis, and some, as well be seen, yield false negative results (cf Table III, group C).

The detection procedure described in this chapter is based on our previous finding that in infected cultures radioactive tracer nucleotides were rapidly incorporated into mycoplasma nucleic acids after pulse-labeling (Levine *et al.*, 1967, 1968). In the same cultures, however, tracer incorporation into host nucleic acids was markedly depressed and, in many instances, no longer detectable, even though synthesis of host nucleic acids was only minimally affected. A suggested explanation for these observations was that tracer uridine, for instance, might be metabolized by mycoplasma, and never reach the sites of host RNA synthesis (Levine *et al.*, 1968). Several lines of evidence support the thesis that conversion of uridine to uracil by mycoplasma accounts for the apparently altered pattern of RNA synthesis in contaminated cultures. Deoxynucleoside cleavage is enhanced in mycoplasma-contaminated cultures (Hakala *et al.*, 1963; Horoszewicz and Grace, 1964; Holland *et al.*, 1967); exogenous uridine added at high concentrations is incorporated into host RNA (Levine *et al.*, 1968; Harley *et al.*, 1970); allyldeoxyuridine, an inhibitor of nucleoside phosphorylase (Hakala *et al.*, 1963; Holland *et al.*, 1967), augments incorporation of tracer uridine into host cell

RNA (Levine, unpublished observations); and radioactive tracer uracil (which would be the product formed by the action of nucleoside phosphorylase on uridine) is rapidly incorporated into mycoplasma RNA in infected cultures but only to a minimal degree in uncontaminated cells (Levine, unpublished observations). In addition, Perez *et al.* (1972) also have demonstrated formation of uracil and thymine when the respective nucleosides were added to the supernatant fluid of infected cultures, and the ratio of uridine-^{3}H to uracil-^{3}H incorporation is markedly decreased in contaminated diploid fibroblasts (Schneider *et al.*, 1973b). All these data suggested that enhanced phosphorylytic cleavage of uridine as well as other nucleosides may be a diagnostic characterisitic of mycoplasma-infected cells. Indeed, as noted earlier, colorimetric determination of deoxyribose (phosphate) liberated from thymidine incubated with mycoplasma-infected cells already has been used to identify some contaminated cultures (Horoszewicz and Grace, 1964).

The recently developed method (Levine, 1972) for mycoplasma detection described in this chapter involves measurement of nucleoside phosphorylase activity by incubation of radioactive uridine with culture lysates and subsequent chromatographic separation of uridine from the reaction product, uracil (see flow diagram, Fig. 1). This procedure has proved to be rapid (1–4 hours), and sensitive enough to detect all the infected cell cultures thus far encountered, including some adjudged negative by other methods. Although as many details as possible have been given below in describing this method, the procedure is inherently quite simple (see Fig. 1). The usefulness of this method is only slightly limited by the finding that a certain few cell lines normally exhibit significant uridine phosphorylase levels, even though they

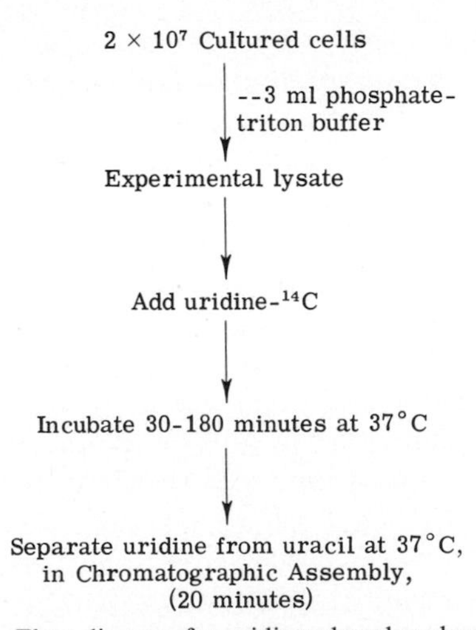

FIG. 1. Flow diagram for uridine phosphorylase assay.

apparently are uncontaminated (false positives). On the other hand, most cell lines including such widely studied cell types as HeLa, 3T3, and human diploid fibroblasts can be satisfactorily monitored for mycoplasma infection by this method, since phosphorylase levels are quite low in uncontaminated cultures of these cells.

II. Uridine Phosphorylase Assay: Facilities, Equipment, Chemicals, and Stock Solutions

Most functioning laboratories will have ready access to the required equipment, or will be able to adapt the detection procedure to their available facilities. The following equipment is used in our laboratory for the procedure here described: scintillation counter, International centrifuge, warm room, water bath, pH meter, hair dryer or heat lamps, ultraviolet lamp (shortwave, 254 mm filter) and darkroom or closet, soldering iron, and 5-μl and 20-μl micropipettes.

The following materials are used: chromatography paper (18×22 inches) numbers 3 and 41, paper clips ($1\frac{5}{16}$ inches), silicone stoppers (No. 4), and test tubes (25×200 mm).

The following chemicals are required: uridine-^{14}C (specific activity approximately 40 mCi/mmole), Triton X-100 (Rohm and Haas, Inc., Philadelphia, Pennsylvania), nonradioactive uridine and uracil, *n*-(1)-butanol, boric acid, dibasic sodium phosphate, and concentrated ammonium hydroxide (specific gravity ca. 0.9).

Stock solutions are prepared as follows:

Concentrates for incubation buffer. A 0.5 *M* solution of Na_2HPO_4 is prepared and adjusted to pH 8.1 (approximately 2.9 ml of 1 N HCl is required per 100 ml final volume of phosphate solution). The second concentrate is a 0.010 *M* aqueous solution of nonradioactive uracil. For the third concentrate, Triton X-100 is diluted with water to yield a 10% (v/v) stock solution. All concentrates are stored cold. Sodium phosphate must be redissolved with warming before use.

Incubation buffer. The above three concentrates are mixed and diluted to prepare the incubation buffer as follows: 10 ml Na_2HPO_4, 10 ml uracil, 5 ml Triton, water to 100 ml. Final concentrations are: Na_2HPO_4, 0.05 *M*; uracil 0.001 *M*; and Triton, 0.5%. Stored cold.

Radioactive uridine stock. Commercial preparations of uridine-^{14}C at a concentration of 10 μCi/ml are used when available; more concentrated solutions are diluted as required. Stored cold.

Chromatography standard solution. Unlabeled uracil and uridine are dissolved together in the same aqueous solution at final concentrations of 0.002 *M* each. The solution is divided into 0.5-ml portions which are stored frozen.

Chromatography solvent. (Adapted from the formula of Rose and Schweigert, 1951.) A 4% (ca. 0.65 *M*) aqueous solution of boric acid is prepared. One milliliter of concentrated ammonium hydroxide is added to 70 ml of boric acid stock solution, after which 430 ml of *n*-(l)-butanol is added to the mixture. At room temperature some separation of phases exists in this solvent. A homogeneous solution is obtained by warming to 37°C, and shaking. For immediate use the solvent is stored at 37°C (tightly stoppered!). Errors in preparation of the chromatography solvent are the most common source of confusing results.

III. Uridine Phosphorylase Assay: Description of Procedure

A. Growth and Harvesting of Cells

Cell cultures are grown routinely in medium supplemented with 10% serum (Eagle, 1959). Culture media devoid of all antibiotics have been employed to ensure that lapses in sterile technique, which might introduce mycoplasma along with bacteria, are not masked (Hayflick and Channock, 1965; Macpherson, 1966; Levine *et al*., 1968). Also, it is possible that widespectum antibiotics, such as Aureomycin, may significantly reduce mycoplasma titers and phosphorylase levels. On the other hand, the presence of penicillin and or streptomycin in the medium has no noticeable effect on assay results. Although assays can be performed on cultures at any stage of growth, waiting for 2–3 days after subculture allows for maximum mycoplasma growth. Cell lysates are prepared (see Section III, B.) from monolayer cultures after decanting the growth medium but without rinsing. Approximately 90% of uridine phosphorylase activity is associated with mycoplasma adherent to the cell monolayer. Suspension cultures are harvested, before addition of fresh medium, by centrifugation at 1000 *g* for 10 minutes in an International PR-2 centrifuge. Approximately 90% of contaminating mycoplasma are sedimented under these conditions.

B. Preparation and Incubation of Lysates

Usually, 2×10^7 cells are assayed although 10^6 cells suffice, if reagent volumes are reduced proportionately. Three milliliters of "incubation buffer" are used to lyse 2×10^7 cells. pH and salt concentration of this buffer have been adapted from previous studies of uridine phosphorylase (Sköld, 1960; Pontis *et al*., 1961; Lindsay *et al*., 1968), although only phosphate concentration appears critical. Lowering the pH from 8.1 to 6.5 enhances enzyme activity somewhat (Pontis *et al.*, 1961), but interferes with the detergent action of the "incubation buffer." Unlabeled uracil in the incubation buffer serves as an effective trap for the labeled product. Conversion of uridine to uracil does not occur in the presence of some mycoplasma strains unless the organisms are disrupted, and Triton (0.5%) is as effective as sonication in "solubilizing" uridine phosphorylase activity. Higher concentrations of Triton do not affect recovery of enzyme activity, and even at 10% Triton, cell nuclei remain intact. Certain cell types are not easily dislodged from glass surfaces with Triton buffer; in such cases, adherent remains of the cell sheet are scraped into buffer using a rubber policeman. In addition, under infrequent circumstances to be discussed later (see p. 237), phosphorylase activity is measured by adding uridine-^{14}C to the supernatant growth medium of living cultures rather than to Triton lysates of the cells.

In some cases, prior to incubation, a portion of the lysate (1.0 to 1.5 ml) is heated in a boiling water bath for 10 minutes, and cooled to 37°C. Uridine-^{14}C (final concentration, μCi/ml) is added to both the original cell lysate and its "boiled control," and each is incubated at 37°C. Samples (0.5 ml) are removed at appropriate times, usually 30 and 180 minutes, chilled in an ice bath and analyzed immediately or stored frozen at −20°C. Freezing and thawing destroys enzyme activity, but does not alter the percentage of uridine already converted to uracil.

C. Chromatographic Analysis of Incubated Lysates

1. Chromatographic Assemblies (See Fig. 2)

Large test tubes (25 × 200 mm) serve as "development tanks." Samples are spotted (see below) on strips measuring 18 × 160 mm, cut from 18 × 22 inch sheets of Whatman No. 41 paper. These strips are supported on one end by fastening them with paper clips to wicks measuring 38 × 190 mm cut from Whatman No. 3 paper and rolled into cylinders 10 mm in diameter. The other ends of the strips are supported by paper clips soldered onto pieces of stiff wire. (Paper clips are a convenient source for such wire.) The clips fastened to wire are inserted near the edge of No. 4 silicone stoppers. Rubber stoppers are not suitable for use with the chromatography solvent. Rusting of the paper clips has no discernible effect on the chromatographic analysis.

2. Chromatographic Analysis

Chromatography strips are spotted at the origin (50 mm from the bottom edge) using micropipettes, with 0.005 ml of the chromatography standard solution (see Section II), and 0.02 ml of incubated lysate or supernatant

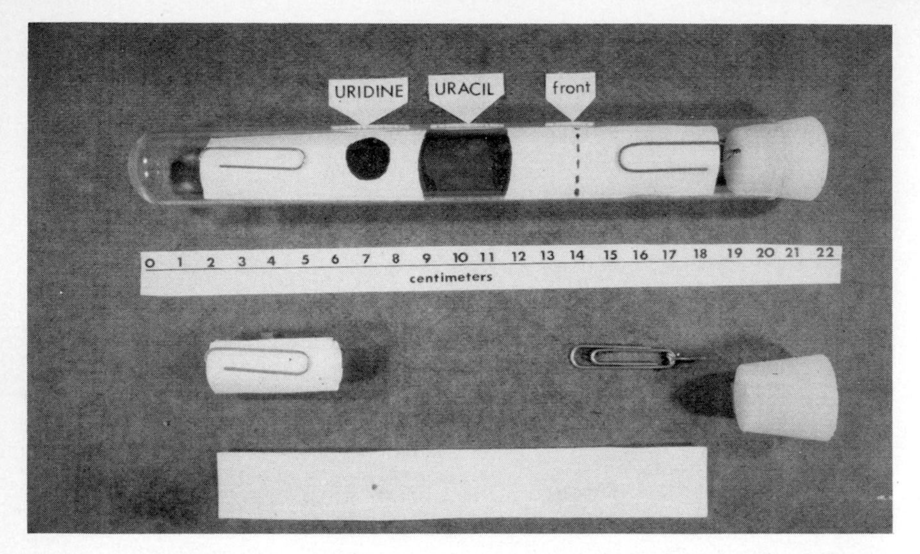

FIG. 2. Chromatographic assembly (see text for details). Migration of uridine ($R_f = O$) and uracil ($R_f = 0.4$) is represented in a "dummy" assembly shown in the top portion of the figure. A wick, strip, supporting clip, and stopper are shown below the centimenter rule.

culture medium (see pp. 235 and 237) (applied in a spot less than 10 mm in diameter with intermittent drying in four or five applications). As indicated above, strips are fastened with paper clips to wicks, leaving a clearance of 10 mm between the origin on the strip and upper edge of the wick. The upper portion of the strip is fastened to the paper clip, which is secured in a silicone stopper. Nine milliliters of chromatography solvent at 37°C is added to the bottom of a prewarmed chromatography test tube. Chromatography at room temperature is slower and requires a solvent with a greater proportion of butanol. The combined wick, strip, and stopper are lowered carefully into the tube, and the stopper is inserted to make a tight seal. The entire assembly is slowly tilted to rest in a horizontal position, with the strips remaining out of the solvent. Once horizontal, tubes must be prevented from rolling over and thereby immersing the chromatography strips directly in the solvent. Chromatograms can be developed in a vertical position, but the solvent advances at a slower rate. Chromatograms are allowed to develop at 37°C for 20 minutes, during which time the solvent front advances 70 to 90 mm. After development, the assembly is held vertically while stopper, strip, and wick are removed and strips are cut away from wicks. Strips are then dried under a heat lamp or in an oven. Carrier spots of uridine and uracil (deriving from the chromatography standard solution) are localized in the dark under UV light and cut out [R_f's: uridine 0.0 (the borate–ribose complex is immobile); uracil 0.4]. Radioactivity in "uridine" and "uracil" areas, as well as a section near the solvent front ("background") is determined using a scintillation counter. Of a total of 20,000–30,000 cpm applied to each strip, no significant amount of radioactivity can be detected other than in the "uridine" and "uracil" areas. Cell lysates sometimes contain a variable amount of nonradioactive UV-absorbing material trailing the uracil, but this seldom obscures the separation between uracil and uridine. Even in the event of apparently poor resolution, an area of demarcation between uridine and uracil can be demonstrated by slightly charring the strip over a bunsen burner.

D. Evaluation of Results

Results are expressed as percent conversion to uracil: "uracil" cmp ÷ ("uracil" cpm + "uridine" cpm) × 100. Uridine phosphorylase activity is calculated by subtracting percentage conversion in the "boiled control" (usually 5 to 10%) from percentage conversion in the original lysate; in some instances, this baseline conversion can be ignored and boiled controls are not run. Commercial preparations of uridine-^{14}C often contain 3 to 5% uracil -^{14}C as a contaminant. Enzyme levels in replicate contaminated cultures are quite similar; in an experiment performed in quadruplicate, the average conversion after 180 minutes was 65.9 ±2.6%. High uridine phosphorylase levels are easily recognized, often after only 30 minutes' incubation (see 3T3, group D, Table III), and a single time point suffices to identify this type of culture. For low rates of conversion, however, the increment in conversion between 30 and 180 minutes is more meaningful than a single time point (Table II). Thus, cultures are presumed positive if conversion to uracil has reached 50% by 30 minutes, or if the corrected percent conversion to uracil increases by more than 10% between 30 and 180 minutes of incubation. In two contaminated cell lines (virus-transformed macrophage and muntjac fibroblast) the interpretation of results was more complicated. In repeated tests on these cell types, Triton lysates exhibited 20–40% conversion of uridine to uracil after 30 minutes incubation but no further increase was noted after 180 minutes incubation. However, when phosphorylase activity was measured by adding uridine-^{14}C to the supernatant growth medium of living cultures of these cell lines (rather than to Triton lysates of the cells), 90% conversion was observed after 30 minutes incubation. The decreased conversion detected using Triton lysates probably stems from instability of the enzyme in these cell lines after cell disruption. A possible explanation might be liberation of "proteases" by Triton treatment, but addition of bovine serum albumin and/or HeLa cell lysates did not increase the enzyme activity observed using Triton lysates from the two lines in question. For borderline cases, more precise values should be obtained from repeated assays and expressed as activity per cell for suspension cultures or per milligram of cell protein for monolayer cultures. Protein content of Triton lysates (diluted 1:50 in reagent "C") can be determined by a modification of the Lowry method (Oyama and Eagle, 1956).

IV. Levels of Uridine Phosphorylase Activity in Various Mycoplasma and Cell Types

A. Uridine Phosphorylase in Cultured Cells and Mycoplasma

Figure 3 illustrates the rate of uridine conversion to uracil by lysates from several types of cell cultures as determined by the uridine phosphorylase assay described above. Mycoplasma-infected "RA" cells clearly exhibit a

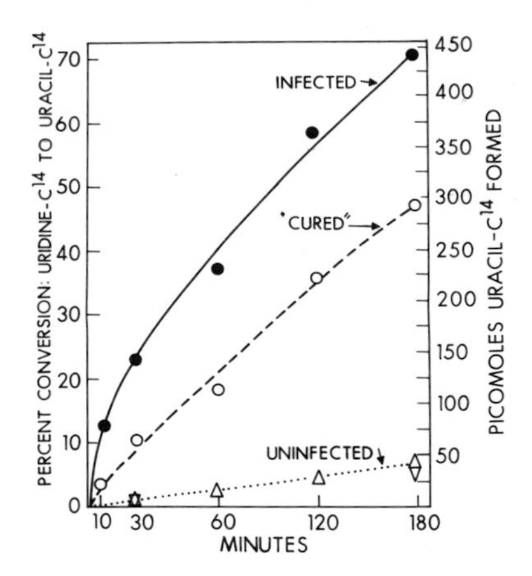

FIG. 3. Uridine phosphorylase activity in infected and uninfected cell cultures. Lysates were prepared as indicated in the text from the following monolayer cultures [origins of cell line are given in Levine (1972)]. ●—●, Infected, "RA" cells contaminated with *Mycoplasma hyorhinis* detected by agar plate cultivation and growth inhibition by appropriate antiserum (Hayflick, 1965; Stanbridge and Hayflick, 1967; Levine *et al.*, 1968). ○—○, Cured, a sister culture of RA cells treated with Aureomycin (0.0025 mg/ml), tylocine (1.5 mg/ml), and kanamycin (0.5 mg/ml) for 4 days and negative by agar plate and broth cultivation. ▽ ··· ▽, Uninfected. HeLa cells uncontaminated as demonstrated by agar plate and broth cultivations and electron microscopical examination. ▽ ··· ▽, Uninfected, KL-2 cells uncontaminated as demonstrated by agar plate and broth cultivations. Lysates were incubated at 37°C, and samples were withdrawn at indicated times. Uridine phosphorylase activity, percentage of conversion to uracil, and picomoles of uracil formed are calculated from a specific activity of 40 mCi/mmole and an approximate counting efficiency of 10^6 cpm/μCi.

level of enzyme activity significantly higher than that in uninfected HeLa or KL-2 cells: a representative contaminated culture converted approximately 70% of the uridine-^{14}C in the incubation mixture to uracil-^{14}C within 180 minutes, while only a 7% conversion was detected in lysates from uninfected cells. As mentioned previously, certain antibiotics have been employed as "antimycoplasma agents." However, treatment of contaminated cultures with Aureomycin, tylocine, and kanamycin reduces, but by no means eliminates uridine phosphorylase activity, even though agar plates and broth cultivations from such cultures are negative. Conversely, the depression of enzyme activity by antibiotic treatment implicates microorganisms as the source of that activity.

Further evidence that uridine phosphorylase activity is associated with mycoplasma is given in Table I. Significant enzymatic activity was easily detected in broth cultures of all ten mycoplasma species tested, seven of which have been reported as tissue cultures contaminants (Hayflick, 1965). *M. hyorhinis, M. laidlawii*, and *M. arginini*, which are the contaminants most often identified in infected cultures, all exhibited high relative activities.

It should be noted, however, that enzymatic activity was not always exactly proportional to mycoplasma titer (colony-forming units, CFU on agar plates). Different mycoplasma isolates exhibited varying enzymatic activities per CFU depending on environmental conditions. For instance, activities were generally higher for mycoplasma grown in cell cultures than for broth-grown mycoplasma. On the other hand, some infected cell cultures exhibited high enzymatic activity and low or apparently negative mycoplasma titers (see Table III), and in other cases antibiotic treatment resulted in a greater decrease in mycoplasma titer than in enzyme activity. Apparently, some of the factors (e.g., antibiotic treatment) that influence colony-forming ability on agar plates do not affect mycoplasma-associated uridine phosphorylase activity to the same extent.

B. Low Levels of Phosphorylase in Uncontaminated Cells

In contrast to the rapid conversion by axenic mycoplasma cultures, most uncontaminated cell cultures assayed under the same conditions showed only minimal uridine phosphorylase activity. In repeated tests of a total of 20 different uncontaminated normal, cancer, and virally transformed cell cultures derived from a variety of animal species and tissues, none showed an increment in conversion greater than 10% between 30 and 180 minutes, and the average was less than 5% (Table II). On the other hand, conversion after 180 minutes incubation by 50 different mycoplasma-infected cultures ranged from 20% to 95%, with an average of 60%.

TABLE I

URIDINE PHOSPHORYLASE ACTIVITY IN REPRESENTATIVE *Mycoplasma* SPECIES

Mycoplasma species	Presumed origin	Relative frequency as tissue culture contaminant[a]	Percent conversion to uracil at: 30 Min	180 Min	Relative uridine phosphorylase activity[b]
M. hyorhinis (GDL)	Porcine	4+	43	90	100
M. laidlawii	Bovine, saprophytic	4+	89	90	313
M. arginini	Ovine, caprine	4+	88	89	312
M. hominis 1	Human	4	12	31	15
M. orale 1	Human	4	19	38	19
M. salivarium	Human	2	28	80	55
M. gallisepticum	Avian	1	13	20	6
M. neurolyticum	Rat	0	30	76	53
M. arthriditis	Rat	0	23	79	49
M. fermentans	Human	0	84	93	280

[a]An approximation derived from frequencies of contamination previously reported by several authors (Hayflick and Channock, 1965; Macpherson, 1966; Levine *et al*., 1968; Hayflick, 1972). In this Laboratory, *M. hyorhinis* has been the most commonly detected contaminant.

[b]Calculated from the reciprocal of the time required for a 50% conversion (uncorrected) to uracil by lysates from 48-hour broth cultures (Hayflick, 1965) containing 10^7 to 10^9 colony-forming units per milliliter. Lysates were prepared by centrifuging 9.5 ml of broth culture at 1000 *g* for 10 minutes and resuspending the pellet in 3 ml of incubation medium. The value for *M. hyorhinis* has been set arbitrarily at 100; this culture effected a 50% conversion of uridine to uracil in 50 minutes.

TABLE II
URIDINE PHOSPHORYLASE ACTIVITY IN UNCONTAMINATED[a] CELL CULTURES

Cell cultures[b]	Origin[c]		Increment[d] in urdine phosphorylase activity
	Species	Tissue or cell type	(%)
HeLa	Human	Cervical carcinoma	4.1[e]
HeLa S3	Human	Cervical carcinoma	4.0
MS-2	Human	Diploid skin fibroblast	0.8
KL-2	Human	Diploid lung fibroblast	6.9
Chang Liver	Human	Liver	6.9
J-111	Human	Monocytic leukemia	3.7
MA160	Human	Prostate[f]	8.8
WI-26-VA	Human	SV40-transformed lung fibroblast	6.4
3T3	Mouse	Embryo	3.9
3T3/Balb c	Mouse	Embryo[g]	0.2
C57	Mouse	Embryo[h]	0.4
45.6	Mouse	Myeloma	1.6
V9	Mouse	Bone marrow	5.3
V10	Mouse	Rauscher-virus infected bone marrow	5.9
Neuro	Mouse	Neuroblastoma	4.3
HTC	Rat	Hepatoma	7.8
E-3	Rat	Liver[i]	3.3
NIL-2	Hamster	Embryo	6.8
B-1	Hamster	BudR-resistant BHK-21/13[j]	2.3
Lens	Rabbit	Lens	9.1

[a]As adjudged by aerobic agar plate cultivation in a 5% CO_2 atmosphere; for other criteria, see footnote *e*.

[b]All cells were grown as monolayer cultures except HeLa S3, which was grown in suspension with mechanical agitation, and the mouse myeloma lines, which were cultured as nonadherent cells in plastic petri dishes.

[c]Unless otherwise indicated, origins of cell lines are given in Levine (1972).

[d]Increase in percentage conversion to uracil between 30 and 180 minutes. With low levels of activity this is a more meaningful value than absolute conversion at a single time point (see text).

[e]Negative by electron microscopic examination, sucrose density gradient analysis (Levine *et al.*, 1967, 1968), agar plate and broth cultivation with cultured cell feeder layers (Zgorniak-Nowosielska *et al.*, 1967), and agar plates incubated anaerobically.

[f]Richman *et al.* (1972).

[g]Aaronson and Todaro (1968).

[h]Initiated by Mrs. M. Solomon, The Wistar Institute.

[i]Initiated by Drs. A. Schwartz (Temple University) and H. Eagle (Albert Einstein College of Medicine).

[j]Littlefield and Basilico (1966).

TABLE III

CORRELATION BETWEEN ELEVATED URIDINE PHOSPHORYLASE ACTIVITY AND MYCOPLASMA IN CONTAMINATED CELL CULTURES

Cell designation[a]	Results of agar plate cultivation[b]	Elevated uridine phosphorylase activity[c] (%)
	A Representative contaminated cultures[d]	
HeLa	+	56
HeLa S3	+	59
L-929	+	74
CV-1	+	44
R. Glia	+	85
	B Contamination with various mycoplasma species[e]	
3T3	+	65
RA	+	82
KL–2–O	+	21
KL-2-H	+	26
HeLa-Ar	+	74
HeLa-La	+	59
	C Antibiotic treatment[f]	
RA	+	22[g]
RA + anti	−	7[g]
3T3	+	21[g]
3T3 + anti	−	9[g]
SV 3T3	+	
SV 3T3 + anti-1	−	16
SV 3T3 + anti-2	−	72
	D "Quasi-negative" cultures	
3T3-"A"	−	4
3T3-"B"	−	46

C. Other Characteristics of Uridine Phosphorylase Activity in Cultured Cells

The correlation between elevated uridine phosphorylase activity and mycoplasma contamination is further considered in Table III. Most cultures with high enzymatic activity were also positive by agar plate cultivation (group A), and more importantly, no culture has yet been encountered with cultivable mycoplasma and a negative result from the uridine phosphorylase method. Several mycoplasma species commonly detected in contaminated cultures cause an elevated enzyme activity in infected cultures (Table III, group B), consistent with the levels of uridine phosphorylase activity present in axenic mycoplasma broth cultures (Table I).

TABLE III (cont'd.)

Cell designation[a]	Results of agar plate cultivation[b]	Elevated uridine phosphorylase activity[c] (%)
HeLa S3-"A"	–	4
HeLa S3-"B"	–	37
Neuro-"A"	–	5
Neuro-"B"	–	60
V9-"A"	–	5
V9-"B"	–	30
V-10-"A"	–	6
V-10-"B"	–	95
	E Effects of various treatments on "False positives" and "Quasi-negatives"	
PK	–	91
PK-heated	–	93
Neuro	–	66
Neuro-heated	–	8
R. Glia	+	85
R. Glia "cured"	–	96
R. Glia "cured" + Heat	–	91
R. Glia "cured" + Heat + anti	–	93

[a]See Levine (1972) for characterization of cell types.
[b]Hayflick (1965); Zgorniak-Nowosielska *et al.* (1967).
[c]Percentage conversion to uracil in 180 minutes.
[d]Species of contaminating mycoplasma not identified, unless otherwise indicated.
[e]3T3 and RA were shown to be infected with *M. hyorhinis* by antiserum growth inhibition tests (Stanbridge and Hayflick, 1967). KL-2-O was deliberately infected with *M. orale I* and KL-2-H with *M. hominis I*, HeLa-Ar with *M. arginini*, and HeLa-La with *M. laidlawii*.
[f]Treated with antibiotics (see Fig. 2) for periods of from 4 days to 2 weeks.
[g]Thirty-minute values.

Although most mycoplasma-infected cultures were able to affect conversion even when intact cells and organisms were tested in their original growth medium, contaminated RA cultures (group B) did not convert uridine to uracil unless disrupted in Triton-containing buffer. Consistent with this indication that the mycoplasma present in RA cultures are "cryptic strains" (i.e., impermeable to nucleosides) is the observation that such cultures pulse labeled with uridine-^{14}C and analyzed by sucrose density gradient analysis did not contain labeled mycoplasma RNA (Levine *et al.*, 1967, 1968). Furthermore, autoradiographs of RA cultures pulse labeled with thymidine-^{3}H did not indicate any cytoplasmic incorporation of the type usually observed in mycoplasma-infected cell cultures.

Data for cultures in group C illustrates the effect of antibiotic treatment on mycoplasma-infected cultures. In agreement with the results presented in Fig. 3, "cured" cultures are negative by agar plate assay, but retain an appreciable measure of phosphorylase activity. In addition, as illustrated by SV3T3, high levels of enzyme activity may reappear in subsequent subcultures even in the presence of antibiotics.

The fact that antibiotic treatment affects the ability of mycoplasma to form visible colonies on agar plates, and to a lesser extent, their capacity to synthesize uridine phosphorylase, bears on the results obtained with cultures in group D, "quasi-negative cultures." When tested initially, most of these cultures (3T3–"A," HeLa–S3–"A," etc.) exhibited low levels of uridine phosphorylase. Subsequently, when cells of the same cell line (derived from these initial stocks or obtained from different sources) were tested, some exhibited high levels of uridine phosphorylase in the absence of cultivable mycoplasma (3T3–"B," HeLa–S3–"B," etc.); i.e., agar plates were "negative." In all the above cases enzyme activity was depressed in sister cultures treated with antibiotics. A similar phenomenon has been noted for cultures of L-929 cells obtained from different sources, with the exception that in some cultures the level of uridine phosphorylase activity was not depressed after antibiotic treatment (cf. Table III, group E). Therefore, unless clonal variations with elevated enzyme levels have occurred in all these lines (Hillcoat, 1971), one may conclude that the enzyme levels in group C cultures are not inherent in those mammalian cell lines, but derive from an extracellular agent. In the case of these quasi-negative cultures, then, this method is able to detect mycoplasmalike organisms which cannot be cultivated on agar plates.

Results presented for cultures in group E are more difficult to interpret. Cultures of line PK consistently exhibited extremely high levels of uridine phosphorylase activity which was not depressed in sister cultures treated with antibiotics. Furthermore, results from other detection methods (Levine *et al.*, 1967, 1968; Todaro *et al.*, 1971) (including the growth of cells in the presence of broth for several weeks) were negative, and electron micrographs did not reveal the presence of extracellular mycoplasma. Further evidence that uridine phosphorylase activity in PK cultures might not be mycoplasma-associated was obtained by incubating such cultures at 42°C for 24 hours, a procedure which mammalian cell cultures can survive under the proper conditions (Levine and Robbins, 1970), but which kills most mycoplasma (Hayflick, 1960). Even after two cycles of such treatment, surviving PK cells still possessed high enzyme levels, as did daughter cultures derived from them. In contrast, when a quasi-negative culture, Neuro, was treated similarly, enzyme levels were depressed to normal levels, indicating that in these cultures uridine phosphorylase activity was mycoplasma-associated. However, certain other mycoplasma-contaminated cultures did not exhibit this characteristic depression of enzyme levels after treatment with either antibiotics or heat. One such cell line, Rat Glia, was demonstrably infected with mycoplasma by agar plate assay and exhibited a high level of uridine phosphorylase when initially tested. After antibiotic "cure" of these cells, agar plate and broth cultivation tests were negative and enzyme levels remained elevated. Subsequently the line was maintained in antibiotic-free medium for a period of months; agar plate and broth cultivation tests performed at this stage were still negative, and enzyme levels were high, but could not be depressed by either antibiotic or heat treatment.

V. Discussion

The importance of maintaining cell cultures free of mycoplasma should be self-evident. Furthermore, although agar plate cultivation has been established as an absolute requirement for the taxonomic characterization of mycoplasma (Hayflick and Chanock, 1965), it is apparent from the data presented here (Table III) and elsewhere (Zeigel and Clark, 1969; Todaro *et al.*, 1971; Clark, 1974; Schneider *et al.*, 1973a,b), that cell cultures can contain contaminants resembling mycoplasma which cannot be cultivated readily on synthetic media. Clearly, multiple approaches are advisable for detecting mycoplasma contamination, and the determination of uridine phosphorylase activity described above represents a simple, rapid, and relatively sensitive procedure which can be used in conjunction with other biochemical and cultivation methods. It is worth noting that the "cryptic" mycoplasma strains reported in these studies cannot be detected by autoradiographic, density gradient, or gel electrophoresis methods if uridine or thymidine is used as a tracer, and additional mycoplasma variants probably exist which cannot transport other nucleic acid precursors.

Nevertheless, certain drawbacks of the phosphorylase method should be kept in mind. No attempt has been made to distinguish between enzyme activity due to contaminating mycoplasma and that which might stem from other microorganisms (Heppel and Hilmoe, 1952; Paege and Schlenk, 1952; Razzell and Kohrana, 1958), although in practical terms such distinction matters little to the investigator. A more important consideration is the occurrence of significant enzyme activity in uncontaminated cell types. Most uncontaminated serially propagated cell types surveyed in the present investigation exhibited low levels of uridine phosphorylase activity, in contrast to the high levels of this enzyme found in many mammalian tissues (see Levine, 1972). This apparent inconsistency may be due to the significant loss of some enzyme activities ("dedifferentiation") which occurs in most serially propagated cells. Conversely, high enzyme levels in PK cells may represent retention of specialized function. In the case of PK and in rat glial cells, however, it may be that enzyme activity found in such "false positives" derives from some sort of cellular response to a past transient infection, and this possibility is currently under investigation. Except in the latter event, cell-associated phosphorylase activity can be distinguished from mycoplasma-associated activity by the depression of the latter in antibiotic and heat-treated cells (see Fig. 2 and Table III, C, E).

The rapidity and sensitivity of the uridine phosphorylase method probably can be improved further. Higher levels and specific activities of isotope and more concentrated cell lysates should allow for shorter incubation times, while, in some cases, it may be more advantageous to use thin-layer chromatography as a separation technique. In addition, under certain circumstances it may be advantageous to determine the conversion of thymidine to thymine as a measure of phosphorylase activity. The present procedure yields results in 1 to 4 hours, is superior to agar plate cultivation and several other methods in detecting most contaminants encountered in this laboratory and should be satisfactory under most circumstances.

ACKNOWLEDGMENTS

The expert technical assistance of Miss Susanne Arelt was invaluable in the development of this detection method. I am grateful also to Mrs. Barbara Becker, who assisted in the more recent experiments. Some of this work was initiated at the Albert Einstein College of Medicine and reported in an earlier publication (Levine, 1972). These investigations were supported by grants from the National Institutes of Health (GM-20306) and the National Science Foundation (GB-35590), and the author was a NIH Career Development Awardee (K3-AI-8532) during the initial phases of this research.

REFERENCES

Aaronson, S. A., and Todaro, G. J. (1968). *J. Cell. Physiol.* **72**, 141–148.
Barile, M. F., Schimke, R. T., and Riggs, D. B. (1966). *J. Bacteriol.* **91**, 189–192.
Carski, T. R., and Shephard, C. C. (1961). *J. Bacteriol.* **81**, 626–635.
Clark, H. F. (1974). *Prog. Med. Virol.* (in press).
Eagle, H. (1959). *Science* **130**, 432–437.
Edward, D. G. (1954). *J. Gen. Microbiol.* **10**, 27–64.
Fogh, J., and Fogh, H. (1964). *Proc. Soc. Exp. Biol. Med.* **117**, 899–901.
Fogh, J., and Fogh, H. (1965a). *Proc. Soc. Exp. Biol. Med.* **119**, 233–238.
Fogh, J., and Fogh, H. (1965b). *Proc. Soc. Exp. Biol. Med.* **119**, 944–950.
Girardi, A. J., Hayflick, L., Lewis, A. M., and Somerson, N. L. (1965). *Nature (London)* **205**, 188–189.
Hakala, M. T., Holland, J. F., and Horoszewicz, J. S. (1963). *Biochem. Biophys. Res. Commun.* **11**, 466–471.
Harley, E. H., Rees, K. R., and Cohen, A. (1970). *Biochim. Biophys. Acta* **213**, 171–182.
Hayflick, L. (1960). *Nature (London)* **185**, 783–784.

Hayflick, L. (1965). *Tex. Rep. Biol. Med.* **23**, 285–303.
Hayflick, L. (1972). *Pathogenic Mycoplasmas, Ciba Found. Symp.* pp. 17–31.
Hayflick, L., and Chanock, R. M. (1965). *Bacteriol. Rev.* **29**, 186–221.
Hayflick, L., and Stanbridge, E. (1967). *Ann. N.Y. Acad. Sci.* **143**, 608–621.
Heppel, L. C., and Hilmoe, R. J. (1952). *J. Biol. Chem.* **198**, 683–694.
Hillcoat, B. L. (1971). *J. Nat. Cancer Inst.* **46**, 75–80.
Holland, J. F., Korn, R., O'Malley, J. O., Minnemeyer, H. J., and Tieckelmann, H. (1967). *Cancer Res.* **27**, 1867–1873.
Horoszewicz, J. S., and Grace, J. T. (1964). *Bacteriol. Proc.* **64**, 131.
House, W., and Waddell, A. (1967). *J. Pathol. Bacteriol* **93**, 125.
Hummler, K., and Armstrong, D. (1967). *Ann. N.Y. Acad. Sci.* **143**, 622–625.
Kraemer, P. M. (1964a). *Proc. Soc. Exp. Biol. Med.* **115**, 206–212.
Kraemer, P. M. (1964b). *Proc. Soc. Exp. Biol. Med.* **117**, 910–918.
Kraemer, P. M., Defendi, V., Hayflick, L., and Manson, L. A. (1963). *Proc. Soc. Exp. Biol. Med.* **112**, 381–387.
Levine, E. M. (1972). *Exp. Cell Res.* **74**, 99–109.
Levine, E. M., and Robbins, E. B. (1970). *J. Cell. Physiol.* **76**, 373–380.
Levine, E. M., Burleigh, I. G., Boone, C. W., and Eagle, H. (1967). *Proc. Nat. Acad. Sci. U.S.* **57**, 431–438.
Levine, E. M., Thomas, L., McGregor, D., Hayflick, L., and Eagle, H. (1968). *Proc. Nat. Acad. Sci. U.S.* **60**, 583–589.
Lindsay, R. H., Wong, M.-Y, Romine, C. J., and Hill, J. B. (1968). *Anal. Biochem.* **24**, 506–514.
Littlefield, J. W., and Basilico, C. (1966). *Nature (London)* **211**, 250–252.
McCarty, K. S., Woodson, B., Amstey, M., and Brown, O. (1964). *J. Biol. Chem.* **239**, 544–549.
Macpherson, I. (1966). *J. Cell. Sci.* **1**, 145–168.
Macpherson, I., and Russell, W. C. (1966). *Nature (London)* **210**, 1343–1345.
Markov, G. G., Bradvarova, I., Mintcheva, A., Petrov, P., Shishkov, N., and Tsanev, R. G. (1969). *Exp. Cell Res.* **57**, 374–384.
Nardone, R. M., Todd, G., Gonzalez, P., and Gaffney, E. V. (1965). *Science* **149**, 1100–1101.
Negroni, G. (1964). *Brit. Med. J.* **1**, 927–929.
Newnham, A. G., and Chu, H. P. (1965). *J. Hyg.* **63**, 1–23.
Oyama, V. I., and Eagle, H. (1956). *Proc. Soc. Exp. Biol. Med.* **91**, 305–307.
Paege, L. M., and Schlenk, F. (1952). *Arch. Biochem.* **40**, 42–49.
Paton, G. R., Jacobs, J. P., and Perkins, F. T. (1965). *Nature (London)* **207**, 43–45.
Perez, A. G., Kim, J. H., Gelbard, A. S., and Djordjevic, B. (1972). *Exp. Cell Res.* **70**, 301–310.
Perlman, D., Rahman, S. B., and Semar, J. B. (1967). *Appl. Microbiol.* **15**, 82–85.
Pollack, M. E., Treadwell, P. E., and Kenny, G. E. (1963). *Exp. Cell Res.* **31**, 321–328.
Pontis, H., Segerstedt, G., and Reichard, P. (1961). *Biochim. Biophys. Acta* **51**, 138–147.
Randall, C. C., Gafford, L. G., Gentry, G. A., and Lawson, L. A. (1965). *Science* **149**, 1098–1099.
Razzell, W. E., and Kohrana, H. G. (1958). *Biochim. Biophys. Acta* **28**, 562–566.
Richman, A. V., Vincent, M. M., Marteno, E. C., and Tauraso, N. M. (1972). *Cancer Res.* **32**, 2186–2189.
Robinson, L. B., Wickelhausen, R. H., and Roizman, B. (1956). *Science* **124**, 1147–1148.
Rose, I. A., and Schweigert, B. S. (1951). *J. Amer. Chem. Soc.* **73**, 5903.
Rothblat, G. H. (1960). *Ann. N. Y. Acad. Sci.* **79**, 430–432.
Rouse, H. C., Bonifas, V. H., and Schlesinger, R. W. (1963). *Virology* **20**, 357–365.

Russell, W. C. (1966). *Nature (London)* **212**, 1537–1540.
Russell, W. C., Niven, J. S. F., and Berman, L. D. (1968). *Int. J. Cancer* **3**, 191–202.
Sabin, A. B. (1968). *Ann. N.Y. Acad. Sci.* **143**, 628–634.
Schimke, R. T. (1967). *Ann. N.Y. Acad. Sci.* **143**, 573–577.
Schimke, R. T., and Barile, M. F. (1963a). *Exp. Cell Res.* **30**, 593–596.
Schimke, R. T., and Barile, M. F. (1963b). *J. Bacteriol.* **86**, 195–206.
Schneider, E. L., Epstein, C. J., Epstein, W. B., Betlach, M., and Abbo-Halbasch, G. (1973a). *Exp. Cell Res.* **79**, 343–349.
Schneider, E. L., Stanbridge, E. J., and Epstein, C. J. (1973b). In preparation.
Shedden, W. I. H., and Cole, B. C. (1966). *Nature (London)* **210**, 868,
Sköld, O. (1960). *Biochim. Biophys. Acta* **44**, 1–12.
Stanbridge, E. (1971). *Bacteriol. Rev.* **35**, 206–227.
Stanbridge, E., and Hayflick, L. (1967). *J. Bacteriol.* **93**, 1392–1396.
Stanbridge, E., Önen, M., Perkins, F. T., and Hayflick, L. (1969). *Exp. Cell Res.* **57**, 397–410.
Todaro, G. J., and Green, H. (1966). *Virology* **28**, 756.
Todaro, G. J., Aaronson, S. A., and Rands, E. (1971). *Exp. Cell Res.* **65**, 256–257.
Wolanski, B., and Maramorosch, K. (1970). *Virology* **42**, 319–327.
Zeigel, R. F., and Clark, H. F. (1969). *J. Cell Biol.* **43**, 163a.
Zgorniak-Nowosielska, I., Sedwick, W. G., Hummler, K., and Koprowski, H. (1967). *J. Virol.* **1**, 1227–1239.
Zucker-Franklin, D., Davidson, M., and Thomas, L. (1966). *J. Exp. Med.* **124**, 531–532.

Chapter 15

The Ultra-Low Temperature Autoradiography of Water and Its Solutes[1]

SAMUEL B. HOROWITZ

Laboratory of Cellular Physiology, Department of Biology, Michigan Cancer Foundation, Detroit, Michigan

I. Introduction

One hardly need emphasize the importance of knowledge of the spatial distribution of metabolites and pharmacological agents at the tissue and cellular levels. The increase in number and sophistication of transport or "membrane" studies in the last decade reflects this importance, and is prob-

[1] This work was supported by grants from the National Institutes of Health, GM 19548, from the National Science Foundation, GB 36005, and by an institutional grant to the Michigan Cancer Foundation from the United Foundation of Greater Detroit.

ably second in magnitude only to those in molecular biology. Nevertheless, with the exception of the area of artificial membranes, there has not yet been a parallel inflorescence in the development of analytical techniques in transport, and a paucity of widely applicable quantitative methods exists. An obstacle to the development of such methods is the inexorability of small solute diffusion. In biological systems, distances are short and transport barriers are unstable. Once the experimenter has intruded, commonly a rapid redistribution of small molecules masks or belies conditions in the living state. Analytical methods are needed which, first, restrain solute redistribution and, second, permit quantitative analysis of solute concentration in spatial arrays.

One promising class of methods is autoradiography under conditions that keep the sample at a sufficiently low temperature to prevent solute redistribution from the time of original freezing and sectioning to the completion of the autoradiograph. In principle, this method allows quantitative localization of any solute that can be radioactively labeled. Such localizations now have been achieved with sufficient resolution and sensitivity to recommend the technique for a wide range of problems in cell physiology. The method, as practiced in our laboratory, is described here.

The autoradiographic method to be described traces its genealogy directly from that of Kinter *et al.* (1960), whence it developed to the "modified sandwich technique" of Kinter and Wilson (1965). Modifications from our laboratory in the technique and its quantitation are described in a series of publications (Horowitz and Fenichel, 1968, 1970; Horowitz, 1972).

Of the autoradiographic techniques now in use, the present method is most like that of Appleton (1964, 1966, 1972) in that from the original quenching in liquid nitrogen to photographic development, the sample is never allowed to thaw, to come in contact with solvents, to be exposed to moisture-laden breath, or to intentionally freeze-dry, all of which may adversely affect resolution. There are, however, marked differences in the two techniques, and these can be appreciated by comparing the present paper with those of Appleton.

The overall method is represented diagrammatically in Fig. 1. The cell or tissue sample is embedded in an appropriate medium, frozen to $-196°C$, and stored until sections are cut. Sectioning is done in a cryostat, usually at $-50°C$. The sections are placed on a Teflon-coated slide, and another slide coated with dry photographic emulsion (the autoradiographic plate) is pressed on top to form a sandwich: slide–emulsion–section–Teflon–slide. This sandwich is maintained at $-85°C$ until exposure is complete. The sandwich then is separated and developed by conventional photographic methods. The final product: an autoradiograph in which the section is above

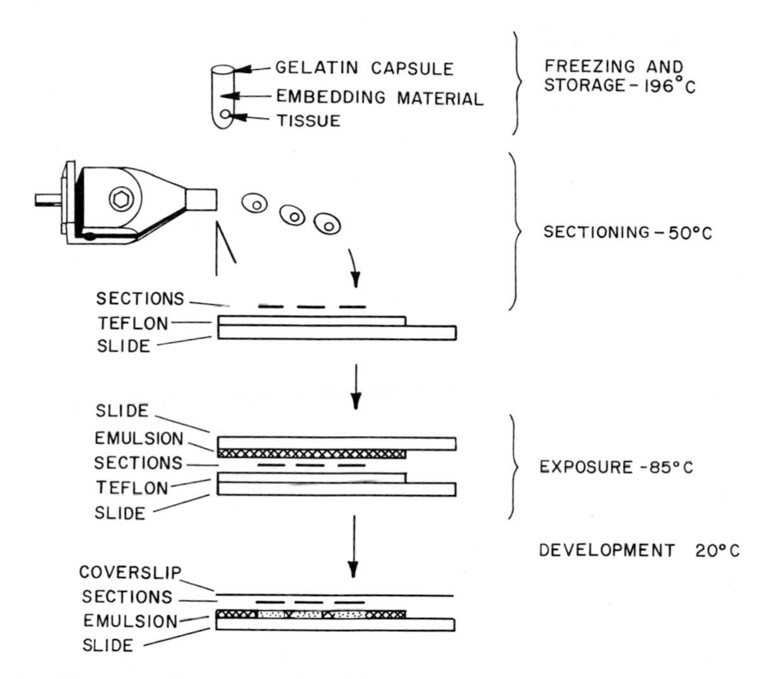

FIG 1. Diagrammatic representation of the steps in ultra-low temperature autoradiography.

the emulsion and in which the silver grain distribution records the quantity and spatial distribution of the radioactivity originally in the section. Grain counting and quantification complete the experimental procedure.

We believe one advantage of the present method is its diminished reliance on technical virtuosity as compared to other available methods. So far as possible we have tried to simplify the various steps and reduce the risk of failure. The techniques involved can be mastered by inexperienced persons with a day or two of instruction and 3 or 4 days of practice.

Keeping in mind that ultra-low temperature autoradiography is new and has been applied to only a limited range of problems, potential users should expect to find situations in which the methods to be described require modification.

II. Sample Embedding and Freezing

The objective of the freezing and embedding process is to reduce the temperature of the tissue sample to that of liquid nitrogen as rapidly as possible to prevent solute redistribution, and to do this in a matrix capable of supporting the sample during the rigors of sectioning.

The method of choice depends on the nature of the sample to be frozen. Two points should be adhered to in freezing. First, sample size should be as small as practical to ensure that cooling at the center of the sample occurs as rapidly as possible. Second, cooling should be performed by rapid immersion in any of a number of liquids cooled to their freezing point with liquid nitrogen. These liquids include propane (f.p. −189°C), isopentane (f.p. −161°C), or dichlorodifluoromethane (f.p. −160°C). We prefer dichlorodifluoromethane because it poses less of a fire hazard than the two alkanes.

Embedding materials vary widely and may include the actual tissue to be frozen, other tissues, specially prepared solutions, or commercially available matrix-forming glycol and resin mixtures such as O.C.T. compound (Ames Co., Elkhart, Indiana) or Cryoform (International Equipment Co., Needham, Massachusetts). We have at various times used 15% w/v bovine serum albumin dissolved in Ringers solution (Horowitz and Fenichel, 1968), minced beef liver, and O.C.T. compound (Horowitz and Fenichel, 1970).

A method of embedding and freezing we find particularly useful relies on the use of gelatin capsules and is described below. For routine work our microtome and entire procedure are set up for this form of embedding.

A. Materials

The following are required for embedding and freezing:

- Gelatin capsules
- Capsule jig
- Hypodermic syringe, 2-ml, containing O.C.T.,[2] with squared-off No. 18 needle
- Vacuum sample-handling device
- Dissecting forceps (11.5 cm) with insulated blades
- Long forceps (29.5 cm)
- Screw-cap vials with internal labels
- Asbestos gloves
- Freezing-bath ensemble
- Liquid nitrogen refrigerator

Capsules. We use size 5, 4, or 3 gelatin capsules. Only the narrow, longer portion of the capsule having an outside diameter of 0.45 to 0.53 cm is used.

Capsule jig. During embedding the capsules are held in a Plexiglas block (Fig. 2) with a graded series of drilled holes, each 0.5 cm deep having diameters appropriate to the capsule sizes.

[2]The O.C.T. compound used in our laboratory is formulated for cutting at −30° to −60°C but is no longer commercially available. A representative of the Ames Co. informs me that the O.C.T. compound currently marketed can be modified for use in the −30° to −60°C range by the addition of 5% propylene glycol w/w.

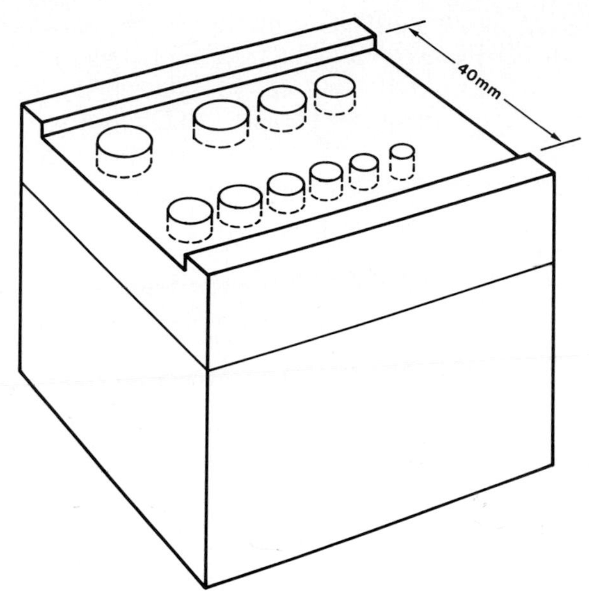

FIG. 2. Drawing of the capsule jig described in the text. Note the parallel rails 40 mm apart on the upper surface.

Vacuum sample-handling device. The tissue sample is manipulated within the small space of the capsule with a Telvac Model K-150 vacuum small parts handling system (Telvac Instruments Co., Van Nuys, California).

Freezing-bath ensemble. Figure 3 includes a diagram of the ensemble used for freezing. A tube containing dichlorodifluoromethane (Freon 12) hangs by braided copper wire from the edge of a Dewar flask containing liquid nitrogen. Adjusting the length of the copper wire keeps about one-half of the Freon 12 solid and the remainder liquid.

B. Procedure

The steps involved in embedding and freezing are diagrammed in Fig. 3.

1. Just before embedding begins, sufficient O.C.T. is added to occupy the curved portion of a capsule, Figs. 3a, b. Care is taken to avoid the walls of the capsule and to exclude air bubbles.

2. The tissue sample is picked up with the vacuum sample-handling device and deposited on the surface of the O.C.T. out of contact with the capsule walls (Figs. 3c, d).

3. Additional O.C.T. is added and allowed to flow over the sample until the capsule is filled to about three-fourths its height (Figs. 3e, f).

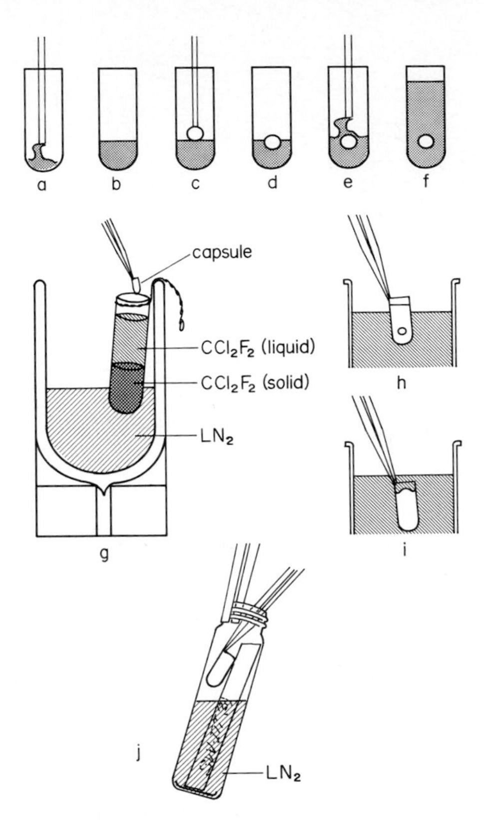

FIG. 3. Diagrammatic representation of the steps in embedding and freezing. (a–f) Embedding procedure; (g) the freezing-bath ensemble; (h–j), the freezing procedure. For additional explanation see text.

4. The capsule is grasped with the dissecting forceps, removed from the capsule jig, and lowered into the freezing bath of Freon 12 (Figs. 3g, h). If the top surface of the O.C.T. freezes before the bulk, the capsule will split and the sample will be unusable. To avoid this the capsule is totally immersed only when a small mound of O.C.T. in the middle of the top surface still is unfrozen (Fig. 3i).

5. When completely frozen, the capsule is placed directly into a labeled screw-cap vial which previously was cooled and filled with liquid nitrogen. The vial is held by long forceps, Fig. 3j. Pressing the capsule firmly against the inside of the vial frees it from the forceps. The vial is grasped with gloved hand, and the cap is put on loosely enough to prevent buildup of pressure within the vial. Vial and capsule are stored in a liquid nitrogen refrigerator.

III. Sectioning, Preparation of the Sandwich, and Exposure

All the steps described in this section are performed at temperatures appropriate to cascade refrigeration systems, −50° to −90°C. In preparation for sectioning, all materials are allowed to reach thermal equilibrium within the cryostat. The temperature of sectioned material must not exceed that of the cryostat.

A. Materials

The cryostat must be located in a darkroom and contain the following objects, illustrated in Fig. 4:

Microtome equipped with antiroll device and appropriate specimen chuck
Autoradiographic plates (in light-proof box)
Teflon-coated slides (in slide box)
Capsule jig
Slide frame
Small sable brush
Dissecting forceps (115 mm) with insulated blades
Curved-tip coverglass forceps
Scalpel (B & P No. 10) with insulated handle
Wooden-handled dissecting needles
Slide boxes containing two bulldog clips each

In addition, a cascade-type freezer or a dry-ice chest is required to store the autoradiograph during exposure.

Cryostat. Any cryostat equipped with a microtome can be used. However, the greater the working space and storage surface within the box, the easier and faster will be the various operations.

If spatial resolution on the order or 3 μm or better for tritiated solutes is required, the cryostat should be capable of maintaining temperatures of −50°C or lower when operating with the normal heat load. In practice this means cooling is by a cascade-type refrigeration system like those installed in microtome cyrostats by Harris Manufacturing Co., Cambridge, Massachusetts, Refrigeration for Science, Island Park, New York, and Slee, London, England.

Our laboratory uses the Wedeen modification of the Coon's type cryostat manufactured by Refrigeration for Science, with modifications in the lighting to conform to the requirements described below, and in the refrigeration

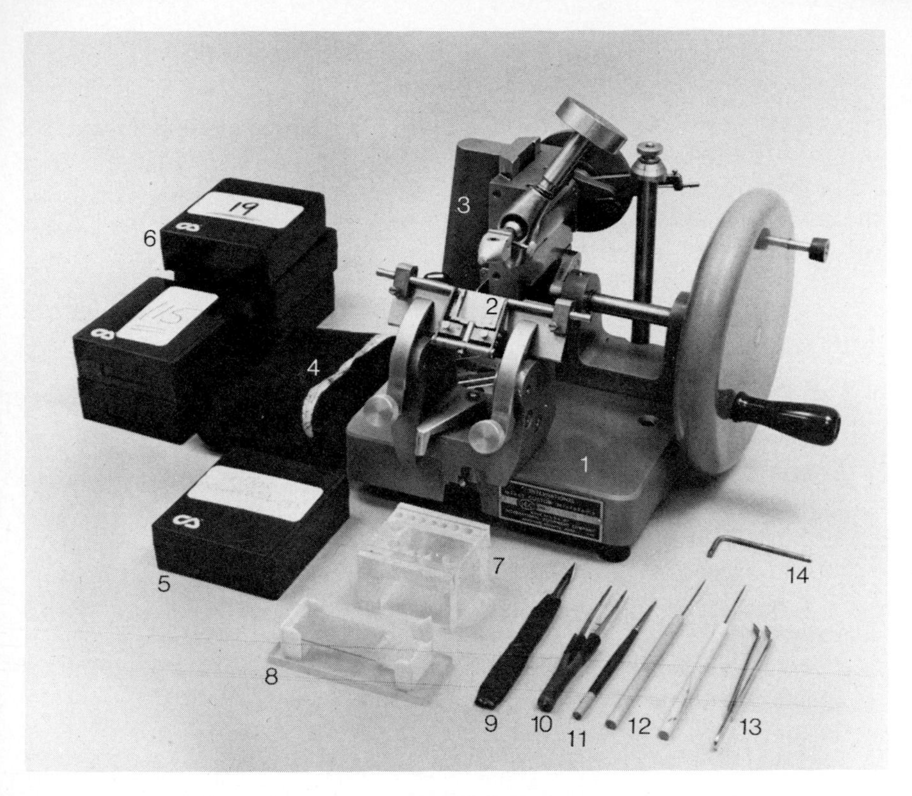

FIG. 4. Photograph of the contents of the cryostat during sectioning. 1, Microtome, with 2, antiroll device, and 3, capsule chuck; 4, lightproof box (in black cloth packet) containing autoradiographic plates; 5, box containing Teflon-coated plates; 6, numbered slide boxes containing 2 bulldog clips each; 7, capsule jig; 8, slide frame; 9, scalpel; 10, dissecting forceps with insulated blades; 11, small sable brush; 12, dissecting needles; 13, coverglass forceps; 14, wrench for capsule chuck.

system to increase reliability in continuous operation. Figure 5 shows the instrument.

The cryostat should have at least two appropriately placed sockets for incandescent bulbs: one for a safelight; the other for a white light source. The latter should be controlled by a footswitch outside the cryostat. Owing to their slow phosphorescent decay at cryostat temperatures, fluorescent lamps are unacceptable. The safelight must be appropriate to the emulsion used. For Kodak NTB emulsions the Gennert, 4 candlepower, dark ruby clear, darkroom bulb is satisfactory. In sufficient space, any of Kodak's Wratten 2 series safelights can be used. Light sources can be outside the box if no unacceptable glare from the surfaces of the cryostat results.

FIG. 5. Photograph of cryostat and operator. Features seen include: 1, a range switch, and 2, potentiometer for measuring temperature with thermocouples in various parts of the system; 3, rheostat for the heated gloves; 4, telephone mounted close to the operator. The panel, 5, controls the operation of the machine. The panel lights are removed in darkroom operation.

Hands and arms must be protected during extended exposure to the low temperatures involved. Lambskin gloves are inadequate at temperatures below about $-25°C$. We have been using capeskin gauntlets with electrically heated liner gloves, both available from the Suchy Division, Reflectronics Corp., New York, New York.

A new product, Arispace Glove Liners (Aris Gloves, New York, New York), woven from aluminized Lurex threads can replace the heated gloves in routine operation. They are much less expensive than heated gloves and require no electrical leads that may on occasion be a nuisance.

As an additional liner layer, it is also desirable to wear latex gloves, especially in closed cryostats. If latex gloves are not worn, frozen water vapor continuously escapes from the hands and can be disconcerting to the operator. The vapor may obscure the view of the cutting area and, more importantly, may coat the microtome blade surface with a tenacious frost. In time, frozen vapor reduces the efficiency of the refrigeration system by coating the evaporator surfaces with an insulating layer of frost.

Although three layers of gloves—capeskin gauntlet, liner, and latex gloves—restrict movement of the fingers, appropriate tools and devices compensate for this restriction.

Microtome. We use the Minot Custom Microtome (International Equipment Co., Needham, Massachusetts), but any microtome designed for cryostat operation probably is adequate. The specimen chuck should be appropriate to the material to be sectioned. In most of our work the tissue sample is embedded in a gelatin capsule, and it is simple to modify the LKB capsule chucks (LKB Instruments Inc., Rockville, Maryland) Models 4820 (B)-(G) (shown in Fig. 7a) to fit the chuck and ball of the Minot microtome.

At ultra-low temperatures, microtome lubrication becomes a problem. After testing a variety of lubricants we settled on Dow Corning 510 fluid, 50 centistoke viscosity (Dow Corning, Midland, Michigan).

An antiroll device is indispensable in frozen sectioning; that of Coons *et al.* (1951) is especially useful. See Fig. 7 for use. These can be obtained from Edward G. Dixon, 6690 Bush Road, Jamesville, New York 13078.

Autoradiographic plates. Kodak NTB and NTB-2 nuclear track emulsions are used for the autoradiographic plates. The property that weighed most heavily in the original choice of these emulsions is their relative freedom from pressure artifacts under our experimental conditions. However, subsequent to the original tests, a number of changes in the method were introduced, and other emulsions, previously rejected, may be useful.

We have used NTB-2 for tritium only; it has about twice the sensitivity to this isotope of NTB. NTB has given acceptable results with ^{3}H, ^{14}C, ^{22}Na and ^{35}S.

The autoradiographic plates are produced as follows. Kodak NTB emulsion, handled in total darkness or under a Kodak Wratten 2 safelight, is melted at 40°C and diluted 1:1 with distilled water, with careful stirring. Clean microscope slides (25 × 75 mm, single-frosted) are dipped to within 0.5 cm of the frosting and allowed to drain briefly. With the slide held firmly by the frosted end, frosting facing the operator, the back of the slide is wiped free of emulsion with a small damp cloth. The slide then is held up to a safelight and examined for air bubbles, which appear in transmitted light as bright spots on an otherwise uniform appearing surface. Slides with bubbles are discarded. Satisfactory slides are placed lengthwise on a rack at an 80° angle, and air-dried over lightly moistened absorbent paper at 20°C for 2 hours. The plates are stored at 4°C in light-tight boxes containing a packet of Drierite, which are sealed with black vinyl tape.

The light-tight boxes are Clay Adams No. A-1604B. By convention the front of the box is the face bearing the label slot. Slides are placed in the box with the frosted edge right and facing forward. Following this convention facilitates operations under the safelight.

Emulsion thickness of the dried slide, determined by weighing and averaged over the entire surface, is 1.9 ± 0.2μm. As first noted by Leblond *et al.* (1963), emulsion thickness is not uniform; it is thinner on the side which is uppermost during dipping and drying. The gradient in thickness causes systematic variation in grain densities when isotopes with energies greater than ^{3}H are used. This will be discussed in Section V, D.

Teflon-coated slides. These are prepared by bonding 0.003 inch bondable skived Teflon tape to single-frosted 25 × 75 mm microscope slides with epoxy Adhesive 85. Tape and adhesive are available from Fluoro-Plastics, Inc., Philadelphia, Pennsylvania. A thin coat of adhesive is spread uniformly over the clear portion of the frosted side of the slide, and a piece of Teflon tape, etched surface downward, is pressed on smoothly. An oversize strip of Parafilm is laid on top of the Teflon, and then another slide. The entire sandwich: slide–adhesive–Teflon–Parafilm–slide is held with two bulldog clips for 24 hours. Polymerized epoxy is removed from glass surfaces with a razor blade. The Teflon surface is wiped with a xylene-dampened cloth. Plates are stored in a slide box, the frosted area facing forward and to the right.

Capsule jig. If embedding is to be done in capsules, a device to hold the frozen capsules during preparation for sectioning is needed. A Plexiglas block with a series of holes, as described previously (Fig. 2), suffices for this purpose.

If two parallel rails about 45 mm apart are fixed to the top of the capsule jig, it can double as a platform to hold the slide frame close to the microtome knife.

Slide frame. The slide frame holds the Teflon-coated slide while sections are being cut, and aligns the autoradiographic plate when the sandwich is

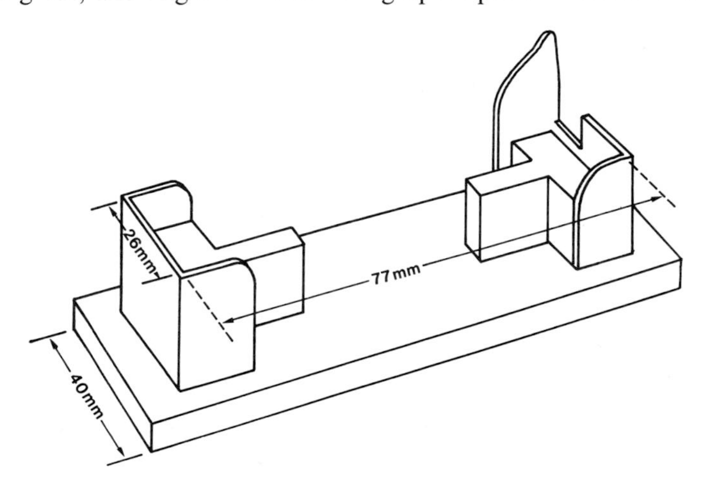

FIG. 6. Drawing of the slide frame, with dimensions.

made and clamped. It is an important aid in the dim light of the safelight. The slide frame is diagrammed, and its dimensions are given, in Fig. 6. The tab in the upper right corner serves to orient the operator.

Slide boxes and bulldog clips. During autoradiographic exposure the sandwich is stored in light-tight boxes. Before use these are kept in the cryostat, each box containing two clips used to clamp the sandwich. The top of each box contains a large label with an easily read code number that identifies the slide for record keeping.

The clips should be "bulldog" clips, Model 1925, of M. Myers & Son Ltd., Oldbury, Birmingham, England. Other clips we tried were less satisfactory. Myers clips can be obtained from stationers in the British Commonwealth.

B. Procedure

The vial containing the frozen sample is removed from the liquid nitrogen refrigerator and placed in the cryostat just before use. Residual liquid nitrogen is allowed to boil off, and with forceps the capsule is placed, rounded end upward, in the capsule jig. The gelatin coat is carefully removed with a scalpel from the exposed two-thirds of the capsule. The sample is placed in the sample chuck and microtome adjusted for sectioning.

To describe the art of sectioning frozen material would overstep the purview of this article. Such variables as temperature, knife blade angle, section thickness, and the angle of the antiroll device all must be adjusted appropriately to the material being sectioned (Baker, 1964; Pearse, 1968). No written description adequately substitutes for practical experience. After the initial frustration, one picks up the knack and can, with only a little experimenting, discover the proper conditions for cutting most kinds of tissue.

The majority of our sections are cut 6 to 10 μm thick, although we have cut as thin as 2 μm and as thick as 20 μm when dealing with special material. When quantitative grain counting is intended, it probably is better to use thick sections to reduce grain density variation due to variation in section thickness. If tritium is used, sections greater than 3 or 4 μm are effectively infinitely thick.

Figures 7a–f show the steps in the sectioning and sandwich-making processes. The Teflon slide sits in the slide frame frosted surface upward and to the right. The slide frame itself is elevated above the cryostat surface between the rails on the capsule jig (Fig. 7b). This facilitates grasping the frame under the safelight and brings the Teflon surface closer to the knife. As the sections are cut, they appear as flat sheets between the knife blade and the face of the antiroll device (Fig. 7a). These are picked up with the tip of a dissecting needle, carried from the knife blade to the slide frame, and carefully laid on the Teflon slide (Figs. 7b, c). An additional dissecting needle or a brush can help with this.

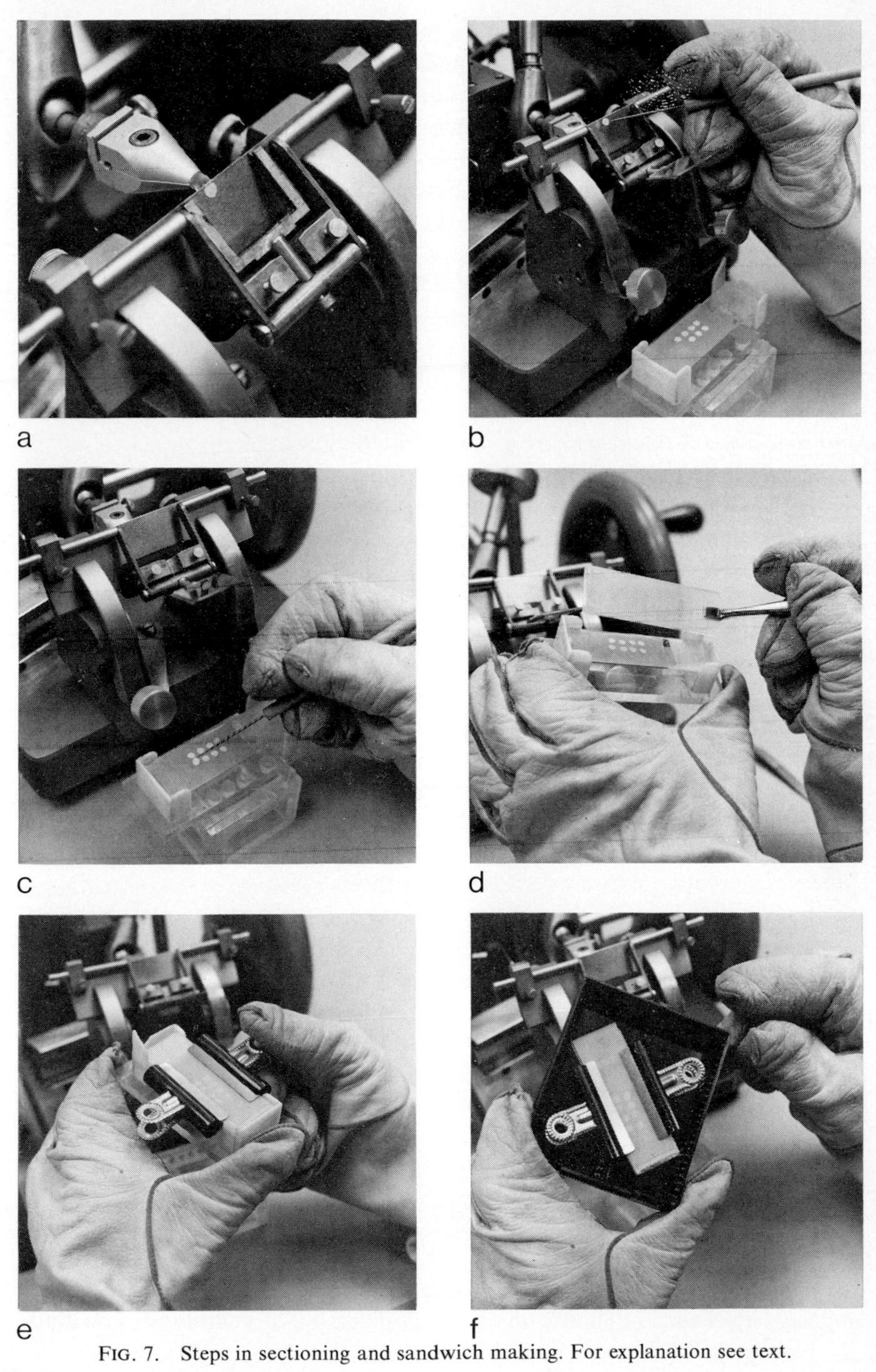

FIG. 7. Steps in sectioning and sandwich making. For explanation see text.

Ideally, the sections will lie flat on the Teflon until the next step is taken. Commonly, however, some curling occurs at the section's edge. If moderate, this can be ignored because the section soon will be flattened by the weight of the autoradiographic plate. When curling is severe, it is treated in one of two ways: (1) The section is allowed to lie between knife and antiroll device, the latter lifted slightly until some curling occurs, and then pressed down on the section again. This process is repeated until the rate of curling enters acceptable limits. (2) The sections are placed on the Teflon and held down with a dissecting needle while a sable brush is used to oppose the curling tendency. This latter method is faster but technically more demanding than the former.

When enough sections have been collected for a single autoradiographic plate, the sandwich is formed. A slide box containing bulldog clips is opened and placed conveniently nearby. The white light is extinguished, leaving the safelight as the only source of illumination.

The box containing autoradiographic plates is opened, and a plate removed with the coverglass forceps is placed emulsion downward on the Teflon and sections. The slide frame is used to guide the operation, as shown in Fig. 7d. The resultant sandwich is clipped with two bulldog clips, as in Fig. 7e. The clipped sandwich is laid diagonally in the slide box (Fig. 7f), and the box is closed.

After checking the security of the plate-bearing boxes, the white light is turned on. The sandwich-bearing box is removed from the cryostat and placed directly into a freezer at $-85°C$. Then sectioning variables and the time of autoradiographic contact are recorded, including blade angle, section thickness, and box temperature as well as notes on cutting behavior of the sample and orientation of the sections. The contact time serves as zero time in calculating exposure, t_e, for quantitative work.

The autoradiographs are kept undisturbed in the freezer until grain densities are in a desirable range for counting. In quantitative work, this is in general the highest grain density consistent with no, or only a small, correction for nonlinearity. The time is chosen from a plot of grain density against activity of the underlying section (expressed as mCi · h/gm). These plots are discussed in Section V, A.

IV. Development, Mounting, and Bleaching

Exposure terminates, and development begins with the removal of the bulldog clips from the sandwich and the separation of Teflon slide and autoradiographic plate, all done under a safelight. The plate is labeled for identification with a soft pencil, and the date and time are noted.

Condensation forms on the plates, and before proceeding they are dried by 20 minutes' exposure to the air.

At this point the sections adhere firmly to the emulsion and normally we proceed to develop. If, however, experience shows an appreciable loss of sections during subsequent steps, it may be useful to dip the autoradiographic plate in a 1% gelatin solution and let it dry before developing (Kinter and Wilson, 1965; see also Appleton, 1972).

The autoradiographic image is developed in staining dishes maintained at 20°C. Development steps are as follows: (1) Kodak D 19 for 6 minutes; (2) 1% acetic acid stop for 10 seconds; (3) Kodak Rapid Fix for 5 minutes; (4) gently running water for 5 minutes; (5) distilled water dip for 1 minute.

The plates are dehydrated by 5-minute passages through graded alcohols (50%, 95%, and two passages through 100%); cleared by a passage through 1:1 absolute alcohol-xylol, two passages through xylol; and mounted in Coverbond (Harleco, Philadelphia, Pennsylvania) under a coverslip.

Our material generally is examined under phase contrast or darkfield illumination, and we have no experience with staining autoradiographs. However, Appleton (1972) described staining techniques for comparable material and reported no special difficulties.

Grain counting on melaninized material often is difficult because melanin granules and silver grains are easily confused. We have found (Horowitz and Fenichel, 1968) that immersion in 30% hydrogen peroxide (Merck Superoxol) at 0°C for 18 to 72 hours (depending on the density of pigment) can lower pigment granule interference to workable levels while having no influence on the emulsion. Some detachment of sections and loss of section quality occur although not sufficiently to be a serious problem.

V. Analysis of Autoradiographs

Grain counting can be done under bright field, phase contrast, or darkfield illumination. The illumination chosen should allow the easy differentiation of silver grains from other particulate matter. Magnifications are from 1200 to 2500 X using a Whipple-Hausser type eyepiece micrometer. The silver grains appear essentially in a single plane of focus: in phase and darkfield as bright refractile spots; in bright field they are black.

Local grain densities, g, are expressed in grains per unit area. For most purposes background is subtracted, and when appropriate (see below) a correction for nonlinearity is made.

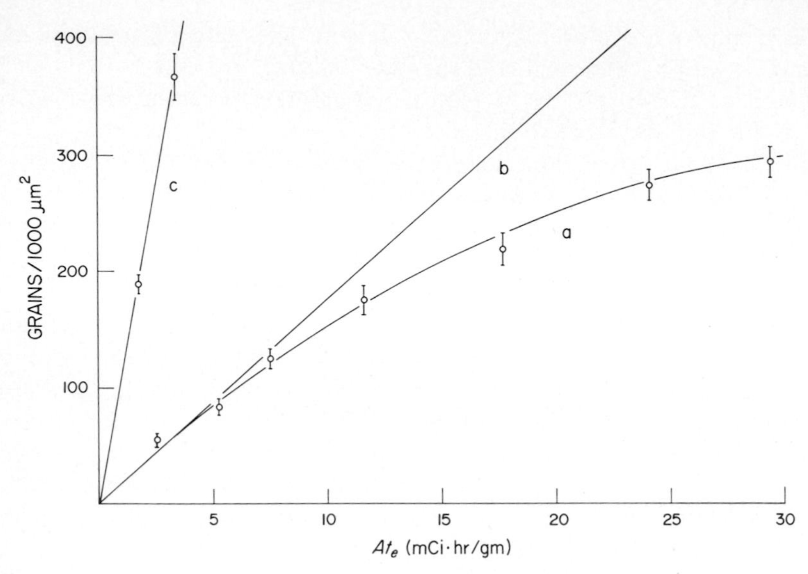

FIG. 8. Relationship between exposure to a radioactive source and grain density in NTB emulsion. The sources were α-aminoisobutyric-^{3}H acid, 0.25 mCi/gm in O.C.T. for curves a and b; and glycine-^{14}C, 0.20 mCi/gm in O.C.T. for curve c. The points show the mean and standard deviation for 30 to 40 determinations. For lines see text.

A. Calibration and Nonlinearity Correction

Figure 8 shows typical curves describing the relation between exposure to radioactive sources and visually determined grain density in NTB autoradiographic plates. These calibration curves are essential for quantitative work. They are prepared by dissolving tracer (in the case of Fig. 8, glycine-^{14}C and α-aminoisobutyric acid-^{3}H) in the embedding material, which then is frozen, sectioned, and autoradiographed as described above. One may vary either the specific activity of the source, A, or the exposure time, t_e, to provide the range of At_e required.[3]

As shown in Fig. 8, a linear relation exists between g and At_e at low exposure levels. As At_e increases, nonlinearity becomes appreciable, as reflected in the deviation of lines a and b. For tritium, it becomes discernible at about 80 grains/1000 μm^2. The points in Fig. 8 are experimental, and the line for tritium (line a) is described by Eq. (1).

$$g = \text{grains}/1000\ \mu\text{m}^2 = 17.6\ At_e - 0.2545\ A^2t_e^2 \qquad (1)$$

[3] The option of varying A or t_e exists because latent image fading is negligible at the temperature and times used for exposure (Pelc *et al.*, 1965).

When counting grain densities in a range of appreciable deviation from linearity, a correction should be made. To do this using Fig. 8, the observed density is found on curve a, and the corresponding theoretical value for the same value of At_e is read off curve b.

Because of the low energy of its β particle, tritium is detected only by a thin hemisphere of emulsion in the immediate vicinity of the source. The particles of the other commonly used isotopes are distributed through a wider energy spectrum and have access to a larger emulsion volume. Grain density-activity plots for isotopes other than tritium, therefore, are linear over a greater range of g, as, for example, the curve for glycine-^{14}C (Fig. 8, line c), which is linear to g greater than 360 grains/1000 μm^2. Hence, no nonlinearity correction is needed for ^{14}C over this range. Counting grain densities above 250 grains/1000 μm^2 requires close, tiring, control of eye movements. It is advisable, therefore, to limit At_e, thereby restricting g to densities below this level. Consequently nonlinearity corrections rarely should be necessary for isotopes other than tritium.

A test of the agreement between the results of ultra-low temperature autoradiography and other quantitative radioisotopic measurements occasionally is possible in cellular systems themselves. Figures 9 and 10 show two examples of this.

In Fig. 9 is plotted the overall concentration of glycerol-^3H in an amphibian oocyte during an efflux experiment. In this relatively simple case, the concentration of tracer could be shown to be uniform over the entire cytoplasm and within the nucleus at all efflux times. The filled circles in Fig. 9 were determined from autoradiographs by correcting for the relative volumes and water contents of the nucleus and cytoplasm and for the differences between the densities of standards and samples. These corrections

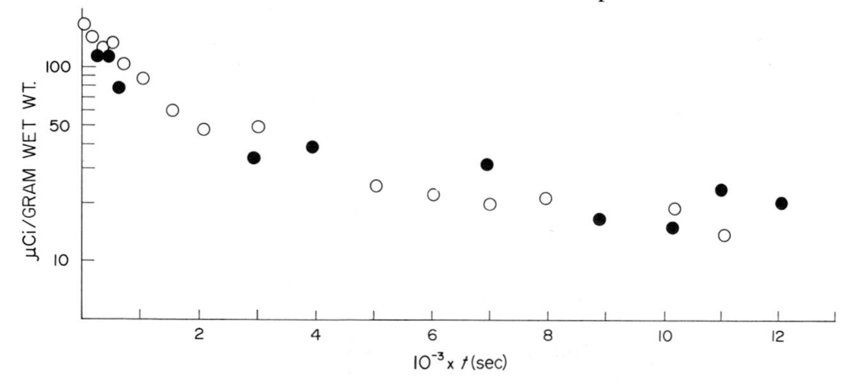

FIG. 9. Oocyte concentration of glycerol-^3H as determined by liquid scintillation analysis (open circles) and autoradiography (filled circles) as a function of efflux time at 13.6°C. Redrawn from Horowitz and Fenichel (1968), with permission.

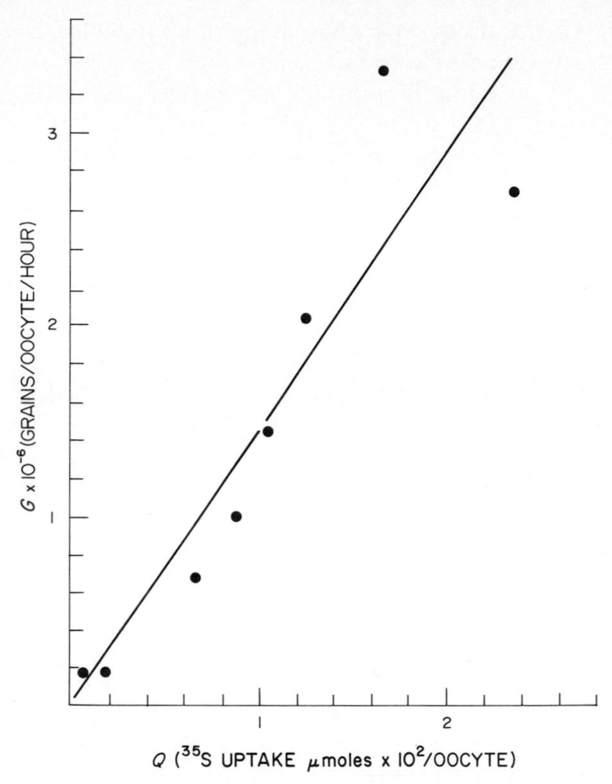

FIG. 10. Relationship between total grains derived from a whole oocyte per hour of autoradiographic exposure, G, and ^{35}S uptake, Q. See text for additional explanation. Points experimental, line fitted. From Horowitz *et al.* (1970), with permission.

are discussed in the original work (Horowitz and Fenichel, 1968). The open circles in Fig. 9 were derived by counting the entire cell content of isotope in a liquid scintillation spectrometer. Obviously there is good agreement between the two sets of data.

Figure 10 shows the relationship between total grains derived from whole oocytes per hour of autoradiographic exposure, G, and ^{35}S uptake, Q, in a cysteamine-^{35}S influx experiment. In this case distribution of intracellular isotopes was not uniform, as seen in Fig. 11. Instead a complex gradient existed, dependent on the permeability of the cell membrane and the diffusional properties of the solute in cytoplasm. To determine G, profiles of g were taken along the radius of sections of oocytes which had been incubated in cysteamine-^{35}S solutions for various periods of time. From these profiles G could be derived from Eq. (2).

$$G = 4\pi \Sigma g_i r_i^2 \, \delta r \tag{2}$$

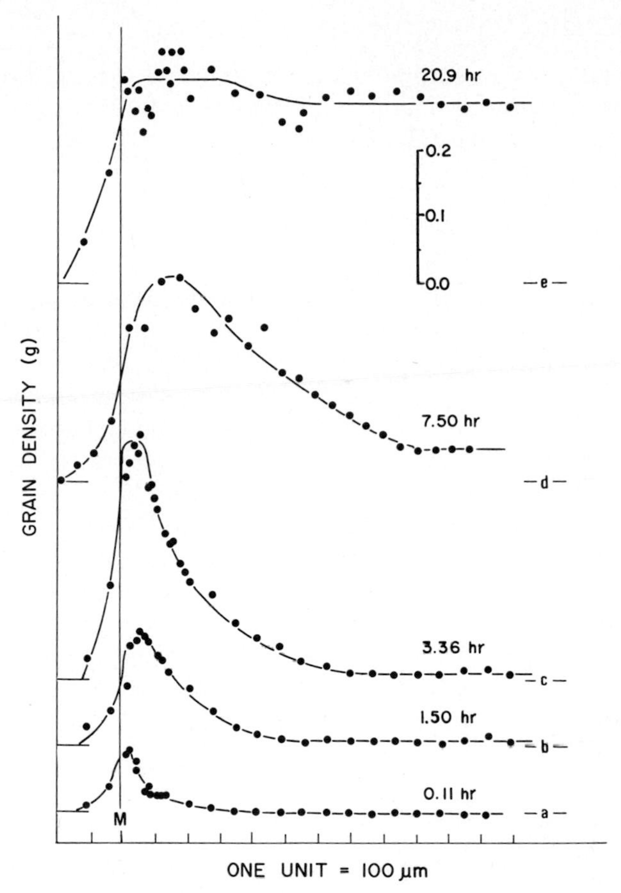

FIG. 11. Radial grain density profiles through the cytoplasm of oocytes incubated in cysteamine-^{35}S phosphate for 0.11 to 20.9 hours. The vertical scale is in grains per 1000 μm^2/hour. Horizontal lines labeled a–e indicate the base line background densities. Position of the cell surface is indicated by vertical line M. From Horowitz *et al.* (1970), with permission.

in which g_i is the local grain density in grains/μm^2/hour; r_i the distance from the center of the section; and δr the measurement interval along the radius, in this case 50 μm. The summation is along the entire radius. Q was determined by liquid scintillation spectroscopy of the entire oocyte content.

We conclude from Figs. 9 and 10 that although the variation in concentration determined by ultra-low temperature autoradiography is greater than by liquid scintillation spectroscopy, sufficient agreement exists between the two methods to assure that autoradiography provides a quantitative measure of radioactivity in biological systems.

B. Resolution

The resolution of an autoradiograph often can be assessed by assuming that the concentration in a large cell previously equilibrated with a labeled substance changes at the cell's surfaces as a step function when the cell is rinsed and embedded. The grain density profile then is determined through the plane of the surface.

Using this method, and defining resolution as the distance over which grain density falls by 50% from its bulk value, we found the resolution for tritium to be about 3 μm (Horowitz, 1972), and for ^{35}S (Horowitz *et al.*, 1970), ^{22}Na (Horowitz and Fenichel, 1970), and ^{14}C (Horowitz, unpublished results) about 45 μm. These values would be substantially smaller if resolution were defined, as is often done, by measuring from the edge of the source to 50% grain density level (cf. Appleton, 1966; Nadler, 1951), but we believe the present definition is more conservative and of greater practical value. By the less stringent criteria, resolution for ^{35}S, ^{22}Na, and ^{14}C is about 7 μm, while for tritium it is too low to determine by our methods.

The similarity in resolution for the more energetic isotopes suggests that not the path length of the emitted particle, but the geometry of the autoradiograph, is the prime determinant of resolution once some maximum energy level is reached. We suspect that improvements in resolution can be achieved for these isotopes simply by cutting thinner sections and making autoradiographic plates from more dilute emulsion.

C. Vaporization

There is a period in the preparation of autoradiographs when water and its volatile solutes are lost from the sample. This occurs during sectioning, when freeze-drying goes on at the cutting face of the sample and from the section itself in the interval between sectioning and being covered by the autoradiographic plate. With the possible exception of sectioning at drastically lower temperatures as, for example, those of liquid nitrogen (Christensen, 1971) or some equally heroic measure, there seems to be no way of avoiding this loss. Hence for quantitative work one must measure loss under the conditions of sectioning and make appropriate corrections. Figure 12 shows the reduction in grain density that occurs in our cryostat when the sections are on Teflon, following sectioning but before sandwich formation. The line is exponential with a half-time of decay of 27.5 minutes. Since 20 to 30 minutes may elapse from the beginning of sectioning to the formation of the sandwich, this rate of loss is appreciable.

The demonstration of water loss during sectioning raises the possibility of another type of systematic error. As water evaporates, a concentration

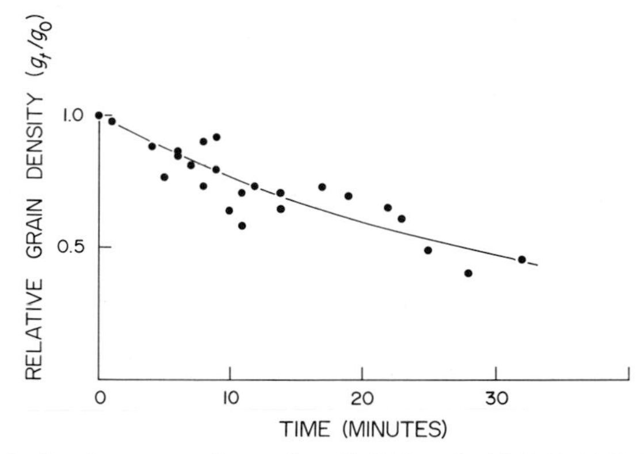

FIG. 12. Grain density over sections of an O.C.T. embedded, G-34 Sephadex bead equilibrated with tritiated water. The reduction in grain density with time is due to freeze-drying in the cryostat. The ordinate ratio, g_t/g_0, is of grain density at time t (on abscissa) to grain density at time 0, taken as 1 minute after sectioning.

and redistribution of nonvolatile solutes must occur in the volume immediately adjacent to the surfaces. If the redistribution involves migration away from the surface into the bulk, we expect a decrease in grain densities for weak emitters. The inverse phenomenon can be imagined; that the redistribution involves increased concentration at the surface, solute being carried by mass flow of water from the bulk of the section to the evaporative surface. In this case an increase in grain density would be expected. At this writing we have not yet investigated these possibilities.

D. Emulsion Gradients

Conventional dipping technique does not yield a uniform emulsion thickness over the entire slide. After dipping, a slide dries vertically, and the flow of emulsion which occurs results in a thinner layer of emulsion at the top than at the bottom. This can be ignored if tritium is used, because at both top and bottom the emulsion layer is effectively infinitely thick to its β particle. When stronger emitters are used, however, the sections closest to the frosted area (the upper side during dipping) will have lower grain densities than more distal sections. Figure 13 depicts the results of grain counts under a constant ^{14}C source as a function of position along the length of three autoradiographic plates. The differences in grain density between the two ends of the plate are substantially greater than the difference between the same position on different plates.

We have controlled this variation in quantitative work by grain counting in a defined narrow band on each slide (Horowitz and Fenichel, 1970). Other solutions to the problem could involve applying more uniform emulsion

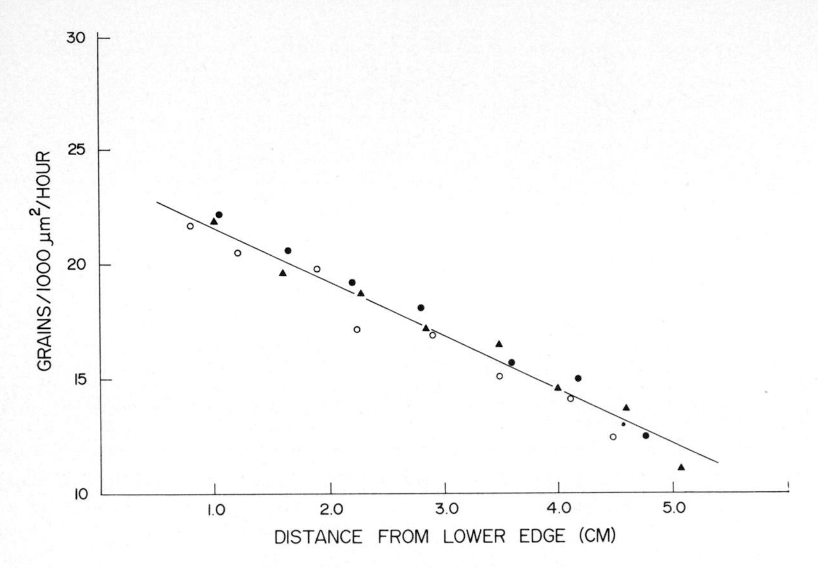

FIG. 13. Relationship between grain density over a uniform ^{14}C source (glycine-^{14}C in an O.C.T. embedded G-34 Sephadex bead) and position on the autoradiographic plate. The abscissa is distance from the distal (contra-frosted) edge of the slide. The data taken from three autoradiographic plates are indicated by different symbols. Each point is the average of five determinations. The line was obtained by the least squares method.

coats, either by using stripping film, or with appropriate instrumentation (Kopriwa, 1966), or by applying sections of a standard source adjacent to each section of experimental material.

E. Artifacts

Two types of artifacts are commonly seen and easily recognized by an experienced observer. Illustrations of both have been published (Horowitz and Fenichel, 1968).

1. A thin line of high grain density may occur at the interface between tissue, or a large cell, and the embedding material. At first this may seem due to a cortical region of high activity, but upon closer examination one finds it does not coincide with a particular zone or structural feature. Furthermore, these lines of grains can be seen at the edge of objects known to be homogenous with respect to tracer distribution; sections of polymerized tritiated methacrylate, or Sephadex beads equilibrated with radioactive solutions, for example. They also have been found in biological material devoid of a radioactive source. We believe the lines are pressure artifacts (Moore, 1951), a consequence of the clamping used to form the sandwich.

2. Occasionally one sees alternate bands of high and low grain density in locations where a uniform distribution of tracer is expected, as in sections of embedding material like O.C.T. in which tracer has been dissolved and carefully mixed. They also are seen beneath the nucleus of amphibian oocytes, although rarely beneath cytoplasm. We do not know their cause, although clearly it is due to a heterogeneity in the section rather than in the emulsion.

VI. Experimental Examples

This section will present briefly examples of the use of ultra-low temperature autoradiography. We are not concerning ourselves with the biological problems under investigation but with suggesting the power of the technique, while offering comments on potential pitfalls in interpreting autoradiographs which may not be immediately apparent.

Figure 14 is a grain density profile through the cytoplasm and nucleus of an amphibian oocyte injected with sucrose-^{3}H and permitted to go to diffusional equilibrium before freezing. The vertical bars mark the cell (C) and nuclear boundaries (N). The grain density of the cytoplasm is uniform throughout, indicating an absence of special pool compartments. The grain density over the nucleus is about 3 times that of the cytoplasm and is also

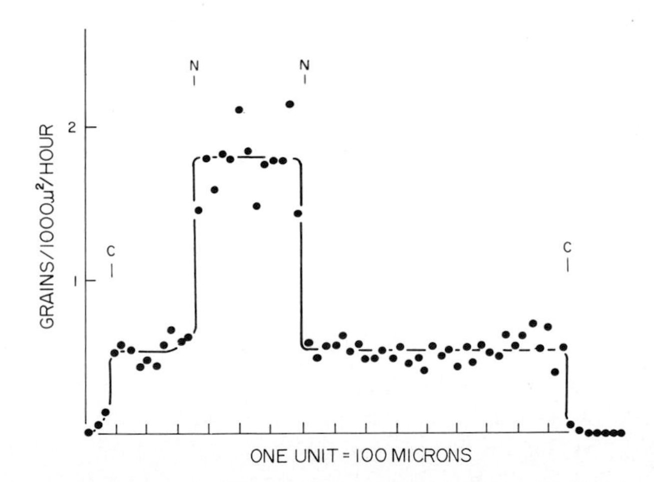

FIG. 14. Grain density profile through the cytoplasm and nucleus of an amphibian oocyte injected with sucrose-^{3}H and permitted to go to diffusion equilibrium before freezing. The vertical bars mark the cell (C) and nuclear boundaries (N). From Horowitz (1972), with permission.

uniform. From this we might conclude that the relative concentration of sucrose in the water of the nucleus is equally high, but this is not the case. Grain density over a radioactive source is proportional to tracer concentration on a volume basis. To determine concentration on a water basis—a more meaningful number—some independent measure of the water content of the source must be available. These data are available for the amphibian oocyte. The ratio of nuclear to cytoplasmic water is 1.88 (Century *et al.*, 1970). The nuclear concentration of sucrose when expressed on a water basis is, therefore, only 1.6, not 3 times that of cytoplasm.

Ultra-low temperature autoradiography can determine diffusion coefficients, D_c, in the cytoplasm of a cell. Figure 15 is a grain density profile through the site of injection, 16 minutes after the microinjection of an aqueous solution of inulin-^{3}H into an oocyte. After injection, inulin diffuses from the site of injection into the surrounding cytoplasm at a rate determined by its diffusion coefficient. The resultant diffusional profile can be compared with theoretical diffusional curves to provide a value for D_c. In Fig. 15 the points are experimental and the solid line a theoretical profile for $D_c = 2 \times 10^{-7}$ cm^2/sec. The concentration peak indicated by the dashed line astride the cytoplasmic gradient marks the injection site. We can see from this peak that the injection solution has not been resorbed totally at 16 minutes.

Another point to note is the deviation (indicated by the dotted line) from

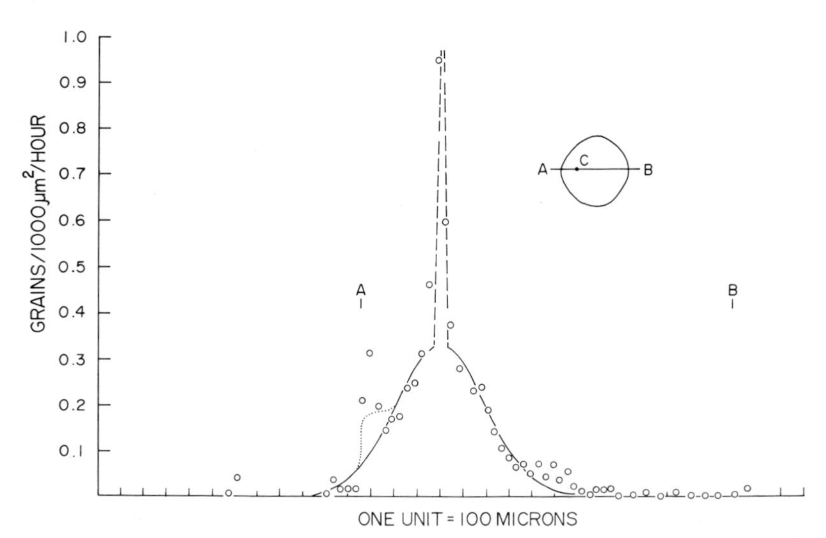

FIG. 15. Grain density profile through the injection site, 16 minutes after the injection of a solution if inulin-^{3}H into the cytoplasm of an amphibian oocyte. Vector of the profile is shown in the inset. A and B mark the cell boundaries, C the position of the injection site. For additional explanation, see text.

the theoretical diffusional gradient at the cell membrane. Because the membrane is impermeable to inulin, there is a backup of inulin. This is seen as a plateau before the grain density falls sharply to background levels just outside the membrane. The anomalous single high point seen above the plateau may be a pressure artifact, as discussed above.

The diffusion of water also can be visualized by ultra-low temperature autoradiography. Large Sephadex beads (Horowitz and Fenichel, 1964), were swollen in O.C.T. and then immersed in O.C.T. containing tritiated water (HTO). Diffusion was permitted for 80 seconds, the system frozen, the bead and O.C.T. sectioned and autoradiographed. Figure 16 shows the grain density profile through the center of a bead 2.2 mm in swollen diameter. The points are experimental. The line is theoretical (Crank, 1957), based on the assumption that the surface concentration of HTO is constant due to a large excess volume of O.C.T. and adequate mixing, and $D = 1.4 \times 10^{-5}$ cm^2/sec (cf. Horowitz and Fenichel, 1964).

Finally, it is well to keep in mind when interpreting autoradiographs of solutes that the intracellular distribution of an exogeneous tracer may not be identical with the nonradioactive cellular form. In the case of a nonmetabolized substance, this may be due to isotope effects or to inadequate time being permitted to achieve exchange equilibrium at every point in the cell. Sometimes equilibrium times for regions within the same cell differ by many orders of magnitude. Such a case is sodium in the amphibian oocyte. Figure 17 shows the distribution of ^{22}Na as determined by autoradiography in a cell which was exposed to a solution of the tracer for 25 minutes and

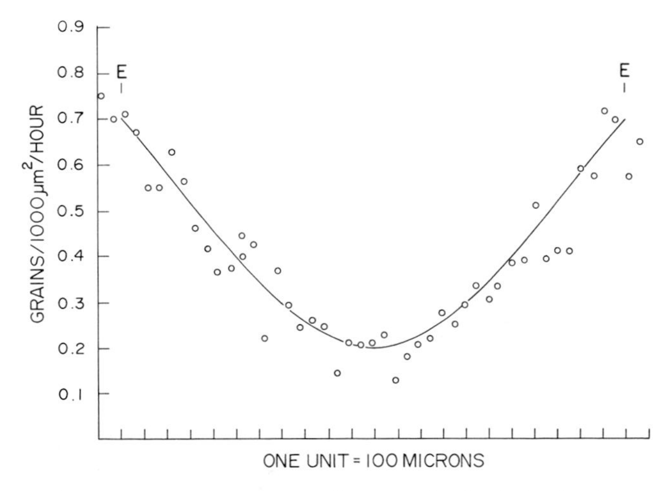

FIG. 16. Grain density profile along the diameter of a Sephadex bead after 80 seconds' exposure to tritiated water. The points are experimental. The bead surfaces are indicated by E. The line is theoretical; for details, see text.

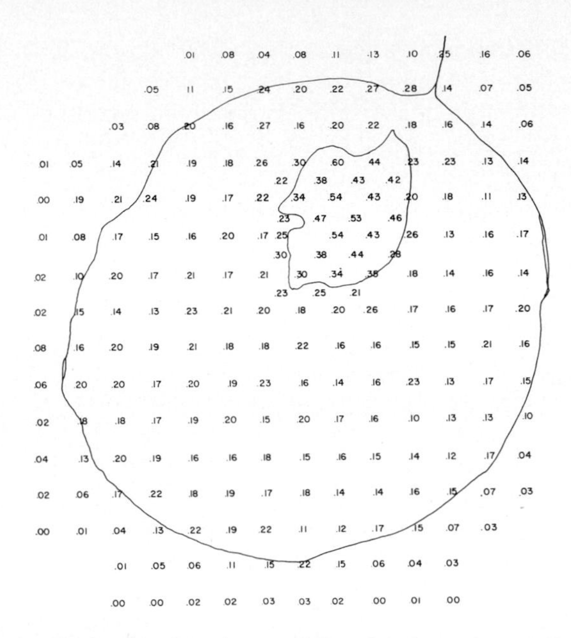

FIG 17. Local grain densities in grains per 1000 μm^2 per hour of autoradiographic exposure over a section of an amphibian oocyte incubated in the presence of ^{22}Na for 25 minutes and rinsed for 1 minute. The nucleus is the enclosed area of high grain density. Numbers over the cytoplasm are 155 μm apart. From Horowitz and Fenichel (1970), with permission.

then briefly rinsed. The nuclear-cytoplasmic grain density ratio is about 2.6. When corrected for the differing water contents, the ratio of ^{22}Na concentration in the two compartments is 1.3. It would seem that a slight excess of Na is found in the nucleus as compared to the cytoplasm, but this is not the case. Analysis by low temperature microdissection and flame photometry (Century *et al.*, 1970) discloses that Na is 5.5 times more concentrated in the cytoplasm than in the nucleus. The cause of the apparent disparity between the results for ^{22}Na and those for total Na has been shown to be a fraction absent from the nucleus but accounting for 85% of the cytoplasmic Na, whose exchange has a half-time of 2.5 days. At 25 minutes less than 1% of this fraction has exchanged; hence, at this time nuclear Na is approaching exchange equilibrium, while cytoplasmic Na is only about 15% exchanged.

ACKNOWLEDGMENTS

I wish to thank Mrs. Jean Reeves and Ms. Marian Herman Horowitz for their contributions to the preparation of this chapter, and especially Mr. Leonard Moore for much of the technical work and grain counting.

REFERENCES

Appleton, T. C. (1964). *J. Roy. Microsc. Soc.* **83**, 277.

Appleton, T. C. (1966). *J. Histochem. Cytochem.* **14**, 414.

Appleton, T. C. (1972). *In* "Autoradiography for Biologists" (P. B. Gahan, ed.), pp. 51–64. Academic Press, London and New York.

Baker, J. R. (1964). "Microtome-cryostat Handbook," 2nd Ed. Int. Equip. Co., Needham, Massachusetts.

Century, T. J., Fenichel, I. R., and Horowitz, S. B. (1970). *J. Cell Sci.* **7**, 5.

Christensen, A. K. (1971). *J. Cell Biol.* **51**, 772.

Coons, A. H., Leduc, E. H., and Kaplan, M. H. (1951). *J. Exp. Med.* **93**, 173.

Crank, J. (1957). "The Mathematics of Diffusion." Oxford Univ. Press, London and New York.

Horowitz, S. B. (1972). *J. Cell Biol.* **54**, 609.

Horowitz, S. B., and Fenichel, I. R. (1964). *J. Phys. Chem.* **68**, 3378.

Horowitz, S. B., and Fenichel, I. R. (1968). *J. Gen. Physiol.* **51**, 703.

Horowitz, S. B., and Fenichel, I. R. (1970). *J. Cell Biol.* **47**, 120.

Horowitz, S. B., Fenichel, I. R., Hoffman, B., Kollmann, G., and Shapiro, B. (1970). *Biophys. J.* **10**, 994.

Kinter, W. B., and Wilson, T. H. (1965). *J. Cell Biol.* **25**, 19.

Kinter, W. B., Leape, L. L., and Cohen, J. J. (1960). *Amer. J. Physiol.* **199**, 931.

Kopriwa, B. M. (1966). *J. Histochem. Cytochem.* **14**, 923.

Leblond, C. P., Kopriwa, B., and Messier, B. (1963). *Proc. Int. Congr. Histochem. Cytochem. 1st.* pp. 1–31. Pergamon Press, London.

Moore, A. C. (1951). *Brit. J. Appl. Phys.* **2**, 20.

Nadler, N. J. (1951). *Can. J. Med. Sci.* **29**, 182.

Pearse, A. G. E. (1968). "Histochemistry, Theoretical and Applied." Little, Brown, Boston, Massachusetts.

Pelc, S. R., Appleton, T. C., and Welton, M. G. E. (1965). *In* "The Use of Radioautography in Investigating Protein Synthesis" (C. P. Leblond and K. B. Warren, eds.), Symposia of the International Society for Cell Biology, Vol. 4, p. 9. Academic Press, New York.

Rogers, A. W. (1967), "Techniques of Autoradiography," Elsevier, Amsterdam.

Chapter 16

Quantitative Light Microscopic Autoradiography[1]

HOLLIS G. BOREN,
EDITH C. WRIGHT, AND
CURTIS C. HARRIS

University of South Florida College of Medicine; Veterans Administration Hospital, Tampa, Florida; Lung Cancer Branch, Carcinogenesis Program, National Cancer Institute, Bethesda, Maryland

---

I. Introduction

Theoretical and technical problems in the quantitation of autoradiography have been extensively reviewed (Maurer and Primbsch, 1964; Perry, 1964; Rogers, 1969; Baserga, 1967; Cleaver, 1967; Baserga and Malamud, 1969; Schultze, 1969; Przybylski, 1970). The methodologies described here offer a practical solution to these problems insofar as tritium-labeled molecules are concerned.

Previous attempts to use autoradiography as a quantitative tool have relied primarily upon the standardization of procedures. Since procedures vary in different laboratories, it has been difficult to compare results from different investigators. Even within a laboratory, multiple variables, for

[1] Supported in part by Public Health Service Research Contract FS 73 206 from the National Cancer Institute.

example, section thickness, temperature of processing solutions, have not been adequately controlled by standardized procedures. Our experience has shown that an internal standard of known specific radioactivity on each individual slide is essential to determine and to correct for the effects of these multiple variables. To this end we have used a tritiated plastic similar to that described by Ritzén (1967).

Quantitation of light microscopic autoradiography can be achieved in a practical manner by: (1) the addition of an internal standard; and (2) the use of 1 μm sections of tissues embedded in plastic, which allows precise identification of cell types as well as cellular compartments. The specific radioactivity of the specimen can be calculated by the methodology described in this chapter.

II. Materials and Procedures

1. Internal Standards

Tritiated poly(*n*-butyl) methacrylate (PBM-^{3}H) in gelatin capsules is obtained from Amersham-Searle (Arlington Heights, Illinois). PBM-^{3}H is available in specific activities of approximately 50 μCi/gm or 500 μCi/gm. The blocks of PBM-^{3}H fit into the chucks of standard ultramicrotomes. The density of these blocks as determined by volume displacement of blocks of measured weight is 1.06 gm/cm^3.

The actual specific activity is determined by dissolving weighed sections in toluene and measuring disintegrations per minute (dpm) with a scintillation counter. Each microcurie gives 2.22×10^6 dpm [3.7×10^4 disintegration per second (dpm) $\times$ 60 sec/min]. Specific activity, μCi/gm, is calculated by dpm/(2.22×10^6 dpm/μCi $\times$ wt, gm). For example, if 0.025 gm of PBM-^{3}H gives 3×10^6 dpm, then specific activity, μCi/gm = 54 μCi/gm.

Internal standards of both available specific activities are useful. PBM-^{3}H (500 μCi/gm) is used for short-term studies in which the exposure time is less than 10 days. PBM-^{3}H (50 μCi/gm) is used when the exposure time is 1–12 weeks.

2. Tissue Fixation

The tracheal epithelium of Syrian golden hamsters is initially fixed *in situ* by intratracheal instillation of cold (4°C) 2% OsO_4 buffered by 0.1 *M s*-collidine, pH 7.4. The tracheas are removed surgically. Cross sections of tracheal rings, which consist of cartilage, submucosa, and epithelium, are immersed in the fixative for 1 hour. After dehydration, the tracheal rings are embedded in Epon 812 containing 5% Araldite 502. By volume displacement of weighed blocks, the density of this polymerized plastic is 1.16 gm/cm^3.

3. Sectioning

One-micrometer sections are cut on either a Huxley LKB or a Porter-Blum (MT-1) ultramicrotome. Section thickness is monitored by interference microscopy using a Zeiss interference microscope and an Ehringhaus quartz compensator.

4. Autoradiography

Kodak NTB 2 nuclear track emulsion is melted in a 45°C water bath. After 15 minutes, bubbles are skimmed off the surface of the emulsion with gauze. Eight prepared slides are placed in a Technicon slide holder. They are thoroughly dried and warmed to 45°C in a slide warmer, then dipped into emulsion and withdrawn at a uniform speed. The slides, held by a slide holder, are hung on a Plexiglas rack to dry at room temperature in the dark for 1.5–3 hours. Slides (eight per box) are then placed in bakelite boxes containing Drierite and sealed with black electrical tape. Slides are stored at 4°C for the duration of dark time, i.e., exposure time.

Photographic development is by Kodak D-19 for 6 minutes at 15°C. Slides are then treated with Kodak acid hardening fixer for 10 minutes and passed through 20 changes of demineralized water at 15°C.

Emulsion is scraped off the back of the slides and they are placed in a clean rack in the refrigerator at 4°C to dry for at least 2 hours.

All procedures through photographic development should take place in complete darkness and at carefully controlled temperatures. Emulsion, developer and fixer must be freshly prepared for each run. The developer and fixer are filtered before use.

5. Organization of the Slide

Stain is applied with an eye dropper to a limited area of the slide (Fig. 1). Both experimental sections and internal standards of PBM-^{3}H are stained for calculation of autoradiographic efficiency. A comparison of grain counts of stained and unstained internal standards demonstrates the magnitude of grain loss, if any, due to staining. Positive or negative chemography is determined by placing on each slide unlabeled sections of tracheal rings which are fixed and processed in the same manner as experimental sections.

6. Staining

Aqueous toluidine blue, 1%, is mixed with 2.5% sodium carbonate in the ratio of 1:20, stored at 4°C and filtered just before use. This cold solution is applied with an eyedropper to the area of the slide to be stained. After 1 hour,

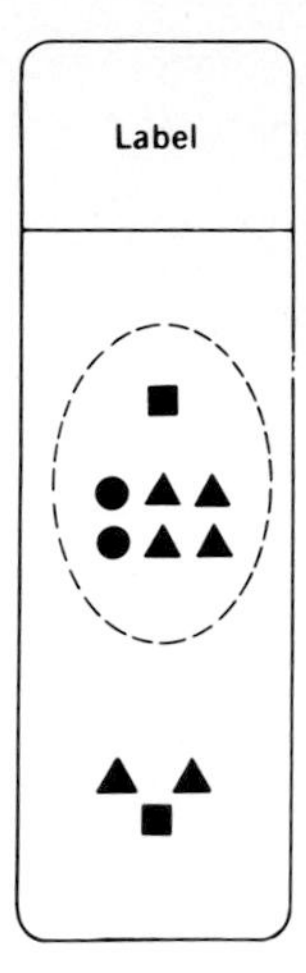

FIG. 1. Organization of the slide. Three types of sections are placed on each slide: ■, internal standards of tritiated poly (*n*-butyl) methacrylate; ●, unlabled tissue sections; and ▲, experimental sections. The area of stain is indicated by the area enclosed by the dashes. Net grain counts over the internal standard are used to calculate quantitative autoradiographic values for the experimental sections. The unlabeled tissue sections demonstrate the presence of either positive or negative chemography. Counts over the unstained internal standard as compared to those over the internal standard in the stained area demonstrate the amount of grain loss due to staining. Unstained experimental sections are compared with both stained experimental sections and unstained internal standards as additional tests of grain loss.

slides are carefully rinsed in 4°C distilled water, then dipped in 0.1% hydrochloric acid to remove excess stain from the emulsion. Slides are rinsed thoroughly, dried at 4°C overnight and brought to room temperature. Coverslips are applied to the slides using a mounting medium of 50% xylene and 50% Permount. Although it is common practice to dip slides into xylene before applying mounting medium, this procedure causes loss of autoradiographic grains.

7. GRAIN COUNTS

For each slide, grain counts are done over five areas: (a) stained internal standard of PBM-^{3}H, (b) unstained internal standard of PBM-^{3}H, (c) background, (d) unlabeled specimen, (e) labeled specimen. Grain counts over the labeled specimen are done separately for each of the three major cell types of tracheal epithelium (basal mucous, and ciliated cells).

Grains overlying the internal standard of PBM-^{3}H are counted from their projected image on a Glarex (Zeiss) moving projection screen which is mounted on the microscope. The number of grains falling within a 100-μm^2 area of the section is counted. For each square section of the PBM-^{3}H standard, ten 100-μm^2 areas are counted: one over each corner, one at the midpoint of each of the four edges, and two areas near the center of the section. Background grain counts are taken over areas of the slide which contain no sections. Net grains per 100 μm^2 are calculated by subtracting the number of background grains per 100 μm^2 from the total number of grains per 100 μm^2 over the PBM-^{3}H standards. This background correction is not significant when the procedures described in this chapter are carefully followed.

Grains over specimens are counted using an insert in the eyepiece. This insert is made from a piece of clear plastic with an outer diameter such that it will be held in position by the diaphragm in the eyepiece tube. The center of the insert is cut out by repeatedly rotating a divider with two sharp metal points. This inner circle gives an unobstructed field as well as sharply defined margins. The diameter of the field is calibrated with a stage micrometer.

Adequate sample size is a critical factor for quantitative autoradiography. In our studies of respiratory epithelium, 6–8 labeled sections are placed on each microscopic slide. Grains are counted over a minimum of 4 experimental sections. For each section, four fields 80 μm in diameter are counted. Each field contains approximately 25 respiratory epithelial cells.

For certain studies grain counts over the experimental sections are made over the nucleus and over the cytoplasm of specific cell types. An example of the application of this approach is the localization of ^{3}H-benzo [*a*] pyrene in hamster respiratory epithelium (Harris *et al.*, 1973).

III. Results and Discussion

A. Autoradiographic Techniques

PBM-^{3}H standards can be used to determine when significant grain loss or displacement occurs during autoradiographic procedures. Using grain count comparisons, we find that the temperature of all solutions placed on the exposed emulsion is critical. Although bulk emulsion is usually heated to 45°C before use, NTB 2 emulsion will melt from 20° to 30°C. If the temperature of photographic and staining solutions is maintained at 15°C or less, grain loss and displacement are greatly diminished. The deleterious effect of staining sections on a hot plate is illustrated in Fig. 2.

Internal standards are also used to determine the time course of photographic development with undiluted Kodak D-19 at 15°C (Fig. 3). Development is complete by 6 minutes and background is not significant until 8 minutes. Omission of a hardener in the photographic fixer causes loss of grains.

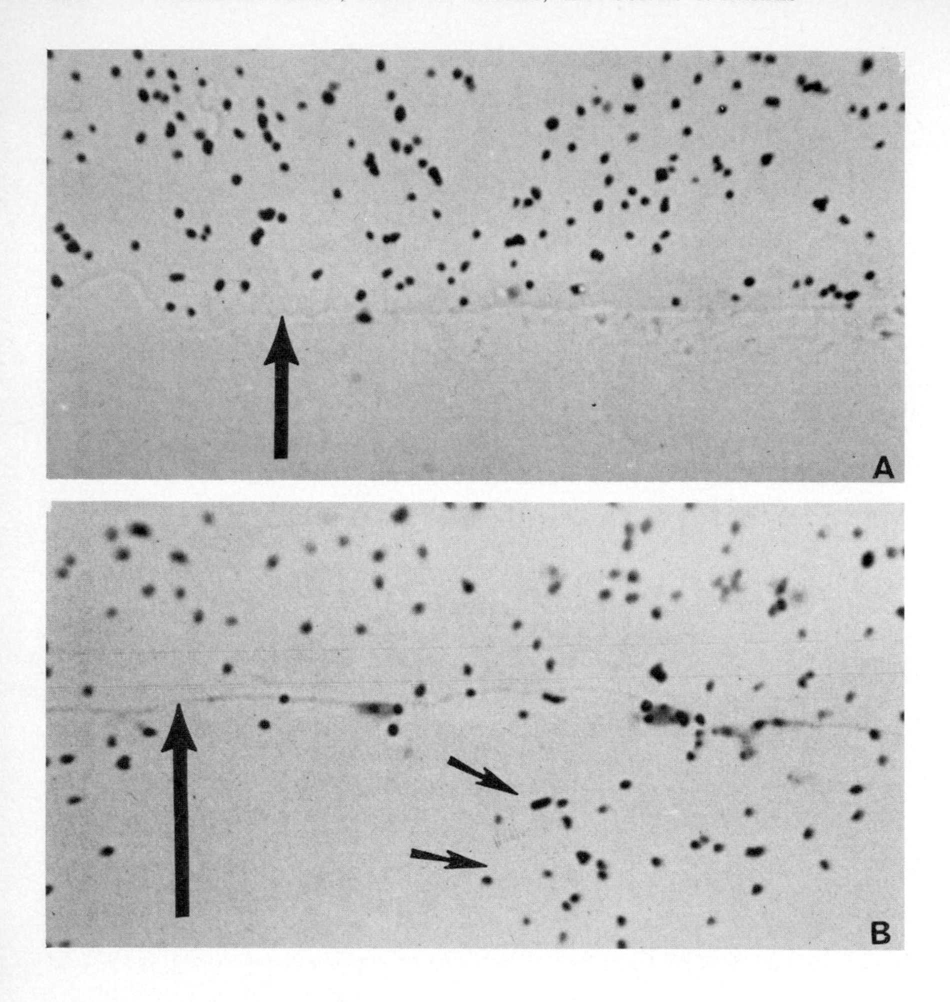

FIG. 2. Two internal standards (A and B) of the same thickness were placed on the same slide. Heavy vertical arrows indicate the edges of the internal standards. The unstained standard (A) shows low background in adjacent areas, no displacement of silver grains, and uniform grains over the internal standard. The internal standard (B) was stained at 45°C on a hot plate with toluidine blue. There was displacement of grains at the edge of the internal standard (small arrows) and irregular loss of grains over the internal standard.

B. Physicochemical Determinants of Quantitation

1. SELF-ABSORPTION OF β-RAYS

The importance of self-absorption of β-rays by molecules between the tritium sources and the emulsion is well recognized (Fitzgerald *et al.*, 1951; Perry, 1964; Maurer and Primbsch, 1964; Pelc and Welton, 1967; Cleaver,

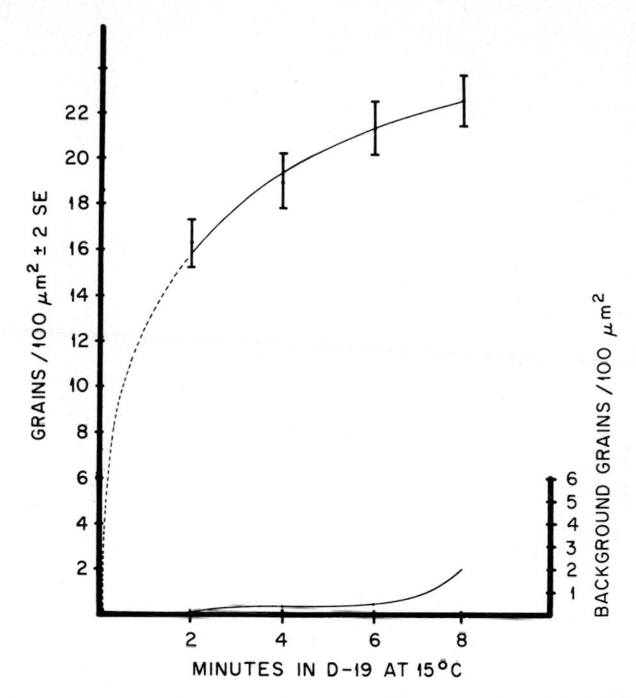

FIG. 3. Photographic development of Kodak NTB 2 by undiluted Kodak D-19 at 15°C. Grain counts per 100 μm^2 were done over unstained internal standards PBM-^{3}H and over background. The mean ± 2 standard errors for the PBM-^{3}H are plotted as vertical bars; mean background counts are plotted as points. Time in minutes is on the abscissa. Grain counts over the internal standard do not increase significantly after 6 minutes, and background is not significant until 8 minutes.

1967). Self-absorption is determined by counting grains over radioactive sections of increasing thickness. Sections of PBM-^{3}H were cut from 0.2 to 1.5 μm in thickness and grains per 100 μm^2 were counted (Fig. 4). No significant increase of grain counts is found as the thickness of the PBM-^{3}H standard increased above 0.8 μm. Absorption of beta radiation at 1 μm of PBM-^{3}H is equivalent to 0.11 mg/cm^2 absorber (mg/cm^2 = cm absorber $\times$ density, mg/cm^3 = 10^{-4} cm $\times$ 1060 mg/cm^3). One micrometer of Epon embedding medium (ρ = 1.16 gm/cm^3) would be 0.12 mg/cm^2. The self-absorption data is applicable to any material of known density.

The number of grains does not increase with sections thicker than 0.8 μm (0.085 mg/cm^2). If sections thicker than 0.8 μm are used, self-absorption is a constant, i.e., the number of β-rays reaching the emulsion from tritium-labeled sections remains the same. One-micrometer experimental sections not only give constant self-absorption ("saturation"), but also allow the precise identification of all types as well as cell compartments.

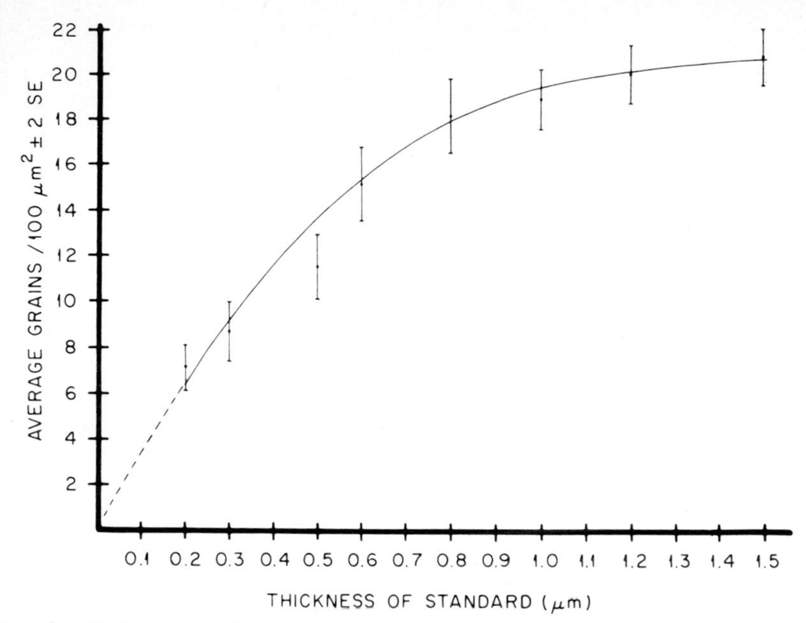

FIG. 4. Grain counts of unstained PBM-^{3}H standards coated with NTB 2 are plotted against the thickness of these standards. Mean net grain counts ± 2 standard errors are indicated by the vertical lines. Thickness varied from 0.2 to 1.5 μm. If there were so self-absorption, grain counts would continue to increase as the thickness, and therefore the amount of radioactivity, increased. Because of self-absorption, grain counts reach a plateau, or saturation value as thickness increases. There was no significant increase of grain counts for thicknesses of BPM-^{3}H greater than 0.8 μm.

2. Latent Image Fading

A major limitation to the continued measurement of radioactivity by autoradiography is latent image fading. With increasing time, silver halide crystals of the emulsion which have been hit by β-rays may lose their ability to become the site of silver grain growth at the time of photographic development. This loss is called latent image fading. This effect has been observed indirectly by keeping PBM-^{3}H standards exposed to NTB 2 emulsion for increasing exposure times. At the time that the rate of latent image fading equals the rate of latent image formation, grain counts would no longer increase (Wheeler and Shaw, 1971). To measure this factor, both 500 μCi/gm and 50 μCi/gm PBM-^{3}H standards are used (Fig. 5). For both specific activities of PBM-^{3}H, grain counts increase linearly with time. No effect of latent image fading is found until after a period of 12 weeks. In addition, the number of silver halide crystals is sufficient to detect β-rays from PBM-^{3}H of high specific activity (500 μCi/gm). If multiple hits or "coincidence" were a

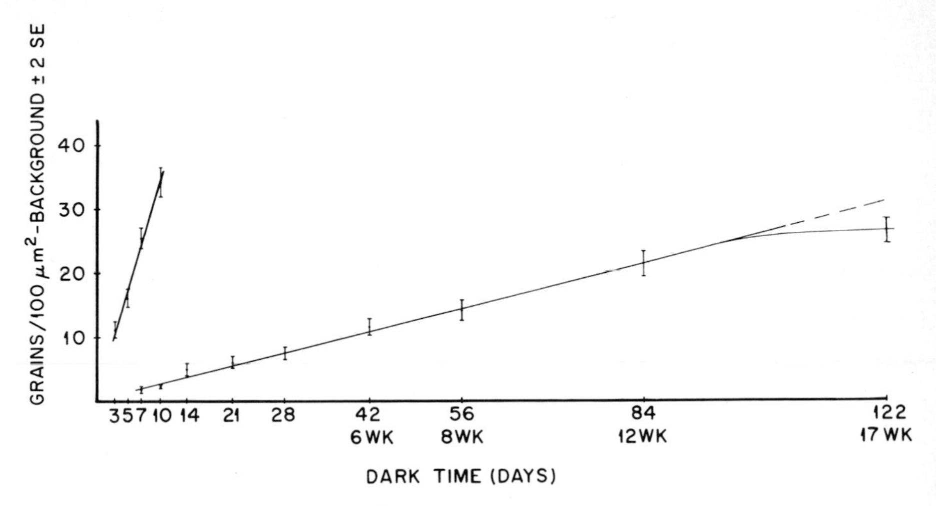

FIG. 5. NTB 2 emulsion was developed in undiluted D-19 at 15°C for 6 minutes. Ordinate gives net mean grain counts per 100 μm^2, and abscissa gives exposure time in days (or weeks). The curve on the left with the steeper slope is for 500 μCi/gm PBM-^{3}H. The curve on the lower right is for 50 μCi/gm PBM-^{3}H. Mean values ± 2 standard errors are indicated by the vertical bars. Both curves extrapolate through zero. The 500 μCi/gm curve has a slope approximately 10 fold that of the 50 μCi/gm curve. For the 500 μCi/gm curve, silver halide crystal packing is still sufficient at 10 days dark time for β-ray detection. Until 84 days (12 weeks) the 50 μCi/gm PBM-^{3}H increases linearly.

significant factor, the ratio of the slopes of the two curves should not be proportional to the ratios of their specific activities.

3. UNIFORMITY OF β-RAY DETECTION

Uniformity of β-ray detection by the emulsion is required if internal standards are to be used to obtain quantitative data from experimental sections. This uniformity is measured by placing 1-μm sections of PBM-^{3}H in three different positions over the 40-mm length of a 25 $\times$ 75-mm glass slide which falls within the boundaries of a coverslip 50 mm long. The slide is then coated with emulsion. Grain counts of over 90 sections demonstrate that regardless of the position of the internal standard on the slide, the number of grains remains the same when undiluted NTB 2 emulsion is used. This emulsion is therefore a uniform detector of β-rays for slides of standard size (25 $\times$ 75 mm) when used by the method described here. This uniformity allows valid comparisons between experimental sections and internal standards of PBM-^{3}H.

IV. Calculations

A. Efficiency Calculations

Falk and King (1963) used ^{3}H-labeled methyl methacrylate to calculate "radioautographic efficiency," and others (Rogers, 1969; Schultze, 1969) have alluded to the use of "grain yield" to get quantitative information from autoradiographs. Ritzén (1967) recognized the advantage of using an "infinite" thickness of the cell, i.e., saturation conditions, for direct comparison to a known reference. Calculations resulting from these considerations are simple (see Eqs. 1–3).

$$\beta = \text{time, days} \times 3.2 \times 10^9\ \beta/(\text{day} \cdot \mu\text{Ci}) \times \mu\text{Ci/gm} \times \rho,\ \text{gm/cm}^3 \times 10^{-10}\ \text{cm}^3 \quad (1)$$

where β = β-rays, ρ = density of standard (1.06), 10^{-10} cm^3 = 100 μm^2 section 1 μm thick.

$$G/\beta = \text{Efficiency} \quad (2)$$

Where G = number of grains per 100 μm^2 overlying the internal standard of PBM-^{3}H.

$$\text{Normalized specimen grain count} = G^E/E \quad (3)$$

when G^E = grain counts over cells in specimen, E = efficiency.

This calculation represents the grain count which would have been observed if efficiency were one, i.e., 100%.

In our experience, the values for "efficiency" vary from 0.034 to 0.064. Significant differences are found between slides processed in the same manner as far as can be determined.

The calculation of efficiency as described above is a slide calibration factor. It is not a measure of the efficiency of the emulsion. It does not allow the calculation either of the amount of radioactivity or of the specific radioactivity of the specimen. It does allow valid comparisons between individual autoradiograms done at different times or performed in different laboratories. When comparing slides, a linear correction can be made for different dark times, i.e., exposure time.

B. Specific Activity Equations

Over the specimen, grain counts per unit area must be made. These counts are then expressed per 100 μm^2.

$$(G/X\ \mu\text{m}^2)^E \cdot 100/X = (G/100\ \mu\text{m}^2)^E \quad (4)$$

where G = number of grains, superscript E = experimental specimen, X = area of the field counted in μm^2.

$$\mu Ci^S = (\mu Ci/gm)^S \cdot \rho^S \cdot 10^{-10}\ [(G/100\ \mu m^2)^E/(G/100\ \mu m^2)^S] \quad (5)$$

where superscript S indicates internal standard, ρ = density.

$$(\mu Ci/gm)^E = (\mu Ci/gm)^S \cdot (\rho^S/\rho^E) \cdot [(G/100\ \mu m^2)^E]/[(G/100\ \mu m^2)^S] \quad (6)$$

Notice that the values for total activity and specific activity differ only by constants.

Data for these calculations are more difficult to obtain because of the area measurements required. However, such an approach is necessary to realize the full potential of autoradiography.

Appendix: Derivation of Specific Activity Equations

Let S = internal standard of PBM-^3H
E = experimental specimen
β = beta rays
D = days
K = 3.2×10^9 disintegrations/(day $\cdot\ \mu$Ci)
G = number of silver grains

$$\beta^S = D \cdot K \cdot \mu Ci^S \quad (a)$$

Let F_S = fraction of β-rays from source reacting with emulsion
F_T = fraction β rays transmitted through source
F_L = fraction latent images remaining at time of development
F_G = fraction silver grains produced from latent images
F_R = fraction silver grains remaining after developing, washing, fixing, and staining.

$$G^S = \beta^S \cdot F_S^S \cdot F_T^S \cdot F_L^S \cdot F_G^S \cdot F_R^S \quad (b)$$

Let $\tau^S = F_S^S \cdot F_T^S \cdot F_L^S \cdot F_G^S \cdot F_R^S$

$$G^S = \beta^S \cdot \tau^S \quad (c)$$

Substituting Eq. (a) in Eq. (c)

$$G^S/(D \cdot K \cdot \mu Ci^S) = \tau^S \quad (d)$$

Similarly for the specimen

$$G^E/(D \cdot K \cdot \mu Ci^E) = \tau^E \quad (e)$$

Dividing Eq. (d) by Eq. (e)

$$\mu\text{Ci}^E = \mu\text{Ci}^S \cdot (G^E/G^S) \cdot (\tau^S/\tau^E) \tag{f}$$

For a single slide $F_L^E = F_L^S$, $F_G^E = F_G^S$, $F_R^E = F_R^S$.

At saturation $F_T^E = F_T^S$.

With Epon not removed and tritium distributed uniformly through specimen, $F_S^E = F_S^S$.

Therefore $\tau^E = \tau^S$.

$$\mu\text{Ci}^E = \mu\text{Ci}^S \cdot (G^E/G^S) \tag{g}$$

Dividing numerator and denominator of last member on right by 100 μm^2

$$\mu\text{Ci}^E = \mu\text{Ci}^S \cdot [(G/100\,\mu\text{m}^2)^E/(G/100\,\mu\text{m}^2)^S] \tag{h}$$

Since $\mu\text{Ci}^S = (\mu\text{Ci/gm})^S \cdot \rho^S\ \text{gm/cm}^3 \cdot 10^{-10}\ \text{cm}^3$

$$\mu\text{Ci}^E = (\mu\text{Ci/gm})^S \cdot \rho^S\ \text{gm/cm}^3 \cdot 10^{-10}\ \text{cm}^3 \cdot [(G/100\ \mu\text{m}^2)^E/(G/100\ \mu\text{m}^2)^S] \tag{i}$$

and

$$(\mu\text{Ci/gm})^E = (\mu\text{Ci/gm})^S \cdot (\rho^S/\rho^E) \cdot [(G/100\ \mu\text{m}^2)^E/(G/100\ \mu\text{m}^2)^S] \tag{j}$$

With conversion of measured specimen area

$$(\mu\text{Ci/gm})^E = (\mu\text{Ci/gm})^S \cdot \rho^S/\rho^E \cdot [(G/\text{X}\ \mu\text{m}^2)\ (100/\text{X})]^E/(G/100\ \mu\text{m}^2)^S \tag{k}$$

REFERENCES

Baserga, R. (1967). *In* "Methods in Cancer Research" (H. Busch, ed.), Vol. 1, pp. 100–108. Academic Press, New York.

Baserga, R., and Malamud, D. (1969). "Autoradiography: Techniques and Application," Ch. 5, pp. 129–145. Harper, New York.

Cleaver, J. E. (1967). "Thymidine Metabolism and Cell Kinetics," Ch. 1, pp. 15–42. North-Holland Publ. Amsterdam.

Falk, G. J., and King, R. C. (1963). *Radiat. Res.* **20**, 466.

Fitzgerald, P. J., Eidinoff, M. L., Knoll, J. E., and Simmel, E. B. (1951). *Science* **114**, 494.

Harris, C. C., Kaufman, D. G., Sporn, M. B., Boren, H. G., Jackson, F., Smith, J. M., Pauley, J., Dedick, P., and Saffiotti, U. (1973). *Cancer Res.* (in press).

Maurer, W., and Primbsch, E. (1964). *Exp. Cell Res.* **33**, 8.

Pelc, S. R., and Welton, M. G. E. (1967). *Nature* (*London*) **216**, 925.

Perry, R. P. (1964). *In* "Methods in Cell Physiology" (D. M. Prescott, ed.), Vol. 1, pp. 305–326. Academic Press, New York.

Przybylski, R. J. (1970). *In* "Introduction to Quantitative Cytochemistry II" (G. L. Wied and G. F. Bahr, eds.), pp. 477–505. Academic Press, New York.

Ritzén, M. (1967). *Exp. Cell Res.* **45**, 250.

Rogers, A. W. (1969) "Techniques of Autoradiography." Elsevier, Amsterdam.

Schultze, B. (1969). *In* "Physical Techniques in Biological Research" (A. W. Pollister, ed.), 2nd Ed., Vol. 3, Part B, pp. 16–24. Academic Press, New York.

Wheeler, K. T., Jr., and Shaw, E. I. (1971). *Int. J. App. Radiat. Isotop.* **22**, 759.

Chapter 17

Ionic Coupling between Nonexcitable Cells in Culture

DIETER F. HÜLSER

Abteilung Physikalische Biologie, Max-Planck-Institut für Virusforschung, Tübingen, Germany

I. Introduction

Since the existence of electrical as opposed to chemical synapses was described by Furshpan and Potter (1959) in abdominal nerve cords of the crayfish, such "low-resistance junctions" have been discovered in numerous types of nonexcitable cells. These junctions allow the exchange of ions and molecules between cells in contact and are, therefore, considered responsible for intercellular communication (Loewenstein and Kanno, 1964; Potter *et al.*, 1966). As far as is known, they are found in all normal *in vivo* cell systems.

Recent findings suggest that these low-resistance junctions are identical with the gap junctions found by electron microscopical observation of membranes after lanthanum impregnation or freeze-cleaving (Revel and Karnovsky, 1967; Revel *et al.*, 1971; Johnson and Sheridan, 1971; Rose, 1971; Gilula *et al.*, 1972; Pinto da Silva and Gilula, 1972; Friend and Gilula, 1972).

In the following sections, the electrophysiological techniques used for the demonstration of low-resistance junctions by the occurrence of ionic coupling pulses will be described. It will also be shown that for the analysis of coupling phenomena between cells, electrical parameters such as potential difference, membrane resistance, and capacitance must be taken into account. In addition, the electrical and mechanical instrumentation necessary for such measurements will be described with special emphasis on electrophysiological studies in cultured cells. Based on experiments with lymphocytes it will be shown that low-resistance junctions can be built up within minutes and that in the case of cells of established lines in culture their existence is related neither to normal nor to malignant cell properties, but appears to be associated with fibroblastoid cells.

II. Technical Equipment

The techniques used for demonstrating low resistance junctions were adapted from neurophysiological methods, and therefore the application of current pulses of millisecond duration and their registration with the oscillograph was understandable. Although this method allows suitable analysis of the transmitted pulse with regard to resistive and capacitative components of the involved membranes, for routine measurements we prefer the continuous registration of the electrical parameters with a pen recorder. This requires the application of current pulses of several seconds' duration and allows a better observation of changes in cell electrical properties during an experiment.

For the determination of ionic coupling between nonexcitable cells it is necessary to measure (1) the potential difference (PD) between the cell interior and the medium, (2) the superimposed pulses resulting from the injection of current pulses into a cell, and (3) the resistances of the cell membrane and junctional membrane (low-resistance junctions) to show the integrity of the cell.

Most of the instruments used for pulse generation, amplification, and recording are commercially available. The instruments used for the measurements described in Section IV are indicated in the following section.

A. Electrical Setup

A schematic drawing of the electrical setup is shown in Fig. 1. Experiments registered by a pen recorder were performed with 6 coupling pulses per minute. In this case the current supplied from the pulse generator G1 (Tektronix 601, modified 161, 162) to the current electrode CE could be directly measured with a nanoammeter (Philips digital multimeter PM 2421). This current flows from CE through the medium and a glass bridge filled with Ringer-agar to a 3 *M* KCl solution into which the indifferent calomel electrode IE is immersed. A symmetrical circuit is desirable, in order to avoid different diffusion potentials which could interfere with the PD measurements (Pfister and Pauly, 1969; Barry and Diamond, 1970).

The current supplied to a cell has to be varied depending on the size of the cell and the degree of coupling. In ionically coupled cells, in which the current spreads through the cell monolayer, a maximum pulse of about 50 nA results in the same hyperpolarization of about 75 mV as in noncoupled cells supplied with a current pulse of about 10 nA. For better time resolution, shorter pulses (20 msec) were supplied via an opto-electronic circuit as described by Baird (1967). The advantage of such an electroluminescent diode combined with a photo-detector (e.g., Monsanto MCD2) is the isolation of the current electrode from ground by no conductive paths other than the preparation. Both the short current and coupling pulses were measured with the oscillograph.

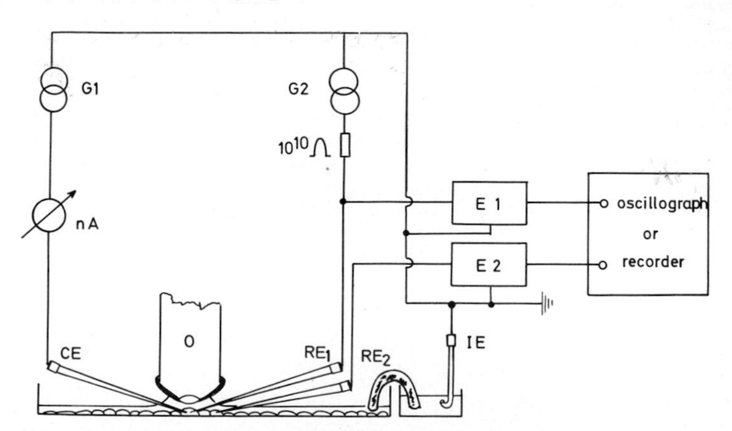

FIG. 1. Scheme of the electrical setup. The current supplied by the generator G-1 to the current electrode CE can be measured with a nanoammeter. The circuit is closed via a Ringer-agar bridge and the indifferent electrode IE. The resulting potential changes can be measured with the recording electrodes RE_1 and RE_2 and the negative capacitance electrometers E 1 and E 2. For continuous electrode resistance control, a current pulse can be supplied from the generator G2. The metallic conus of the water-immersion objective, O, is insulated by a beeswax-colophonium cover.

The PDs were measured with the recording electrodes RE_1 and RE_2 using negative capacitance electrometers E 1 and E 2 (Keithley 605) and observed at an oscillograph (Tektronix 564) or registered with a pen recorder (Graphirac, Sefram). RE_1 could be connected via a 10^{10} Ω resistor to a second pulse generator (0.1 Hz), thus permitting continuous measurements of the electrode resistance (see Section III). Two types of coupling measurements were performed with this setup.

1. In cells which could be impaled with two electrodes, both CE and RE_1 were inserted into one cell. The resulting voltage pulse V_1 superimposed onto the PD line of this cell was compared with the voltage pulse V_2 in an ionically coupled cell impaled with RE_2. If no pulse could be detected with RE_2, the cells were not considered to be ionically coupled. The ratio $V_2 : V_1$—defined by Loewenstein and Kanno (1967) as "communication ratio"—indicates the degree of ionic coupling and the existence of low-resistance junctions (see Section III).

2. In the case of small cells, such as lymphocytes, single cells cannot be routinely impaled with two electrodes. Therefore, ionic coupling was in this case determined with two electrodes only. The small surface area of these cells results in a high ohmic resistance, Therefore, the successful insertion of CE was indicated by a decreased current, whereas, when RE_1 was successfully inserted, an increased resistance was indicated by the continuous resistance pulse. It was necessary to register these parameters, because the PD values in lymphocytes varied around zero.

Since changes in pH over a range between pH 7 and pH 8.4, as well as temperature changes between 20°C and 37°C did not noticeably influence the PD (Hülser, 1971), the measurements were performed at room temperature under microscopical observation (Zeiss Standard RA). We used a water-immersion objective (40×), electrically insulated with a beeswax-colophonium mixture coating the metallic conus. No demonstrable toxic effect on the cell cultures was observed during the measurements. This procedure permitted an overall 400 × magnification, sufficient to observe any morphological alteration of the cells which might occur during impalement.

B. Mechanical Setup

The measurements in cells of established cell lines in monolayer culture were performed in surface-treated plastic petri dishes (Falcon or Greiner). Prior to the measurements, 2 cm of the wall of the dish on opposite sides had to be removed with a flat soldering iron, permitting a low-angle approach of the electrodes. This was necessary because of the short working distance of the insulated water-immersion objective. It was not necessary to cover the petri dish bottom with a soft resin, as was described by Borek *et al.* (1969), since this low-angle approach prevented the electrode from being so easily broken.

For positioning the electrodes correctly under the microscope objective, we used Leitz micromanipulators (Leitz, Wetzlar, Germany) which allow electrode movement in three dimensions with different transmissions and without a backlash. Various other micromanipulators are commercially available that will meet the same requirements (see Kopac, 1964).

For measurements with lymphocytes and explants of tumors and tissues, special preparation techniques were used, they are described in Section IV.

A serious problem for measurements with micro-manipulators is the absorption of vibrations which may result from different sources. Among a number of possibilities to overcome this problem, we chose a castiron plate of about 150 kg situated on two inner tubes in a table frame. On this plate, the microscope and the micromanipulators can be balanced by lead blocks. Most of the vibrations that occur in a laboratory are thus completely absorbed or transposed into low oscillations of the plate, which are then followed synchronously by the microscope, the micromanipulators, and the electrodes without damaging the preparation.

The transition from the electrolyte solution to the connective cable can either be performed using an Ag/AgCl wire with the AgCl in the electrolyte solution or with a calomel electrode. We preferred calomel electrodes because of their high stability with regard to drifts of potentials. Calomel electrodes (Fig. 2) can be prepared by first fixing a Pt wire into the end of a 30 mm-long glass tube (5 mm outer diameter, 3 mm inner diameter) by melting the glass. The other end is narrowed so that a magnesia rod (Merck, Germany, No. 5809) will just pass. The part of the tube with the Pt wire is then filled with Hg and the other half with Hg_2Cl_2 in saturated KCl. A 10 mm-long magnesia rod is then pressed into the calomel and sealed with a quick-acting glue. The whole calomel unit is fixed with a Plexiglas glue into a Plexi-

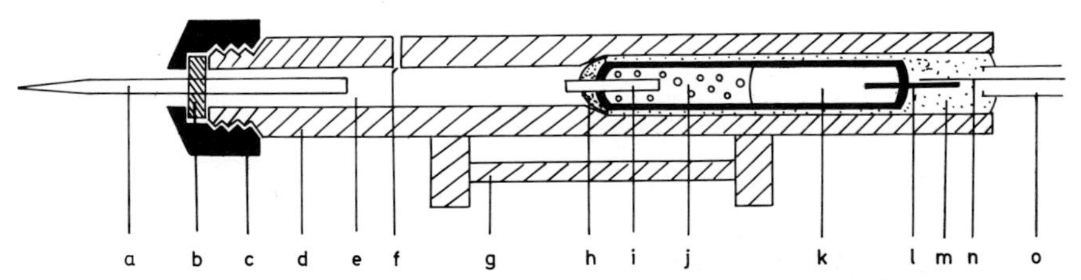

FIG. 2. Plexiglas electrode-holder with a calomel unit. a, glass microelectrode; b, rubber washer; c, stainless steel screw; d, Plexiglas tube; e, electrolyte solution; f, hole for pressure compensation, g, Plexiglas bar for attachment to a micromanipulator; h, quick-acting glue, i, magnesia rod; j, Hg_2Cl_2 in saturated KCl; k, Hg; l, Pt; m, Plexiglas glue; n, connective cable; o, shield of the connective cable.

glas holder. The protruding end of the magnesia rod must be covered with the electrolyte solution to prevent the calomel from drying.

The Plexiglas holder should have a small hole on top to allow pressure compensation. Otherwise a flow of electrolyte through the tip will result, which may affect the electrical properties of the electrode. A small Plexiglas bar on the bottom of the holder enables it to be attached to a micromanipulator.

C. Glass Microelectrodes

For the intracellular recording of potentials, glass electrodes are widely used. Since their introduction by Ling and Gerard (1949), there has been considerable progress concerning the optimal properties of the glass and the technical equipment used for pulling and filling the pipettes. Electrodes with a tip diameter of the order of 0.5 μm, which can impale a cell without damaging it, have been termed "microelectrodes." Even though numerous metal electrodes are now available, the open-tip glass microelectrodes are preferred for several reasons: they are easily and cheaply prepared in a sufficiently reproducible manner; their resistances are low compared with metal electrodes; they are nonpolarizable and they can be used for iontophoretic injection of different molecules or ions (Chowdhury, 1969).

In this section the preparation of glass microelectrodes is described as carried out in our laboratory. This procedure results in a sufficient number of electrodes having the desired properties needed for our investigations. However, other methods have been described for different types of investigations, especially for filling the pipettes with electrolytes (Tasaki *et al.*, 1954; Burkhardt, 1959; Frank and Becker, 1964; Oliveira-Castro and Machado, 1969; Machemer, 1970; Zettler, 1970).

Some of the procedures used for glass microelectrode production are entirely empirical and may sometimes appear alchemistic; but nevertheless, most of the proposed variations have resulted in certain advantages.

1. Preparation of the Electrodes

Glass capillaries (Corning Pyrex glass 7740/234400) with an outer diameter of 2 mm are cut into 8-cm lengths, and the ends are rounded in the Bunsen flame without narrowing the lumen. These capillaries are soaked in distilled water for 24 hours and then sonicated in a test tube with dilute detergent (Mucasol, Merz & Co.) by inserting the tip of a sonicator (Branson-Sonifier B-12) just above the capillaries for about 3 minutes. The detergent is washed off under tap water, followed by distilled water in which the capillaries are sonicated a second time before washing in acetone and drying at 180°C for 20 minutes.

The cooled capillaries are pulled into micro pipettes by a vertical pipette puller (David Kopf Instruments, Model 700C). Such a puller allows the preparation of pipettes with different shapes and lengths of the shank and different sizes of the tip by variation of the temperature of the heater coil and the pulling force. Both pipettes pulled from one capillary can be used for filling with the electrolyte; significant differences in their electrical properties have not been observed.

The pipette puller should be cleaned with alcohol or acetone before use, and the capillaries and pipettes should be handled only with clean forceps to avoid contamination of the electrode tips. The pipettes are placed with their tips down into a Plexiglas disk with drill holes, where they are kept secure with a rubber band situated in a groove. This disk will hold 30 elec-

FIG. 3. Plexiglas disk with glass microelectrodes, fixed by a Plexiglas bar in the lid of the Witt flask.

trodes, as shown in Fig. 3. The disk is screwed onto a Plexiglas bar fixed in the neck of the lid of a Witt filter apparatus containing a side tube. The Plexiglas bar has a central bore hole from outside to its middle corresponding to the normal filling level for the solutions.

The electrodes are shaken once in methanol in the cleaned Witt flask, and after the methanol is poured through the side tube, the flask is filled with methanol again, so that all the pipettes are covered. The side tube is then closed, and by evacuating through the Plexiglas bar with an aspirator, the pipettes are filled with methanol.

Now the methanol in the flask is removed and enough solution of 3 *M* KCl + 2 m*M* potassium citrate is added to cover the pipettes. This electrolyte solution should be freshly filtered through a 0.45-μm Millipore filter. The methanol in the pipettes is displaced by the electrolyte solution. To speed up this diffusion, the flask is placed in a water bath at 40°C for about 1 hour; it is then evacuated once more to remove the remaining methanol. The Plexiglas disk with the electrodes is now unscrewed from the Plexiglas bar and turned, so that the electrode tips point upward. In this position the electrodes can be stored up to 1 week or longer. Normally they are not used within the first 2 days after preparation, because of their unstable electrical properties during this period.

2. Electrical Properties of the Electrodes

There are two electrical properties by which electrodes are generally selected: Ohmic resistance and tip potential. Electrodes with resistances between 20 and 50 MΩ, and tip potentials $\leq$ 5mV are preferred for most experiments. Both resistance and tip potential depend on the tip diameter, and they change with the age of the electrode.

a. Electrode Resistance. The resistance changes with the direction, amplitude, and duration of the current carried by the tip (Frank and Becker, 1964; Zettler, 1970), as well as with the ion concentration of the fluid in which the electrode is immersed (see Table I). Therefore, electrode resistances can be compared only if determined in solutions of the same ion content. The changes caused by the current will not influence the determination of resistances if the calibration is performed with identical methods, but they can be of importance when current is supplied to the cells for the determination of cell membrane resistances.

The electrode resistance is easily determined by differentiation of a ramp signal with a RC combination formed by the electrode resistance (or, for calibration, with known resistors) and a constant capacitor. Figure 4 shows the setup used to determine the electrode resistance. The calibrator of the oscillograph supplies square pulses (a) which can be integrated (b) by a 200 kΩ resistor and a 0.47 μF capacitor into ramp signals (c). These ramp

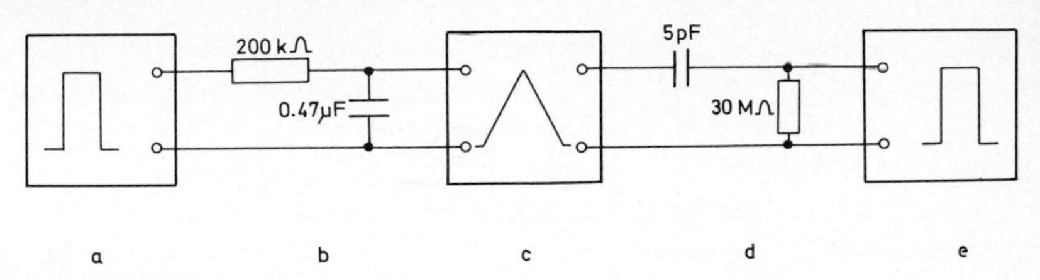

FIG. 4. Determination of the electrode resistance. A square pulse (a) is integrated (b) into a ramp signal (c). This ramp signal can be differentiated by the *RC* combination of the electrode resistance and a constant capacitor (d). The amplitude of the resulting square pulse (e) depends on the electrode resistance.

signals are differentiated (d) into square pulses (e) by a constant capacitor in the Keithley 605 electrometer and the resistance of the electrode. The amplitude of these square pulses (e) increases with increasing resistance. They are, therefore, used for the determination of the unknown electrode resistances by comparing their height with calibration values obtained from the differentiation with known resistors between 10 and 100 MΩ by the same setup. Furthermore, the risetime of the square pulse (e) is an indicator for the electrode capacitance and every other capacitance introduced into the circuit.

b. Electrode Tip Potential. Much more important for the potential difference (PD) measurements in cells is the tip potential (TP) of the electrodes. The possible origin of TPs has been discussed in several publications (Adrian, 1956; Bernhardt and Pauly, 1967; Redmann and Kalkoff, 1968; Agin, 1969; Lavallée and Szabo, 1969; Snell, 1969); their influence on PD measurements will be demonstrated. The TP is defined as the PD which appears in a closed circuit when the electrode tip is broken off. With a broken electrode tip, only symmetrical diffusion potentials are found in the circuit, and the resulting potential can be considered as the zero potential. When a fine-tipped electrode is introduced in place of a broken one, a potential change will be observed which indicates the height of the TP of the introduced electrode. For electrodes filled with 3 *M* KCl, this TP is always negative when the tip is inserted into a cell culture medium and can reach values of > 40 mV. When other electrodes with tips broken in the same way are introduced and the same electrode holder is used in the same circuit, small changes of about 1–2 mV can be observed, indicating that the real TP of a glass microelectrode can only be determined by breaking the tip after the experiment. At the end of the experiment, however, the TP may have been altered by a film of cellular substances adhering to the tip after withdrawal of the electrode from a cell. This would result in a change of the base line which may escape observation, if experiments performed with the same electrode are not recorded continuously.

TABLE I

VARIATIONS OF TIP POTENTIAL [mV] AND RESISTANCE [MΩ] OF A SINGLE ELECTRODE WHEN IMMERSED IN EAGLE-DULBECCO MEDIUM OR DIFFERENT CONCENTRATED SALT SOLUTIONS

Solution	Eagle-Dulbecco medium		Concentration							
			0.003 *M*		0.03 *M*		0.3 *M*		3 *M*	
	[mV]	[MΩ]	[mV]	[MΩ]	[mV]	[MΩ]	[mV[	[MΩ]	[mV]	[MΩ]
Potassium chloride	−4	26	−44	61	−18	35	−3	20	0	9
Sodium chloride	−5	28	−43	53	−17	36	−2	21	+1	11
Choline chloride	−4	25	−50	73	−22	50	−1	28	+10	16
Potassium citrate	−6	28	−35	57	−23	30	−25	18	−13	20
Sodium cyclamate	−1	29	−42	83	−16	43	−15	26		
Sodium bicarbonate	+1	25	−48	100	−18	40	−11	21		

A single electrode changes its TP when inserted into different salt solutions with different ion concentrations, as can be seen from Table I. The use of a solution buffered with 2 m*M* Hepes buffer to a pH of about 8 did not result in a different TP value. The experiment started with the measurement of the TP and resistance of one electrode in the medium and continued with measurements in increasing concentrations of the solutions. After the measurement in 3 *M* KCl solution, the values were once more determined in the medium before continuing with the NaCl solutions. It will be noticed that the electrode TPs and resistances depend to a greater part on the ion concentration, but also on the "history" of the electrode shown by the different values in Eagle-Dulbecco medium. After the measurements in the bicarbonate solutions, the tip was immersed in the medium for a short period before being broken to reveal the originally determined TP of −4 mV.

The different intracellular ion concentration, as compared with that of the medium, means that the tip potential of an electrode will also change when the electrode is inserted into a cell. This can be demonstrated using two glass microelectrodes. The tip potential decreases after cell impalement resulting in PD values lower than the actual membrane potential of the cell. This is illustrated in the schematic drawing of Fig. 5. The zero line is obtained with broken glass microelectrodes as recording and indifferent electrodes (see

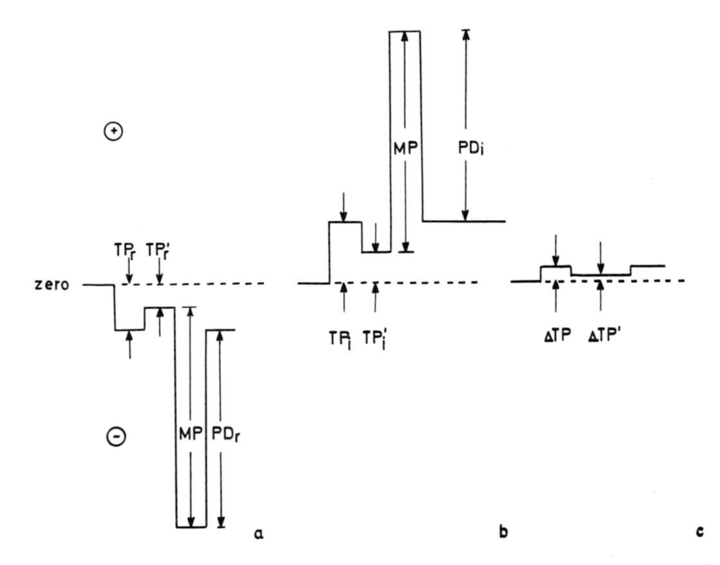

FIG. 5. Membrane potential measurements: influence of microelectrode tip potential. TP_r: tip potential of the recording electrode in medium; TP_r': tip potential of the recording electrode in a cell; TP_i: tip potential of the indifferent electrode in medium; TP_i': tip potential of the indifferent electrode in a cell; MP: membrane potential; PD: Potential difference measured between cell interior and medium; $\Delta TP = TP_i - TP_r$; $\Delta TP' = TP_i' - TP_r'$.

Fig. 1). After introducing an intact recording electrode RE, its tip potential TP_r is measured (Fig. 5a). When inserted into a cell, the tip potential decreases (TP'_r), but at the same time the membrane potential (MP) is recorded, so that this change of the tip potential cannot be observed, and only a potential difference (PD_r) is registered. If a glass microelectrode had first been chosen as the indifferent electrode IE, a situation similar to that found with RE will be observed, but the voltage deflections would have the opposite direction (Fig. 5b). When both IE and RE are intact glass microelectrodes in the medium, a change of the zero line for ΔTP is observed which corresponds to the difference of the two tip potentials (Fig. 5c). With the insertion of both electrodes into one cell, only the difference of the two tip potentials in the cytoplasm ($\Delta TP'$) will be recorded. $\Delta TP'$ is always less than ΔTP, indicating that the electrode tip potential has decreased after insertion into a cell. Upon withdrawal of the electrodes, the original tip potential is recorded.

In Section IV some examples are presented which show the precautions required for evaluating the membrane potential of different cell types.

III. Determination of Cell Membrane and Junctional Membrane Resistance

It should be emphasized that the electrode resistance can change simply by insertion of the electrode into another solution. Therefore, the measurement of cell membrane resistance with single-electrode techniques requires careful interpretation. According to the values of Table I, the resistance change should not be important if only small TP changes occur, but as different cytoplasmic resistivities for different cells have been measured (Schanne, 1969; Gradmann, 1972), it must be taken into account. The differentiation method allows at least a very careful demonstration of resistance and capacitance changes of cell membranes and, furthermore, provides a way of distinguishing between coupled and noncoupled cell types (see Section IV, Fig. 12). The cell membrane resistance determined by this method will always be less than the actual value, owing to the decrease of the electrode resistance when impaling a cell.

A convenient method for continuous resistance control of the electrode is the application of a current pulse to the recording electrode via a high ohmic resistor as shown in Fig. 1 (Woodbury and Woodbury, 1963; Hegel and Frömter, 1966). This current pulse causes a voltage drop across the electrode resistance which is superimposed onto the PD measurement and can be registered in order to control the electrode stability. In the case of small lymphocytes which have a high ohmic membrane resistance owing to their small surface area, this method could also be used as an indication of successful impalement of the cells (see Section IV, B). Another method, the single-electrode bridge technique, enables the injection of current through the voltage recording electrode. This may be useful for some experiments, but requires cautious interpretation of the recorded current voltage relations, as shown recently by Engel *et al.* (1972).

The influence of the described change of electrode resistance is diminished if the measurements of the resistance of the cell surface membrane and the junctional membrane are performed with two electrodes, as described by Loewenstein and Kanno (1964, 1967).

By the two-electrode method, the current and the voltage of the circuit are measured directly, and the ohmic resistance can then be calculated. For better understanding of the conductive paths, a schematic drawing of an

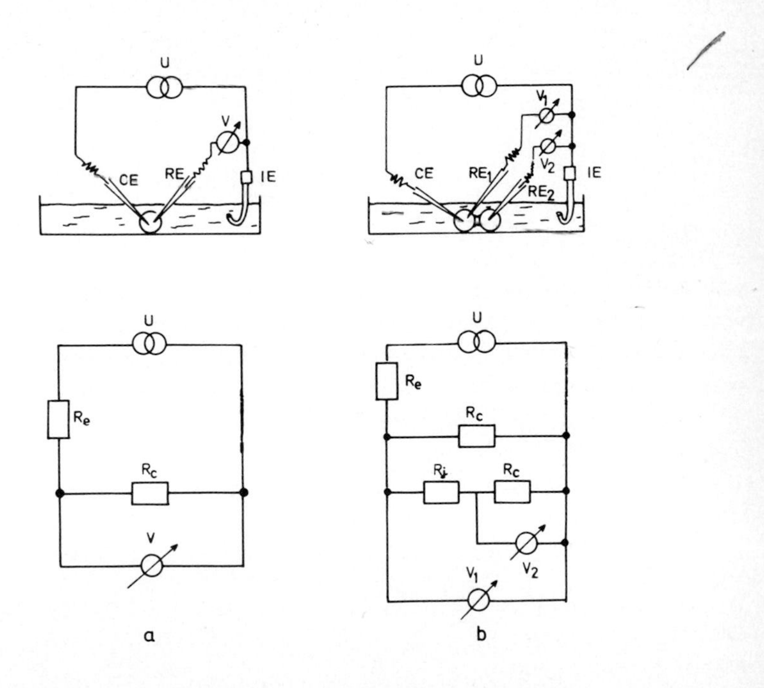

FIG. 6. (a) Determination of the ohmic cell membrane resistance. (b) Determination of the junctional membrane resistance. U, Voltage pulse from the generator; V, voltage measurement using the electrometer and the recording electrodes RE; CE, current electrode; IE, indifferent electrode; R_e, resistance of the current electrode; R_c, resistance of the cell membrane; R_j, resistance of the junctional membrane.

impalement of a single cell is shown in Fig. 6a, together with the respective electrical circuit. The voltage pulse V is recorded superimposed onto the PD and is due to the voltage drop across the resistance of the cell surface membrane R_c:

$$V = U\frac{R_c}{R_e + R_c} \tag{1}$$

(U: voltage pulse supplied by the generator: R_e: resistance of the current electrode CE; the resistance of the medium plus the circuitry has been neglected because it is less than 0.1% of the total resistance).

The situation with one attached cell is also very simply demonstrated, whereas that for a monolayer or tissue becomes more complex and requires mathematical evaluation (Cole, 1968; Shiba, 1970a,b; Shiba and Kanno, 1971; Siegenbeek van Heukelom *et al.*, 1972a,b).

The schematic drawing in Fig. 6b shows the electrical circuit of two coupled cells isolated from other cells. R_j represents the resistance of the junctional membrane. The voltage drop across the cell membrane resistance is now given by

$$V_1 = U\frac{1/2\,R_c}{R_e + 1/2\,R_c} \tag{2}$$

where the resistance of the medium plus the circuitry has again been neglected for the determination of the parallel cell resistance, and R_j can be neglected in a first approximation for the case of coupling. The coupling pulse V_2 is then given by

$$V_2 = V_1\frac{R_c}{R_c + R_j} \tag{3}$$

If $R_j \gg R_c$ the coupling pulse $V_2 \ll V_1$. This is the situation of noncoupling. With $R_j \ll R_c$ the coupling pulse $V_2 = V_1$, indicating complete coupling. The communication ratio $V_2 : V_1 \leq 1$, and is also an indicator for the extent of the junctional resistance.

Measurement of the resistance of cell membranes with only one electrode by the differentiation method gives values of the same order as the two-electrode measurement. For comparing different cell membranes, the specific resistance ($\Omega \times cm^2$) of the cells must be calculated by determining the cell surface area.

IV. Measurements of Ionic Coupling in Different Cells

A. Established Mammalian Cell Lines in Monolayer Culture

Cells of so-called established cell lines in culture, i.e., cells that have acquired the capacity for indefinite proliferation, are generally aneuploid and, morphologically, either of the fibroblastoid or the epithelioid type. As wc have observed in 14 different cell lines (Hülser, 1971; Hülser and Webb, 1973) and as suggested by the data of a number of authors (Azarnia and Loewenstein, 1971; Borek *et al.*, 1969; Furshpan and Potter, 1968; O'Lague *et al.*, 1970; Gilula *et al.*, 1972), the ionic coupling between cells in a monolayer generally appears to be associated with their morphological appearance: fibroblastoid cells are ionically coupled, epithelioid cells are not.

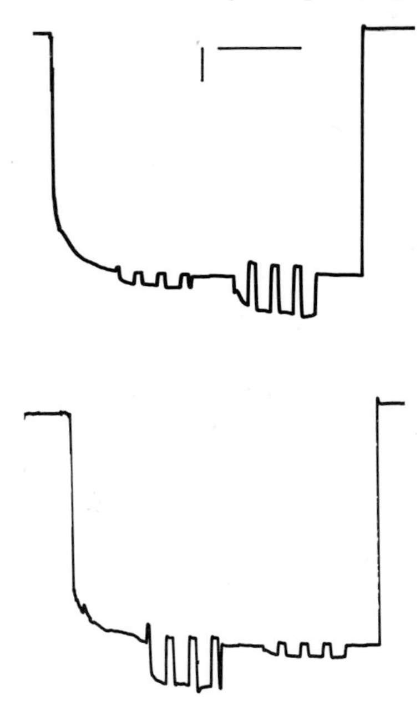

FIG. 7. Recorder diagram of an experiment with ionically coupled fibroblastoid BICR/M1R-K tumor cells. Starting on the left side, one observes the initial voltage deflections after insertion of both recording electrodes into two cells, separated by three other cells. The final PDs of about 70 mV are reached after a short time interval. The insertion of the current electrode into one of the two cells is indicated by the beginning of the coupling pulse, which also indicates the displacement of both traces due to the recorder. After about 1 minute, the current electrode is withdrawn and inserted into the other cell. Note that the communication ratio is the same in both cases, i.e., about 0.3. Upon final withdrawal of the current electrode, the PDs remain stable. Retraction of both recording electrodes results in the original base lines. Horizontal bar, 1 minute; vertical bar, 10 mV.

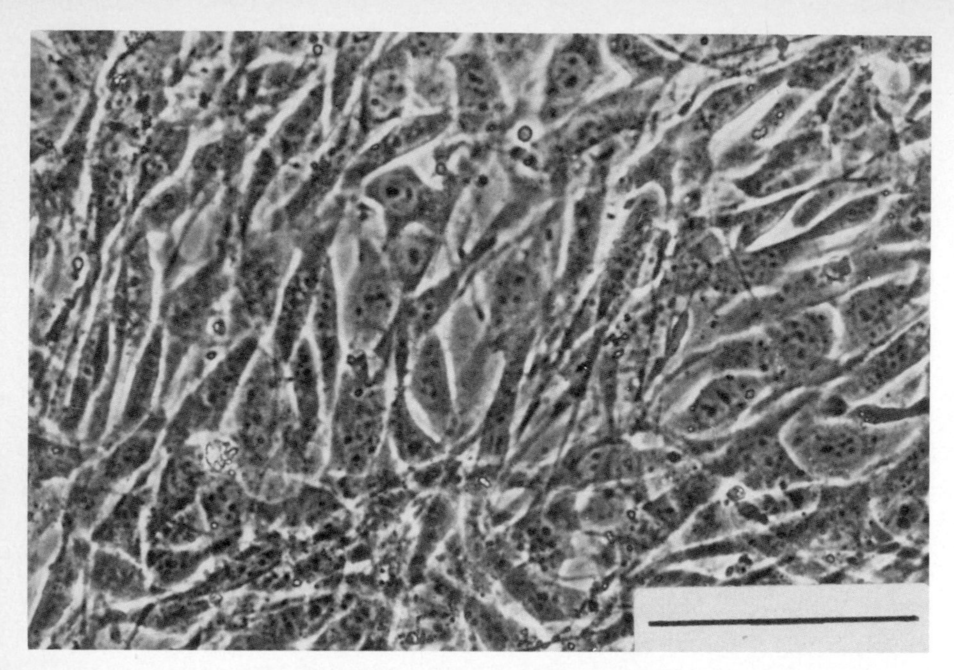

FIG. 8. 3T3 cells, an example of ionically coupled fibroblastoid cells. Bar indicates 100 μm.

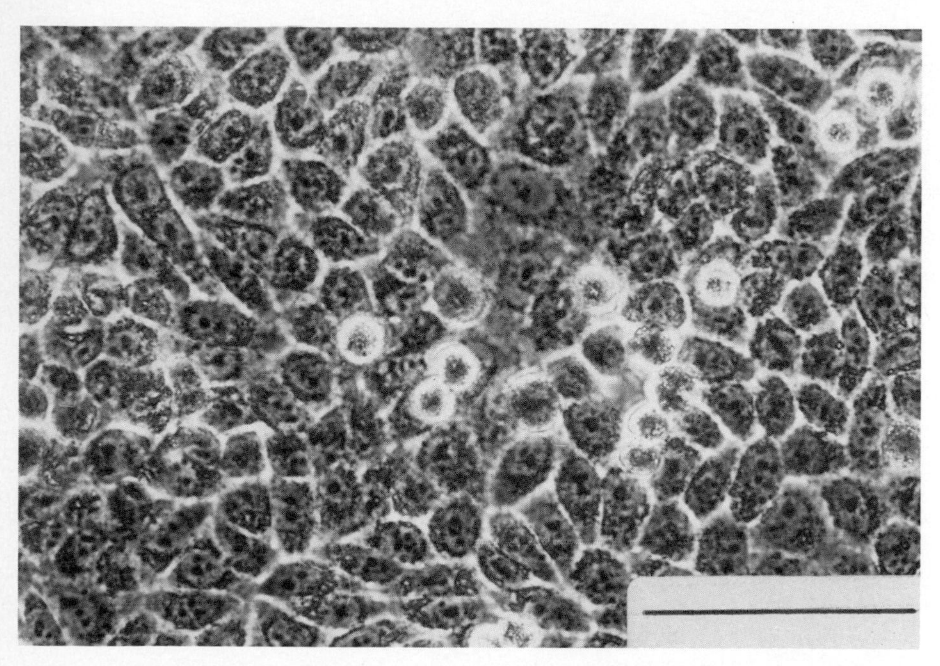

FIG. 9. HeLa cells, an example of ionically noncoupled epithelioid cells. Bar indicates 100 μm.

The presence or absence of ionic coupling is apparently not related to the origin of the cell lines or to their tumorigenicity. An epithelioid cell line (Hülser and Frank, 1971; Frank *et al.*, 1972) derived from embryonic rat cells failed to produce tumors even when 2×10^6 cells were injected into 10-day-old isogeneic rats. In these cells we were unable to detect ionic coupling. A fibroblastoid cell line derived from a transplantable BICR/M1R tumor gave rise to tumors within 14 days when 1×10^6 cells were injected into isogeneic baby rats. These cells were found to be ionically coupled (Fig. 7). A further example of coupled fibroblastoid cells is the nonmalignant 3T3 cell line shown in Fig. 8, whereas the HeLa tumor cell line (Fig. 9) is representative for the noncoupled epithelioid cells. The respective oscillograph tracings of the current pulses are demonstrated in Fig. 10.

The communication ratio varies in different fibroblastoid cell lines but has been detectable in every case investigated. In epithelioid cells faint ionic coupling may sometimes be observed several seconds to minutes after the

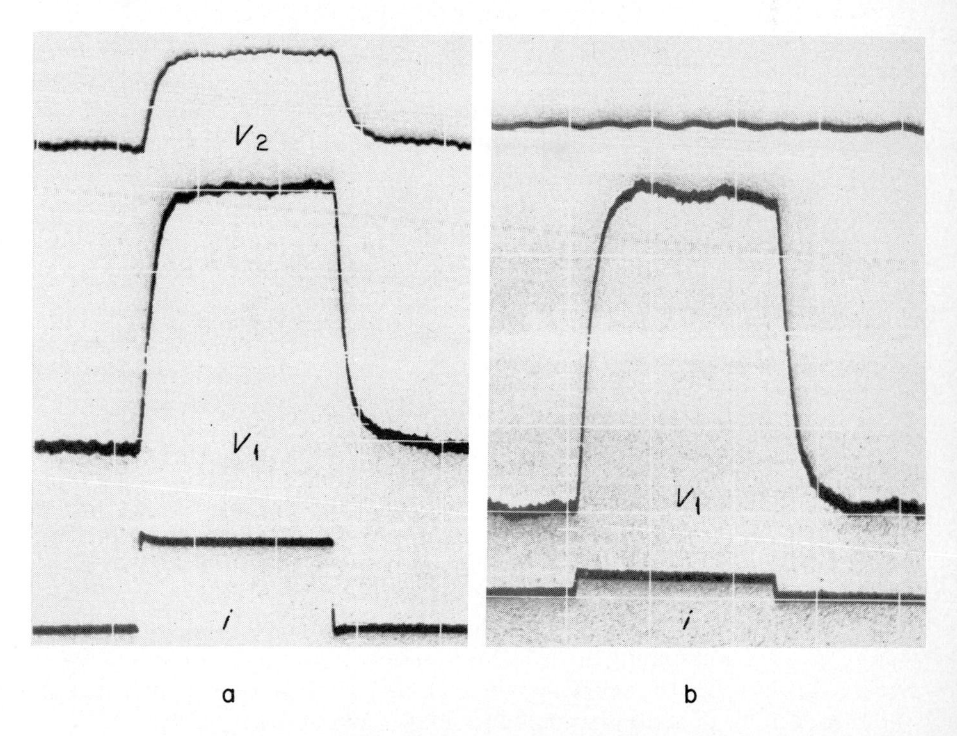

FIG. 10 Oscillograph tracings of experiments with coupled and noncoupled cells. (a) ionically coupled fibroblastoid 3T3 cells: a 50 nA current pulse (i) of 20 msec duration results in a 75 mV hyperpolarization pulse V_1 and a 30 mV coupling pulse V_2. (b) Noncoupled epithelioid HeLa cells: a 15 nA current pulse (i) of 20 msec duration results in a 90 mV hyperpolarization pulse V_1. A coupling pulse V_2 is not recorded.

insertion of the electrodes into two neighboring cells. This effect is due to an artifact, not to the recovery from transient uncoupling caused by the insertion of the electrodes, since at the same time one or more small blebs can usually be seen budding from the impaled cells. These blebs are more clearly seen at the border of a group of cells than in a dense monolayer. In Fig. 11 this is demonstrated for the epithelioid HeLa cells. Probably because their outer membrane has properties different from that of the intact cell, these blebs may lead to an artificial type of coupling between cells which normally possess no low-resistance junctions. The blebs also cause a drastic change in the membrane resistance and capacitance (Hülser and Webb, 1973) as demonstrated in Fig. 12 for noncoupled HeLa cells. After insertion of an electrode into a cell, the resistance increased from about 30 MΩ to about 45 MΩ, as indicated by the increased pulse height. With the appearance of blebs, there was not only a further increase in the resistance, but also in the capacitance, as can be seen from the longer rise time.

True absence of coupling is not easily demonstrated, since it could be due to an uncoupling of coupled cells by the insertion of three electrodes into two neighboring cells; i.e., the probability of disrupting junctional membranes must be taken into account. The likelihood of uncoupling is reduced by insertion of one electrode only and determination of the membrane resistance by the differentiation method as described in Section II,C. This procedure also provides an indication of the presence or the absence of low-resistance junctions: noncoupled cells have a similar ohmic membrane resistance if they are isolated or in contact with each other; coupled cells have a lower ohmic membrane resistance when contacting each other compared with isolated cells. The resistance of the junctions is some orders of magnitude lower than the membrane resistance, so that the resistance of the coupled cells should decrease in proportion to the increase in the common surface area due to the attached cells.

When coupled cell lines are cocultivated, cells of one line are also coupled to cells of other lines (Michalke and Loewenstein, 1971; Gilula *et al.*, 1972). This phenomenon can be used to demonstrate coupling or noncoupling by manipulating the cell of interest in such a way that it provides the only link between the two groups of coupled cells (Azarnia and Loewenstein, 1971). This "cell-bridge method" allows the demonstration of coupling or noncoupling without impaling the cell in question with an electrode, since only the ion flow through the cell is observed, whereas the injection of current and the recording of the coupling pulse are performed in other cells.

Another method for demonstrating coupling or noncoupling is the injection of fluorescein by iontophoresis. The spread of the dye into neighboring cells can be taken as an indication for low-resistance junctions, as it is reported to be associated with the finding of ionic coupling (Loewenstein

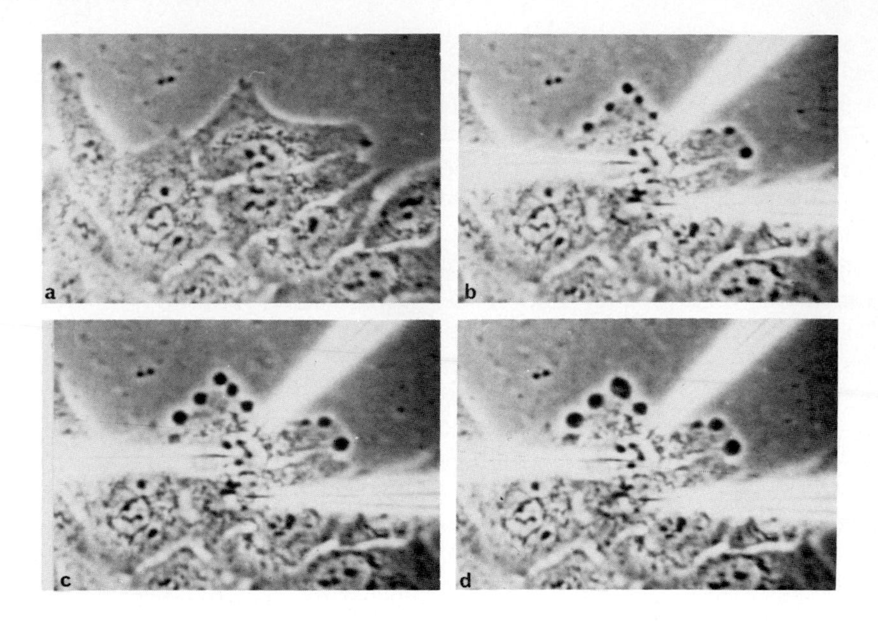

FIG. 11. Appearance of blebs during a coupling experiment with HeLa cells. All pictures were taken within 90 seconds from a television screen. (a) Cells before electrode insertion. (b)–(d) Situation 30, 60, and 90 seconds after insertion of two recording electrodes and one current electrode into two neighboring cells. The appearance of these blebs was accompanied by the onset of slight coupling (communication ratio 0.003–0.005).

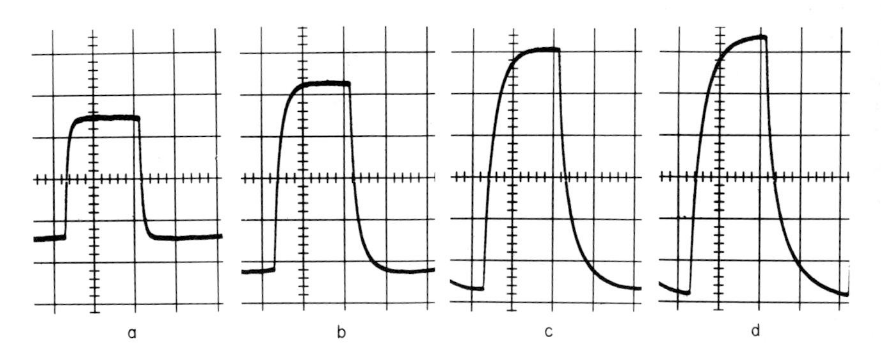

FIG. 12. Membrane resistance measurement by the differentiation method in a HeLa cell in a monolayer culture. (a) Pulse height (30 MΩ) indicates the electrode resistance in Eagle-Dulbecco medium. (b) Pulse height (45 MΩ) indicates the electrode resistance of the inserted electrode plus the ohmic resistance of a HeLa cell, immediately after impalement. (c) Forty seconds after insertion of the electrode where blebs could be observed, not only the ohmic membrane resistance has increased, but also the capacitance, as indicated by the longer rise time of the pulse. (d) Ninety seconds after insertion of the electrode. Further increase of the resistance and the capacitance can be observed.

and Kanno, 1964; Potter *et al.*, 1966; Kanno and Loewenstein, 1966; Furshpan and Potter, 1968; Loewenstein, 1967; Sheridan, 1970, 1971a,b; Rose, 1971; Rose and Loewenstein, 1971; Oliveira-Castro and Loewenstein, 1971). Using this method it could be shown that molecules with a molecular weight of at least 10,000 could pass from one cell to another in the salivary glands of *Drosophila* (Kanno and Loewenstein, 1966); however, it is not yet known whether this value also applies to mammalian cells. Since the iontophoretic application of substances also requires the impalement of cells, which may cause uncoupling, further improvement should be forthcoming from the cell-bridge method, using cells which have, for example, accumulated fluorescein by hydrolysis of fluorescein esters (Sellin *et al.*, 1971).

B. Lymphocytes

The presence of ionic coupling has been interpreted as an indication of information exchange between coupled cells (Loewenstein, 1968a,b; Furshpan and Potter, 1968). It was, therefore, of interest to measure the ionic coupling between cells that can change from one physiological state to another. For example, small lymphocytes can be induced, by nonspecific as well as immunologically specific stimulants, to shift from a resting state to an activated state with high metabolic activity. We used the nonspecific stimulation with phytohemagglutinin, which activates a high proportion of lymphocytes and is accompanied by cell agglutination, which allows the measurement of ionic coupling (Hülser and Peters, 1971, 1972).

The lymphocytes were obtained from bovine lymph nodes, dissociated and suspended in Eagle medium supplemented with 10% inactivated calf serum, purified by passage through a glass fibre column, and maintained in Eagle's medium (Hausen *et al.*, 1969). This procedure results in a more than 95% pure lymphocyte preparation. Stimulation was performed by adding 1.0 μl of phytohemagglutinin-P (Difco) to 1 ml medium containing 5×10^6 cells. To agglutinate lymphocytes without a stimulating effect, 10% horse antipig-thymocyte serum was used instead of inactivated calf serum. Immobilization of the lymphocytes was achieved by a special embedding method performed at 37°C with glassware of the same temperature (Hülser and Peters, 1972). A stock solution of 2% agar (Agar Noble Special, Difco) in physiological saline was heated to 100°C. When the agar had cooled down to about 60°C, 0.8 ml was mixed with 1.2 ml of a 37°C solution of Eagle's medium supplemented with 10% inactivated calf serum and 0.6% gelatin (Merck, Germany). Sedimented lymphocytes were suspended in this agar–gelatin mixture at a concentration of about 1 to 1.5×10^8 lymphocytes per milliliter. From this cell suspension, 0.15 ml was pipetted into a scratched petri dish and smoothed down to provide a thin film which immobilized the lymphocytes after a 3- to 5-minute gelatinization period at room temperature. With 2.5 ml of Eagle's medium + 10% calf serum, the cells could be kept in a humidified CO_2 incubator for more than 24 hours. Several test procedures confirmed that cells prepared in this way were in the same viable state as cells kept under normal culture conditions. Less than 10% of the cells were trypanblue positive, these cells appearing dark under phase contrast compared with the bright viable cells. Resting lymphocytes prepared in this way can be stimulated with phytohemagglutinin or pokeweed mitogen (Peters, 1972).

An indication of successful impalement of a cell is normally the PD measured between the interior of the cell and the medium. During the measurements with lymphocytes, we observed variations in the positive PD values which were related to the electrode TP. Regression analysis of 1200 PD measurements of resting lymphocytes with electrodes of TPs between −1 and −40 mV showed a linear relation between TP and PD expressed by Eq. (4)

$$PD = -0.55\, TP + 0.009 \tag{4}$$

The intercept (0.009) indicates that a resting lymphocyte has practically no membrane potential, the measured PD being due only to the change of the TP on insertion of the electrode into another solution (see Section II,C). Electrodes that had been used for several impalements occasionally produced changes in the base line, which were sometimes followed by negative PD values. This was probably caused by adherence of cellular substances to the electrode tip as described in Section II,C. However, even when measured PD values are not due to a membrane potential, they are, together with continuous membrane resistance measurements as described in Section II,C, a useful indicator of the successful impalement of a cell.

As long as the PD and the resistance values remained stable and the cell kept its bright appearance, we considered it to be viable. Sometimes an impaled cell suddenly became dark; this was always associated with a decrease of the potential to the base line and a decrease of the resistance to the value of the electrode resistance, as if the electrode had been withdrawn from the cell.

A high sensitivity is required to measure small PDs and to avoid unnecessary hyperpolarization during coupling experiments. This led to the registration of the bath coupling pulse arising from the electrical circuitry and to a less extent from the resistivity of the medium. With the recording and current electrodes in different cells, the bath coupling pulse must be less than the measured coupling pulse. To test this, the electrodes were placed in the medium at a similar distance as they had been in the cells. If the cells had been agglutinated and stimulated by phytohemagglutinin, the coupling pulse exceeded the bath coupling pulse in most cases, and communication ratios between 0.1 and 0.5 could be observed. To determine the communication ratio $V_2:V_1$, the bath coupling pulse u is subtracted from the coupling pulse v. Thus the maximum pulse height V_1 which is obtained by insertion of both electrodes into one cell is given by

$$V_1 = v_1 - u \tag{5}$$

whereas the pulse height V_2 obtained by insertion of the electrodes into different cells of an agglutinate is given by

$$V_2 = v_2 - u \tag{6}$$

In nonstimulated lymphocyte agglutinates, no coupling could be detected, suggesting that ionic coupling is not necessarily linked to lymphocyte agglutination. The onset of ionic coupling in these agglutinates after the addition of phytohemagglutinin could be demonstrated. Figure 13 shows an example of such an experiment, although the fully established coupling pulse is not demonstrated. This experiment was performed with a high TP electrode which resulted in a positive PD. Starting on the left side, the short pulse

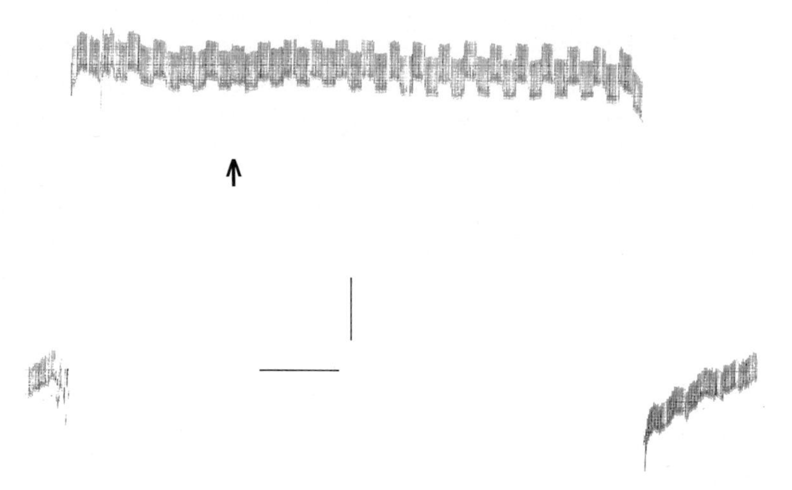

FIG. 13. The onset of ionic coupling between lymphocytes after the addition of phytohemagglutinin. Starting on the left side, the impalement of a cell by the recording electrode is noted by the positive potential change and the increase of the short resistance measuring pulse. The insertion of the current electrode is indicated by the appearance of the small coupling pulse which is identical to the bath coupling pulse. The arrow indicates the moment of addition of phytohemagglutinin to the medium, which results in an increased coupling pulse after about 1 minute. Horizontal bar, 1 minute; vertical bar, 1 mV.

superimposed onto the base line indicates the electrode resistance. No bath coupling pulse can be detected. The irregularities before the voltage drop are due to electrode movements on the cell surface. After impalement of the cell, not only is a PD of about + 5 mV recorded, but also an increase in the resistance pulse, preferentially due to the surface membrane. The impalement of a more distant cell by the current electrode is indicated by the slight irregularity and the beginning of a coupling pulse, which is identical with the bath coupling pulse. The arrow indicates the moment of phytohemagglutinin addition to the medium, which results in an increased coupling pulse after about 1 minute. Upon withdrawal of both electrodes, a small overshoot of about 0.5 mV occurs before the base line is reached. The resistance pulse again indicates the original electrode resitance. In control experiments, the coupling pulse remained stable for up to 1 hour while the electrodes were inserted into cells of nonstimulated agglutinates.

These experiments indicate that different sites may exist at the cell surface: unspecific sites, where the cells can be connected without triggering cellular processes, and more specific sites, where an agglutination of lymphocytes is associated with ionic coupling and the stimulation of cells. Upon addition of phytohemagglutinin, these specific sites react within minutes simultaneously producing agglutination, ionic coupling and—as an early sign of stimulation—an increase in the rate of certain transport processes (Quastel and Kaplan, 1970; Peters and Hausen, 1971a,b; Allwood *et al.*, 1971; van den Berg and Betel, 1971; Whitney and Sutherland, 1972; Resch *et al.*, 1972). It was demonstrated by Peters (1972) that lymphocytes need cell contact to become stimulated. Therefore, this cell activation may be a multicellular process rather than a single-cell phenomenon. The hypothesis has been advanced that the establishment of ionic coupling between lymphocytes marks the beginning of stimulation and mediates cellular cooperation (Hülser and Peters, 1972). Interestingly, using the fluorescein method, Sellin *et al.* (1971) have demonstrated a flow of this dye from one cell to another in the course of specific immune reactions.

C. Explants from Normal and Malignant Tissues

Loewenstein and Kanno (1967) reported that normal liver cells showed good ionic coupling, whereas hepatoma cells did not. Morphologically unaltered liver cells in the neighborhood of hepatoma cells had a diminished communication ratio. Together with the results of Jamakosmanović and Loewenstein (1968), these findings suggested a possibility of distinguishing between normal and malignant cells and especially for the detection of early stages in the expression of malignant cell properties on the basis of the presence or the absence of ionic coupling. In the meantime not only were cell lines in culture described that were both malignant and ionically coupled (Furshpan and Potter, 1968; Borek *et al.*, 1969; Johnson and Sheridan, 1971; Hülser and Rajewsky, 1971; Hülser, 1971), but also certain tumors showed ionic coupling under *in vivo* conditions (Boitsova *et al.*, 1970; Sheridan, 1970; Hülser and Rajewsky, 1971), indicating that this concept could not be generalized.

It is known (Loewenstein *et al.*, 1967; Loewenstein and Penn, 1967) that ionic coupling may disappear if cells are injured. Therefore, in our experiments the preparation of explants was performed carefully, in order to avoid uncoupling. Whole tumors or liver lobes were removed from rats and immediately put into Eagle-Dulbecco medium supplemented with 10% calf serum. For immobilization of the explants in the petri dish, we prepared 1-cm-long threads soaked in hot beeswax. Pieces of about 5-mm length were carefully cut from the tissues and secured on a dry dish by melting the ends of a thread laid across the explant before adding the medium. Four different rat tumors, as well as intact lobes of rat liver, were investigated (Hülser and Rajewsky, 1971).

The cells of the transplantable rat mammary tumor BICR/MIR (Rajewsky, 1970; Rajewsky and Grüneisen, 1972) were found to be ionically coupled, as were the cells of the respective malignant culture cell line BICR/MIR-K (see Fig. 7) which originated from the BICR/MIR tumor. Between cells of ethylnitrosourea- induced cystic anaplastic neurinomas of the BD IX rat and in a kidney metastasis of one of these tumors, no ionic coupling could be found, whereas in one of the respective culture cell lines the cells were coupled. In subcutaneous transplanted neurinomas ionic coupling with a low communication ratio was detected.

Control experiments with intact rat liver lobes revealed a high communication ratio, which decreased rapidly after the tissue was damaged with a needle. In every explant the measured PDs were considerably lower (about 15–30 mV) than usually found in cells cultured *in vitro* (about 50 mV). This result coincides with the finding of Penn (1966), who described lower PDs in isolated liver cells as compared with liver cells *in situ*, and with Schanne and Coraboeuf (1966), who showed that the high PD of about 50 mV in liver *in situ* decreases to about 25 mV within 30 minutes after the death of the animal.

D. Mammalian Cells Cultured in Intraperitoneal Diffusion Chambers

The preparation of tissue or tumor explants is often unsatisfactory because of the damage that may occur owing to the dissection procedures. A method which avoids such difficulties, and which also facilitates the culturing of cells under *in vivo* conditions, is the diffusion chamber method described by several authors (Algire *et al.*, 1958; Capalbo *et al.*, 1964; Benestad and Breivik, 1972; Boyum *et al.*, 1972; Laerum *et al.*, 1973).

We used diffusion chambers consisting of a Plexiglas ring of 10 mm i.d. and 13 mm o.d. containing a hole which could be closed by a plastic stopper. A square piece of a glass coverslip was glued into such a ring by "Tensol" cement (ICI). Then Millipore filters (GSWPO 1300) were glued on both sides of the ring. The tightness of the chambers was checked before sterilization overnight at 80°C. Two chambers with about 5×10^5 cells each were implanted into the peritoneum of mice or rats, and after 2 to 5 days the cells were harvested. One filter was removed, then the glass was taken from the plastic ring and transferred into a petri dish for the measurements. Since a coagulum is formed around the cells within the chamber, the removal of the filter and the glass has to be performed carefully. Usually, cells were attached to both sides of the glass and could be used for electrophysiological measurements.

Comparative investigations with established cell lines in tissue culture showed that the morphological and electrical properties of cells cultured in diffusion chambers or *in vitro* are identical: fibroblastoid cells also exhibit fibroblastoid morphology in the chambers and are ionically coupled, whereas epithelioid cells preserve their epithelioid morphology and are not ionically coupled (Laerum and Hülser, in preparation).

E. Nonmammalian Cells

As an example for nonmammalian cells we describe some results obtained with a cellular slime mold (*Dictyostelium discoideum*). When aggregating, these cells communicate by an intercellular system which consists of chemotaxis, cell contact formation, and periodic stimuli from cell to cell (see Gerisch, 1971). It was of interest to investigate the possibility of an intercellular ion transport by measuring the ionic coupling between aggregating cells.

The ax-2 strain of *Dictyostelium* was grown in axenic medium (Watts and Ashworth, 1970) up to a density of approximately 10^7/ml in suspension culture. The cells were washed three times and resuspended in 16.7 m*M* phosphate buffer pH 6.0. When seeded into plastic petri dishes, the cells adhere to the bottom and can easily be impaled with electrodes. The lower ion concentration of the phosphate buffer, as compared with Eagle-Dulbecco medium, results in higher TPs of the electrodes. As in the case of lymphocytes, a linear relation was found between the TP and the PD.

Linear regression results in Eq. (7)

$$PD = -0.80\ TP - 9.11 \tag{7}$$

where the intercept indicates that the membrane potential of these slime mold cells is of the order of -9 mV. The slope is different from that obtained for the lymphocyte Eq. (4) and may indicate a different intracellular ion concentration in these cells.

The PD measured after impalement of slime mold cells was stable for only 30 seconds. A vacuole could then be seen forming at the electrode tip. At the same time, there was usually a change in the PD. In most cases the vacuole soon disappeared and the original PD was reestablished. This often occurred several times before the cell finally retracted from the elecrode (Hülser and Malchow, unpublished results). The appearance of vacuoles can be interpreted as unsuccessful attempts by the *Dictyostelium* cells to phagocytose the electrodes. These phenomena precluded an unequivocal demonstration of ionic coupling between aggregating cells, however by the use of inhibitors of contractile proteins it may be possible to perform coupling experiments with this type of cell.

V. Conclusion

Electron microscopic investigations on different mammalian tissues by Friend and Gilula (1972) have shown that all known interconnections between nonexcitable cells can be subdivided into two general categories: cell junctions sharing a structural component and cell contacts which do not. It seems obvious that low-resistance junctions should be represented by cell junctions rather than by cell contacts. Indeed, recent evidence tends to establish gap junctions as the ultrastructural equivalent of low-resistance junctions (Revel *et al.*, 1971; Johnson and Sheridan, 1971; Rose, 1971; Gilula *et al.*, 1972; Pinto da Silva and Gilula, 1972; Hülser and Demsey, 1973). The question whether these junctions are always present or only transient must still be regarded as unanswered. The fact that lymphocytes need only a transient contact to become stimulated (Peters, 1972), together with our demonstration of ionic coupling immediately after stimulation (Hülser and Peters, 1971, 1972), suggests that low-resistance junctions may be established rapidly and may only exist transiently. Therefore, it is understandable that only after careful preparation could gap junctions be detected in cells reported to possess low-resistance junctions. Furthermore this indicates the necessity of measuring cell membrane properties, such as PD, resistance, and capacitance to detect possible changes of the cell viability that may occur during an experiment. In this context experiments with metabolically coupled cells (Pitts, 1971; Cox *et al.*, 1972; Gilula *et al.*, 1972) are informative, since they allow the demonstration of junctional membranes without the interference of probes, such as electrodes.

The lack of coupling in cultured cells with epithelioid morphology shows that a mere close attachment of the cells is not sufficient for ionic coupling. Membranes of such cells are apparently unable to build up low-resistance junctions, even when the cells are in close contact and have proliferative characteristics similar to ionically coupled fibroblastoid cells. Since cells of different coupled cell lines can be interconnected by low-resistance junctions, it seems that these junctions are not cell-type-specific but are related to membrane properties common to these cells. The varying morphological appearance of cultured cells may be due to differences in membrane structure and composition which are reflected by different properties revealed with electrophysiological methods. For instance, the PD in cultured epithelioid and fibroblastoid cells is about 50 mV in Eagle-Dulbecco medium containing 44 m*M* bicarbonate. In medium with a reduced bicarbonate concentration the permeability of fibroblastoid cells is changed, resulting in a PD of about 35 mV, whereas epithelioid cells are unaffected (Hülser, 1971).

The described electrophysiological methods not only allow the demonstration of common membrane properties of different cell lines, but by the determination of ionic coupling they also offer a useful tool for the investigation of cell interactions and their role in biological processes requiring cellular cooperation.

Acknowledgments

The author is indebted to Professor Dr. H. Friedrich-Freksa for valuable discussion and encouragement and to Drs. O. D. Laerum, D. Malchow, J. H. Peters, M. F. Rajewsky and Mr. D. J. Webb, who collaborated in the experiments described in Section IV. The work was partially supported by the Deutsche Forschungsgemeinschaft.

References

Adrian, R. H. (1956). *J. Physiol. (London)* **133**, 631–658.

Agin, D. P. (1969). *In* "Glass Microelectrodes" (M. Lavallée, O. F. Schanne, and N. C. Hébert, eds.), pp. 62–75. Wiley, New York.

Algire, G. H., Borders, M. L., and Evans, V. J. (1958). *J. Nat. Cancer Inst.* **20**, 1187–1201.

Allwood, G., Asherson, G. L., Davey, M. J., and Goodford, P. J. (1971). *Immunology* **21**, 509–516.

Azarnia, R., and Loewenstein, W. R. (1971). *J. Membrane Biol.* **6**, 368–385.

Baird, I. (1967). *Med. Biol. Eng.* **5**, 295–298.

Barry, P. H., and Diamond, J. M. (1970). *J. Membrane Biol.* **3**, 93–122.

Benestad, H. B., and Breivik, H. (1972). *Norw. Def. Res. Estab. Rep.* No. 61.

Bernhardt, J., and Pauly, H. (1967). *Biophysik* **4**, 101–112.
Boitsova, L. Y., Kovalev, S. A., Chailakhyan, L. M., and Sharovskaya, Y. Y. (1970). *Tsitologiya* **12**, 1255–1265.
Borek, C., Higashino, S., and Loewenstein, W. R. (1969). *J. Membrane Biol.* **1**, 274–293.
Boyum, A., Carsten, A. L., Laerum, O. D., and Cronkite, E. P. (1972). *Blood* **40**, 174–188.
Burkhardt, D. (1959). *Glas-Instrum.-Tech.* **3**, 115–122.
Capalbo, E. E., Albright, J. F., and Bennet, W. E. (1964). *J. Immunol.* **92**, 243–251.
Chowdhury, T. K. (1969). *In* "Glass Microelectrodes" (M. Lavallée, O. F. Schanne, and N. C. Hébert, eds.), pp. 404–423. Wiley, New York.
Cole, K. S. (1968). "Membranes, Ions and Impulses." Univ. of California Press, Berkeley.
Cox, R. P., Krauss, M. R., Balis, M. E., and Dancis, J. (1972). *Exp. Cell Res.* **74**, 251–268.
Engel, E., Barcilon, V., and Eisenberg, R. S. (1972). *Biophys. J.* **12**, 384–403.
Frank, K., and Becker, M. C. (1964). *In* "Physical Techniques in Biological Research" (W. L. Nastuk, ed.), Vol. 5, pp. 22–87. Academic Press, New York.
Frank, W., Ristow, H.-J., and Schwalb, S. (1972). *Exp. Cell Res.* **70**, 390–396.
Friend, D. S., and Gilula, N. B. (1972). *J. Cell Biol.* **53**, 758–776.
Furshpan, E. J., and Potter, D. D. (1959). *J. Physiol. (London)* **145**, 289–325.
Furshpan, E. J., and Potter, D. D. (1968). *In* "Current Topics in Developmental Biology" (A. A. Moscona and A. Monroy, eds.), Vol. 3, pp. 95–127. Academic Press, New York.
Gerisch, G. (1971). *Naturwissenschaften* **58**, 430–438.
Gilula, N. B., Reeves, O. R., and Steinbach, A. (1972). *Nature (London)* **235**, 262–265.
Gradmann, D. (1972). *Can. J. Physiol. Pharmacol.* **50**, 817–823.
Hausen, P., Stein, H., and Peters, J. H. (1969). *Eur. J. Biochem.* **9**, 542–549.
Hegel, U., and Frömter, E. (1966). *Pfluegers Arch. Gesamte Physiol. Menschen Tiere* **291**, 121–128.
Hülser, D. F. (1971). *Pfluegers Arch.* **325**, 174–187.
Hülser, D. F., and Demsey, A. (1973). *Z. Naturforsch. C* **28**, 603–606.
Hülser, D. F., and Frank, W. (1971). *Z. Naturforsch. B* **26**, 1045–1048.
Hülser, D. F., and Peters, J. H. (1971). *Eur. J. Immunol.* **1**, 494–495.
Hülser, D. F., and Peters, J. H. (1972). *Exp. Cell Res.* **74**, 319–326.
Hülser, D. F., and Rajewsky, M. F. (1971). *Deut. Krebskongr. Hannover* pp. 22–23. (Abstr.)
Hülser, D. F., and Webb, D. J. (1973). *Exp. Cell Res.* **80**, 210–222.
Jamakosmanović, A., and Loewenstein, W. R. (1968). *J. Cell. Biol.* **38**, 556–561.
Johnson, R. G., and Sheridan, J. D. (1971). *Science* **174**, 717–719.
Kanno, Y., and Loewenstein, W. R. (1966). *Nature (London)* **212**, 629–630.
Kopac, M. J. (1964). *In* "Physical Techniques in Biological Research" (W. L. Nastuk, ed.), Vol. 5, pp. 191–233. Academic Press, New York.
Laerum, O. D., Grüneisen, A., and Rajewsky, M. F. (1973). *Eur. J. Cancer* **9**, 533–541.
Lavallée, M., and Szabo, G. (1969). *In* "Glass Microelectrodes" (M. Lavallée, O. F. Schanne, and N. C. Hébert, eds.), pp. 95–110. Wiley, New York.
Ling, G., and Gerard, R. W. (1949). *J. Cell. Physiol.* **34**, 383–396.
Loewenstein, W. R. (1967). *Develop. Biol.* **15**, 503–520.
Loewenstein, W. R. (1968a). *Perspect. Biol. Med.* **11**, 260–272.
Loewenstein, W. R. (1968b). *Proc. Can. Cancer Conf.* **8**, 162–170.
Loewenstein, W. R., and Kanno, Y. (1964). *J. Cell. Biol.* **22**, 565–586.
Loewenstein, W. R., and Kanno, Y. (1967). *J. Cell Biol.* **33**, 225–234.
Loewenstein, W. R., and Penn, R. D. (1967). *J. Cell Biol.* **33**, 235–242.
Loewenstein, W. R., Nakas, M., and Socolar, S. J. (1967). *J. Gen. Physiol.* **50**, 1865–1891.
Machemer, H. (1970). *Z. Naturforsch. B* **25**, 895.

Michalke, W., and Loewenstein, W. R. (1971). *Nature (London)* **232**, 121–122.
O'Lague, P., Dalen, H., Rubin, H., and Tobias, C. (1970). *Science* **170**, 464–466.
Oliveira-Castro, G. M., and Loewenstein, W. R. (1971). *J. Membrane Biol.* **5**, 51–77.
Oliveira-Castro, G. M., and Machado, R. D. (1969). *Experientia* **25**, 556–558.
Penn, R. D. (1966). *J. Cell Biol.* **29**, 171–174.
Peters, J. H. (1972). *Exp. Cell Res.* **74**, 179–186.
Peters, J. H., and Hausen, P. (1971a). *Eur. J. Biochem.* **19**, 502–508.
Peters, J. H., and Hausen, P. (1971b). *Eur. J. Biochem.* **19**, 509–513.
Pfister, H., and Pauly, H. (1969). *Biophysik* **6**, 94–112.
Pinto da Silva, P., and Gilula, N. B. (1972). *Exp. Cell Res.* **71**, 393–401.
Pitts, J. D. (1971). Ciba Found. Symp. "Growth Control in Cell Cultures" (G. E. W. Wolstenholme and J. Knight, eds.), pp. 89–105., Churchill, London.
Potter, D. D., Furshpan, E. J., and Lennox, E. S. (1966). *Proc. Nat. Acad. Sci. U.S.* **55**, 328–336.
Quastel, M. R., and Kaplan, J. G. (1970). *Exp. Cell Res.* **63**, 230–233.
Rajewsky, M. F. (1970). *Exp. Cell Res.* **60**, 269–276.
Rajewsky, M. F., and Grüneisen, A. (1972). *Eur. J. Immunol.* **2**, 445–447.
Redmann, K., and Kalkoff, W. (1968). *Experientia* **24**, 975–976.
Resch, K., Gelfand, E. W., Hansen, K., and Ferber, E. (1972). *Eur. J. Immunol.* **2**, 598–601.
Revel, J. P., and Karnovsky, M. J. (1967). *J. Cell Biol.* **33**, C7–C12.
Revel, J. P., Yee, A. G., and Hudspeth, A. J. (1971). *Proc. Nat. Acad. Sci. U.S.* **68**, 2924–2927.
Rose, B. (1971). *J. Membrane Biol.* **5**, 1–19.
Rose, B., and Loewenstein, W. R. (1971). *J. Membrane Biol.* **5**, 20–50.
Schanne, O. (1969). *In* "Glass Microelectrodes" (M. Lavallée, O. F. Schanne, and N. C. Hébert, eds.), pp. 299–321. Wiley, New York.
Schanne, O., and Coraboeuf, E. (1966). *Nature (London)* **210**, 1390–1391.
Sellin, D., Wallach, D. F. H., and Fischer, H. (1971). *Eur. J. Immunol.* **1**, 453–458.
Sheridan, J. D. (1970). *J. Cell Biol.* **45**, 91–99.
Sheridan, J. D. (1971a). *Develop. Biol.* **26**, 627–636.
Sheridan, J. D. (1971b). *J. Cell Biol.* **50**, 795–803.
Shiba, H. (1970a). *Jap. J. Appl. Phys.* **9**, 1405–1409.
Shiba, H. (1970b). *J. Theor. Biol.* **30**, 59–68.
Shiba, H., and Kanno, Y. (1971). *Biophysik* **7**, 295–301.
Siegenbeek van Heukelom, J., Denier van der Gon, J. J., and Prop, F. J. A. (1972a). *J. Membrane Biol.* **7**, 88–110.
Siegenbeek van Heukelom, J., Slaff, D. W., and van der Leun, J. C. (1972b). *Biophys. J.* **12**, 1266–1284.
Snell, F. M. (1969). *In* "Glass Microelectrodes" (M. Lavallée, O. F. Schanne, and N. C. Hébert, eds.), pp. 111–123. Wiley, New York.
Tasaki, I., Polley, E. H., and Orrego, F. (1954). *J. Neurophysiol.* **17**, 454–474.
van den Berg, K. J., and Betel. I. (1971). *Exp. Cell Res.* **66**, 257–259.
Watts, D. J., and Ashworth, J. M. (1970). *Biochem. J.* **119**, 171–174.
Whitney, R. B., and Sutherland, R. M. (1972). *Cell Immunol.* **5**, 137–147.
Woodbury, D. M., and Woodbury, J. W. (1963). *J. Physiol. (London)* **169**, 553–567.
Zettler, F. (1970). *Z. Vergl. Physiol.* **67**, 423–435.

Chapter 18

Methods in the Cellular and Molecular Biology of Paramecium[1]

EARL D. HANSON[2]

Shanklin Laboratory, Wesleyan University, Middletown, Connecticut

This review follows those by Sonneborn (1970) and van Wagtendonk and Soldo (1970) published in this series. It is, therefore, unnecessary to add further introductory or general comments at this point other than to state the goals of this particular review.

The very broad variety of techniques covered by cellular and molecular techniques necessitated the following treatment: Throughout, each section is introduced by a brief summary of the biological problems being pursued. Then the essentials of the techniques used in pursuit of these problems is presented to the point that the reader can determine what basic equipment

[1]Supported by NSF Research Grant GB-22861 and ACS Grant E-526.

[2]With contributed sections by I. Finger, Biology Department, Haverford College, Haverford, Pennsylvania; L. Hufnagel, Department of Pathology, Wayne State University, College of Medicine, Detroit, Michigan; A. Reisner, CSIRO, Division of Animal Genetics, P.O. Box 70, Epping, N.S.W., Australia; and J. Wille, Department of Theoretical Biology, University of Chicago, Chicago, Illinois.

is needed, and should be able to decide whether his or her laboratory is set up to tackle a given problem. In most cases details are left to reading of the original literature. Throughout, different laboratories inevitably report minor variations in the salt or buffer solutions used, in centrifugational protocols, etc. The text alerts the reader to such varieties in experimental procedures. This review is, then, an *introduction* to current methods in the cellular and molecular biology of *Paramecium*.

I. Cellular Biology

To elucidate cellular life there must be operations that manipulate living systems in qualitative and quantitative terms. We shall start with microscopy of the whole cell and proceed to fine structure and then to isolation of cell parts. Then ways to study these parts as they relate to the functioning of the whole cell will be examined.

A. Light Microscopy and Related Techniques

The use of the light microscope, itself, needs no comment but its applications in the context of special staining techniques and special viewing problems do deserve consideration.

1. Staining and Sectioning

General procedures for fixing and staining are given in Sonneborn (1948). Of the more recently used techniques, that of silver impregnation has become extremely important in revealing the distribution of surface structures (Fig. 1). The original technique is that of Klein (1926), who discovered a "dry" method which still has limited applications (Kozloff, 1964). The so-called "wet" method is somewhat more complex but avoids distortion of the cell due to air-drying and can result in beautiful, clear preparations which, when done optimally, provide information on all surfaces of the cell rather than only the upper one. The basic English reference is that of Corliss (1953) which follows the earlier work of Chatton and Lwoff (1930, 1935) (see also von Gelei, 1932.) A modification applicable specifically to paramecia is that of Gillies (Gillies and Hanson, 1968).

There are, of course, endless variations to standard nuclear and cytoplasmic stains. The basic techniques are given in Sonneborn (1950). The individual worker can adapt these to his or her needs. However, two useful staining techniques not given in previous summaries of methods are those

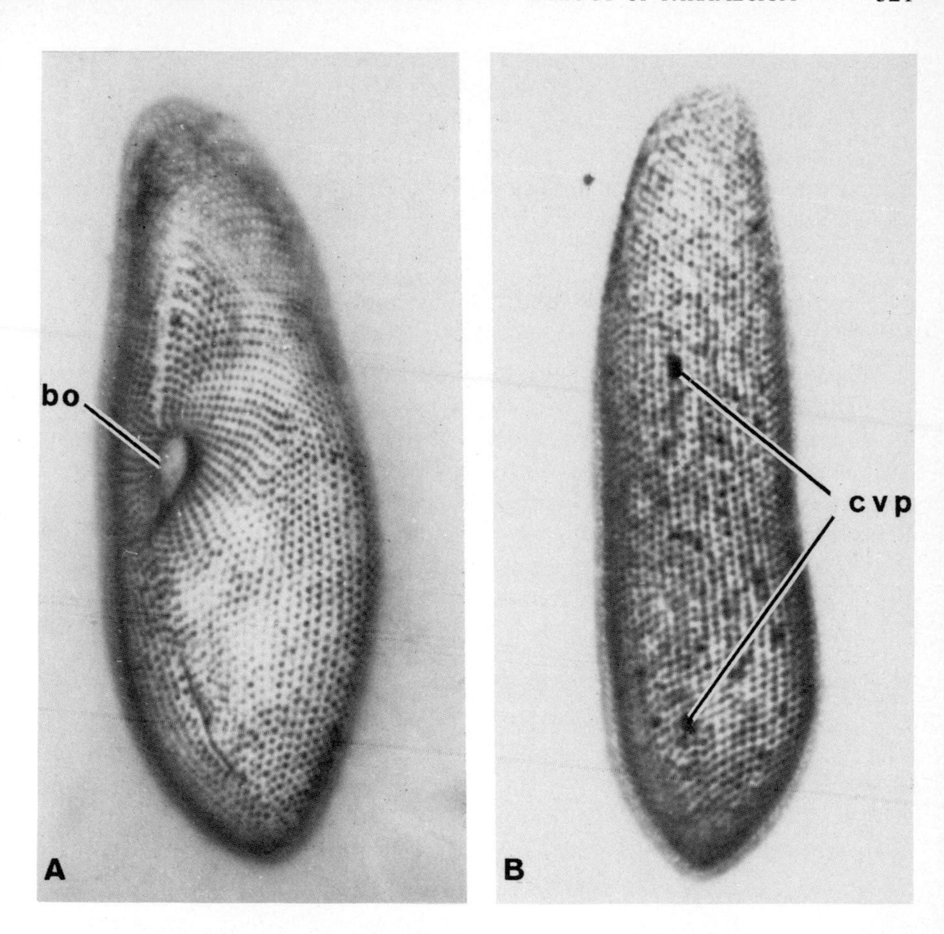

FIG. 1. Silver-impregnated specimens of *Paramecium aurelia*. (A) Ventral surface showing mid-ventral buccal opening (bo) and characteristic array of cortical units around it. Cytoproct (c) lies on the posterior suture line. (B) Dorsal surface showing longitudinal orientation of cortical units into ciliary rows or kineties. Two contractile vesicle pores (cvp) are also stained.

for a quick nuclear stain, used, for example, when checking whole cells for autogamy (Dippell, 1955) and a ciliary stain (Grębecki, 1964).

Increasingly, there are occasions for examining sections of paramecia. Many techniques are available in the general literature but specific references to paramecium are not common for work with light microscopy. The essential problem especially in work with single cells, is to be able to orient the cell as desired. This can be achieved by mounting the fixed cell in agar—sandwiching it between two layers of agar by placing it on solidified agar lying on a glass surface and then adding molten agar on top of the cell. A block of agar containing the cell in known orientation is then cut out and treated as a very small piece of tissue by embedding in paraffin, esterwax, or other desirable material and sectioning it. Results of such a technique (Figs. 11–13 of Hanson *et al.*, 1969) provide structural details and a three-dimensional view of the cell not gained from whole-cell preparations and are somewhat easier to prepare than sections for electron microscopy.

2. Dark-Field, Phase Contrast, and Interference Contrast Microscopy

The special benefit of dark field viewing of cells has been to gain records of cellular behavior. By using low magnifications and time exposures, the swimming paths of paramecia are easily and accurately recording (Dryl, 1958, 1961; Cooper, 1965). This is also useful for determining the behavioral response of cells exposed to various chemicals and salt concentrations (see Section I, C, 3).

Phase contrast is of course especially useful for observing live cells. The problem of immobilization is readily solved by physical compression in a

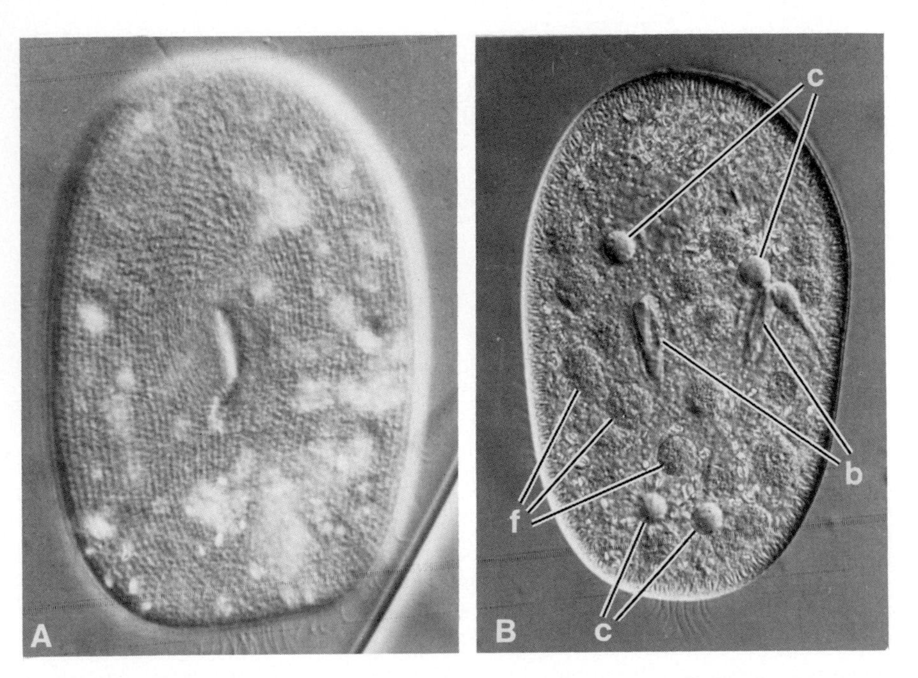

Fig. 2. Interference contrast pictures of live homopolar doublets of *Paramecium aurelia*. (A) Lower surface of the cell (note reversal of kinety patterns from Fig. 1A). (B) Internal structures of cell showing both sets of buccal organelles (b), four contractile vesicles (c), and food vacuoles (f). The caudal tuft of cilia is clear in both pictures.

rotocompressor (obtainable from Biological Institute, 2018 N. Broad Street, Philadelphia, Pennsylvania) which flattens the cell between parallel glass surfaces. This flattening, though distorting the normal shape of the cell, significantly improves the viewing of the specimen since phase contrast is most effective on relatively thin specimens (ca. 2 μm thick). Suhama (personal communication) has been able to follow development of cilia in live *P. trichium* by phase contrast optics.

Interference contrast, developed by Nomarsky, has the real advantage of allowing detailed resolution of surface cortical patterns (Hanson and Ungerleider, 1973) as well as internal structures of live cells (Fig. 2). Used with the rotocompressor, as in phase contrast viewing, it seems destined to replace the phase system as the method of choice when dealing with living material. It has been used by Allen (1970) for cinematography of conjugation and other processes in *P. aurelia*.

B. Electron Microscopy

This section was contributed by L. Hufnagel.

Almost from its inception, the electron microscope has been utilized to investigate *Paramecium* (Jakus *et al*., 1942; Jakus, 1945; Jakus and Hall, 1946; Wohlfarth-Bottermann, 1950, 1953; Knoch and Konig, 1951; Sedar, 1952; Metz *et al*., 1953). However, many early studies, while informative in their time, have become antiquated through technical advancement. Most of this progress has been through improvements in preparation of the material for examination in the electron microscope, especially improved fixation and development of methods for increasing specimen contrast. Most procedures in use today have appeared since the late 1950's.

Those procedures which have proved to be of use in studies on *Paramecium* are referenced below. Of course many techniques and procedures exist which have not been applied to *Paramecium*. For these, the reader is referred to the general books on electron microscopy by Pease (1964), Kay (1965), Sjöstrand (1967), and Hayat (1970). It is assumed that the reader has a general knowledge of the principles and uses of electron microscopy. If not, consult the books cited above.

1. General Procedures: Transmission Electron Microscopy

a. Whole Cells and Parts of Cells. Originally, whole cells and their parts were mounted on electron microscope grids and viewed directly in the microscope without further treatment (Potts, 1955) or after being shadow-cast with palladium or chromium (Jakus and Hall, 1946; Metz *et al*., 1953; Potts and Tomkin, 1955). The resolution obtained was not very great (> 30Å according to Anderson, 1966). Improved resolution has since been effected by a procedure called "negative staining," in which the structures are embedded on the grid in a thin layer of electron dense material, such as phosphotungstic acid, silicotungstic acid, or uranyl acetate (Horne, 1965). Since these substances can combine with organic molecules, some positive staining can also occur. Negative staining allows resolution to about 10–20 Å (Anderson, 1966). It has proved to be very useful in the study of trichocysts (Jakus, 1945; Steers *et al.*, 1969; Stocken and Wohlfarth-Bottermann, 1970), isolated cortical organelles such as membranes and fibres (Pitelka, 1965b; Hufnagel, 1966, 1969), and bacterial endosymbionts (Anderson *et al.*, 1964; Preer *et al.*, 1966; Preer and Preer, 1967; Preer and Jurand, 1968; Jurand and Preer, 1968; Preer, 1969). Negative staining with uranyl acetate also was applied to isolated nuclei and nuclear chromatin which had been spread on water prior to staining (Wolfe, 1967a). Negative staining also can be used to characterize some macromolecules and was applied to purified immobilization antigen by Reisner *et al.* (1969b).

b. Sectioned Cells: Fixation. When it is not necessary to study selected cells, paramecia are fixed *en masse* following concentration by centrifugation or filtration. Satisfactory fixation has been achieved either with buffered osmium tetroxide (OsO_4) or with buffered glutaraldehyde followed by buffered OsO_4. An early success in fixation with OsO_4 was obtained by Sedar and Porter (1955), using Palade's (1952) Veronal-acetate buffer, and many later investigations have utilized variations on this fixation, at pH's usually around 7.2–7.4 but ranging from 6.0 (Ehret and de Haller, 1963) to 8.6 (Sommerville and Sinden, 1968; Sinden, 1971). Other buffers utilized with OsO_4 are Millonig's (1961) phosphate buffer (Pitelka, 1965b), McLlvaine's citrate-phosphate buffer (Powers *et al.*, 1955; Roth, 1958; Rudenberg, 1962) and collidine buffer (Soldo *et al.*, 1970).

A variation of the buffered osmium fixative which appears to give generally good fixation in *P. aurelia* and *P. caudatum* is a mixture of osmium tetroxide and potassium dichromate developed by Wohlfarth-Bottermann (1957a,b) and also used extensively by Schneider (1959a, 1960a, 1961, 1963, 1964a), Yusa (1963) and Stockem and Wohlfarth-Bottermann (1970).

Fixation in osmic acid vapors also has been reported to give satisfactory preservation of fine structure without excessive shrinkage in *P. multimicronucleatum* (Jenkins, 1964, 1970).

Excellent visualization of fibers and microtubules in the cytoplasm of *P. caudatum* has been reported after fixation in buffered glutaraldehyde, followed by OsO_4 fixation (Allen, 1971). Fixation with glutaraldehyde also has been cited for the preservation of a certain type of food vacuole in *P. caudatum* (Esteve, 1967), and has been said to give a different appearance of pellicular alveoli than does fixation with osmium alone (Pitelka, 1965b). Buffers used with glutaraldehyde include Veronal-acetate (Dippell, 1968), phosphate (Pitelka, 1965a,b; Soldo *et al.*, 1970; Sinden, 1971), cacodylate (Esteve, 1970; Allen, 1971), and collidine (Allen, 1971).

In some cases, the presence of $CaCl_2$ or KCl during fixation of paramecia has been reported to be beneficial, e.g., for preserving nuclear microtubules (Jurand and Selman, 1970) and membranes and granular elements (Jurand and Selman, 1969).

It may be of value to control tonicity (Hayat, 1970) during fixation of *Paramecium*. Some investigators have done so by using sucrose or saline (Jurand *et al*., 1962; Mott, 1963a,b; Pitelka, 1965b; Jurand and Preer, 1968; Kennedy and Brittingham, 1968; Sommerville and Sinden, 1968; Sinden, 1971).

Fixation is usually carried out for 0.5–1 hour at temperatures ranging from 0°C (Soldo *et al*., 1970) to 37°C (Jurand and Preer, 1968), room temperature being used most generally.

The relative effectiveness of various fixation procedures has been compared within the same species of *Paramecium* in a few cases (Beale and Jurand, 1960; Jurand *et al*., 1962; Pitelka, 1965b; Jurand and Preer, 1968; Jurand and Selman, 1969; Allen, 1971). It is clear that more information about general morphology can be obtained when two or more different fixation procedures are used in a single study. However, it is not clear whether the same procedure will work equally well on different species of *Paramecium*.

c. Sectioned Cells: Staining. For increased contrast, sections usually are stained with uranyl acetate, followed by lead citrate (e.g., Allen, 1971). However, several authors have found useful a mixture of uranyl acetate and potassium permanganate (Mott, 1963, 1965; Jurand and Selman, 1970; Preer and Jurand, 1968; Beale and Jurand, 1966). Wohlfarth-Bottermann (1957a,b) developed a method for staining fixed *Paramecia* during dehydration with a mixture of phosphotungstic acid and uranyl acetate in 70% ethanol. This stain has since been used effectively by others (Yusa, 1963; Schneider, 1964a,b; Stockem and Wohlfarth-Bottermann, 1970).

2. General Procedures: Scanning Electron Microscopy

The scanning electron microscope provides high resolution, three-dimensional images of the surfaces of cells or parts of cells (Oatley *et al*., 1965; Hayes and Pease, 1968; Boyde and Wood, 1969; Carr, 1971). Methods for examining protozoa, including *Paramecium*, have been developed by Small and Marzalek (1969) and were applied to *Paramecium* by Wessenberg and Antipa (1970). Apparently, the potential of this microscope for characterizing large areas of and charting developmental events in the cortex of *Paramecium* has not yet been realized (Antipa, 1971).

3. Special Procedures

In order to apply the electron microscope to physiological questions, a variety of special procedures has been developed for use with *Paramecium*.

Many of these are designed to identify specific chemical substances in cells of known physiological state.

a. Processing of Single Cells. Procedures for handling single cells during fixation and embedding for orientation of cells prior to sectioning have been described for *Paramecium* by Esteve (1970), de Haller *et al.* (1961), Jurand *et al.* (1962), and Janisch (1967). Also see Flickinger (1966) for a broad treatment of methods available for handling small numbers of cells for electron for electron microscopy.

b. EM Histochemistry. Although a wide variety of histochemical procedures for use with thin sections now exists, few such procedures have been used on *Paramecium*. The sites of activity of the enzyme, acid phosphatase, have been identified in *P. caudatum* by Esteve (1970). Dippell (1963, 1968), Dippell and Sinton (1963), and Beale and Jurand (1960) have adapted to *Paramecium* procedures for locating the nucleic acids, DNA and RNA, in thin sections.

c. EM Autoradiography. Techniques for autoradiography at the electron microscopic level are now readily available (Caro, 1964; Salpeter, 1966; Stevens, 1966). However, these have been applied to *Paramecium* in only two cases, after incubation, of the cells in thymine-^{3}H (Jurand and Jacob, 1969) and in leucine-^{3}H (Ehret *et al.*, 1964).

d. Staining with Antibodies. Techniques for location of immobilization antigens in *P. aurelia*, using ferritin-conjugated antibodies, were developed by Mott (1963, 1965) and later refined by Sinden (1971).

e. Staining with Colloidal Thorium Oxide (Thorotrast). Apparently, the electron-dense compound thorium oxide is taken up selectively into food vacuoles of *Paramecium*. It has been used to mark food vacuoles during their transit through the cytoplasm by Schneider (1964b) and by Jurand *et al.* (1971).

f. Techniques for Specific Organelles and Inclusions.

(1) *Membranes of P. aurelia.* These were reported to be preserved more satisfactorily in the presence of 0.03% $CaCl_2$ (Jurand and Selman, 1969) during fixation.

(2) *Granular elements.* The addition of $CaCl_2$ to the fixative also was reported to enhance preservation of granular constituents in sections of whole cells (Jurand and Selman, 1969). Techniques have also been described for obtaining thin sections of isolated ribosomes of *P. aurelia* (Sommerville and Sinden, 1968).

(3) *Fibrous structures.* Jurand and Selman (1969) report the best visualization of fibrous structures in *P. aurelia* when fixation is conducted in the absence of $CaCl_2$. In *P. caudatum*, Allen (1971) obtained the best visualization of fibers and microtubules after fixation in glutaraldehyde buffered with collidine.

(4) *Mitochondria.* These were studied in *P. caudatum* after fixation in an

osmium dichromate mixture (Wohlfarth-Bottermann, 1957a, 1958; Wohlfarth-Bottermann and Schneider, 1961).
(5) *Trichocysts*. According to Selman and Jurand (1970), in *P. aurelia* mature trichocysts tend to fire during fixation, unless carried out at 37°C.
(6) *Nuclear structure*. Procedures that give good preservation of macronuclear organization have been described by Jurand *et al.* (1962) and Jurand and Selman (1970). Microtubules are evident in published photographs of sections through micronuclei of *P. multimicronucleatum* fixed in vapors of osmium tetroxide (Jenkins, 1970).
(7) *Contractile vacuole system*. Schneider (1959a, 1960a) has been able to study the structure of the contractile vacuoles or nephridial system in *Paramecium* after fixation in an osmium–dichromate mixture.
(8) *Symbionts*. As mentioned above, good preservation of the internal structure of symbionts, *in situ*, appears to require special techniques for fixation of *Paramecium* (Preer and Jurand, 1968). Negative staining has also been particularly useful in revealing the internal organization of endosymbionts (Anderson *et al.*, 1964; Preer *et al.*, 1966; Preer and Preer, 1967; Preer and Jurand, 1968; Jurand and Preer, 1968; Preer, 1969).

C. Organelle Systems

Being eukaryotes, paramecia have the expected array of cellular organelles. Additionally, being free-swimming, ciliated protozoans, they have some special organelles. These include organization of cilia and basal bodies along with associated filaments, microtubules, and membranes into kinetosomal territories. These in turn are compounded into predominantly longitudinal rows or kineties in which trichocysts alternate with the unit territories. Also associated with the surface structure are the following: a single feeding organelle (termed a buccal apparatus by most ciliatologists or a gullet by parameciologists); two contractile vacuole or vesicle pores which lead internally to the vesicles themselves and, in all species except *P. trichium*, there are also radial canals; and a cytoproct which is a slitlike structure through which undigested materials from food vacuoles are egested. Internally, the special organelles include the food vacuoles and the macronucleus which occurs in vegetative cells as a highly polyploid (ca. 860n) membrane-bound mass of chromatin. Depending on the species there is one, two, or more diploid micronuclei. Some strains of paramecia contain symbionts, such as kappa. These have been dealt with earlier (Sonneborn, 1970).

The isolation of these organelles along with techniques for studying their function and development will be reviewed, starting from the more external structures and moving internally.

1. Cilia

Isolated cilia have been used to obtain purified antigens and to analyze the chemical induction of mating. The study of antigens is given more fully below (see Section II,B,3). Here we simply note that concentrated cells (Sonneborn, 1970; van Wagtendonk and Soldo, 1970; Aylmer and Reisner, 1971) are deciliated in a salt solution (Preer and Preer, 1959). Mild centrifugation (700–900 *g*) leaves the cilia in the supernatant, which is purified further by stronger centrifugation (24,000 *g*).

For studying chemical induction of mating (Miyake, 1964), cilia are removed by suspending reactive, concentrated cells in normal culture fluid to which $K_2Cr_2O_7$ is added (final concentration 20–30 m*M*, higher concentrations are to be avoided). When the cells die, the cilia are released, the dead cell bodies are precipitated by mild centrifugation, and the supernatant is passed through a single filtration (Toyo filter paper No. 2) to remove the remaining dead cells and discharged trichocysts. The filtered cilia are washed by repeated centrifugation or by dialysis to remove the $K_2Cr_2O_7$. This technique, especially because of the filtration, results in considerable loss of cilia. Also a very few trichocysts and bacteria remain in the preparation along with some unidentified granules thought to be derived from the culture medium.

Cilia attached to the cell have been examined to determine patterns of metachronal beat and to observe development of new cilia. The study of ciliary beating took on a wholly new aspect with Parducz's development of a method for instantaneous fixation of cilia with O_sO_4 as the chief fixing agent. [The original reference is Párducz (1952). This is in Hungarian with a German summary. A brief but adequate English account is in the appendix to Párducz (1967).] Grębecki (1964) offers a much improved technique for staining the immobolized cilia. The combined Párducz–Grebecki method provides large numbers of beautifully fixed and stained cells (e.g., Machemer, 1969). This has obviated certain earlier criticisms of the Párducz method which had relied on searching through large numbers of cells to find well-fixed and well-stained specimens. The search implied selection of atypical cells. As being used currently, the technique is a source of many useful data on ciliary action (for further comments see Section I,D,3.)

The alternative to looking at fixed material is to look at high-speed cinematography of live material. This has been accomplished (Kuźnicki, 1969; Kuźnicki *et al.*, 1970) and has resulted in analyses of the pattern of motion of individual cilia.

2. Kinetosomes

These organelles have been studied from three points of view: fine structure, mode of formation, and possible presence of DNA. Fine structure studies have necessarily depended on electron microscopy, and suitable techniques are those mentioned above (see Section I,B,1 and 3). Similarly, certain aspects of the development of new basal bodies have also been studied by electron microscopy. Dippell (1968) in particular, in a brilliant series of micrographs, has been able to describe kinetosome formation starting from an electron dense mass of material lying anterior to an already differentiated kinetosome and following, step by step, the apparent epigenetic assembly of the full organelle.

At a different level of resolution, the distribution of old and new kinetosomes on the cell surface have been studied by the use of silver impregnation on cells at known times in their interfission cycle (Gillies and Hanson, 1968; Sonneborn, 1970). These have revealed patterns of proliferation of the kinetosomal territories that demonstrate duplication, triplication, and quadruplication to occur.

The question of DNA being present in kinetosomes arose initially from claims that kinetosomes are self-reproducing (Lwoff, 1950), a conclusion based on inferences from silver staining. Dippell's (1968) work now shows that duplication in the sense of a template-copy mechanism is highly unlikely (see also Fulton, 1971). In paramecia, attempts to label the kinetosomes with tritiated thymidine and to stain them with acridine orange, which gives a green fluorescence when bound to DNA, were both used (Smith-Sonneborn and Plaut, 1967). Positive results were reported. Whole cells were disrupted to release pellicular fractions, which were then washed; autoradiographs and stains were made of this material. This frees the surface membranes and the attached kinetosomes from the bulk of the rest of the cell, but it does not exclude the possibility of adsorption of DNA released from the ruptured macronucleus onto the basal bodies. Smith-Sonneborn and Plaut attempted to minimize this possibility by showing that deliberate contamination by DNA had little effect. Their work contrasts with that of Hufnagel (1966, 1969), who reports that pellicular material has a strong affinity for DNA, and hence the presence of DNA in these preparations is thought to be a technical artifact. Fractions were prepared differently by Hufnagel and by Smith-Sonneborn and Plaut, as pointed out by the latter (Smith-Sonneborn and Plaut, 1967, p. 231) and may account for the different interpretations.

3. Trichocysts

These enigmatic cortical structures have attracted a steady flow of attention over the years with the result that their fine structure, formation, chemical composition, and something about their mechanism of discharge is known. But their use to paramecia is still a source of debate.

Many of the studies depend on obtaining trichocysts free from the rest of the cell, and the fact that they are extrusible is exploited in obtaining pure preparations. Discharge of trichocysts can be caused by a variety of agents (Jennings, 1908) including mechanical injury, electric stimulation, desication, and various chemicals including weak solutions of acids or salts.

Electric stimulation has probably been the most common agent for eliciting discharge. Fine platinum electrodes connected to a 5V ac source, are touched to a small drop of culture fluid containing cells. Short, repeated dips of the electrodes into the fluid elicits a heavy discharge (Wohlfarth-Bottermann, 1950; Yusa, 1963). Such material can be used for electron microscopy (Schmitt *et al.*, 1943; Jakus, 1945). Furthermore, the stimulated cells can be followed for study of trichocyst regeneration by electron microscopy (Yusa, 1963).

Air-dried cells tend to discharge trichocysts just before they die, and such material can also be used for electron microscopy (Ehret *et al.*, 1964) and for light-microscope work. Trichocysts labeled with leucine-^{3}H have been shown to conserve their label; the detection of the label being done by high-resolution autoradiography on undischarged and discharged trichocysts. Procedures for introducing the label and preparing the material for electron microscopical examination are given in Ehret *et al.* (1964).

The chemical composition of trichocysts has been reported by Steers *et al.* (1969), but this first report carries only an outline of their methods of trichocyst isolation and purification. Greater detail is promised for a subsequent publication. Basically, chemical stimulation using dimethyl sulfoxide and then ethanol, resulting "in at least 90% of the trichocysts being rapidly extruded" (Steers *et al.*, 1969, p. 393), is followed by differential centrifugation. Amino acid analyses on the trichocyst preparation were performed using polyacrylamide disc gels.

4. Cortex

The cortex has been defined through light microscopy, electron microscopy, and through developmental and genetic studies. These definitions are *not* precisely coextensive (Kaneda and Hanson, 1973). Here we shall follow Hufnagel's (1969) definition based on electron microscopy which includes the following as components of the cortex: cilia, kinetosomes, surface unit membranes, subsurface ciliary vesicles which are bounded by a unit membrane, parasomal sac, kinetodesmatas, various sets of microtubules, and trichocysts. There should be added, however, the presence of contractile vesicle pores, the cytoproct, and the buccal opening with attached buccal parts. The basic unit is the ciliary corpuscle (Ehret and Powers, 1959) or kinetosomal territory (Pitelka, 1969). These are organized into linear rows called kineties, and the kineties themselves are organized into a highly distinctive pattern (Fig. 1). It is the development and perpetuation of this organization, whether at the level of the smallest organelle (e.g., a microtubule in a kinetosome) or at the level of the whole cell (e.g., the kinety patterns) that has caught the interest of various cell biologists. Their studies include fine structure, biochemical, developmental, or genetic questions. Since the preceeding sections have already dealt with cilia, kinetosomes, and trichocysts, and in that the remaining membranous and fibrous units have only been studied in detail with electron microscopy and that too has been dealt with above, we shall here focus on the unit territories, kineties, and kinety patterns.

Isolated cortical fragments from the cell surface (also termed pellicular fragments) have been used for fine structure studies on the relationships between various cortical components (see Section I, B,1.) Isolated fragments have also been used for autoradiography and acridin orange staining (Smith-Sonneborn and Plaut, 1967). These latter fragments appear to be devoid of cilia and trichocysts and consist largely, if not exclusively, of patches of membrane and attached basal bodies and kinetodesmal fibers.

Cortical structures are studied in the intact cell by examining live material under phase or interference optical systems or by fixing and staining. The silver-impregnation technique (Section I, A,1) provides excellent preparations for following the behavior of basal bodies in particular. The silver is deposited (Dippell, 1962) around the base of cilium and at the basal plate region of the kinetosome or basal body. It is also deposited on kinetosomes which lack cilia, on the opening of the parasomal sac, and can be deposited on the tip of the trichocyst and in the folds at the edge of a unit territory. It does not reveal cilia (see Grębecki, 1964, for a good ciliary stain) or other parts of unit territories. In good preparations, the silver impregnation technique clearly demarcates the cytoproct and contractile vesicle pores (Fig. 1). This stain, when applied critically, can demonstrate unit territories which have one or two basal bodies plus parasomal sac and at times of basal body proliferation can show up to four basal bodies per unit (Gillies and Hanson, 1968; Sonneborn, 1970). It can also be used to identify inverted kineties, in that the characteristic array of basal bodies and parasomal sac can be seen to be reversed (Beisson and Sonneborn, 1965; Sonneborn, 1963; Gillies and Hanson, 1968). All of this adds up to a technique of central importance in visualizing certain details of cellular differentiation and morphogenesis.

Finally, there should be mentioned techniques for introducing changes into the cortex. These can be done by grafting, cutting, or ultraviolet microbeam irradiation. The latter two approaches have been described and referenced previously (Sonneborn, 1970). The grafting technique is dependent on conjugation and aberrant separation of conjugants (Sonneborn, 1963). It is not precisely controllable.

5. Oral or Buccal Structures

The buccal parts or gullets have been isolated, too. Using single cells, the buccal structures can be released by compressing the cell rapidly under a coverslip and then releasing the compression (Ehret and Powers, 1957). The lysed cell leaves much debris, but among it there is often an intact buccal apparatus. Another method (Sibley and Hanson, unpublished) is to expose a cell to 1.0 *M* hexylene glycol. The cell starts to lyse and further breakdown is achieved by mechanical disruption. A glass needle is vibrated electronically (Fig. 3) and placed beside the cell, which is held in place by pressure from another needle held in a simple micromanipulator; this results in almost instantaneous dissolution except that the oral parts remain and without visible damage. Such a preparation can be air-dried for autoradiography or examined as a wet mount by phase or interference contrast microscopy.

Isolation of buccal parts from mass cultures is also possible. The following account provided by J. Wille (personal communication) has been used with *P. aurelia*. Four grams of packed cells, harvested by low speed centrifugation, are resuspended in gullet homogenization and stabilization medium (GHSM), which consists of the following ingredients: 0.01 *M* Tris (Trizma base, Sigma Biochem, Inc.), 1×10^{-3} *M* EDTA (acid form) in 0.25 *M* sucrose, pH = 8.3. The cell suspension is left for at least 3 hours at 4°C and then

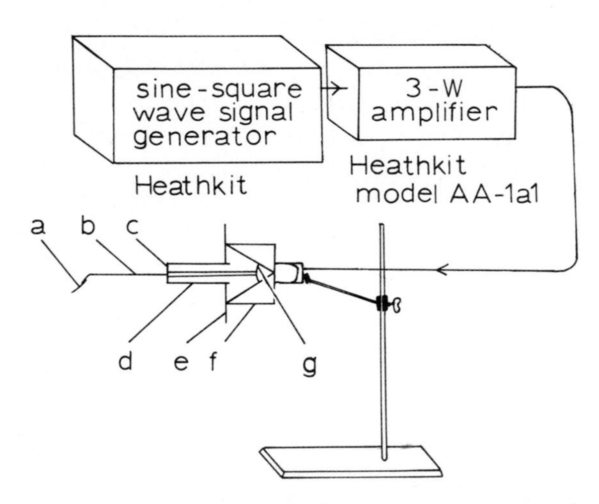

FIG. 3. Sonicator or mechanical disrupter of single cells. a. Fine glass needle, attached by rubber cement to b.b: Steel needle of dissecting probe with wood handle. Base of needle (where it enters wood handle) is supported by c. End of wood handle is glued (epoxy) to g. c: Cloth support for b. d: Tube (diameter ca. 3 cm, length ca. 7 cm). e: Rigid sheet attached to d and f. f: Small speaker cut down to desired size. g: Point of attachment of dissecting probe handle to speaker.

homogenized in a Mickle tissue disintegrator for 5 minutes at maximum vibration. The resulting homogenate is then forcibly extruded through a glass syringe (2.0 ml, hypodermic syringe, 200 strokes). Phase contrast microscopic examination of the brei during this latter treatment will indicate the progress of separation of cortical fragments from the intact gullet, and allow an estimate of the yield of gullets prior to sucrose gradient purification. One-step sucrose gradient purification: The entire homogenate is layered on top of a 40% sucrose solution containing GHSM medium, and spun at 3000 *g*, for 5 minutes (5.0 ml of homogenate on top of 25 ml of gradient, 40-ml polyethylene centrifuge tubes, in an SS 34 rotor in an RC-28 refrigerated centrifuge). The gullets are recovered in the pellet, and are only contaminated with crystals. The crystals can be removed by re-

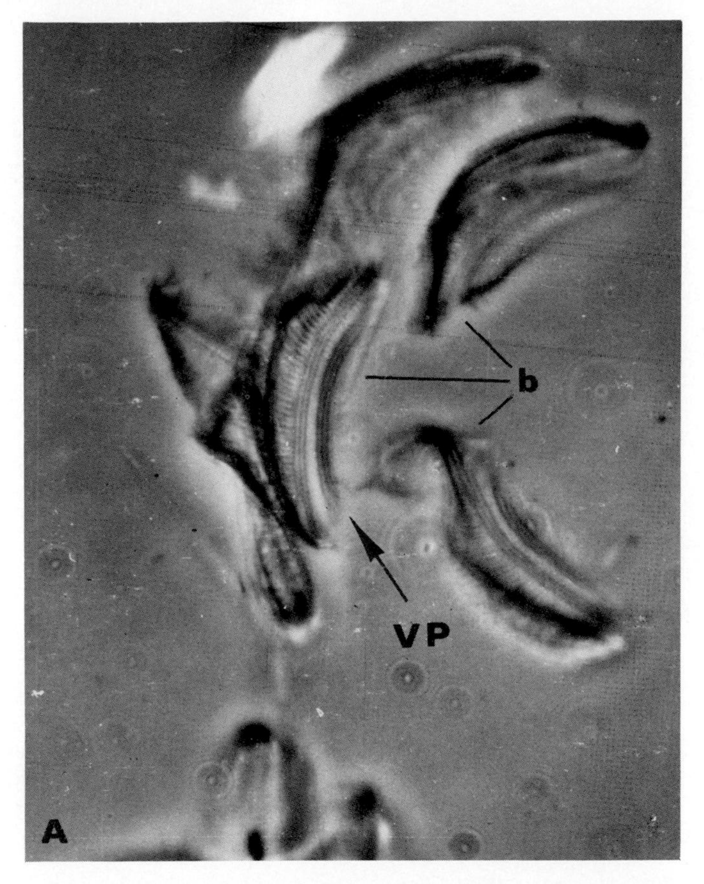

FIG. 4. Isolated buccal structures (b). (A) Ventral peniculus (vp, arrow) and other internal details are clearly visible. (B) Almost intact buccal structure (b) with attached vestibular cortex (c). See p. 334 for (B). (Photographs by courtesy of Dr. J. J. Wille, Jr.)

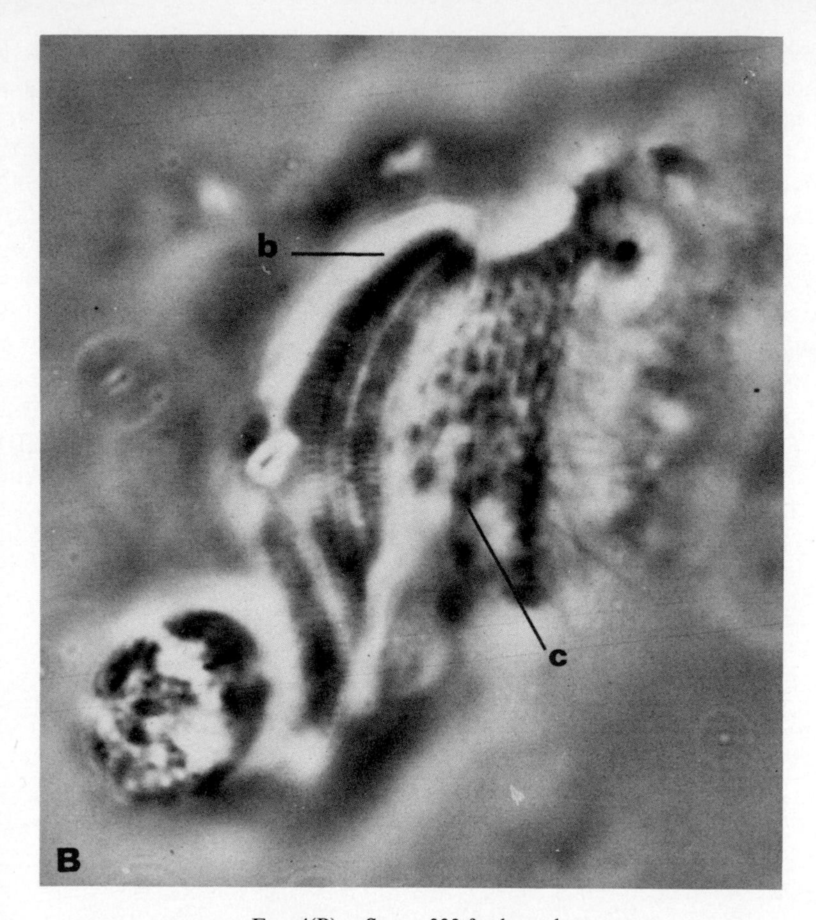

FIG. 4(B). See p. 333 for legend.

suspending the pellet in GHSM and using low speed centrifugation (1000 *g*, for 2 minutes). If further purification is required, the one-step gradient step should be repeated. Figure 4 is a phase contrast micrograph of a preparation of purified gullets. One notes the apparent structural integrity of the three main membranelles—dorsal and ventral peniculi, and quadrulus—and the nonciliated ribbed wall. Also present and conspicuous are the fused cilia along the edges of the membranelles, and the cytopharyngeal fibers at the base of the gullet. Of interest is the occasional partial, or almost complete, detachment of the ventral peniculus (e.g., note the partial unraveling of the rows of this peniculus, arrow in Fig. 4).

6. MITOCHONDRIA

These organelles have been studied from three points of view: their fine structure, cytoplasmic inheritance, and DNA.

Fine structure work reveals the tubular cristae typical of protozoan mitochondria. Techniques for electron microscopy are those given above (Section I, B,1 and 3).

Cytoplasmic inheritance has focused thus far on erythromycin and chloramphenicol resistance. Analyses show that these traits are associated with the mitochondria. Techniques used here include selection for resistance and subsequent crosses to show that segregation patterns are non-Mendelian (Beale, 1969; Adoutte and Beisson, 1970). Furthermore, electron microscopy reveals differences in the mitochondria of cells transforming from erythromycin sensitive to erythromycin resistant. These mixed populations of mitochondria allow the investigator to determine the subsequent phenotype of the population through appropriate selection (Adoutte and Beisson, 1972). Also, by special microinjection techniques it is possible to transfer mitochondria of different phenotypes from one cell to another and follow the fate of the injection on the recipient cell (Beale *et al*., 1972). Thus, there are now available a variety of approaches for manipulating intracellular populations of mitochondria in *P. aurelia*.

Mitochondrial DNA has been isolated from *P. aurelia* and various of its physical features were characterized. Suyama and Preer (1965) suspended the cells in 0.2 *M* raffinose, 0.25% bovine serum albumin (fraction V, Calbiochem), and 1 m*M* potassium phosphate buffer at pH 6.2. Flavell and Jones (1970) used 0.3 *M* sucrose, 2 m*M* EDTA, 10 m*M* Tris-HCl buffer at pH 7.2. For homogenization, the former workers used a milk homogenizer, and the latter a tissue disintegrator. The homogenate was then centrifuged, first at 500 *g* to remove nuclei and pellicular fragments, and then at 5000 *g* to bring down the mitochondria. See the original references for details regarding repeated cycles of centrifugation and resuspension.

Suyama and Preer described a technique for quantifying the mitochondrial suspension by using yeast cells of known concentration added to the mitochondrial suspension.

Isolation of DNA from the purified mitochondria follows, in essential detail, that of Luck and Reich (1964). Suyama (1966) has published details that he subsequently used for further work on *Tetrahymena*, and this technique was followed successfully by Flavell and Jones (1970), who went on to repeat the cesium chloride density measurements of Suyama and Preer and determined, also, the sedimentation velocity, thermal denaturation and renaturation rates, and degradation kinetics of the purified mitochondrial DNA.

7. NUCLEI

A macronucleus and, depending on the species, one to several micronuclei are present in any normal paramecium (Wichterman, 1953; Raikov, 1968). Most studies have concentrated on the macronucleus for a variety of reasons:

it is large and readily observed; its high ploidy—as much as 860*n* (Woodard *et al.*, 1961), special behavior at fission (amitosis), and reorganization during sexual processes are of interest to cytogeneticists; and its essential role in survival of the cell, combined with the above points, make it a genuinely intriguing organelle.

Most of the work on nuclei has been done on intact or sectioned cells. Macronuclei have been isolated, but usually as a prelude to obtaining DNA or other nuclear macromolecules. Therefore, the methods that follow will be devoted largely to intracellular nuclei—their visualization, measurement, and characterization of contents. Isolation of nuclei will be briefly mentioned but discussed further under the section devoted to molecular biology, wherein the study of various macromolecules is described.

Microscopic observation, *in vivo*, of micro- and macronuclei depend on phase or interference contrast microscopy for best results. The behavior of nuclei during fission and conjugation have been reported, both under normal and experimental conditions (Wichterman, 1953; Vivier and Andre, 1961; Gill and Hanson, 1968; Hanson *et al* 1969). The only serious technical problem is immobilization of the cells, especially for photography, and flattering to optimize the contrast need for critical viewing. Both ends are achieved by use of the rotocompressor (see Section I,A,2).

Observations *in vitro* have been described by Sonneborn (1950), and further details are also given above.

The volume of macronuclei has been determined by measuring length and width of the organelle in fixed, unflattened cells and assuming that the shape approximated a prolate spheroid (Berger, 1971). Namely, $V \equiv 4/3\ \pi ab^2$, where V is the volume and a and b are the major and minor axes, respectively. Such measurements go back to the early work of Popoff and Jennings and many other workers (full references in Wichterman, 1953).

The qualitative characterization of nuclear contents can be done cytochemically. Sometimes it is useful to use an enzymatic digestion followed by the appropriate staining or other techniques which can demonstrate the loss of the digested component. For example, absence of staining with the Feulgen reagent following DNase treatment effectively demonstrates loss of DNA, and absence of purple-blue staining with azure B bromide demonstrates loss of RNA following RNase treatment. Both nucleic acids can be removed by hot (90°C) 5% TCA (30-minute exposure) as demonstrated by loss of all staining by azure bromide. Electron microscopy of digested and undigested sections readily shows the selective effect of the enzymes (Jurand *et al.*, 1962, 1964).

Enzymatic digestion can be carried out on whole, unmounted cells (Berger, 1971) or on cells that are air-dried on glass slides (Kimball *et al.*, 1959; Woodard *et al.*, 1961). In the latter case it is strongly recommended that the slides be subbed with a chrom-alum gel (Fitzgerald, 1959) before use. As is common experience, DNase preparations must be carefully buffered and provided with Mg ions, whereas use of RNase is more straightforward. [Many variations are available for the preparation and use of the enzyme solutions. The directions given by Berger (1971) will provide consistently reliable results.] In all cases undigested and digested control cells, stained appropriately, should be used to check on the efficacy of the enzymatic solutions.

Quantitative determinations of the nucleic acid and protein content of macronuclei have been achieved by several techniques. Microspectrophotometry of Feulgen-stained macronuclei has been used to determine DNA levels (Moses, 1950; Walker and Mitchison, 1957; Kimball and Barka, 1959; Woodard *et al.*, 1961). Absorption of monochromatic ultraviolet light has also been used (Raikov *et al.*, 1963; Kimball *et al.*, 1960). This latter technique also measured total nucleic acid and, by inference, the difference obtained by subtracting the DNA values from total nucleic acid provide data for RNA, too. Moses (1950) measured the RNA content by determining loss of UV absorption at 254 nm following ribonuclease digestion. And others have determined amounts of RNA by microspectrophotometric measurements of azure B staining (Woodard *et al.*, 1961).

Measurements of protein has been done on *P. caudatum* in terms of total protein and nonhistone protein (Moses, 1950). In both cases photometric analysis of critically stained material was used (Million reaction for total protein and sulfuric acid Millon reaction for nonhistone protein). The difference between these two sets of data was thought to represent histones. Woodard *et al.* (1961) measured total protein (Naphthol Yellow S) and tyrosine (Millon reaction) in single cells.

The total dry weight of the macronucleus was measured by scanning interference microphotometry (Kimball *et al.*, 1960). Their calculations indicate that the macronucleus comprises 6% of the total dry weight of a recently divided cell, which is 2×10^{-8} gm.

In the foregoing instances of microspectrophotometry, there appear to be differences in the design of photometric devices and the treatment of cells in terms of fixation and of staining, which may affect the data obtained. Overall there seems to be quite good agreement among the various reports, and, therefore, a researcher entering the field can expect to obtain reliable results from these techniques but is well-advised to examine carefully the original literature to find that particular approach that best suits the question he or she wishes to examine.

The foregoing studies have often had as their main purpose the study of quantitative changes in amounts of the various macromolecular constituents during the cell cycle. Timing of the cells for such studies is given below (Section I, D).

Isolation of macronuclei is necessary when using the UV absorption technique on whole macronuclei because the effect of the surrounding cytoplasm has to be avoided (Kimball *et al.*, 1960). Macronuclei have been isolated by various methods. Schwartz (1956), using hand-held needles, removed them from *P. bursaria* immobilized in the presence of cold CO_2. Kimball *et al.* (1960) describe two osmotic shock methods which they say work well for well-fed paramecia (*P. aurelia*) and an ethyl alcohol–acetic acid–water (63:17:20 parts, respectively) method for starved cells.

8. Ribosomes

Reisner and his co-workers as well as Sommerville and Sinden have been able to isolate ribosomes and use them as a system for *in vitro* protein synthesis, as well as proceeding to characterization of subunits within the 80 S units of the initial ribosomal fraction, and isolation of ribosomal RNA and protein. The isolation of the ribosomal fraction will be outlined here (Reisner and Macindoe, 1968; Sommerville and Sinden, 1968), and in Section II, on molecular biology, there will be discussed the methods for more detailed study of the ribosomes and their role in protein synthesis.

Both groups of workers start with cell populations concentrated by mild centrifugation to ca. 10^6 or 2×10^7 cells/ml. [For another method for concentrating cells see Aylmer and Reisner (1971).] If the culture is taken from paramecia fed on bacteria, the food-organism is eliminated by washing and centrifugation at this stage. Concentrated cells are suspended in an ice-cold Tris buffer (see original papers for slight differences in the solution) and homogenized by hand using a Teflon homogenizer. This homogenate is then fractionated by centrifugation. Reisner's group centrifuges at 3000 *g* for 15 minutes followed by decantation of the supernatant and further centrifugation for 15 minutes at 10,000 *g*. This supernatant is then layered on 25 ml of 15 to 30% linear sucrose gradients and centrifuged for 3 hours at 25,000 rpm in a Spinco SW 23.1 rotor or its equivalent. Samples are removed and absorbance at 260 nm is recorded. The four samples closest to the 80 S peak of each gradient are pooled and centrifuged at 150,000 *g* for 1 hour. The resulting pellet is surface rinsed with the following solution: 50 m*M* Tris, 25 m*M* KCl, 3 m*M* $MgCl_2$, pH 7.6. It is resuspended in that medium and centrifuged for 10 minutes at 3000 *g*. This final preparation is reported to contain no polysomes and is 57% RNA and 43% protein.

A somewhat different procedure is used by Somerville and Sinden (1968) and summarized in Sommerville (1970a) (Fig. 5). Here, there is no use of sucrose gradients to produce the ribosomal fraction, which is the pellet produced by the final centrifugation.

The ribosomal fractions prepared by Reisner and his co-workers have been used to study the dissociation of the ribosome through exposure to varying concentrations of Mg^{2+} (Reisner *et al.*, 1968). This treatment plus

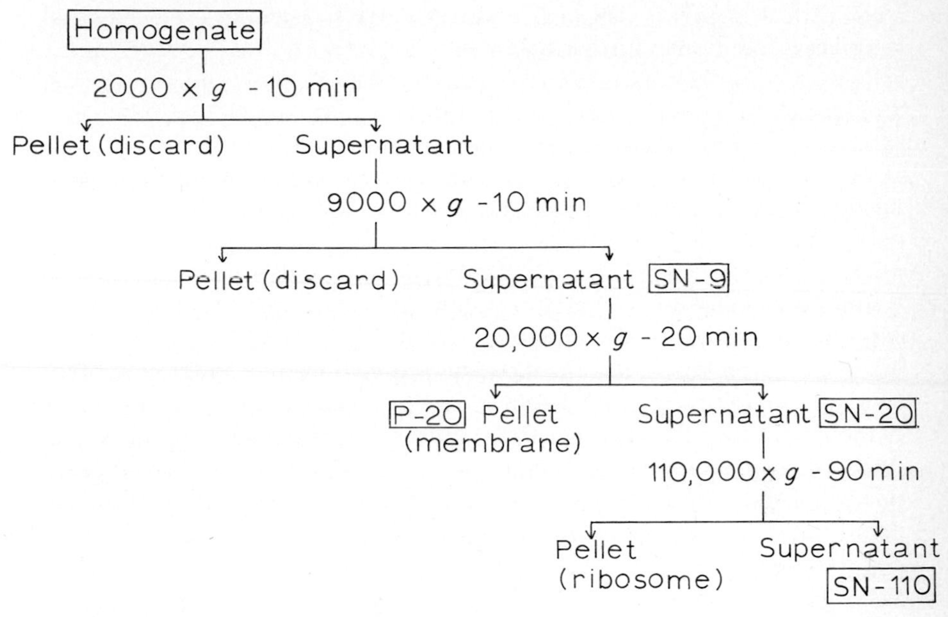

FIG. 5. Outline of procedure for obtaining selected cell fractions from a paramecium homogenate. Redrawn from Sommerville (1970a).

the appropriate use of sucrose density gradients has led to the recovery of 28 S and 8 S subunits. This laboratory (Reisner *et al.*, 1972) has also developed techniques using sedimentation coefficients for determining the size of polyribosomes.

The further use of these ribosomal preparations for isolation of RNA and protein and as part of the system used for *in vitro* protein synthesis will be discussed later (see Sections I, A,2; and 4).

9. OTHER ORGANELLES

The remaining organelle systems of paramecia, i.e., contractile vacuoles or vesicles, food vacuoles, pinocytotic vesicles, mitotic apparatus, endoplasmic reticulum, Golgi apparatus, and lysosomes have all been studied to some degree either by light microscopy (Wichterman, 1953) or electron microscopy (Jurand and Selman, 1969) or both. The function of the contractile vesicles has received significant attention from a physiological point of view. These studies have used intact cells [see review by Kitching (1967)]. The concern has often been with environmental factors which control the rate of pulsation, such as temperature, oxygen tension, and salt concentration. Earlier, very general and largely descriptive studies on the physiology of feeding are given in Wichterman (1953) and a more recent experimental study (effect of pH) is that of Brutkowska (1963). Feeding as a behavioral problem is mentioned further below (Section I,D, 4).

D. Cell Cycle

The historical continuity of life depends, ultimately, on the ability of one cell to make more cells. Paramecia, with their richness of differentiated detail, have engaged the efforts of many workers interested in the biochemical, physiological, genetic, and developmental mechanisms which control cellular reproduction. To carry out such studies, various parameters of the cell must, of necessity, be measured, ranging from simple metric measurements to automatized, simultaneous determinations of certain major biosynthetic functions. It will be convenient to review first the measurements made on whole cells, then examine various manipulative techniques, and finally to look at special techniques in cell physiology and behavior. More purely genetic techniques are found in Sonneborn (1951, 1970) and culture techniques in van Wagtendonk and Soldo (1970).

1. Whole Cell Measurements

Determinations on one or another feature of a cell are most meaningful when the interfission age of the cell being measured is accurately established.

a. Interfission Age. At constant temperatures, given strains of paramecia divide regularly provided food and pH are also controlled. However, the regularity just referred to is not much better than ±30 minutes for an interfission period of about 6 hours, at room temperature, for genetically indentical (isogenic) lines of *P. aurelia*. Kimball *et al.* (1959) showed that sister cells tend to be more alike in the duration of their interfission period than nonsister cells. [They used well-fed cells taken at random from a stock culture and obtained a correlation coefficient, r, of 0.6. Using cells 3–4 fissions after autogamy Hanson and Kaneda (1968) obtained r values of 0.93.] Using this finding Kimball and co-workers formulated a simple method for determining the interfission period of experimental cells whose normal cell cycle might be interrupted by experimental treatments and either shortened or prolonged. For this determination, a comparison to the expected cell cycle is needed. The expected interfission time, $\hat{T}$, is given as follows

$$\hat{T} = rT + (1-r)\,\overline{T}$$

where r is the correlation coefficient, T the interfission period of the control sister cell, and $\overline{T}$ the average interfission period of all the sister cell controls. Interfission age is then estimated by taking the time elapsed from the preceding fission to the time for experimentation divided by $\overline{T}$. Interfission age is expressed than as a decimal portion of one interfission period (Gill and Hanson, 1968).

b. Cell Size. Length and width measurements were used by early workers (e.g., Jennings and Popoff, see Wichterman, 1953, for full references) to follow cell growth at stated intervals after a given fission. These measurements were made on both live and fixed cells. Some attempts were also made to calculate cell volume using the formula of the prolate spheroid (see Section I,C,7). However, paramecia are not regular geometric figures, largely because of the oral groove, and volume measurements in the foregoing terms are only approximations.

In addition to measurements in metric units, the number of cortical units present along a given surface meridian have been used to follow changes in the cell surface (Kaneda and Hanson, 1973). The problem here is that the rows of cilia, the kineties, are rarely true pole-to-pole longitudinal meridians. They curve around the oral opening on the ventral side, and on the lateral and dorsal sides they may or may not reach the cell apices (Fig. 1). What must be done is to follow the meridian formed by a focal plane or the edge of the cell, or one determined by other surface features, such as the contractile vesicle pores or the suture lines, and use these reference lines for counting units when kineties are not available. One advantage of using the cortical units is that their regular spacing and packing pattern provides a set of detailed landmarks for mapping the cell surface, much as longitudinal and latitudinal lines can apply to the earth's surface. Few cells are as cooperative as ciliates in this matter. However, the regularity of this patterning can be overstressed, for the latitudinal lines [or paratenes (Ehret, 1967)] do not continue around the cell with genuine regularity (Kaneda and Hanson, 1973).

c. Cell Mass. This parameter has been measured by Kimball *et al.* (1959) and Kimball *et al.* (1960) using either X-ray absorption or a scanning interference microphotometer. They calculated the dry weight of a recently divided *P. aurelia* as being 2×10^{-8} gm.

Rates of synthesis of cellular constituents such as DNA, RNA, protein, and so on, have been determined by application of techniques already described for identifying and measuring these molecular components (Section I, C,7) with the added proviso that the measurement be made at known points in the cell cycle.

2. Whole Cell Manipulations

a. Microsurgery. Cutting operations can be carried out on paramecia by means of fine needles or by a microbeam of ultraviolet light. Sonneborn (1970) gives details on both of these techniques. Additionally, papers by

Chen-Shan (1969, 1970) have now been published which describe precise amputation of anterior or posterior portions of *P. aurelia.* Tartar's (1964) earlier paper on methods should also be consulted. This describes single-cell and mass culture techniques for operating on paramecia. The latter provides large numbers of cell fragments. A somewhat similar apparatus (basically, rapid rotation of razor blades) has been reported for use on *Urostyla* but can surely also apply to paramecia (Jerka-Dziadosz, 1968).

b. Micromanipulation. See Sonneborn (1970) for reference to Koizumi's techniques. Recently Beale *et al.* (1972) have described transfer of mitochondria by microinjection.

c. Radioactive Labelling. Radioactive tracers have been used extensively in paramecia, primarily to determine rates of synthesis of nucleic acids but also to locate the fate of this material in degenerating nuclei and to label proteins and follow the behavior of organelles assembled from this material.

It is most convenient to comment, in order, on labeling of DNA, RNA, and proteins, and then describe labeling of organelles. The most intensive study of nucleic acid labeling in paramecia has been done by Berger on *P. aurelia.* Much of this work is still in the form of a Ph. D. thesis (Berger, 1969). The kinetics of incorporation of DNA precursors using monoxenic and axenic cultures has been published (Berger and Kimball, 1964) and recently extended (Berger, 1971). It was found that *Escherichia coli*, strain 15 T^-, which requires thymine, fed on tritiated thymine and then used as the food organism for *P. aurelia* results in entry of the labeled cell into paramecia in a matter of minutes, followed by label in a cytoplasmic DNA precursor pool, and then rapid incorporation into the macronucleus, if that organelle is in its S period. It appears that autoradiographic detection of incorporated material is a more sensitive measure of the onset of DNA synthesis than are photometric techniques. The latter show synthesis beginning at about interfission age 0.5, Berger's technique shows the S period starting at age 0.25–0.30. Others using this approach had evidence of similar results but did not show the marked incorporation reported by Berger (Kimball and Perdue, 1962; Pasternak, 1967; Rao and Prescott, 1967).

Berger (1971) also shows that using "exogenously labelled cells," i.e., paramecia in a bacterized medium to which tritiated thymidine was added directly, there is significantly less incorporation of the label into the macronucleus, and more label not removable by DNase. Clearly the "bacterially labelled cells" appear as the better experimental material. Neither Berger nor Berger and Kimball compared their results with those obtainable from axenic culture techniques (van Wagtendonk and Soldo, 1970).

In addition to thymidine-^{3}H, cytidine-^{3}H can be used to label DNA (Kimball and Perdue, 1962). The specificity of these precursors needs special comment. Thymidine-^{3}H always results in some cytoplasmic labeling which cannot be removed by nuclease or hot TCA. It is thought to be labeling in protein (Kimball and Perdue, 1962). The amount of this nonspecific labeling is negligible in short pulse experiments. Cytidine goes predominantly into RNA even though administered late in the interfission period when DNA synthesis is at its peak. Kimball and Perdue stated that the specificity of cytidine for RNA is better than thymidine for DNA. Uridine-^{3}H goes predominantly into RNA, but some label may become removable by DNase. This agrees with Soldo and van Wagtendonk's (1961) report that uridine-^{14}C and cytidine-^{14}C can be precursors for thymidine in axenically cultured *P. aurelia*.

All the foregoing warns the experimenter to use appropriate controls with nuclease and hot TCA digestions in preparing slides for grain counting.

Labeling of nucleic acids with ^{14}C- containing precursors has been done by van Wagendonk's group (Soldo and van Wagtendonk, 1961; Butzel and van Wagendonk, 1963) using extractions and counters of radioactivity. Autoradiographic techniques appear to be simpler, more sensitive, and when ^{3}H is used, provide high intracellular resolution.

Proteins have been radioactively labeled in most cases by incorporation of tritiated amino acids and the disposition of the label followed by autoradiography. Butzel and van Wagtendonk (1963) described using glycine-^{14}C and the radioactivity of the recovered protein fraction was measured by a scintillation counter. Their culture medium was axenic. All other reports have used a bacterized medium for routine culturing and for exposure to the label or else the latter was placed in a sterile medium to which the paramecia were added for the period of exposure. Kimball and Prescott (1964) found that tritiated histidine, isoleucine, leucine, lysine, phenylalanine, proline, or valine could result in labeling of protein. Here, the sterile lettuce medium which contained the labeled amino acid also contained "a small amount" of egg albumin "to promote uptake," that is, presumably to stimulate food vacuole formation. A similar technique (Kimball and Perdue, 1965) used the baked lettuce medium with *Aerobacter aerogenes* (the normal food organism), leucine-^{3}H, and egg albumin and also achieved good labeling of cellular proteins. Finally, Ehret *et al.* (1964) also using labeling by leucine-^{3}H in a bacterized medium, obtained good labeling of cytoplasmic proteins and organelles, the trichocysts in particular. The latter two papers concur in stating that the incorporated label is extremely stable (unless, in case of the trichocysts, they are discharged) and persists for many cell generations.

One procedure for autoradiography (Kimball and Prescott, 1964) is the following. The cells are air-dried as individuals or glass slides and extracted with ice-cold 5% TCA (w/v) for 5 minutes in work with nucleosides but extracted with 5% TCA at 25°C for 90°C for 5 minutes in work with labeled amino acids. (Digestion with RNase or DNase can be inserted at this point if desired. See Section IC,7 for details.) The slides are washed free of TCA and dipped in Kodak NTB3 liquid emulsion. They are stored in a dry container in the refrigerator and exposed for the desired length of time before being developed. The exposure time can be highly variable and is best determined by running some test slides along with the experimentals and developing these at appropriate intervals to determine the optimal exposure.

In our laboratory it is found useful to use slides marked on the *underside*, by a diamond- or carbide-tipped pencil, as shown in Fig. 6. Such slides are subbed (see Section I,C,7) and then the cells to be studied are air dried on the unmarked, upper surface, between the parallel, vertical arms of the L. In this way they are easily located after the autoradiogram is developed. Also the asymmetry of the L always allows one, even in the darkroom to know which side of the slide is up.

The foregoing techniques can be used not only to label certain macromolecular components of the cell, but also certain of its organelles. The specificity of the label is variable in all cases. Labeling of DNA results in heavy nuclear labeling. There is DNA in the mitochondria of paramecia (Suyama and Preer, 1965), but labeling of cytoplasmic DNA on this account is very light or negligible, especially when short pulses (up to 10 minutes' duration) are used. Labeling of RNA results in an early accumulation of label in the nucleus (within 10 minutes of starting the exposure) and then the label moves rapidly to all parts of the cytoplasm. Ehret *et al.* (1964) were able to label trichocysts heavily by inducing trichocyst regeneration in cells exposed to leucine-^{3}H. (See Section I,C,3 for induction of trichocyst extrusion.) Leucine-^{3}H is also found in the cells throughout the cytoplasm.

d. Antimetabolites and Other Chemicals. A certain range of chemical inhibitors of metabolic activities have been used in *P. aurelia*. Mitomycin at 100 μg/ml inhibits detectable incorporation of cytidine-^{3}H or thymidine-^{3}H into DNA (Hanson, unpublished). Actinomycin D, a powerful inhibitor of DNA-dependent RNA synthesis, effectively blocks incorporation of cytidine-^{3}H or uridine-^{3}H into RNA when cells are exposed to 40 μg/ml for as briefly as 10 minutes. There is also a delay in the next fission (Hanson and Kaneda, 1968). Rasmussen (1967) also reports a fission delay following pulse exposures, but whereas Kaneda and Hanson found the delay to equal to the addition of an extra fission, Rasmussen reported the delay to decrease as

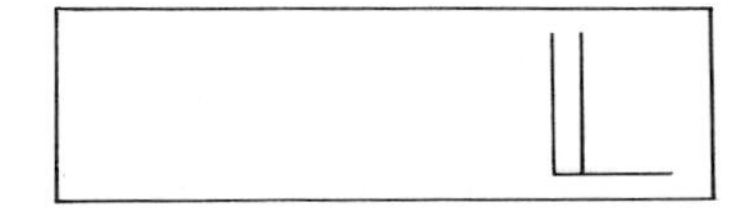

FIG. 6. Diagram of a standard glass slide marked for us with autoradiograms. Note that the L-shaped figure is scratched into the under side of the slide, and materials to be studied are oriented relative to it on the upper side of the slide.

exposures occurred later and later into the interfission periods. The reason for these different results is not clear. Extended exposures to concentrations of 10 μg/ml resulted in no fission and eventual death, whereas 2.5 μg/ml simply caused a depression of the fission rate (Gill and Hanson, 1968).

The methods used for timing the pulsed exposures is that mentioned earlier (Section I,D,1) which uses sister cells to calculate the interfission age at which the experimental cell is exposed to the antimetabolite.

Protein synthesis has been inhibited by treatment with cycloheximide (Suhama and Hanson, 1971). However, the stock of *P. aurelia* used in this work (stock 51) proved to be highly resistant to the inhibitor. Exposure to 1.8 mg/ml blocked only 70% of the incorporation of leucine-^{3}H found in normal cells. By comparison, 20 μg/ml completely inhibits protein synthesis in *Tetrahymena* (Frankel, 1969). Dr. Mary Austin (personal communication) has found several strains of *P. aurelia* which are resistant to cycloheximide and apparently Suhuma and Hanson used one of these in their work.

Austin *et al.* (1967b) have used a variety of chemicals to examine their effects on the stabilization or transformation of the immobilization antigens of *P. aurelia*. These include actinomycin D, puromycin, and chloramphenicol as well as patulin and acetamide.

Colchicine and colcemid have been used (Clark and Hanson, unpublished) on *P. aurelia* to inhibit fission or morphogenesis of organelles, or both. The results have been hard to interpret in that very little is observed other than total blockage of fission, with no morphogenesis and subsequent death, at higher concentrations (ca. 4–5 μg/ml) or, at slightly lower concentrations, only slight disruption of oral structures was observed.

3. The Physiology of Locomotion

There has been a continuing interest in paramecia regarding the mechanism of ciliary action and the synchronization of ciliary beating over the cell surface. The former has involved exposure of cells to media containing different ions, notably Ca^{2+}, K^{+}, and Ba^{2+} and exposure to electrical stimuli, both as discharges through the medium and by means of intracellular microelectrodes. The pattern of ciliary beating has necessitated, first, a careful description of normal beating in live or specially fixed cells, complemented by observations on specially manipulated cells—cut by glass needles, electrically stimulated, or exposed to various chemicals. Most recently, behavioral mutants are being examined from an electrophysiological point of view. All of these are contributing to a broadened understanding of the basic mechanisms underlying ciliary action and transmission of coordinative impulses in single cells.

The electrophysiological studies started with the early observations of Verworn, Ludloff, and Jennings (reviewed by Jennings 1923, 1966). Here

Alfa-Laval (or de Laval) cream separator technique (Preer and Preer, 1959). This is almost the standard first step now; it is used in almost all the reports cited in this section. Various techniques were used to disrupt the cells by osmotic shock using 0.3 *M* sucrose. Stabilization of the nuclear membrane by calcium (0.0006 *M* $CaCl_2$) was found to be best. Using further refinements of this technique Stevenson (1967) isolated macronuclei capable of incorporating labeled ATP and UTP into an acid-insoluble residue believed to be RNA, and measured amounts of DNA, RNA, and protein present in these isolated nuclei.

Isaacks *et al.* (1969) reported isolation of macronuclei following collection, concentration, and disruption at 0–4°C with a Waring Blendor. Again calcium ions were needed to ensure obtaining intact macronuclei. The nuclei were purified by sucrose density gradient centrifugation. These nuclei were used to isolate DNA, RNA, and protein, but details on these latter techniques are not given.

Methods for characterizing the macromolecules from isolated macronuclei are given by Cummings (1972). Special attention is given to stabilizing the nuclei (both macro- and micronuclei were isolated and as separate fractions) by use of calcium ions and spermidine following rupture of the cells by detergents. DNA and RNA were separated from proteins in both the macro- and micronuclear material.

Other workers have omitted the step of nuclear isolation and have lysed cells and extracted the DNA from the lysate. Allen and Gibson (1971) describe two methods using basically a phenol extraction. (Outline given here; consult original paper for details.) Starting from tightly packed cells, in the first method lysis is brought about by addition of 10% sodium lauryl sulfate (final concentration 0.5%) whereas in the second method cells are frozen in liquid nitrogen, ground by means of a mortar and pestle, and then lysed in 6% sodium 4-aminosalicylate with 1% NaCl. In both methods DNA was collected from the aqueous layer of the phenol extractions by precipitating with cold absolute ethanol and winding onto glass rods. Then follows treatment with enzymes to remove only RNA in the second method, and RNA plus carbohydrates and proteins in the first method, followed by further washings and precipitations with propanol, overnight dialysis, and storage over chloroform at 4°C. Estimated yields were 50% for the first method and 65% for the second. Various tests for purity of the DNA were run which revealed a maximum possible contamination by RNA of 15–20% (orcinol-positive material), 23% carbohydrate, and little or no protein. The purified DNA was double-stranded and had a molecular weight range from 7×10^5 to 2×10^6 daltons indicating that considerable shearing had taken place.

These workers emphasize the great importance of removing all possible bacteria from cells grown in a bacterized medium. Axenically cultured cells give constant density values whereas cells fed on bacteria are variable and only starved cells provide values like those obtained from axenic cultures. This work and that which follows has used *P. aurelia*.

Reisner and Bucholtz (personal communication) have provided the following unpublished outline for preparing DNA, starting from monoxenically cultured paramecia. The food organism is one whose DNA bouyant density is 1.731 while that of *P. aurelia* is 1.685 and, therefore, the two are easily distinguished, thus obviating part of the important problem cited by Gibson and Allen.

The food organism *Micrococcus lysodeikticus* is grown to stationary phase in "P-broth," consisting of 1% Bacto tryptone, 0.5% Bacto-yeast, 0.1% glucose, 0.5% NaCl (all w/v percentages), at pH 7.5. The paramecia are grown in the following medium: To a 2-liter Erlenmeyer flask is added 1500 ml of water, 12.0 ml of a $33\frac{1}{3}$% (w/v) Vegenite suspension, and 30.0 ml of "P-broth" concentrate (15% Bacto tryptone, 7.5% Bacto-yeast, 1.5% glucose, 7.5% NaCl). This solution is autoclaved at 15 psi for 35 minutes. To each such flask, 1 ml of *M. lysodeikticus* suspension and 10 ml of paramecia grown in 200-ml starting cultures of the above medium are added. Flasks are incubated for 7 days at 25°C.

The DNA preparation is a modification of the phenol extraction method of Saito and Miura (1963). Solutions used are:

A. Tris–SDS–NaCl buffer: 0.1 *M* Tris; 1% (w/v) sodium dodecyl sulfate; 0.1 *M* NaCl. Adjust to pH 9.0 with HCl.

B. Tris–NaCl buffer: 0.1 *M* Tris, 0.1 *M* NaCl. Adjust to pH 9.0.

C. SDS-buffered phenol (make up fresh). To each 100 of phenol, add 13 ml of solution A.

D. Buffered phenol (make it fresh). To each 100 ml of phenol add 13 ml of solution B.

E. 10× SSC: 0.15 *M* NaCl, 0.15 *M* sodium citrate. Make up SSC and 0.1× SSC by dilution of 10× SSC.

P. aurelia grown monoxenically with *M. lysodeikticus* are harvested using a cream separator and then pear-shaped centrifuge tubes. The further procedural steps are as follows.

1. The packed organisms (80 to 100 $\times$ 10^6) are suspended in 25 ml of solution A (0°C) and transferred to a 100-ml beaker in an ice bath. The suspension is stirred for about 5 minutes, and it becomes quite viscous.

2. The viscous lysate is transferred to a 100-ml glass-stoppered flask, and 25 ml of solution C (room temperature) is added. The flask is shaken vigorously by hand for about 5 seconds, it is then placed in a New Brunswick Gyratory water-bath shaker which contains an ice-water mixture (0°C) and is shaken another 20 minutes at 400 cycles/min (top speed).

3. The emulsion is then centrifuged at 20°C in 25-ml tubes for 15 minutes at 1500 *g*. The aqueous layer is carefully pipetted off and held in ice.

4. The phenol layer is then reintroduced into the glass-stoppered flask, 25 ml of solution B is added, and steps 2 and 3 are repeated, except that shaking is done for 10 minutes.

5. To the combined aqueous layer is added 2 volumes of cold 95% EtOH with stirring and the precipitate centrifuged down at 1500 *g* for 10 minutes.

6. The suspension is decanted and the centrifuge tubes are wiped, then a total of 25 ml of solution A (room temperature) is used to suspend the precipitate and transfer it to a 100-ml glass-stoppered flask. It is dissolved at room temperature by swirling the suspension by hand.

7. Once dissolved, step 2 is repeated but with 10 minutes of shaking at 0°C.

8. Step 3 is performed and followed by step 5.

9. The precipitate is dissolved in 20 ml of 0.1 × SSC by transfer to a 100-ml glass-stoppered flask and swirling by hand at room temperature. Two milliliters of 10 × SSC is then added to make the solution 1 × SSC. Sufficient T_1 RNase solution (10,000 units/ml) is added to give a final concentration of 100 units/ml, as well as pancreatic RNase (0.2% solution) to give a final concentration of 50 μg/ml. This mixture is incubated for 30 minutes at 37°C.

10. The flask is cooled to room temperature (swirled 1 minute in an ice bath), and an equal volume of solution D is added. The flask is then shaken at 400 cycles/min in the 0°C water bath for 10 minutes.

11. The mixture is transferred to a Spinco SW 25.1 rotor and spun for 30 minutes at 2°C and 25,000 rpm. This hard spin ensures breaking of the emulsion.

12. The aqueous layers in the centrifuge tubes are carefully removed, pooled, and transferred to polyallomer tubes for the Spinco SW 50.1 rotor. The rotor is spun for 45 minutes at 40,000 rpm and 2°C in order to separate the glycogen.

13. The supernatant is carefully pipetted off leaving the loose glycogen pellet behind (if the supernatant is cloudy, step 12 is repeated). Two volumes of cold EtOH are then added and the DNA precipitate spooled up on a glass rod. After pressing excess fluid from the spooled precipitate, it is dipped successively in 70%, 80%, and finally 90% EtOh, and the excess solvent is again pressed from the precipitate.

14. The precipitate is then shaken into 10 ml of 0.1 × SSC and dissolved by holding at 4°C over night and then if necessary warming it in a 37°C water bath.

Using *P. multimicronucleatum*, Barnett *et al.* (1972) developed techniques for whole-cell lysis and phenol extraction of labeled DNA and RNA for use in DNA–RNA hybridization experiments.

The DNA from certain symbionts of paramecia have been studied (Smith-Sonneborn *et al.*, 1963; Behme, 1964; Soldo and Reid, 1968; Gibson *et al.*, 1971).

2. Isolation and Characterization of RNA

Thus far, except for the special case of metagon RNA and new work on mate-killer particles (Baker, 1970), only rRNA has been successfully purified in *Paramecium*, and all studies, so far have dealt with *P. aurelia*.

Initially Gibson (Gibson and Beale, 1964; Gibson and Sonneborn, 1964) extracted ribosomal RNA after disrupting the organisms by one of three methods, i.e., freeze thawing, grinding in a mortar with a glass pestle, or shaking in sodium lauryl sulfate, followed by centrifugation at increasingly higher speeds until after spinning at 105,000 *g* for 90 minutes, a presumed "ribosome fraction" was obtained. Subsequently he (Gibson, 1967) followed the method of Barlow and Mathias (1966). This depends on phenol extraction of the ribosomal pellet obtained at 150,000 *g*.

Reisner *et al.* (1968) found that phenol extraction caused degradation of 18 S rRNA and therefore used the guanidinium chloride method of Cox (1966) with the following modification (Reisner, personal communication). When precipitated rRNA is redissolved in the 4 *M* guanidinium chloride, 10 m*M* EDTA solution, the temperature of the solution should be kept as cold as possible while still allowing the rRNA to dissolve. This is achieved by placing the preparative container in a beaker of 0°C water, which then is allowed to warm up. The 4 *M* guanidium chloride RNA preparation is seen to dissolve before the water bath reaches 6°C. [See also Reisner *et al.* (1972).]

Somerville and Sinden (1968) reported successful use of an orcinol technique developed by Schneider (1957) for measuring the amount of isolated rRNA.

3. DNA–DNA and DNA–RNA Hybridization

Gibson (1966, 1967) has pioneered in the use of nucleic acid hybridization techniques to study intergeneric homologies within the ciliated protozoans.

The technique is essentially that of McCarthy and his co-workers (McCarthy and Bolton, 1963; McCarthy and Hoyer, 1964). This is the method in which single-strand (denatured) DNA from one source is trapped in agar. Labeled, sheared, and denatured DNA is added. Hybridization between complementary sequences then takes place. Nonhybridized, labeled DNA is removed. The radioactivity is measured, thus allowing an estimate of the degree of homology (details in Gibson, 1966).

Barnett *et al.* (1972) used DNA–RNA hybridization techniques for testing the chronon theory of circadian time-keeping. This approach for analyzing biological clocks is a new approach to the problem, and many of the tech-

it and in comparison to related stocks. Given below are techniques not described in the Sonneborn and Margolin references.

a. Maintenance. The stability of a line showing a given antigenic type can be affected by a great variety of factors, and different lines may behave differently in response to the same factors.

A widely used physical factor affecting stability is temperature (Beale, 1952a; Sonneborn *et al.*, 1953; Margolin, 1956). Chemical factors, such as salt concentration (Beale, 1954) and nature of the culture medium (reviews by Beale, 1954, 1957), can play a role. Biological factors of importance include exposure to homologous antisera (Jollos, 1921; Skaar, 1956), growth rates (Sonneborn, 1950; Margolin, 1956); genetic constitution (Beale, 1952b; Sonneborn *et al.*, 1953); previous antigenic history of the line (Skaar, 1956), and release into the medium of as yet unidentified stabilizing factors (Finger *et al.*, 1967).

b. Transformation. Transfer of lines from one temperature to another can induce a change of antigenic type within a matter of hours, in some cases it takes longer—up to days or weeks. The direction of the change is predictable in many cases and can involve all cells in the culture (Sonneborn, 1947; Beale, 1952a, 1954). Also X-irradiation and exposure to ultraviolet light can induce nonmutational antigenic changes (Sonneborn, 1947; Sonneborn and Schneller, 1955). Exposing cells to a variety of chemical agents is also effective. Such agents include salts (Beale, 1954), proteolytic enzymes (Kimball, 1947; van Wagtendonk, 1951), antibiotics (Austin *et al.*, 1956, 1967a,b), and others. Biological factors that affect stability also relate to transformation. The references given above regarding exposure to homologous sera, growth rate, genetic constitution, prior history of a clone, and release of conditioning factors, all apply here, too.

c. Obtaining antisera. This section was contributed by I. Finger. Rabbits are still the mammal of choice for development of antisera. In addition to the basic techniques already in the literature (Sonneborn, 1950) the use of an adjuvant has been successfully employed (Finger and Heller, 1962). Here three injections are given subcutaneously or intramuscularly, 5 days apart, each consisting of 5.0 ml of equal volumes of the antigen and Freund's complete adjuvant (Difco Laboratories, Detroit). When titers have dropped to one fourth or less of the highest previous titer a booster injection of 2×10^5 cells plus adjuvant can be given and bleeding taken after 4 or 5 days. The adjuvant-antigen emulsion (total, 4–5 ml) can be injected into two or four footpads of a rabbit (Leskowitz and Waksman, 1960). A booster injection into the marginal ear vein can be given a month later and does not use adjuvant.

Drawing of blood and removal of serum are done by techniques given earlier (Sonneborn, 1950).

d. Antigens from Cell Suspensions. Concentrated cells can be freeze-dried in their culture medium and then left overnight in a 0.9% NaCl solution at 40°C. After centrifugation at 24,000 *g* for 2 minutes the supernatant is kept as an antigen source. Similarly, a concentrated culture can be frozen in dry ice or the deep freeze, and after thawing the resulting homogenate can be used as antigen source or purified further (Finger, 1956).

e. Ciliary Antigens. Finger (1956) and Preer and Preer (1959) have described methods for obtaining highly purified ciliary preparations. In outline, their procedures use the following steps. Concentrated cells (van Wagtendonk and Soldo, 1970; or references given above) are placed in a buffered salt solution (Preer and Preer also add ethanol) to remove the cilia. Mild centrifugation (700–900 *g*) leaves the cilia in the supernatant, which is carefully removed and purified by further centrifugation (24,000 *g*) and steps given in the next section.

There are now available detailed methods for assay, purification and determination of various properties of paramecium antigens.

f. Assay. Double diffusion in agar is the major technique applied by Preer and co-workers (Preer, 1956, 1959a,b, 1970; Preer and Telfer, 1957; Preer and Preer, 1959). Another approach is to determine absorption of the antibody by an antigen extract. The reduced ability of the antiserum to immobilize paramecia is then measured (Preer, 1959a; Jones, 1965). The former technique is preferred by Preer as "easier and much more precise."

Finger *et al.* (1972) have developed a radioimmunoassay of surface antigens. Labeling was achieved by feeding paramecia leucine-^{14}C-labeled *Escherichia coli*. The antigen is purified (Preer, 1959a,b; modified by Finger and Heller, 1962) and a quantitative precipitin technique is used to obtain materials that are counted in a Packard Tri-Carb liquid scintillation spectrometer. Not only does the technique allow quantitative determinations of very small amouts of antigen, but it also allows identification of closely related antigens.

A semiquantitative assay for both soluble and bound antigen is given by Macindoe and Reisner (1967). Here cell fractions are used to adsorb antibody, and then the remaining antibody is measured by its ability to immobilize paramecia. Used in conjunction with gel-diffusion techniques (detects soluble antigen), it is possible to determine the relative amounts of soluble and bound antigen.

g. Purification. Preer (1959b) isolated the antigenic material from ciliary preparations using the techniques just mentioned plus the added steps of centrifugation at low pH followed by fractional precipitation using ammonium sulfate. The purity of the material was checked by tests on its solubility, electrophoretic pattern, and behavior in the ultracentrifuge. In all tests it behaved as a single protein molecule (but see below for characterization of constituent parts). And, finally, tests of its antigenicity showed that the purified fraction was a potent stimulus to antibody production.

A further modification of this technique has been developed by Macindoe and Reisner (1967) and used subsequently (Reisner and Rowe, 1969) to isolate a very large polypeptide—"the largest presently known," having an estimated 2930 amino acid residues and a molecular weight of 301, 500 ± 4500. In a subsequent publication other, comparable proteins were also reported (Reisner *et al.*, 1969a,b).

h. Physical Properties. Present thinking (reviewed by Preer, 1968; Sommerville, 1970b) takes the view that a molecular weight of 310,000 is best present value for the size of the antigen molecule. Earlier, Preer (1959b), using sedimentation and diffusion techniques, had argued for a molecular weight of 240,000. And others, using the Archibold equilibrium method (Bishop, 1961) and a short column, low speed, equilibrium technique (Jones, 1965) reported similar values. The higher values comes from Yphantis high speed method used by Steers (1965), which provides a figure for a "zero concentration" and using this a more accurate estimate of the molecular weight is provided. The work from Reisner's group (cited in the previous section) agrees with these larger molecular weights.

As to the structure of the molecule itself, suggestions range from a dimer to a nonamer made up of three different types of chains, each of which is present in triplicate, i.e., $\alpha_3\beta_3\gamma_3$ (Jones, 1965; Steers, 1962, 1965). Here it has been necessary to break disulfide bonds and alkylate them and then collect the fragments. Both Jones and Steers have followed techniques used on material other than paramecium antigens and have adapted them as needed [Steers followed Anfinsen and Haber (1961); Jones followed Crestfield *et al.* (1963)]. In a convincingly critical review of the structure of six of the serotype antigens Reisner *et al.* (1969b) again conclude that all are single polypeptides and suggest that, since there is little evidence of helices in the molecules, "that they might resemble double bimolecular leaflets."

Furthermore, the monomer units have been digested—Jones (1965) and Steers (1965)—to their constituent amino acids, and the latter were identified by the means of the basic technique of fingerprinting by two-dimensional chromotography and electrophoresis (Katz *et al.*, 1959). Reisner *et al.* (1969b) carried out amino acid analyses of three serotypic antigens using a Beckman 120C amino acid analyzer.

i. Immunochemistry. Preer's double diffusion technique (see Section II, B,3 f) has been used to characterize antigens. Details of the method, its variations, and their applications were given by Finger (1970). These procedures probably provide the most sensitive method available for discovering different antigenic members present in a mixture and for identifying the pattern of cross-reaction of one antigen to different antibodies. Great care must be taken in determining the concentration of the reactants since this affects the position of the precipitate in the agar gel and care must also be taken in setting up and reading the gels [see especially the following papers: Finger (1964), Finger and Heller (1962, 1963, 1964), Finger *et al.* 1965, 1972); Macindoe and Reisner (1967)].

j. Cellular Localization. Various successful attempts have been made to determine the cellular location of the paramecium antigens. Work with homogenates and cell fractions shows that most of the antigen important in the immobilization reaction with antibodies is isolated from the cilia and other surface structures of the cell (Preer and Preer, 1959). But there is also a significant amount of soluble antigen (Seed *et al.*, 1964; Macindoe and Reisner, 1967). Direct observation by light microscopy showed that fluorescein-labeled antibody (Beale and Kacser, 1957; Beale and Mott, 1962) becomes adsorbed to cilia and surface membranes. The latter work was especially convincing since sections of cells were exposed to the antibody and its deposition was limited to surface structures. Further work using ferritin-conjugated antibodies and observing localization by electron microscopy (Mott, 1963, 1965) has been able to show the distribution of surface antigens in normal and in transforming cells.

4. *In Vitro* Protein Synthesis

It is now possible to obtain *in vitro* incorporation of amino acids into polypeptides (Reisner and Macindoe, 1967) and even into specific proteins, such as the immobilization antigens (Sommerville, 1970a). The first steps of the procedure are those needed to concentrate cells and homogenize them (see Section I,C,7). Reisner and Macindoe take this homogenate and centrifuge it at 3000 *g* for 15 minutes, decant the supernatant and centrifuge again at 10,000 *g* for 15 minutes. This second supernatant can be used for incorporation experiments. To this supernatant (10 parts) there is added 0.25 μM GTP (1 part), 10 μM creatine phosphate (1 part) 100 μg of creatine phosphokinase (1 part), and 0.5 μC labeled amino acid (1 part). Incubation at 35°C for 30 minutes is followed by addition of an equal volume of 10% (w/v) trichloroacetic acid. Samples are then prepared for counting as appropriate to the counting technique to be employed.

Sommerville's (1970a) technique starts with the supernatant obtained after initial homogenization and two rounds of centrifugation (2000 *g*, 10 minutes; 9000 *g*, 10 minutes) with retention of supernatants. Then a third round at 20,000 *g* for 20 minutes provides a supernatant of cell sap, ribosomes, and ribosomal aggregates. This supension was maintained at 0°C in a solution of 0.05 *M* Tris · HCl (pH 7.6), 0.025 *M* KCl, 0.005 *M* $MgCl_2$, and 0.005 *M* β-mercaptoethanol and was dialyzed against this solution for

2–4 hours before incubation as a cell-free system. Amino acids were used in a final isotope concentration of 1 μCi/ml. The assay for labeled antigenic material used either of two immunological procedures: autoradiography of immunoelectrophoretic arcs or precipitation of the antigen with specific antisera followed by scintillation counting.

III. Perspectives

Representatives of the genus *Paramecium* are surely among the most thoroughly explored eukaryote cells. Much of this is due to their large size and ease of culture, a good deal is due also to the special opportunities they provide for attacking specific research questions. It is not common to be able to carry out Mendelian analyses on behavioral mutants at the cellular level, or to separate cleanly nuclear and cytoplasmic factors in development and heredity, or to have the choice of working on single cells or mass cultures and to be able to induce nonrandom, phenotypic changes in either one. Despite this ability to conjoin molecular, fine structural, genetic, developmental, physiological, and behavioral techniques in one or many cells, there are still certain limitations that need to be recognized in the use of paramecia as experimental organisms.

In terms of their molecular biology it seems fair to say that the work has only begun. The work with immobilization antigens is at a stage in its technical refinement where *in vitro* analysis of control of protein synthesis is a reality. To extend this approach to other developmental problems, such as the control of events in the cell cycle may have to depend on synchronizing techniques to provide large numbers of cells at known interfission ages. So far all attempts at mass induction of cell synchrony in paramecia have failed.

Coupled with these considerations is that of axenic culture. Although a reliable medium is now available (van Wagtendonk and Soldo, 1970) for mass cultures, experience with it for single cells is not entirely encouraging (Hanson, unpublished). Single cell isolations may or may not survive, and the interfission period is close to 24 hours rather than the 6 hours found with a bacterized medium. Certain developmental problems that would be ideally studied with single cells in axenic medium are, therefore, not readily solved. However, further improvement of axenic culturing techniques is surely only a matter of time.

Looking to future studies, two comments seem appropriate. With the variety of approaches now available for studying paramecia, and ciliates in general, it seems that these cells might well be turned to certain technological ends. One such effect is that of Weiss and Weiss (1966), who examined the toxicity of cigarette smoke on paramecia. This work, in fact, showed that bubbling tobacco smoke through a medium to which paramecia were then exposed was toxic to them. To the degree that paramecia are representative of eukaryotes, they could be used as screening agents for various chemical substances suspected of being mutagenic or capable of inducing abnormal development (e.g., Hutner, 1964).

Furthermore, paramecia might also be used as monitors of pollution of inland waters. Noland and Gojdics (1967) discuss some of these problems, and Fauré-Fremiet (1967) discusses the ecology of Protozoa in broad, chemical terms. With regard to specific studies, Finley and McLaughlin (1961) have done pioneering work using the peritrich ciliates as indicators of pollutants. These indicate the possibilities in this field.

Acknowledgments

The following have contributed detailed information in preparation of various sections: A. Adoutte, Laboratoire de Génétique, Bot. 400, Faculté des Sciences, 91-Orsay, France; A. Tait, Institute of Animal Genetics, West Mains Road, Edinburgh EH9 3JN, Scotland. And many other others have contributed through informal advice and comment; to all of these I express my deepest thanks.

References

Adoutte, A., and Beisson, J. (1970). *Mol. Gen. Genet.* **108**, 70.

Adoutte, A., and Beisson, J. (1972). *Nature* (*London*) **235**, 393.

Allen R. D. (1968). *J. Protozool.* **15**, 7. (Abstr.)

Allen, R. D. (1970). Protists (Film Loop, 81–5928), Ealing Corp. Cambridge, Massachusetts.

Allen, R. D. (1971). *J. Cell Biol.* **49**, 1.

Allen, S. L. (1964). *J. Exp. Zool.* **155**, 349.

Allen, S. L., and Gibson, I. (1971). *J. Protozool.* **18**, 518.

Allen, S. L., and Weremiuk, S. L. (1971). *J. Protozool.* **18**, 509.

Allen, S. L., Byrne, B. C., and Cronkite, D. L. (1971). *Biochem. Genet.* **5**, 135.

Anderson, T. F. (1966). *In* "Physical Techniques in Biological Research" (A. W. Pollister, ed.), 2nd Ed., Vol. 3, Part A, pp. 319–387. Academic Press, New York.

Anderson, T. F., Preer, J. R., Jr., Preer, L. B., and Bray, M. (1964). *J. Microsc.* (*Paris*) **3**, 395.

Anfinsen, C. B., and Haber, E. (1961). *J. Biol. Chem.* **236**, 1361.

Antipa, G. (1971). *J. Protozool.* **18**, Suppl., 17.

Austin, M. L., Widmayer, D., and Walker, L. M. (1956). *Physiol. Zool.* **29** (4), 261.

Austin, M. L., Pasternak, J., and Rudman, B. M. (1967a). *Exp. Cell Res.* **45**, 289.

Austin, M. L., Pasternak, J., and Rudman, B. M. (1967b). *Exp. Cell Res.* **45**, 306.

Aylmer, C., and Reisner, A. H. (1971). *J. Gen. Microbiol.* **67**, 57.

Baker, R. (1970). *Heredity* **25**, 657.

Barlow, J., and Mathias, A. P. (1966). *In* "Procedures in Nucleic Acid Research" (G. L. Cantoni and R. D. Davies, eds.), pp. 444–454. Harper, New York.

Barnett, A., Wille, J. J., Jr., and Ehret, C. F. (1971). *Biochim. Biophys. Acta* **247**, 243.

Barnett, A., Ehret, C. F., and Wille, J. J., Jr. (1972). *In* "Biochronometry" (M. Menaker, ed.), pp. 637–651: Nat. Acad. Sci., Washington, D.C.
Beale, G. H. (1952a). *Science* **115**, 480.
Beale, G. H. (1952b). *Genetics* **37**, 62.
Beale, G. H. (1954). "The Genetics of *Paramecium aurelia*." Cambridge Univ. Press, London and New York.
Beale, G. H. (1957). *Int. Rev. Cytol.* **6**, 1.
Beale, G. H. (1969). *Genet. Res.* **14**, 341.
Beale, G. H., and Jurand, A. (1960). *J. Gen. Microbiol.* **23**, 243.
Beale, G. H., and Jurand, A. (1966). *J. Cell Sci.* **1**, 31.
Beale, G. H., and Kacser, H. (1957). *J. Gen. Microbiol.* **17**, 68.
Beale, G. H., and Mott, M. R. (1962). *J. Gen. Microbiol.* **28**, 617.
Beale, G. H., Knowles, J. K. C., and Tait, A. (1972). *Nature* (*London*) **235**, 396.
Behme, R. J. (1964). *Genetics* **50**, 235.
Beisson, J., and Sonneborn, T. M. (1965). *Proc. Nat. Acad. Sci. U.S.* **53**, 275.
Berger, J. D. (1969). Thesis, Indiana Univ. Bloomington, Indiana.
Berger, J. D. (1971). *J. Protozool.* **18**, 419.
Berger, J. D., and Kimball, R. F. (1964). *J. Protozool.* **11** (4), 534.
Bishop, J. M. (1961). *Biochim. Biophys. Acta* **50**, 471.
Boyde, A., and Wood, C. (1969). *J. Microsc.* **90**, 221.
Brutkowska, M. (1963). *Acta Protozool.* **1**, 71.
Butzel, H. M., and van Wagtendonk, W. J. (1963). *J. Gen. Microbiol.* **30**, 503.
Caro, L. G. (1964). *Methods Cell Physiol.* **1**, 327.
Carr, K. E. (1971). *Int. Rev. Suisse Zool.* **75**, 583.
Chatton, E., and Lwoff, A. (1930). *C. R. Soc. Biol.* **104**, 834.
Chatton, E., and Lwoff, A. (1935). *Arch. Zool. Exp. Gen.* **77**, 1.
Chen-Shan, L. (1969). *J. Exp. Zool.* **170**, 205.
Chen-Shan, L. (1970). *J. Exp. Zool.* **174**, 363.
Cooper, J. E. (1965). *J. Protozool.* **12**, 381.
Corliss, J. O. (1953). *Stain Technol.* **28**, 97.
Cox, R. A. (1966). *Biochem. Prep.* **11**, 104.
Crestfield, A. M., Moore, S., and Stein, W. H. (1963). *J. Biol. Chem.* **238**, 622.
Cummings, D. J. (1972). *J. Cell. Biol.* **53**, 105.
de Haller, G., Ehret, C. F., and Naef, R. (1961). *Experientia* **17**, 524.
Dippell, R. V. (1955). *Stain Technol.* **30**, 69.
Dippell, R. V. (1962). *J. Protozool.* **9** (suppl.), 24.
Dippell, R. V. (1963). *J. Cell Biol.* **19**, 20A.
Dippell, R. V. (1968). *Proc. Natl. Acad. Sci. U.S.* **6**, 461.
Dippell, R. V., and Sinton, S. E. (1963). *J. Protozool.* **10** *Suppl.*, 22.
Dryl, S. (1958). *Bull. Acad. Pol. Sci. Ser. Sci. Biol.* **6**, 429.
Dryl, S. (1961). *Bull. Acad. Pol. Sci. Ser. Sci. Biol.* **9**, 71.
Dryl, S., and Grębecki, A. (1966). *Protoplasma* **62**, 255.
Eckert, R., and Naitoh, Y. (1970). *J. Gen. Physiol.* **55**, 467.
Ehret, C. F. (1967). *J. Theor. Biol.* **15**, 263.
Ehret, C. F. (1960). *Cold Spring Harbor Symp. Quant. Biol.* **25**, 149.
Ehret, C. F., and de Haller, G. (1963). *J. Ultrastruct. Res., Suppl.* **6**, 1.
Ehret, C. F., and Powers, E. L. (1957). *J. Protozool.* **4**, 55.
Ehret, C. F., and Powers, E. L. (1959). *Int. Rev. Cytol.* **8**, 97.
Ehret, C. F., Savage, N., and Albinger, J. (1964). *Z. Zellforsch. Mikrosk. Anat.* **64**, 129.
Ehret, C. F., Barnes, J. H., and Zichal, K. E. (1973a). *In* "Chronobiology: Proc. Int. Soc. Study of Biol. Rhythms" (L. E. Scheving, S. Helberg, and J. E. Pauly, eds.), pp. 44–50. Igakushoin Ltd., Tokyo.

Ehret, C. F., Wille, J. J., and Trucco, E. (1973b). *In* "Biological and Biochemical Oscillators" (B. Chance, E. K. Pye, A. K. Ghosh, and B. Hess, eds.), pp. 503–512. Academic Press, New York.
Esteve, J. C. (1967). *C. R. Acad. Sci., Ser. D.* **265**, 1991.
Esteve, J. C. (1970). *J. Protozool.* **17**, 24.
Fauré-Fremiet, E. (1967). *In* "The Protozoa" (G. Kidder, ed.), Chemical Zoology, Vol. 1, pp. 21–54. Academic Press, New York.
Finger, I. (1956). *Biol. Bull.* **111**, 358.
Finger, I. (1964). *Nature* (*London*) **203**, 1035.
Finger, I. (1971). *In* "Methods in Immunology and Immunochemistry" (C. A. Williams and M. W. Chase, eds.), Vol. 3, pp. 138–146. Academic Press, New York.
Finger, I., and Heller, C. (1962). *Genetics* **47**, 223.
Finger, I., and Heller, C. (1963). *J. Mol. Biol.* **6**, 190.
Finger, I., and Heller, C. (1964). *Genet. Res.* **5**, 127.
Finger, I., Onorato, F., Heller, C., and Wilcox, H. B., III (1965). *Progr. Protozool., Proc. Int, Congr. Protozool.*, 2nd, *Excerpta Med. Found. Int. Congr. Ser.* No. **91**, p. 244.
Finger, I., and Heller, C. and Larkin, D. (1967). Genetics **56** 793.
Finger, I., Fishbein, G. P., Spray, T., White, R., and Dilworth, L. (1972). *Immunology* **22**, 1051.
Finley, H., and McLaughlin, D. (1961). *Progr. Protozool., Proc. Int. Congr. Protozool., 1st, Prague* pp. 308–320.
Fitzgerald, P. J. (1959). *In* "Analytical Cytology" (R. C. Mellors, ed.), 2nd Ed., pp. 381–429. McGraw-Hill, New York.
Flavell, R. A., and Jones, I. G. (1970). *Biochim. Biophys. Acta* **232**, 255.
Flickinger, C. J. (1966). *Methods Cell Physiol.* **2**, 311.
Frankel, J. (1969). *J. Cell. Physiol.* **74**, 135.
Fulton, C. (1971). *In* "Results and Problems in Cell Differentiation" (J. Reinert and H. Ursprung, eds.), pp. 170—221. Springer Publ., New York.
Gibson, I. (1964). *Proc. Roy. Soc., Ser. B* **161**, 538.
Gibson, I. (1966). *J. Protozool.* **13**, 650.
Gibson, I. (1967). *J. Protozool.* **14**, 687.
Gibson, I., and Beale, G. H. (1964). *Genet. Res.* **5**, 85.
Gibson, I., and Sonneborn, T. M. (1964). *Genetics* **50**, 249.
Gibson, I., Chance, M., and Williams, J. (1971). *Nature* (*London*), *New Biol.* **234**, 75.
Gill, K. S., and Hanson, E. D. (1968). *J. Exp. Zool.* **167** (2), 219.
Gillies, C., and Hanson, E. D. (1968). *Acta Protozool.* **6**, 13.
Grębecki, A. (1964). *Acta Protozool.* **2**, 69.
Hanson, E. D., and Kaneda, M. (1968). *Genetics* **60**, 793.
Hanson, E. D., and Ungerleider, R. (1973). *J. Exp. Zool.* **185**, 175.
Hanson, E. D., Gillies, C., and Kaneda, M. (1969). *J. Protozool.* **16** (1), 197.
Hayat, M. A. (1970). "Principles and Techniques of Electron Microscopy: Biological Applications." Vol. 1. Van Nostrand-Reinhold, Princeton, New Jersey.
Hayes, T. L., and Pease, R. F. W. (1968). *Advan. Biol. Med. Phys.* **12**, 85.
Horne, R. W. (1965). *In* "Techniques for Electron Microscopy" (D. H. Kay, ed.), 2nd Ed., pp. 311–327. Davis, Philadelphia, Pennsylvania.
Hufnagel, L. A. (1966). *Electron Microsc., Proc. Int. Congr., 6th, Kyoto* **2**, 235. (Abstr.)
Hufnagel, L. A. (1969). *J. Cell Sci.* **5**, 561.
Hutner, S. H. (1964). *J. Protozool.* **11**, 1.
Isaacks, R. E., Santos, B. G., and van Wagtendonk, W. J. (1969). *Biochim. Biophys. Acta* **195**, 268.
Jahn, T. L. (1961). *J. Protozool.* **8**, 369.
Jahn, T. L. (1967). *J. Cell. Physiol.* **70** (1), 79.

Jahn, T. L., and Bovee, E. C. (1967). *In* "Research in Protozoology" T.-T. Chen, ed.), Vol. 1, pp. 40–200. Pergamon, Oxford.
Jakus, M. A. (1945). *J. Exp. Zool.* **100**, 457.
Jakus, M. A., and Hall, C. E. (1946). *Biol. Bull.* **91**, 141.
Jakus, M. A., Hall, C. E., and Schmitt, F. O. (1942). *Anat. Rec.* **84**, 474.
Janisch, R. (1967). *Folia Biol.* (*Prague*) **13**, 386.
Jenkins, R. A. (1964). *J. Cell. Biol.* **23**, 46A.
Jenkins, R. A. (1970). *J. Gen. Microbiol.* **61**, 355.
Jennings, H. S. (1899). *Amer. J. Physiol.* **2**, 311.
Jennings, H. S. (1908). *J. Exp. Zool.* **5**, 577.
Jennings, H. S. (1923). "Behavior of the Lower Organisms." Columbia Univ. Press, New York.
Jennings, H. S. (1966). "Behavior of Lower Organisms" (Reprint). Indiana Univ. Press, Bloomington, Indiana.
Jensen, D. D. (1959). *Behaviour* **15**, 82.
Jerka-Dziadosz, M. (1968). *Acta Protozool.* **6** (35), 383.
Jollos, V. (1921). *Arch. Protistenk.* **43**, 1.
Jones, I. G. (1965). *Biochem. J.* **96**, 17.
Jurand, A., and Jacob, J. (1969). *Chromosoma* **26**, 355.
Jurand, A., and Preer, L. B. (1968). *J. Gen. Microbiol.* **54**, 359.
Jurand, A., and Selman, G. G. (1969). "The Anatomy of *Paremecium aurelia*." St. Martin's Press, New York.
Jurand, A., and Selman, G. G. (1970). *J. Gen. Microbiol.* **60**, 357.
Jurand, A., Beale, G. H., and Young, M. R. (1962). *J. Protozool.* **9**, 122.
Jurand, A., Beale, G. H., and Young, M. R. (1964). *J. Protozool.* **11**, 491.
Jurand, A., Rudman, B. M., and Preer, J. R., Jr. (1971). *J. Exp. Zool.* **177**, 365.
Kaneda, M., and Hanson, E. D. (1973). *In* "Paramecium: A Current Survey" (W. J. van Wagtendonk, ed.), pp. 221–265, Elsevier, Amsterdam.
Katz, A. M., Dreyer, W. J., and Anfinsen, C. B. (1959). *J. Biol. Chem.* **234**, 2897.
Kay, D. H. (1965). "Techniques for Electron Microscopy," 2nd Ed. Davis, Philadelphia, Pennsylvania.
Kennedy, J. R., Jr., and Brittingham, E. (1968). *J. Ultrastruct. Res.* **22**, 530.
Kimball, R. F. (1947). *Genetics* **32**, 486.
Kimball, R. F., and Barka, T. (1959). *Exp. Cell Res.* **17**, 173.
Kimball, R. F., and Perdue, S. W. (1962). *Exp. Cell Res.* **27**, 405.
Kimball, R. F., and Perdue, S. W. (1965). *Exp. Cell Res.* **38**, 660.
Kimball, R. F., and Prescott, D. M. (1964). *J. Cell Biol.* **21**, 496.
Kimball, R. F., Caspersson, T. D., Svenson, G., and Carlson, I. (1959). *Exp. Cell Res.* **17**, 160.
Kimball, R. F., Vogt-Köhne, L., and Casperson, T. D. (1960). *Exp. Cell Res.* **20**, 368.
Kinosita, H., and Murakami, A. (1967). *Physiol. Rev.* **47**, 53.
Kitching, J. A. (1956a). *Protoplasmatologia* **IIID**, 3a, 1.
Kitching, J. A. (1956b). *Protoplasmatologia* **IIID**, 3b, 1.
Kitching, J. A. (1967). *In* "Research in Protozoology (T.-T. Chen, ed.), Vol. 1, pp. 307–336, Pergamon, Oxford.
Klein, B. (1926). *Arch. Protistenk.* **56**, 243.
Knoch, M., and Konig, H. (1951). *Naturwissenschaften* **38**, 531.
Kozloff, E. N. (1964). *Carolina Tips* **27**, 9. (Carolina Biol. Supply Co., Burlington, North Carolina.)
Kung, C. (1969). *J. Cell Biol.* **43**, 75a.
Kung, C. (1971a). *Z. Vergl. Physiol.* **71**, 142.
Kung, C. (1971b). *Genetics* **69**, 29.

Kung, C., and Eckert, R. (1972). *Proc. Nat. Acad. Sci. U.S.* **69**, 93.
Kuźnicki, L. (1969). *Progr. Protozool., Proc. Int. Congr. Protozool., 3rd, Leningrad* p. 171. (Abstr.)
Kuźnicki, L. (1970). *Acta Protozool.* **8**, 83.
Kuźnicki, L., Jahn, T. L., and Fonseca, J. R. (1970). *J. Protozool.* **17**, 16.
Leboy, P. S., Cox, E. C., and Flaks, J. G. (1964). *Proc. Nat. Acad. Sci. U.S.* **52**, 1367.
Leskowitz, S., and Waksman, B. H. (1960). *J. Immunol.* **84**, 56.
Lowry, O. H., Rosebrough, N. J., Farr, A. L., and Randall, R. J. (1951). *J. Biol. Chem.* **193**, 265.
Luck, D. J. L., and Reich, E. (1964). *Proc. Nat. Acad. Sci. U.S.* **52**, 931.
Lwoff, A. (1950). "Problems of Morphogenesis in Ciliates." Wiley, New York.
McCarthy, B. J., and Bolton, R. J. (1963). *Proc. Nat. Acad. Sci. U.S.* **50**, 156.
McCarthy, B. J., and Hoyer, B. H. (1964). *Proc. Nat. Acad. Sci. U.S.* **52**, 915.
Machemer, H. (1969). *J. Protozool.* **16** (4), 764.
Machemer, H. (1970a). *Z. Naturforsch. B* **25**, 895.
Machemer, H. (1970b). *Acta Protozool.* **7**, 531.
Macindoe, H., and Reisner, A. H. (1967). *Aust. J. Biol. Sci.* **20**, 141.
Margolin, P. (1956). *Genetics* **41**, 685.
Metz, C. B., Pitelka, D. R., and Westfall, J. A. (1953). *Biol. Bull.* **104**, 408.
Millonig, G. (1961). *J. Appl. Phys.* **32**, 1637.
Miyake, A. (1964). *Science* **146**, 1583.
Moses, M. J. (1950). *J. Morphol.* **87** (3), 493.
Mott, M. R. (1963). *J. Roy. Microsc. Soc.* **81**, 159.
Mott, M. R. (1965). *J. Gen. Microbiol.* **41**, 251.
Muramatsu, M. (1970). *Methods Cell Physiol.* **4**, 195.
Murray, K. (1966). *J. Mol. Biol.* **15**, 409
Naitoh, Y., and Eckert, R. (1969). *Science* **164**, 963.
Naitoh, Y., and Eckert, R. (1972). *In* "Experiments in Physiology and Biochemistry" (G. A. Kerkut, ed.), Vol. 5, pp. 17–38. Academic Press, New York.
Noland, L. E., and Gojdics, M. (1967). *In* "Research in Protozoology" (T.-T. Chen, ed.), Vol. 2, pp. 215–266. Pergamon, Oxford.
Oatley, C. W., Nixon, W. C., and Pease, R. F. W. (1965). *Electron. Electron Phys.* **21**, 181.
Palade, G. E. (1952). *J. Exp. Med.* **95**, 285.
Párducz, B. (1952). *Ann. Histo.-Natur. Musei Nat. Hung.* **2**, 5.
Párducz, B. (1967). *Int. Rev. Cytol.* **21**, 91.
Pasternak, J. J. (1967). *J. Exp. Zool.* **165**, 395.
Pease, D. C. (1964). "Histological Techniques for Electron Microscopy," 2nd Ed. Academic Press, New York.
Pitelka, D. R. (1965a). *Progr. Protozool., Proc. Int. Congr. Protozool., 2nd, Excerpta Med. Found. Int. Congr. Ser. No.* **91**, p. 220.
Pitelka, D. R. (1965b). *J. Microsc.* (*Paris*) **4**, 373.
Pitelka, D. R. (1969). *Progr. Protozool., Proc. Int. Congr. Protozool., 3rd, Leningrad* pp. 44–46.
Potts, B. P. (1955). Biochim. Biophys. Acta **16**, 464.
Potts, B. P., and Tomkin, S. G. (1955). *Biochim. Biophys. Acta* **16**, 66.
Powers, E. L., Ehret, C. F., and Roth, L. E. (1955). *Biol. Bull.* **108**, 182.
Preer, J. R., Jr. (1956). *J. Immunol.* **77**, 52.
Preer, J. R., Jr. (1959a). *J. Immunol.* **83**, 385.
Preer, J. R., Jr. (1959b). *Genetics* **44**, 803.
Preer, J. R., Jr. (1968). *In* "Research in Protozoology" (T.-T. Chen, ed.), Vol. 3, pp. 127–278. Pergamon, Oxford.
Preer, J. R., Jr. (1969). *J. Protozool.* **16**, 570.

Preer, J. R., Jr. (1971). *In* "Methods in Immunology and Immunochemistry" (C. A. Williams and M. W. Chase, eds.), Vol. 3, pp. 225–229. Academic Press, New York.
Preer, J. R., Jr., and Jurand, A. (1968). *Genet. Res.* **12**, 331.
Preer, J. R., Jr., and Preer, L.B. (1959). *J. Protozool.* **6**, 88.
Preer, J. R., Jr., and Preer, L. B. (1967). *Proc. Nat. Acad. Sci. U.S.* **58**, 1774.
Preer, J. R., Jr., and Telfer, W. H. (1957). *J. Immunol.* **79**, 288.
Preer, J. R., Jr., Hufnagel, L. A., and Preer, L. B. (1966). *J. Ultrastruct. Res.* **15**, 131.
Raikov, I. B. (1968). *In* "Research in Protozoology" (T.-T. Chen, ed.), Vol. 3, pp. 1–128. Pergamon, Oxford.
Raikov, I. B., Cheissin, E. M., and Buze, E. G. (1963). *Acta Protozool.* **1**, 285.
Rao, M. V. N., and Prescott, D. M. (1967). *J. Cell Biol.* **28**, 281.
Rasmussen, L. (1967). *Exp. Cell Res.* **45**, 501.
Reisner, A. H., and Bucholtz, C. (1972). *Exp. Cell Res.* **73**, 441.
Reisner, A. H., and Macindoe, H. (1967). J. Gen. Microbiol. **47**, 1.
Reisner, A. H., and Macindoe, H. (1968). *J. Mol. Biol.* **32**, 587.
Reisner, A. H., and Rowe, J. (1969). *Nature* (*London*) **222**, 558.
Reisner, A. H., Rowe, J., and Macindoe, H. M. (1968). *J. Mol. Biol.* **32**, 587.
Reisner, A., Rowe, J., and Macindoe, H. (1969a). *Biochim. Biophys. Acta* **188**, 196.
Reisner, A. H., Rowe, J., and Sleigh, R. W. (1969b). *Biochemistry* **8**, 4637.
Reisner, A. H., Askey, C., and Aylmer, C. (1972). *Anal. Biochem.* **46**, 365.
Roth, L. E. (1958). *J. Ultrastruct. Res.* **1**, 233.
Rowe, J., Gibson, I., and Cavill, A. (1971). *Biochem. Genet.* **5**, 151.
Rundenberg, F. H. (1962). *Tex. Rep. Biol. Med.* **20**, 105.
Saito, H., and Miura, K. (1963). *Biochim. Biophys. Acta* **72**, 619.
Salpeter, M. M. (1966). *Methods Cell Physiol.* **2**, 229.
Schmitt, F. O., Hall, C. E., and Jakus, M. A. (1943). *Biol. Symp.* **10**, 261.
Schneider, L. (1959a). *Zool. Anz., Suppl.* **23**, 457.
Schneider, L. (1959b). *Z. Zellforsch. Mikrosk. Anat.* **50**, 61.
Schneider, L. (1960a). *J. Protozool.* **7**, 75.
Schneider, L. (1960b). *Naturwissenschaften* **47**, 543.
Schneider, L. (1961). *Protoplasma* **53**, 530.
Schneider, L. (1963). *Protoplasma* **56**, 109.
Schneider, L. (1964a). *Z. Zellforsch. Mikrosk. Anat.* **62**, 198.
Schneider, L. (1964b). *Z. Zellforsch. Mikrosk. Anat.* **62**, 225.
Schneider, W. C. (1957). *In* "Preparation and Assay of Substrates" (S. P. Colowick and N. O. Kaplan, eds.), Methods in Enzymology, Vol. 3, pp. 680–684. Academic Press, New York.
Schwartz, V. (1956). *Biol. Zentralbl.* **75**, 1.
Sedar, A. W. (1952). *Proc. Soc. Protozool.* **3**, 12.
Sedar, A. W., and Porter, K. R. (1955). *J. Biophys. Biochem. Cytol.* **1**, 583.
Seed, J. R., Shafer, S., Finger, I., and Heller, C. (1964) *Genet. Res.* **5**, 137.
Selman, G. G., and Jurand, A. W. (1970). *J. Gen. Microbiol.* **60**, 365.
Sinden, R. E. (1971). *J. Microsc.* **93**, 129.
Sjöstrand, F. S. (1967). "Electron Microscopy of Cell and Tissues," Vol. 1. Academic Press, New York.
Skaar, P. D. (1956). *Exp. Cell Res.* **10**, 646.
Skoczylas, B., Panusz, H., and Gross, M. (1963). *Acta Protozool.* **1**, 411.
Small, E. B., and Marzalek, D. S. (1969). *Science* **163**, 1064.
Smith-Sonneborn, J., and Plaut, W. (1967). *J. Cell Sci.* **2**, 225.
Smith-Sonneborn, J., Green, L., and Marmur, J. (1963). *Nature* (*London*) **127**, 385.

Soldo, A. T., and Reid, S. J. (1968). *J. Protozool.* **15**, *Suppl.*, 15.
Soldo, A. T., and van Wagtendonk, W. J. (1961). *J. Protozool.* **841**.
Soldo, A. T., Musil, G., and Godoy, M. (1970). *J. Bacteriol.* **104**, 966.
Sommerville, J. (1970a). *Biochim. Biophys. Acta* **209**, 240.
Sommerville, J. (1970b). *Advan. Microbial Physiol.* **4**.
Sommerville, J., and Sinden, R. (1968). *J. Protozool.* **15**, 644.
Sonneborn, T. M. (1947). *Growth* **11**, 291.
Sonneborn, T. M. (1948). *Proc. Nat. Acad. Sci. U.S.* **34**, 413.
Sonneborn, T. M. (1950). *J. Exp. Zool.* **113**, 87.
Sonneborn, T. M. (1951). *In* "Science in Progress" (G. A. Baitsell, ed.), 7th Ser. Ch. 5, p. 167. Yale Univ. Press, New Haven, Connecticut.
Sonneborn, T. M. (1963). *In* "The Nature of Biological Diversity" (J. M. Allen, ed.), pp. 165–221. McGraw-Hill, New York.
Sonneborn, T. M. (1970). *Methods Cell Physiol.* **4**, 241.
Sonneborn, T. M., and Schneller, M. (1955). *Rec. Genet. Soc. Amer.* **24**, 596.
Sonneborn, T. M., Ogasawara, F., and Balbinder, E. (1953). *Microbiol. Genet. Bull.* **7**, 27.
Steers, E., Jr. (1962). *Proc. Nat. Acad. Sci. U.S.* **48**, 867.
Steers, E., Jr. (1965). *Biochemistry* **4**, 1896.
Steers, E., Jr., Beisson, J., and Marchesi, V. T. (1969). *Exp. Cell Res.* **57**, 302.
Stevens, A. R. (1966). *Methods Cell Physiol.* **2**, 253.
Stevenson, I. (1967). *J. Protozool.* **14**, 412.
Stockem, W., and Wohlfarth-Bottermann, K. E. (1970). *Cytobiologie* **1**, 420.
Suhama, M., and Hanson, E. D. (1971). *J. Exp. Zool.* **177** (4), 463.
Suyama, Y. (1966). *Biochemistry.* **5**, 2214.
Suyama, Y., and Preer, J. R., Jr. (1965). *Genetics* **52**, 1051.
Tait, A. (1968). *Nature (London)* **219**, 941.
Tait, A. (1970a). *Nature (London)* **225**, 181.
Tait, A. (1970b). *Biochem. Genet.* **4**, 461.
Tartar, V. (1964). *Methods Cell Physiol.* **1**, 109.
van Wagtendonk, W. J. (1951). *Exp. Cell. Res.* **2**, 615.
van Wagtendonk, W. J., and Soldo, A. T. (1970). *Methods Cell Physiol.* **4**, 117.
Vivier, E., and Andre, J; (1961). *J. Protozool.* **8**, 416.
Walker, P. M. B., and Mitchison, J. M. (1957). *Exp. Cell Res.* **13**, 167.
Weiss, W., and Weiss, W. (1966). *Arch. Environ. Health* **12**, 227.
Wessenberg, H., and Antipa, G. (1970). *J. Protozool.* **17**, 250.
Wichterman, R. (1953). "The Biology of Paramecium." McGraw-Hill (Blakiston), New York.
Wille, J. J., Barnett, A., and Ehret, C. F. (1972). *Biochem. Biophys. Res. Commun.* **46**, 685.
von Gelei, J. (1932). *Arch. Protistenk.* **77**, 152.
Wohlfarth-Bottermann, K. E. (1950). *Naturwissenschaften* **37**, 562.
Wohlfarth-Bottermann, K. E. (1953). *Arch. Protistenk.* **98**, 169.
Wohlfarth-Bottermann, K. E. (1957a). *Z. Naturforsch. B* **12**, 164.
Wohlfarth-Bottermann, K. E. (1957b). *Naturwissenschaften* **44**, 287.
Wohlfarth-Bottermann, K. E. (1958). *Protoplasma* **50**, 82.
Wohlfarth-Botterman, K. E., and Schneider, L. (1961). *Strahlentherapie* **116**, 25.
Wolfe, J. (1967a). *Chromosoma* **23**, 59.
Wolfe, J. (1967b). *J. Cell Sci.* **6**, 579.
Woodard, J., Gelber, B., and Swift, H. (1961). *Exp. Cell Res.* **23**, 258.
Yusa, A. (1963). *J. Protozool.* **10**, 253.

Chapter 19

Methods of Cell Transformation by Tumor Viruses

THOMAS L. BENJAMIN

Department of Pathology, Harvard Medical School, Boston, Massachusetts

I. Introduction and Scope

Viruses of at least five major groups have now been implicated in naturally occurring or experimentally induced cancers of animals (Gross, 1970). A number of these viruses have been studied extensively in their interactions with animal cells in culture. This chapter will describe some of the methods of observation which have been used to detect and quantitate changes in cultured cells infected by known tumor viruses.

Except as they pertain to transformation studies, general virological and cell culture techniques will not be dealt with. Methods for transformation described in the original research literature will be found to vary according to the particular virus–cell system under study, and also for a given system as studied in different laboratories. The reasons for the variations in methodology are partly historical, but they also reflect the rapidly expanding knowledge of tumor viruses and new techniques which are directed toward particular aspects of the many changes still lumped together under the term "transformation." Here an effort will be made to generalize the methods as much as possible, emphasizing those that apply most broadly to the various systems and are based on common virus-induced cellular changes. The basic techniques can readily be adapted to serve particular needs of the investigator interested in various processes accompanying the conversion of normal to transformed cells, as well as in studying the behavior of the converted cells in physiological, biochemical, or immunological terms.

This chapter is intended, at best, to be useful as a guideline; omissions of sometimes useful and perhaps essential details are inevitable. Additional information can be sought in the original research literature, or in any of several books on methods, such as "Fundamental Techniques in Virology" (Habel and Salzman, 1969). "The Biology of Animal Viruses" (Fenner, 1968) and "The Molecular Biology of Tumor Viruses" (Tooze, 1973) offer many pertinent facts and interesting perspectives.

II. Materials and General Methods

Many of the commonly used viruses and cell strains can be obtained from the American Type Culture Collection or other biological supply company. Special strains, virus or cell mutants, etc., may be sought from individual investigators. Various tissue culture media with different mixtures of salts, vitamins, and amino acids are in common use and are available commercially as liquid concentrates or in powdered form; exact formulations are given by the suppliers or can be found in the original literature. Animal sera are also available commercially. To help ensure reproducibility over a reasonable period of time, it is advisable to obtain samples of several lots of sera from one or more suppliers and to test them in a particular system before ordering a large quantity.

Given the appropriate virus and cell materials, the basic manipulations in virus transformation experiments can easily be carried out in the standardly equipped tissue culture laboratory. Minimal essential pieces of equipment are a temperature-controlled humidified CO_2 incubator, a standard low-power inverted microscope (overall magnification in the range of 10–100×), clinical centrifuge (up to 500 gm), and 37°C and 45°C water baths. Special applications requiring additional equipment are fluorescence or time-lapse microscopy and autoradiography.

General laboratory operations involving work on viruses or cell cultures containing potentially harmful agents should be conducted with appropriate safety regulations. Guidelines for personnel as well as recommendations for special equipment and laboratory designs can be obtained from the Office of Biohazard and Environmental Control, National Institutes of Health; these are also discussed in "Biohazards in Biological Research" (Hellman, Oxman, and Pollack, 1973).

III. A Brief Survey of Tumor Viruses

The longest and perhaps best known examples of tumor viruses are found among the C-type RNA-containing group. Structurally, these viruses consist of an inner nucleoprotein core, containing approximately 10 million daltons of RNA, and an outer membrane envelope. Avian and murine viruses of this group are well established as agents causing various leukemias and sarcomas in their natural hosts. More recently C-type RNA tumor viruses from several other mammalian species have been found. Viruses of this class have indeed been found in most vertebrate species examined so far, although not all are associated with or been found capable of inducing cancer. Complex and as yet poorly understood genetic and immunological factors in the virus–host relationship may account for the fact that these viruses are found far more frequently than the neoplastic disease which they sometimes cause; viruses of this class which are fundamentally non-oncogenic may also exist.

In contrast to the RNA tumor viruses which form a single structural class, DNA-containing tumor viruses are known from several distinct families

TABLE 1

SOME TUMOR VIRUSES AND THEIR HOSTS[a]

Virus group (viral strain, disease)	Host of origin	Hosts susceptible to tumor induction by virus	Host cells susceptible to transformation by virus
I. RNA, C-type			
A. Avian leukosis–sarcoma complex	Chicken	Chicken, duck, pheasant, turkey, pigeon, rat, mouse, hamster, guinea pig, rabbit, monkey	Chicken, duck, pheasant, quail, turkey, rat, mouse, hamster, monkey, human
Rous sarcoma virus (RSV-Bryan, Schmidt-Ruppin, Carr-Zilber, Prague, Fujinami)			
Rous-associated viruses (RAV-1, RAV-2, RAV-50, etc.)			
Avian myeloblastosis virus (AMV)			
Avian lymphomatosis virus (RPL-12)			
Avian erythroblastosis			
Avian myelocytomatosis (MC-29)			
B. Murine leukosis–sarcoma complex			
Murine sarcoma virus (MSV-Moloney, Harvey, Kirsten)	Mouse	Mouse, rat, hamster	Mouse, rat, hamster, human
Mouse osteosarcoma			
Murine leukemia viruses			
Friend (erthroleukemia)			
Moloney (lympoid leukemia)			
Rauscher (erythroblastosis; lymphoid leukemia)			
Gross (lymphoid leukemia)			
Kirsten (erythroblastosis)			
Graffi (myeloid leukemias)			
C. Others: leukemias and sarcomas of cat, hamster, dog, guinea pig, cattle		(Species of origin and others)	(Species of origin and others)
II. RNA, B-type			
Mouse mammary tumor (Bittner) virus	Mouse	Mouse	

III. DNA Viruses			
A. Papova viruses			
Polyoma	Mouse	Mouse, rat, hamster, rabbit, ferret, guinea pig, Mastomys	Mouse, rat, hamster, bovine
SV40	Monkey	Hamster, Mastomys	Hamster, mouse, rat, pig, rabbit, human, monkey, bovine, guinea pig
Papilloma viruses: rabbit (Shope), dog, bovine, horse, human (warts)		(same as species of origin)	(Species of origin and others? reports of studies limited)
B. Adenoviruses			
Human adenoviruses: types 12, 18, 31 and others	Human	Hamster, rat, mouse, Mastomys	Hamster, rat, rabbit, human
Simian adenoviruses: SA7 and others	Monkey	Hamster, rat, mouse	Hamster
Avian adenoviruses (CELO)	Chicken	Hamster	
Bovine adenoviruses, type 3	Cattle	Hamster	
C. Herpes viruses[b]			
Lucké virus (frog renal carcinoma)	Frog	Frog, salamander	
Marek's disease virus (MDV) (neurolymphomatosis)	Chicken	Chicken, turkey	Chicken?
Herpes sylvilagus (lymphomas)	Rabbit	Rabbit	
Herpes Saimiri (HVS) (lymphomas)	Monkey	Monkey, rabbit	
Herpes ateles (HVS) (lymphomas)	Monkey	Monkey	
Equine herpes (EH-3)	Horse	Hamster	
Epstein-Barr virus (EBV) (Burkitt lymphoma; infectious mononucleosis)	Human		
Herpes simplex type 2 (cervical carcinoma)	Human	Hamster	Hamster
Nasopharyngeal carcinoma (?)	Human		
D. Pox viruses			
Rabbit myxoma-fibroma (Shope)	Rabbit	Rabbit	Rabbit
Squirrel fibroma virus	Squirrel	Squirrel	
Deer fibroma virus	Deer	Deer	
Molluscum contagiosum	Man	Man	
Yaba monkey virus	Monkey	Monkey	

[a] See Gross (1970).
[b] See Biggs *et al.* (1972).

possessing marked differences in size and structure. Examples of tumor viruses may in fact be found in each of the major classes of DNA animal viruses—papova, adeno, herpes, and pox. The first two classes of DNA viruses have a crystalline (icosahedral) structure. Papova viruses are among the simplest of all known animal viruses and have a DNA content of 3 million daltons, while the adenoviruses have DNA contents in the range of 20–30 million daltons. A variety of cancers are induced in laboratory animals by certain members of the papova and adeno groups; however, the oncogenic potential of these viruses is frequently not demonstrable in the natural host, i.e., the species in which the viruses are found. In contrast to the cancers induced by RNA viruses, which can readily be passaged by inoculations of cell-free filtrates of tumor tissue into normal animals, the cancers induced by papova and adenoviruses usually cannot be so passaged.

Herpes and poxviruses form the two major classes of enveloped DNA viruses. The herpes group has roughly 100 million daltons of DNA which replicates in the nuclei of infected cells. Poxviruses are the largest and most complex animal viruses, containing upward of 160 million daltons of DNA; these viruses replicate in the cytoplasm. Viruses belonging to these two groups have been found in association with naturally occurring cancers in a variety of animals; some of these cancers can be transmitted by passage of virus derived from tumors. A listing of some of the commonly known tumor viruses of the various classes and their hosts is given in Table I.

IV. Neoplastic Transformation

When cells are exposed *in vitro* to known tumor viruses, they may become "neoplastically transformed"—i.e., they may acquire the ability to form malignant tumors upon injection back into the original host. Neoplastic transformation of cultured cells is well established with various C-type, papova and adeno viruses. Neoplastic transformation of cultured cells has also recently been described for Herpes *simplex* type 2 (Rapp and Duff, 1972), although not so far for other known or suspected oncogenic herpes viruses. Transformations of cultured cells by viruses of the pox group have not been extensively reported; however, the rabbit fibroma virus has been reported to transform cultured rabbit kidney cells which form tumors that grow and then regress in the hamster cheek pouch (Hinze and Walker, 1964), perhaps reflecting the usually benign and regressing primary growths induced by the virus in rabbits.

The basic finding with several of the well known tumor viruses, reproducing the process of malignant alteration outside the animal, has led to further, and by now extensive, investigations into both the nature of the cellular changes accompanying neoplastic transformation and possible mechanisms of action by the virus. When compared to normal uninfected cells from which they derive, virus-induced neoplastically transformed cells manifest a variety of new properties. Changes are seen at various levels of cell structure and behavior, and these are detected by a corresponding variety of experimental techniques and methods of observation. A partial list of these changes is given in Table II.

Some of the newly acquired properties are easily discerned with a low power microscope—alterations in cell morphology, in the patterns of cell–cell interaction, and in cell growth behavior. The most interesting of the macroscopic changes are those which are manifested in loss of control of cell growth: the growth of virus-transformed cells becomes relatively insensitive to population density, to the presence of growth factors in serum, and to other conditions to which normal cells respond by ceasing mitosis. Other changes, occurring at the subcellular level, can be detected by appropriate genetic, immunological, or biochemical methods.

At present, there is no general agreement on which particular property or set of properties of virus-transformed cells underlies their malignant growth potential. Nevertheless, general support for the relevance of the *in vitro* alterations to cancer in the animal comes from the demonstrations of tumorigenicity of transformed cells, and conversely (though not as systematically studied) from the general similarity of behavior between virus-transformed cells and cultured cells derived from the corresponding primary virus-induced tumors.

Cells growing in culture may spontaneously undergo neoplastic transformation. They may also become transformed following exposure to X-irradiation (Borek and Sachs, 1966, 1967, 1968), to any of a variety of carcinogenic and mutagenic chemicals, such as benzopyrene, methylcholanthrene, or dimethylnitrosamine (Berwald and Sachs, 1965; Huberman *et al.*, 1968), or to extracts of smog (Freeman *et al.*, 1971). The *in vitro* growth properties of such transformed cells overlap to a large extent with those of virus-transformed cells. Viruses and chemicals (Todaro and Green, 1964a; Freeman *et al.*, 1970; Stich *et al.*, 1972) or viruses and X-rays (Stoker, 1963; Pollock and Todaro, 1968; Coggin, 1969) can act together to produce enhanced transformation frequencies. In general, however, viruses are the most rapid and efficient agents for inducing neoplastic transformation of cultured cells.

TABLE II

SOME CHANGES INDUCED IN CELLS BY KNOWN TUMOR VIRUSES

Class of change	Description
I. Loss of growth control	
(a) Tumorigenicity	Cells form tumors on injection into isologous host
(b) Density-dependent inhibition of growth	Cells grow past monolayer stage, form multilayers
(c) Growth in suspension; loss of anchorage dependence for growth	Cells divide in semisolid suspension media; can be readily adapted to liquid spinner culture
(d) Serum independence	Growth is relatively independent of factor(s) in serum
(e) Cloning	Cells form clones when seeded on top of normal cell monolayers
II. Cell movement	
(a) Contact inhibition	Cells lose contact inhibition of motion
(b) Wound serum requirement	Cells migrate from monolayer to empty area ("wound") with no (or low) serum requirement
III. Morphology	
(a) Cell shape	Changes depend on virus; transformed cells are generally less adhesive
(b) Cell growth pattern	Cells are randomly oriented and multilayered
IV. Nutritional and metabolic	
(a) Viability	Cell survival is more serum dependent
(b) Glycolysis	Cells usually have higher rates of glycolysis
(c) Uptake	Cells have enhanced uptake of sugars and possibly other metabolites
(d) Drug, hormone sensitivities	Various
V. Genetic	
(a) Karyotype	Cells usually become aneuploid
(b) Addition	Integration of viral genetic material
(c) Gene expression	Virus-specific RNA and protein synthesis; altered patterns of expression of cellular genes
VI. Immunologic	
(a) Intracellular	New virus-specific antigens
(b) Cell surface	New virus-specific and tumor-specific transplantation antigens; quantitative and qualitative changes in expression of cell-specified antigens
VII. Other surface changes	
(a) Agglutinability	Altered patterns of agglutinability by lectins such as wheat germ agglutinin and concanavalin A
(b) Membrane junctions	Decreased junctional coupling and intercellular passage of small molecules

V. Cells in Culture

The term "transformation" has most often been applied to changes in cultured cells brought about by tumor viruses; however, in the broadest sense, it refers to the acquisition of any heritable change by cells in culture. Cells taken from an animal and put into culture invariably undergo change. In the absence of viral infection, changes can occur in cell morphology, karyotype, various biosynthetic capabilities, and most important, in the various parameters of cell growth control used to characterize virus transformed cells, including the acquisition of tumorigenicity. The tendency for cells to vary spontaneously and progressively while growing in culture raises obvious uncertainties in assessing viral transformation, both with respect to the role of the virus in observed changes and to the relationship of the *in vitro* virus–cell interaction to that occurring in the animal. The nature of the variants and the frequency with which they arise and are maintained in populations of cultured cells depends to a considerable degree on the cells' history in culture, the general culture conditions and manner of handling of the cells. In order to assess the changes induced by viruses, it is therefore necessary to know the nature of the spontaneous variations which may occur and the extent to which they can be controlled.

Two kinds of cell culture are used in assaying transformation by tumor viruses—short-term cultures and permanent cell lines. For short-term culture, cells are derived usually from embryonic tissues and grown for a limited number of cell generations before being exposed to virus. Cultures of this type can be maintained through a number of serial low-dilution passages (usually 1:2 to 1:6), during which fibroblastic cells tend to predominate. Typically, there is little cell death after the first few passages; surviving cells grow rapidly and are maximally sensitive to viral infections. Such low passage cultures of chicken and mouse embryo cells are commonly used to assay transformation by avian (Rous) and murine sarcoma viruses. After about 10–15 passages, or more depending on the culture conditions and species of cell, the rate of cell division slows and the cells enter a degenerative phase (Hayflick and Moorhead, 1961; Saksela and Moorhead, 1963). A majority of the cells fail to survive further passage. With variable frequency, cells may survive this period of "crisis" and become permanently adapted tissue culture lines. However, cells passing through crisis often fail to maintain the normal diploid chromosome constitution; at this point, or after further passage, other changes can occur leading to "spontaneous" neoplastic transformation. Permanently adapted cell lines must therefore be thoroughly characterized with respect to the particular properties being assayed in viral transformation experiments. Two permanent lines, the

3T3 line from mouse embryo (Todaro and Green, 1963) and the BHK line from baby hamster kidney (Macpherson and Stoker, 1962), are widely used in transformation studies with the papova viruses, polyoma, and SV40, as well as murine sarcoma virus.

Given access to the appropriate animal material, the routine (weekly or biweekly) preparation of short-term cultures has the distinct advantage over the use of continuous cell lines of providing essentially normal cells in amounts more than adequate to meet usual requirements, and without risk of having accumulated spontaneous transformants of the kind and number which would interfere with viral transformation assays. Although more normal than cells of established lines, cells in short-term culture do, nevertheless, undergo detectable physiological transformations compared to their cells of origin *in vivo*. Conditions of culture induce stable physiological changes in newly cultured cells resulting in the gain (Holland and Hoyer, 1962), or loss (Weiler, 1959; Holtzer *et al.*, 1960) of specific functions. Furthermore, growth rates and general viability vary with age in culture (Todaro and Green, 1963; Jensen *et al.*, 1963). A potential disadvantage of short-term cultures resides in their lack of homogeneity; despite the immediate and strong selection imposed by placing cells into culture, such cell populations are not necessarily homogeneous with respect to cell type or tissue of origin; furthermore, since nonestablished cells grow poorly at high dilution, they cannot effectively be cloned to obtain homogeneity. Heterogeneity may exist in uncloned cell populations with respect to virus susceptibility and expression of transformed cell properties; this may pose difficulties in detection and in quantitative analyses of certain kinds of experiments; however, such factors if present are usually not of sufficient weight to interfere with routine transformation assays. Genetic differences among embryos affecting virus susceptibility, or the variable presence of interfering viruses, are possible factors that must be taken into account in the routine preparation of short-term cultures.

Cells of established lines differ in at least two important respects from cells of short-term cultures: first, since normal cells taken from the animal have only a finite life-span in culture, the ability of established cell lines to grow continuously in culture constitutes, per se, a transformation in growth properties; and second, they have usually undergone heteroploid transformation. These two changes aside, it has been possible to derive continuous cell lines which retain aspects of normal growth regulation and which are nontumorigenic, and these have been widely used to study the loss of growth regulation following infection by tumor viruses.

The manner in which continuous lines are derived affects their properties of growth regulation once they become established. As first shown by Todaro and Green (1963), continuous passage of mouse embryo cells in the absence of extensive cell-cell contacts leads to established cells which show strong sensitivity to contact inhibition of growth. This was achieved by means of a subculture routine in which 3×10^5 cells were transferred to 50-mm diameter petri dishes and allowed to grow for only 3 days; at each transfer cells were again diluted, new cultures inoculated with 3×10^5 cells/dish, and incubated for 3 more days. Under these conditions, cells were continuously maintained at subconfluent densities. The 3T3 line of cells which emerged on this routine grew rapidly in the absence of cell–cell contact, but ceased dividing when a complete monolayer formed. Under similar subculturing routines, but using inocula of 6 or 12×10^5 cells at each transfer, lines 3T6 and 3T12 were obtained; these lines, selected for growth in the presence of cell–cell contacts, failed to show the sensitivity to contact displayed by 3T3 and grew to correspondingly higher saturation densities. Aaronson and Todaro (1968) studied the tumorigenicity of lines derived from inbred Balb/c mice by the 3T3, 3T6, and 3T12 routines, and showed a positive correlation between saturation density *in vitro* and tumorigenicity. Furthermore, the nontumorigenic Balb 3T3 line could be transformed by SV40 or mouse sarcoma virus; these transformants grew in culture to high saturation densities and were tumorigenic in Balb/c mice.

The use of continuous cell lines with normal growth regulation has obvious advantages over short-term cultures for viral transformation studies. Although the properties of growth control can easily be lost through spontaneous variation and selection (if, for example, 3T3 cells are allowed to remain confluent for long periods of time), the cells of established lines have a high efficiency of cloning and can be subcloned repeatedly to ensure maximum homogeneity with retention of the original desirable properties. The use of established lines eliminates the need for a steady source of animal material, and at the same time allows for the indefinite propagation of any number of transformed derivatives along with their matched controls. With perhaps minor adjustments in technique, cell lines can be derived from other desired species or particular inbred strains of animal for use in viral transformation experiments. A notable exception occurs, however, with avian cells which have never been successfully adapted to continuous culture despite repeated efforts.

Cell Freezing

To have a supply of cells of an established line over an extended period of time and to minimize the effects of variation in cell properties, it is advisable to freeze cells in a large number of aliquots; samples of frozen cells of a given passage number can be thawed periodically and used for a defined number of passages. To freeze, cells are suspended in growth medium containing serum and 5–10% of either dimethyl sulfoxide or glycerol. Aliquots of 1–2ml of cell suspension at 10^6 to 10^7 cells/ml are placed in small ampoules and sealed. Freezing must take place slowly. This may be done conveniently by placing the ampoules in a beaker with ethanol at room temperature and transferring the beaker to a −70°C (or lower) freezer overnite. Cells remain viable at these temperatures for several months to a year, or even longer if transferred and maintained in liquid nitrogen. To thaw, the ampoules are placed directly in water at 37°C and shaken gently. The cells are transferred to a culture dish and growth medium is added; the cultures can be fluid changed to remove dimethyl sulfoxide or glycerol as soon as cells are attached, usually after 6–12 hours.

VI. General Properties of Transformed Cells

Five general properties reflecting alterations in regulation and patterns of cell growth have been particularly useful in detecting, isolating, and characterizing cells transformed by tumor viruses: (1) morphological alterations, (2) decreased density dependent inhibition of growth, (3) decreased serum requirement for growth, (4) ability to grow in semisolid suspension cultures, and (5) tumorigenicity. Changes in one or some combination of the first four properties form the bases for essentially all viral transformation assays with DNA and RNA viruses.

These properties are not entirely independent of one another; distinctions between them reside more in the operational methods involved than in their being manifestations of independent and fundamentally different alterations. Virus-transformed cells often acquire several and sometimes all these properties together. In certain cases, however, it is possible with appropriate methods to select for changes in one property with little or no change in the others. It is also clear from studies on reversion of transformed cells that various properties may be lost as well as acquired independently (see Vogel and Pollack, 1973). The properties themselves are not discrete or "all or none"; they are quantitative with ranges of more or less continuous variation. Independently derived transformed cell lines, produced by infection of the same cells with a given virus and selected for a given property may when tested for other properties display them to different degrees. (see Risser, 1973).

A. Morphological Alterations

Morphological changes accompanying viral transformation may be discerned both at the level of single cells and at the level of multicellular patterns of growth which emerge from the interactions between cells. Figure 1 shows examples of such changes in cells of several species transformed by DNA or RNA viruses. Normal fibroblasts of either short-term cultures or established lines typically form monolayered growth patterns in which the cells become aligned, the long axes of neighboring cells lying largely parallel to one another. Transformed cells, in contrast, form multilayered growth patterns and the cells are randomly oriented with respect to their neighbors. Changes in cell shape accompanying these alterations in growth pattern may

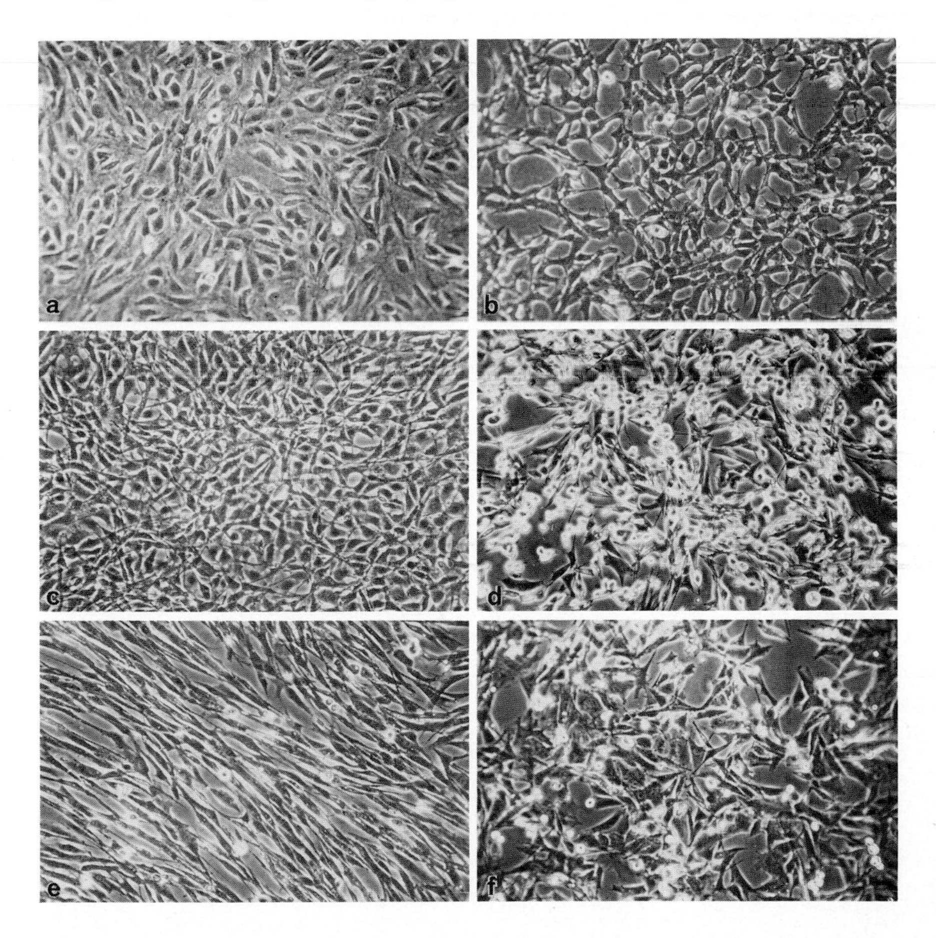

FIG. 1. Growth patterns and morphological changes in normal and transformed established cell lines. (a) Balb-3T3 cells, nearly confluent; (b) Balb-3T3 transformed by polyoma virus at cell density similar to Balb-3T3 in (a); (c) same as (b), 24 hours later; (d) Balb-3T3 transformed by mouse sarcoma virus; (e) BHK cells nearly confluent; (f) BHK cells transformed by polyoma virus; (g) $RECL_3$; (h) $RECL_3$ transformed by polyoma virus; (i) 3T3 (Swiss), confluent; (j) 3T3 transformed by polyoma virus, low density; (k) same as (j), high density. For (g–k) see p. 380. Phase contrast, ×75.

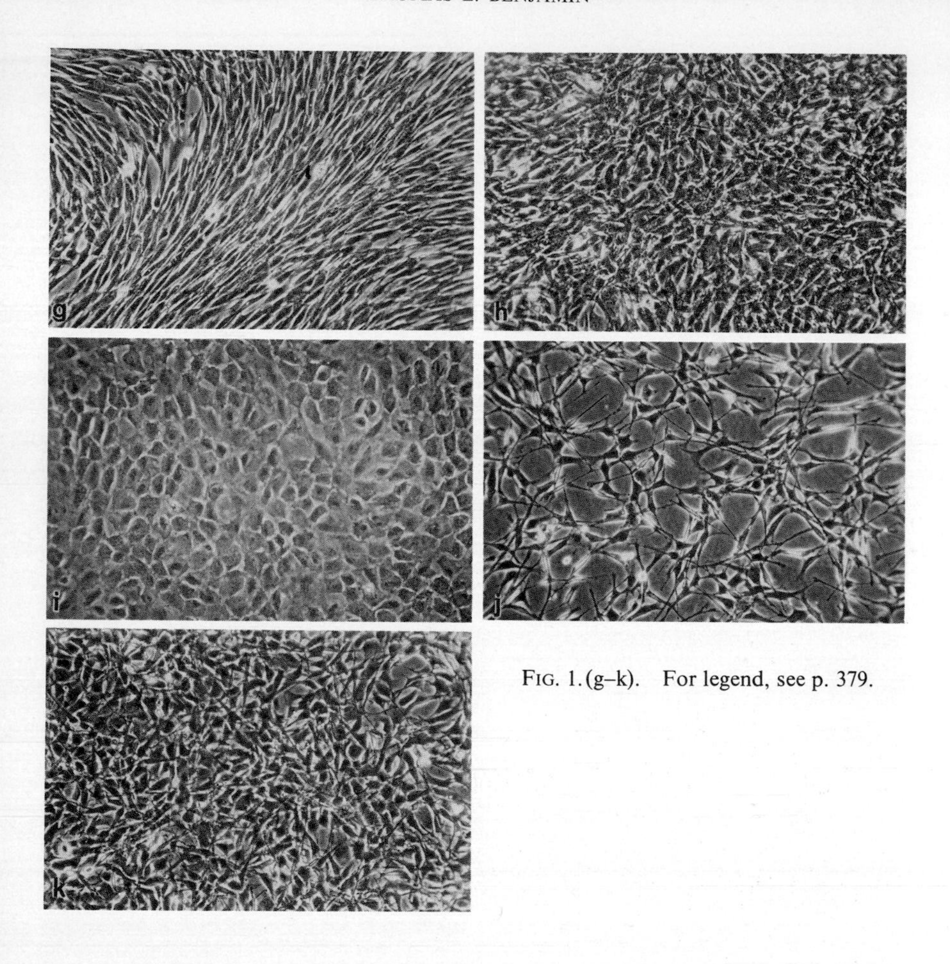

FIG. 1.(g–k). For legend, see p. 379.

be more or less pronounced, depending on the cell and transforming virus.

BHK and $RECl_3$ are established lines of fibroblasts from hamster and rat, respectively, which show the normal pattern of growth (Fig. 1 e and g). When transformed by polyoma virus, they lose their parallel alignment and produce a disordered and multilayered growth (Fig. 1 f and h). The 3T3 lines from Balb or Swiss mice at subconfluent cell densities show elongated or fusiform cells; as they reach confluence, their shape tends to become more polygonal or epithelioid, producing a "cobblestone" appearance (Fig. 1 a and i). Polyoma transformants of these lines retain their fibroblastic shapes as they grow past confluency, producing crisscross multilayered patterns (Fig. 1 b, c, j, and k). Balb 3T3 cells transformed by mouse sarcoma virus also pile up, and the cells tend to be more rounded (Fig. 1d). The changes in morphology and growth pattern observed in mass cultures of normal and transformed cells can be discerned readily in isolated colonies (Fig. 2A). Colony morphology has been used in clonal analysis and assays of transformation of continuous lines, such as BHK or other susceptible fibroblasts by polyoma, as well as 3T3 by SV 40 (see Section VIII). Morphological criteria have also been used to distinguish between poloma and SV 40 transformants of 3T3 cells (Todaro *et al.* 1965a).

Changes in cell shape and growth patterns form the bases of the traditional focus assays for avian and murine sarcoma viruses (see Section X). Foci of altered cells easily stand out against the background of normal cells (Fig. 2B). Such foci result both from the virus-induced alteration of cell shape and subsequent proliferation of the altered cells. Whereas a typical focus assay requires about a week, morphological changes in the initially infected cells are complete in about one day. A rapid transformation assay based solely on morphological changes without cell proliferation has been described for Rous sarcoma virus (Hanafusa, 1969). Within 24 hours after

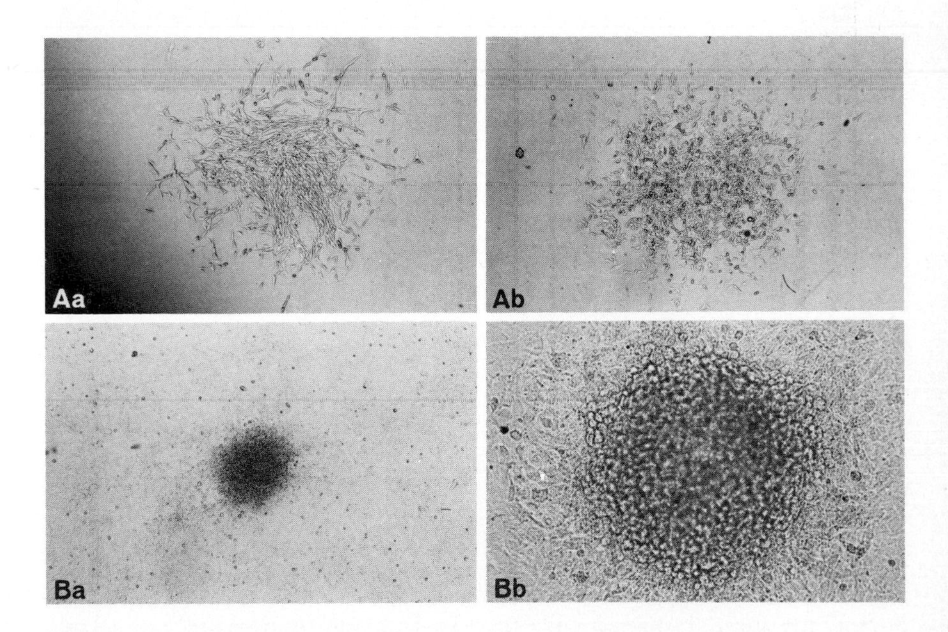

FIG. 2. Tranformation Assays. (A) Colony morphology: (a) colony of normal BHK; (b) colony of polyoma-transformed BHK. (B) Focus assay: (a) and (b) Rous sarcoma virus-induced focus on chick embryo fibroblasts. (C) Agar suspension assay: (a) 3T3; (b) polyoma-transformed 3T3; (c) normal $RECl_3$; (d) polyoma-transformed $RECl_3$, (e) polyoma-transformed BHK. For (C), see p. 382. (A) ×18; (B): (a) ×18; (b) ×75; (C) ×75.

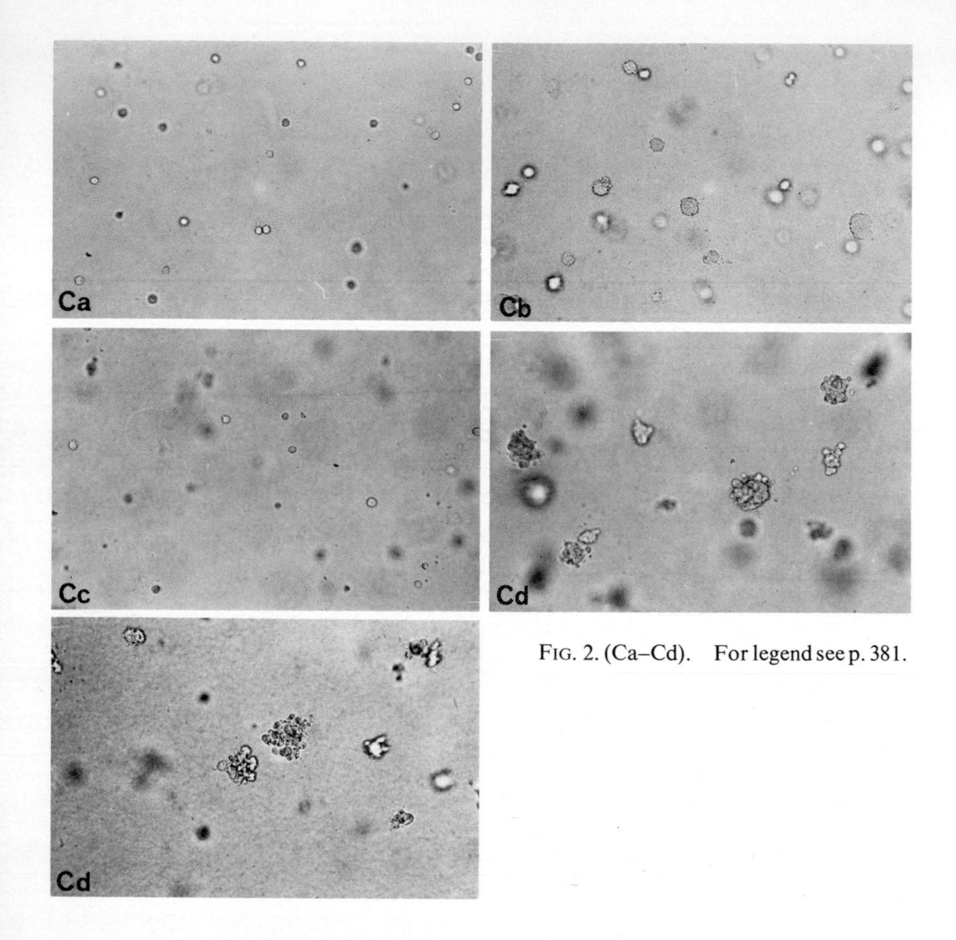

FIG. 2. (Ca–Cd). For legend see p. 381.

infection of chick embryo fibroblasts by Rous sarcoma virus, there is alteration in cell morphology and disruption of the normal pattern of cell orientation. Beginning 14–16 hours after infection, cells begin to lose their elongated fusiform appearance and to assume a more rounded spindle shape. Decreased adhesion to the culture dish is also characteristic of these altered cells. Polyoma infection of confluent BHK cells leads to pronounced alterations in morphology as indicated by the progressive change from predominantly bipolar shaped cells to cells with multiple pseudopodia; this change occurs in over 80% of the cells within 24 hours after infection as determined by counting pseudopodia of cells recorded in time-lapse films (Stoker *et al.*, 1972).

Mutational differences in Rous sarcoma virus can affect the morphology of the transformed cell (Temin, 1960; Martin, 1971); in chick embryo fibroblasts, morph-r strains of virus produce round or spindle-shaped cells, and morph-f strains produce elongated fibroblasts which form bundles of refractile cells. Transformation of epithelial cells from chick iris by the two kinds of "morph" mutants again show a direct affect of the virus in controlling cell shape (Ephrussi and Temin, 1960). Viral genotype does not act entirely independently of cell type, however, since infection of fibroblasts of different tissue or species origin with a given virus strain produces transformed cells with detectably different morphologies (Temin, 1960). It is also clear that cells derived from a single embryo and infected with a given strain of virus produce a minority of foci in which the cells show different morphologies. Since the virus recovered from these foci is not genetically altered, the morphological variations are most likely due to differences in the population of cells prior to infection (Hanafusa, unpublished).

Cells of established lines can undergo spontaneous neoplastic transformation while maintaining normal growth morphology (Defendi *et al.*, 1963; Sanford *et al.*, 1967). Thus morphological transformation of cells by viruses, although usually accompanied by neoplastic transformation, is not, per se, a necessary condition for tumorigenicity. Highly tumorigenic variants have been shown to be present in low frequency in populations of "normal" BHK cells (Defendi, 1963, Jarrett and Macpherson, 1968); such variants have flat parallel growth patterns and can be further transformed by polyoma virus with respect to morphological and other growth criteria.

Associated Changes

The alterations in growth pattern and morphology used to detect cell transformation by virus clearly represent composite changes, and these can be broken down into more discrete and operationally distinct terms. For example, the role of contact in regulating locomotory behavior of normal and transformed cells has been extensively studied by time-lapse and interference photomicroscopy; results of these studies have been described in a series of publications by Abercrombie and his colleagues and discussed by Martz and Steinberg (1973). The loss of contact inhibition of cell movement which accompanies viral transformation would, by itself, be expected to result in the observed change in cell orientation from parallel to crisscross in confluent or near confluent cultures of normal and transformed cells (Abercrombie and Ambrose, 1958; Abercrombie and Heaysman, 1954; Abercrombie, 1962). A separate aspect, to be discussed in detail below, determines the change from monolayering to multilayering, and has been described in terms of the loss in transformed cells of density-dependent inhibition of cell division (Stoker and Rubin, 1967).

Relative strengths of cell–cell and cell–substrate adhesion underlie such apparent morphological changes as the rounding and partial detachment from the monolayer in cells transformed by the RNA sarcoma viruses, and in the loss of the highly flattened and spread appearance of 3T3 cells when transformed by papova viruses (see Figs. 1 and 2). Studies directed specifically toward the formation of cell–cell adhesions by cells in suspension have shown a decreased ability on the part of polyoma-transformed BHK cells to form such adhesions compared to normal BHK cells (Edwards and Campbell, 1971; Edwards *et al.*, 1971). In addition to affecting cell shape, the differences in adhesive properties between normal and transformed cells may also play a role underlying differences in locomotory behavior, overlapping, and relative tendencies toward monolayering and multilayering of cells (see Martz and Steinberg, 1972).

B. Decreased Density-Dependent Inhibition of Growth

Local cell density affects both the speed of movement (Martz, 1972) and the growth rate of normal cells (Zetterberg and Auer, 1970; Martz and Steinberg, 1972). As growth in a normal cell monolayer approaches confluency, there is a rapid decrease in these parameters as well as a falling off in DNA and net RNA and protein synthesis (Levine *et al.*, 1965). The ability of normal cells to regulate their numbers in culture and to achieve a stable cell density is usually lost in virus-transformed cells. The parameter used to measure this regulation of cell number is the "saturation density" or "stationary density"; it is obtained directly by removing cells from the monolayer with trypsin or Versene (EDTA), counting in a hemacytometer, and dividing the total number of cells by the surface area available for cell growth. The numbers obtained in this way are constant for normal cells under defined conditions of growth and media renewal, but are strongly dependent on the concentration and frequency of renewal of serum; continuous perfusion of normal cells, for example, leads to pronounced multilayering (Kruse and Miedema, 1965). For transformed cells, the terms are somewhat misleading, since a stable saturation density is not always attained. Transformed cells may continue to grow for several generations past the saturation density of their normal counterparts under the same conditions; although the growth rate may slow somewhat, the cells will often slough and die without having attained a steady saturation density. The measurements as defined operationally are nevertheless valid in demonstrating the ability of transformed cells to continue dividing after confluency is reached.

Some saturation density values of normal and various virus-transformed cells of the 3T3 lines are shown in Table III. The conditions employed here were to seed 60-mm petri dishes with 2×10^5 cells in 6 ml of minimal essential

TABLE III

SATURATION DENSITIES OF NORMAL AND VIRUS TRANSFORMED 3T3 CELLS[a]

Cell line	Number of cells/cm^2 × 10^{-4}
3T3 (Swiss)	1.7
Py-3T3 (Swiss)	17.4
Balb-3T3	2.3
MSV-Balb-3T3	14.9
SV-Balb-3T3	13.7
RSV-Balb-3T3	16.5

[a]See Section VI,B for conditions.

medium (MEM) with 10% calf serum and 15 m*M* Hepes buffer (see below). Cells were removed and counted 5 days later with no intervening fluid change; subsequent counts taken on days 6 and 7 showed no further increase in cell numbers. It is clear that viral transformants grow to cell densities 5- to 10-fold higher than their normal counterparts under these conditions. Estimates in the literature, using repeated fluid changes or other modifications in procedure, vary in the absolute values but give similar differences in saturation density between normal and transformed cells. Differences in cell size can be a factor in comparisons of saturation density between normal and transformed cells; when such differences exist, it is useful to measure the density in terms of total cell mass (protein or RNA) per unit surface area rather than cell number.

Another valid measure of density-dependent inhibition of growth is the cell (or mass) doubling time as a function of cell density. Normal and transformed cells can be maintained for a number of passages at widely different initial cell densities, and the average doubling time measured (Todaro *et al.*, 1964). The results of this type of experiment show that normal and transformed cells have nearly equal doubling times at subconfluent densities (in the range of 18–24 hours for less than 10^4 cells/cm^2). The doubling time increases sharply for normal cells growing near or at confluency while remaining nearly the same as or only slightly longer than the subconfluent rate for transformed cells. Normal cells in confluent monolayers are arrested in the G-1 phase of the cell cycle (Nilhausen and Green, 1965). Density-dependent inhibition of growth is therefore reflected in declines in rates of DNA synthesis and mitosis as cells reach confluency; direct estimates of mitotic indices or autoradiographic determinations of numbers of labeled nuclei after exposure to thymidine-^{3}H provide alternate and useful ways of studying effects of cell density on normal and transformed cells. Normal

3T3 cells in stationary phase have low but measurable "turnover rates" measured by autoradiography of DNA-synthesizing cells; serum level and frequency of serum (media) changes have a pronounced effect (Todaro *et al.*, 1965b).

For cells of established lines and their transformed derivatives, the increase in saturation density is generally a good index of tumorigenicity (Aaronson and Todaro, 1968). Phenotypic revertants of SV-40 transformed 3T3 cells which have regained density-dependent inhibition of growth have also been shown to be less tumorigenic (Pollack *et al.*, 1968; Pollack and Teebor, 1969). When 3T3 cells are inoculated into petri dishes at cell densities in excess of their saturation density, there is no net growth, and the artifically high saturation density is not maintained; most of the cells initially attach, forming multilayers; cells are then gradually lost over a period of days until the culture approaches its normal saturation value (Todaro *et al.*, 1964). If normal 3T3 cells are added to confluent monolayers of the same cells, they will not form clones; spontaneous or viral transformants of 3T3, having higher saturation densities, are able to form colonies on 3T3 monolayers. This parameter of efficiency of cloning on monolayers of normal cells also correlates well with tumorigenicity (Aaronson and Todaro, 1968).

Possible Mechanisms

The mechanism of density-dependent inhibition of growth is not clear. Factors such as exhaustion of nutrients from the media, cell–cell contact, and utilization of serum factors have been considered. Nutrient depletion is not a major factor under normal conditions. This is shown by the fact that media removed from confluent stationary 3T3 cells can support the growth of subconfluent cells (Todaro *et al.*, 1964). The role of contact in regulating cell division has been much discussed, and the phenomenon of density-dependent inhibition has often been referred to as contact inhibition of growth. Time-lapse studies (Martz and Steinberg, 1972) of 3T3 cells indicate that neither a high level of cell-cell contact experienced by cells in G-1 nor the local cell density environment necessarily prevent a subsequent round of cell division. The effects of contact and density on cell division may be exerted with a delay of approximately one generation time. On the other hand, contact does appear to lead directly to a reduction in the speed of normal cells (Martz, 1972). Although "contact inhibited" monolayers of 3T3 show little nuclear overlap, electron micrographs of vertical sections through such monolayers show evidence of extensive overlapping of cytoplasm. This contrasts with similar pictures of transformed cells which show multilayers with extensive nuclear as well as cytoplasmic overlap (Todaro *et al.*, 1964). Thus the amount of surface contact may be nearly the same in cultures of normal and transformed cells at maximum growth. More efficient utilization of serum growth factors by transformed cells may be a major factor in controlling cell density (see section below).

The pH of the medium has a pronounced affect on the rate and extent of cell growth (Ceccarini and Eagle, 1971; Ceccarini, 1971). Comparisons of saturation densities of normal and transformed cells should thus be carried out under controlled conditions of pH; this is conveniently done by adding organic buffers in various combinations to the growth media (Eagle, 1971); at appropriate concentrations, these buffers are nontoxic and serve to control the fluctuations in pH which normally occur toward the alkaline side in sparse cultures and the acid side in dense cultures using the usual CO_2–bicarbonate buffering system. Within the physiological range of pH, the saturation densities of a given cell may vary by a factor of 4 or more, with the optimum pH varying for different cells.

In addition to serum and pH, factors synthesized by cells may influence the extent and rate of cell growth. Conditioning factors and overgrowth stimulating factors have been described (Rubin, 1970); also "inhibitors" and "potentiators" of multiplication stimulating activity of serum (see Temin *et al.*, 1971). Exogenously added proteolytic enzymes can cause a transient loss of density-dependent inhibition of growth in 3T3 cells (Burger, 1970) and chick embryo fibroblasts (Sefton and Rubin, 1970).

When cells are scraped and removed from a portion of a confluent monolayer of normal cells, the cells adjoining the "wound" will divide and move into the area (Todaro *et al.*, 1965b; Gurney, 1969; Kolodny and Gross, 1969; Dulbecco and Stoker, 1970). Repopulation of the "wound" by migration and cell division strongly suggests effects due to local cell density, though contact per se is not necessarily the causative factor. Highly localized gradients of growth inhibitory substances which are rapidly inactivated on diffusing away from the cell, or an indirect effect of local high density on the cells' ability to utilize serum cannot be ruled out. The wounded experiment lends itself to quantitative studies which are useful in evaluating relative contributions of topographic factors and serum in the density controlled behavior of normal and transformed cells (see following section).

Viral transformation of cells of short-term cultures may not result in as dramatic a loss of density-dependent inhibition of growth as does the transformation of normal cells of continuous lines such as 3T3 which have been selected for their low saturation densities. Primary or secondary mouse embryo fibroblasts have a higher saturation density than 3T3 cells; the difference is accountable only in part by the near-tetrapoloid chromosome constitution and larger cell volume of 3T3. Chick embryo fibroblasts also do not show dramatic differences in density-dependent inhibition of growth after transformation by Rous sarcoma virus under usual conditions of growth (Hanafusa, 1969), but do differ significantly with respect to growth under agar (Hanafusa, 1969) or in serum-depleted media (Temin, 1967).

C. Decreased Serum Requirements for Growth

Serum and occasionally other high molecular weight compounds (such as those present in embryo extracts) are essential to the growth of cells in culture, with few exceptions (e.g., see Ham, 1965). Cells that are selected to grow in the absence of serum or other added macromolecular components may themselves be capable of producing and secreting growth factor(s) into the medium (Shodell, 1972). Besides the effect on growth, serum contains factors essential for cells to adhere to and spread on solid surfaces, to move, and to survive in the absence of growth; normal and transformed cells have been shown to differ in their requirements for, or abilities to utilize, various serum factors; they also may differ in synthesizing "potentiators" and "inhibitors" of growth-stimulating activity (for reviews, see Temin *et al.*, 1971, and various articles in Wolstenholme and Knight, 1971).

A major effect of serum is on the extent of growth or saturation density of normal and transformed cells. Cells transformed by murine and avian sarcoma viruses and papova viruses from both short-term cultures and established lines, have been extensively studied in this regard. The growth of normal cells is strongly dependent on serum, and the growth-promoting substance(s) in serum is removed in the course of supporting cell growth (Temin, 1966; Holley and Kiernan, 1968; Jainchill and Todaro, 1970). There is a marked species specificity; mouse serum, for example, is roughly ten times richer in growth-stimulating activity for 3T3 cells than calf serum, which is normally used. Growth-promoting substances have also been found in human urine and other sources (Holley and Kiernan, 1968). The factor(s) in serum responsible for the promotion of growth are nondialyzable, heat labile, and sensitive to proteolytic enzymes. They are removed from sera with the γ-globulin fraction by precipitation with alcohol (modified Cohn procedure) or 0.4 to 0.5 saturated ammonium sulfate (Jainchill and Todaro, 1970; Taylor-Papadimitriou *et al.*, 1971). Reconstitution with the alcohol precipitate restores growth promoting activity to γ-globulin-depleted serum; the active factor appears not to be a γ-globulin, however, since naturally occurring a-gamma sera of bovine or human origin possess growth-stimulating activity (Jainchill and Todaro, 1970).

Media depleted of serum growth factors may be obtained in either of two ways: (1) medium containing whole serum (unusually 20%) is incubated for 3 days with confluent or near-confluent normal cells to exhaust the factor(s); this medium is removed and diluted 1:1 with fresh medium without serum to restore low molecular weight nutrients; (2) a-gamma serum is prepared by alcohol or ammonium sulfate precipitation, or obtained from a commercial supply; residual growth factor may be removed either by incubating a-gamma medium with monolayers of normal cells as described, or by heating the a-gamma serum to 70°C for 30 minutes; heated sera should be diluted in medium and stored at a concentration of 20% to avoid gel formation (Jainchill and Todaro, 1970). The heating procedure is quicker and is preferable in that it avoids the possibility of introducing toxic or inhibitory substances from cells into the medium.

Growth of transformed cells is also dependent on serum, but to a lesser degree. Table IV shows the effects of the concentration of serum, either whole or depleted, on the growth of normal, polyoma, and mouse sarcoma virus-transformed Balb-3T3 cells. Cells were seeded at 8×10^4 cells in 3.5-mm petri dishes in 2.5 m 1 of MEM buffered with 10 m*M* Hepes and containing the indicated amounts of serum; cell counts were made 5 days later with no intervening fluid changes. It is clear that normal cell growth is limited by the concentration of serum, even up to 15%; both types of transformed cells grow better than the normal cell at all serum concentrations tested, and are limited apparently only at the lower serum concentrations (0.6 to 3%). Normal cell growth is limited in conditioned medium or in medium containing a-gamma serum, while transformed cells grow nearly as well under these conditions as in media containing whole fresh serum.

More extensive data of a similar kind have been reported for 3T3 cells transformed by SV40 (Holley and Kiernan, 1968) and Rous sarcoma virus (Smith *et al.*, 1971), as well as by polyoma and mouse sarcoma virus (Jainchill and Todaro, 1970). Earlier studies by Temin on chick embryo fibroblasts transformed by Rous sarcoma virus (Temin, 1967) and rat cells transformed by murine sarcoma virus (Temin, 1970), and by Stanners *et al.* (1963) on hamster embryo fibroblasts transformed by polyoma, have shown that reductions in serum requirement for growth also accompany viral transformation of normal cells of short-term cultures. In addition to net growth

TABLE IV

EFFECT OF SERUM ON GROWTH OF NORMAL AND VIRUS TRANSFORMED 3T3 CELLS[a]

	Number of cells $\times 10^{-5}$ *(increment)*[b]					
Serum concentrations	Balb-3T3		Py-Balb-3T3		MSV-Balb-3T3	
0.60% Calf (whole)	1.05	*(1.3)*	4.6	*(5.8)*	4.3	*(5.4)*
3.0% Calf (whole)	1.9	*(2.4)*	15.0	*(19.0)*	11	*(14.0)*
15.0% Calf (whole)	6.7	*(8.4)*	16.0	*(20.0)*	12	*(14.0)*
3% Calf (a-gamma)	1.1	*(1.4)*	9.3	*(12.0)*	7	*(8.8)*
15% Calf (a-gamma)	2.1	*(2.6)*	9.4	*(12.0)*	6	*(7.5)*
5% Calf (conditioned)	1.4	*(1.8)*	6.3	*(8.0)*	5.9	*(7.4)*

[a]See Section VI, C for conditions.
[b]Number of cells per plate on day 5 per number of cells seeded.

of cells in mass culture, the effect of serum on growth can be measured in terms of the efficiency of cloning of cells in media containing different kinds and amounts of serum. Normal nonestablished hamster embryo cells, for example, in the presence of irradiated feeder cells, are reduced in their cloning efficiency 100-fold by a reduction in the concentration of serum from 10% to 3.5%; polyoma-transformed hamster embryo cells under the same conditions are either unaffected or reduced by a factor of only 3 to 4 (Stanners *et al.*, 1963). When the cells shown in Table IV were compared for their cloning efficiencies in media with whole or a-gamma serum at 5%, a 200-fold difference was observed for Balb-3T3 cells, while the two viral transformants had efficiences in a-gamma medium of 0.4–0.8 of those in medium with whole serum.

1. Abortive, Cryptic, and Stable Transformation

The ability to grow in a-gamma or conditioned medium has been extensively studied in the SV40/Balb-3T3 system (Smith *et al.*, 1971; Scher and Nelson-Rees, 1971; Todaro *et al.*, 1971). These studies have shown that the ability to grow in serum factor-free media may be acquired independently of changes affecting morphological growth patterns and saturation density. When Balb 3T3 cells are infected with a high multiplicity of SV40 and plated at high dilution in a factor-free medium, many cells are able to initiate a few rounds of cell division; in a majority of clones, this ability is maintained for several generations (up to 6 or 7), after which growth ceases; such cells are said to have undergone abortive transformation (Smith *et al.*, 1971; Todaro *et al.*, 1971). Cell growth can be reinitiated in these cells by the addition of whole serum. The descendants of abortively transformed cells become phenotypically normal and are indistinguishable by the usual criteria from uninfected cells; however, in some cases these apparently normal cells show evidence of persisting SV40 DNA sequences (see Sections VII and VIII) and are called "cryptic transformants" (Smith *et al.*, 1972).

Among the many clones which initiate growth in factor-free medium a small fraction will continue to grow, maintaining the property of serum independence indefinitely. When such stable transformants are tested in complete medium, many pile up and grow to high saturation densities; some, however, maintain flat growth patterns and show little or no change in saturation density compared to normal Balb 3T3 cells (Scher and Nelson-Rees, 1971). These studies make it clear that the ability to grow in factor-free medium is not by itself sufficient for the loss of density-dependent inhibition of growth. Stable transformants which are serum independent while maintaining a low saturation density are not sensitive to density inhibition of DNA synthesis as shown by autoradiography (Smith *et al.*, 1971). A similar phenotype can be obtained by selections of "flat" revertants with FudR starting with "fully" transformed SV-3T3 cells (Pollack *et al.*, 1968); "flat" serum-independent transformants probably also arise after infection of hamster embryo cells by polyoma virus (Stanners, 1963).

2. Wound Serum Requirement and Topoinhibition

Wounding experiments (Todaro *et al.*, 1965b; Kolodny and Gross, 1969; Gurney, 1969) have been useful in studying serum as well as density controlled properties in normal and transformed cells (Dulbecco and Stoker, 1970; Dulbecco, 1970). The ability of cells to initiate DNA synthesis in the wound or monolayer as a function of serum concentration can be followed by autoradiography; normal or transformed cells are first grown to confluency, depleted of serum factors by incubation in serum-free medium, and a wound introduced into the monolayer by scraping with a bent Pasteur pipette or other instrument. Media with various concentrations of serum are added, the cells are exposed to thymidine-^{3}H, and the proportions of labeled nuclei in cells in the monolayer and wound determined by autoradiography (Dulbecco, 1970). Two parameters emerging from such studies have been useful in differentiating normal and transformed cells. One is the "wound serum requirement," defined as $(W_H - W_L)/W_H$, where W_H is the proportion of DNA-synthesizing cells in the wound at high serum concentrations (usually 10% or higher), and W_L the proportion of DNA-synthesizing cells in the wound at low (or zero) serum concentrations. Observed values range from 0.75 to 1.0 for normal fibroblasts, such as BHK or 3T3, to 0 to 0.25, approximately, for SV40 or polyoma-transformed derivatives. Normal cells of epithelial origin resemble transformed cells, however, in their having no wound serum requirement. The second parameter, topoinhibition, measures restrictions due to top topographical factors; it is defined as $(W_L - M_L)/W_L$, where W_L and M_L are the proportions of DNA-synthesizing cells in the wound and monolayer, respectively, measured at low or zero serum concentrations. This parameter approaches 1.0 for normal fibroblasts or epithelial cells and approaches zero for transformed fibroblasts. It is interesting that the FUdR "flat" revertants (Pollack *et al.*, 1968; Vogel and Pollack, 1973) of SV-3T3 cells, and presumably the serum-independent flat transformants selected directly (Smith *et al.*, 1971; Scher and Nelson-Rees, 1971), show low wound serum requirement and topoinhibition parameters (Dulbecco, 1970).

3. Mechanisms

Multiple factors in serum with separate mechanisms of action are most likely involved in the various serum effects described (Dulbecco, 1970; also see Temin *et al.*, 1971; Wolstenholme and Knight, 1971; Defendi and Stoker,

1967; Holley, 1973). Certain polyanions, such as dextran sulfate, may antagonize growth stimulation factors in serum (Temin, 1966; Clarke and Stoker, 1971). Insulin added to serum-free medium allows growth of both normal and transformed chick cells, but transformed cells are more efficient in utilizing added insulin or insulinlike factors in serum (Temin, 1967, 1968).

Serum fractionation experiments have shown that components with growth-promoting activity are separable from those that stimulate migration into a wound (Lipton *et al.*, 1971); furthermore, separate factors may be involved in promoting the growth of normal or SV40 transformed 3T3 cells (Paul *et al.*, 1971). Survival factors in serum are both operationally and physically distinguishable from growth-promoting and migration-stimulating factors (Dulbecco, 1970; Lipton *et al.*, 1971; Paul *et al.*, 1971). In addition to these apparently specific factors, serum and other macromolecules may act in a nonspecific way to block or neutralize the effects of growth inhibitors synthesized by cells (Burk, 1966; Temin, 1966).

Serum causes a selective enhancement of transport of small molecules into stationary 3T3 cells (Cunningham and Pardee, 1969); increases in uptake of uridine and phosphate, but not of amino acids or 3-O-methylglucose, occur in confluent cells within 15 minutes after addition of fresh serum. Little or no effect of fresh serum is seen in subconfluent, actively dividing 3T3 cells or in polyoma-transformed 3T3 cells, even at confluent densities. Increases in rates of incorporation of uridine-^{3}H into RNA are due not solely to enhanced uptake, but also to a rapid turn-on of ribosomal RNA synthesis after serum stimulation (Tsai and Green, 1973; Green, 1973).

A role of serum in making nutrients available intracellularly could be a critical factor in controlling cell growth; this could occur by a variety of mechanisms which indirectly affect permeability, or by a direct role of serum proteins as carriers of vitamins and other small molecules. Alterations in the cell surface brought about by tumor viruses (or other means) which increase nutrient availability could be a fundamental cause of malignant change (see Holley, 1972).

D. Ability to Grow in Semisolid Suspension Culture

Normal fibroblasts or epithelial cells must attach and spread on a solid surface before being able to divide; this requirement is easily demonstrated by the failure of normal cells maintained in liquid suspension (spinner culture) or immobilized in semisolid suspension to increase their numbers despite the presence of the same nutrients and serum factors that support their growth in monolayer cultures; inclusion of small glass beads or other solid material in a semisolid suspension medium allows attachment of cells and results in cell growth (Stoker *et al.*, 1968; Maroudas, 1972). The loss of this property of "anchorage dependence" for cell growth (Stoker *et al.*, 1968) has been shown to accompany the transformation of normal cells by polyoma (Macpherson and Montagnier, 1964; Stoker, 1968), SV40 (Black, 1966), human adenovirus type 12 (McAllister and Macpherson, 1969), Rous sarcoma virus (Rubin, 1966; Kawai and Yamamoto, 1970), hamster and murine sarcoma viruses (Zavada and Macpherson, 1970; Bassin *et al.*, 1970). Semisolid suspension cultures have been especially useful in the quantitative assay of transformation by polyoma (Macpherson and Montagnier, 1964; Stoker, 1968). Since the assay imposes strongly selective conditions allowing the growth only of transformed cells, it permits the screening of large numbers of infected cells and detection of rare transformants. With virus–cell systems which for various reasons have a low efficiency of transformation, this type of selective assay may be particularly advantageous; the limits of detection and reliability are determined by the frequency in the uninfected cell populations of spontaneous transformants capable of growth in suspension.

Assays for the loss of anchorage dependence are conveniently carried out using semisolid media consisting of the usual nutrients and serum plus either agar at 0.3 to 0.4% (see Macpherson, 1969) or carboxymethyl cellulose (Methocel from Dow Chemical Corporation, or methylcellulose from Fisher Chemicals, 4000 centipoises) at 1.0 to 1.2% (Stoker, 1968). As many as 10^4 to 10^5 cells can be suspended in 1.5 to 2 ml of semisolid media and overlaid onto a 6-ml nutrient agar base in a 60-mm petri dish. Under these conditions, normal cells remain in suspension, predominantly as single cells, for up to several weeks with little or no loss of viability. Depending on the cell type and precise culture conditions, one or a few rounds of cell division may occur, giving rise to clones of normal cells up to 8 cells in size, along with rare variants which form larger clones. Transformed cells, on the other hand, will divide repeatedly and form clones that may be upward of 100 cells in size. Figure 2C shows the appearance of normal and polyoma-transformed cells after 6 days in agar suspension cultures. One week to 10 days after plating, the numbers of transformed clones can easily be counted with the naked eye or with a low power microscope. Clones can be recovered from suspension cultures by picking with a Pasteur pipette and transferring to liquid media; before plating, the resuspended clone is vigorously pipetted or mixed on a vortex mixer to release the cells from agar. Because of its water solubility, methyl cellulose is preferable to agar for experiments requiring recovery of clones from the suspension media.

1. Qualifying Factors

While normal fibroblasts of short-term cultures have an essentially absolute anchorage dependence for growth, various factors and conditions can strongly affect the ability of normal established cells such as BHK to grow in suspension (Clarke and Stoker, 1971; Montagnier, 1968, 1971). For example, the use of agarose (or washed agar) instead of regular agar results in some growth of BHK cells. This partial loss of anchorage dependence by normal cells is apparently due to the elimination of polyanionic substances in agar which act to inhibit the growth in suspension of normal, but not of transformed, cells. Use of conditioned medium (taken from normal cell monolayers) further enhances the growth of BHK in agarose, and growth of normal cells in semisolid suspension can be stimulated at high cell densities presumably by a cross-feeding effect. Use of dialyzed serum does not allow growth in agarose, and this can be corrected by addition of serine or possibly other amino acids.

Addition of DEAE-Dextran (100μg/ml) to unwashed agar leads to some growth of BHK cells, due most likely to the complexing of polyanionic inhibitors. Addition of high molecular weight dextran sulfate (5×10^5 to 2×10^6 MW, 5 to 15μg/ml) to agarose again establishes an inhibition of normal, but not of virus-transformed, BHK cells (Montagnier, 1968). Besides dextran sulfate, heparin and, to a lesser extent, chondroitin sulfate act to depress growth of BHK cells in agarose or methyl cellulose, while collagen acts to stimulate growth of BHK in methyl cellulose; these same factors have no affect on the ability of transformed cells to grow in suspension (Sanders and Smith, 1970).

The inhibition of growth of normal BHK in methyl cellulose is not readily overcome by increasing the concentration of serum, although some stimulation of growth is observed at very high (50%) serum levels or with added insulin (Clarke and Stoker, 1971). Adenine or adenine-containing nucleotides added to agar leads to the growth of a class of viral transformants that would not otherwise grow (Montagnier, 1971). It is therefore clear that a number of macromolecular and low molecular weight components affect the growth responses of both normal and transformed cells in semisolid suspension.

Infection of normal BHK by various species of mycoplasmas leads to both morphological transformation and the ability to grow in agar (Macpherson and Russell, 1966). These alterations persist even after apparent curing of the cells. An enzyme assay based on phosphorlysis of pryimidine nucleosides is useful in monitoring cultures for the presence of these organisms (Levine, 1972).

2. Mechanisms

The inhibition imposed by sulfated polysaccharides in agar is selective for normal cells, and specific for the suspended state (rounded shape) since incorporation of glass spicules into the agar layer leads to attachment, spreading, and subsequent growth of normal cells (Stoker *et al.*, 1968). When glass beads of various sizes are tested for their ability to support the growth of BHK in suspension, a lower threshold size of around 50 μm in diameter is found; this size limit seems to correspond approximately to what is required for a cell to assume its natural shape in spreading on a two-dimensional surface (Maroudas, 1972). Other experiments have shown a dependence on the degree of spreading of cells in monolayer culture in order to allow cell division; as cell crowding occurs, the growth rate declines approximately in proportion to the amount of surface occupied per cell (Zetterberg and Auer, 1970).

The alterations in growth requirements and the sensitivities to inhibitors which characterize normal cells in suspension may be related to cell surface changes accompanying the change in cell shape. The relative indifference of transformed cells to the same conditions may also plausibly be linked to cell surface alterations which accompany transformation. The presence of ruthenium red-staining material (acid mucopolysaccharides) in greater amounts on the surfaces of transformed cells than of normal cells (Martinez-Palomo and Brailovsky, 1968), and the apparent relative inaccessibility to phospholipase C of plasma membrane phospholipids in transformed cells, both of which may depend on growth in the presence of exogenous adenine-containing nucleotides, are examples of possibly relevant surface changes (see Montagnier, 1971).

Polyoma transformants of BHK, isolated by the nonselective colony morphology assay (Macpherson and Stoker, 1962), as well as by the selective assay of growth in agar (Macpherson and Montagnier, 1964) are tumorigenic as shown by inoculation into adult hamsters (Jarrett and Macpherson, 1968). The acquisition of tumorigenicity is a direct result of viral transformation, not of a selection by the virus of preexisting, spontaneous tumorigenic variants of BHK. Conversely, a variant of normal (uninfected) BHK cells which has been selected to grow as an ascites tumor in hamsters has also been shown to acquire the ability to grow in agar (Sanders and Burford, 1964).

E. Tumorigenicity

Immunological, hormonal, and nutritional factors operating in the host pose selective barriers which can mitigate either for or against malignant growth of injected cells. Since these factors are complex and only partially understood, it is not surprising that no single property or set of properties manifested by transformed cells *in vitro* can be used to predict their tumorigenicity with certainty. The general properties of transformed cells described above correlate reasonably well with malignant growth but are not absolute indices (see Eagle *et al.*, 1970). Exceptions to most, if not all, correlations between individual properties of transformed cells and tumorigenicity can be found in the literature. Thus, transplantability remains the only valid test of the malignant growth potential of transformed cells.

A major source of difficulty encountered in assessing transplantability is the presence of histocompatability differences between donor cells and host recipient. The use of highly inbred animals for deriving continuous cell lines and short-term cultures and for testing transplantability of transformed cells largely overcomes this difficulty, and permits the direct assessment of various spontaneous or viral induced cellular changes with respect to malignant growth (Aaronson and Todaro, 1968). Virus-specific and tumor-specific transplantation antigens of transformed cells would remain as intrinsic factors and would be expected to play a role in the overall host response (see Section VII,B).

Isogenic hosts are not always available, and various immunosuppressive treatments such as X-irradiation, cortisone, antilymphocytic sera, etc., may be used to overcome strong histocompatability barriers. Use of newborn animals as recipients or "immunological sanctuaries" like the hamster cheek pouch have also been heavily relied upon. Transformed cells which produce infectious virus (as is commonly found with the C-type sarcoma viruses) pose additional problems in assessing tumorigenicity; the use of sex chromosomes or other cellular markers may be used to distinguish between the growth of injected cells and induction of primary tumors by cell-associated virus.

Like other properties of transformed cells, tumorigenicity is a quantitative character. Evaluation is based on statistical observations of injected animals. The most common method is based on the median tumor dose; groups of identical or randomized hosts are injected with serial $\log_{10}$ dilutions of cells, and the proportions of animals developing tumors of a given size after a given amount of time are recorded for each group. The dose of cells required to give tumors to 50% of the animals ("median tumor dose" or "tumor dose $_{50}$") is then calculated. Alternatively, a "time to tumor" test can be carried out by injecting a fixed number of cells in excess of the median tumor dose, and recording the mean time required to produce a tumor of given size. An example of the application of these methods can be found in Jarrett and Macpherson (1968).

F. Surface Properties

The morphological and growth criteria described above have been widely used in detecting and characterizing changes at the level either of the whole cell or of the interactions between cells. The ensemble of these properties suggest that a major subcellular site of alteration operative in transformed cell behavior is likely to be the cell surface. This view receives considerable support from genetic and biochemical studies of both DNA and RNA tumor viruses (see Section XI). Two kinds of studies have revealed structural and functional differences between normal and transformed cell surface membranes: cell agglutination studies using lectins, and transport studies of various small molecules, such as glucose and amino acids.

1. Agglutination Studies

Wheat germ agglutinin and concanavalin A have been widely used in agglutination studies with a variety of normal cells and their viral transformed derivatives. The techniques for these studies have been carefully described (Burger, 1972). These agglutinins are able to cause the agglutination of viral transformed cells and tumor cells induced by a variety of means, while being much less effective at agglutinating normal cells or "phenotypic revertants" of transformed cells. On the basis of hapten inhibition and other studies (Burger, 1973) these two agglutinins appear to bind to different carbohydrate-containing structures on the cell surface. The binding sites themselves are present in normal cells; mild protease treatment of normal cells makes them as agglutinable as their transformed counterparts; the interaction of lectin with binding sites of normal cells changes during the cell cycle—during mitosis the cells bind the agglutinins strongly whereas in interphase they do not (Fox *et al.*, 1971). Electron microscopic studies of the distribution of binding sites in normal and transformed cells have revealed that concanavalin A, under appropriate conditions, can induce a rearrangement (clustering) of sites on the transformed or trypsinized normal cell surface, but not on untreated normal cells (Nicolson, 1971, 1972; Rosenblith *et al.*, 1973).

The relationship between transformed cell behavior and the manner of interaction with lectins is based primarily on studies of 3T3 and its transformants and revertants. The general finding is that the agglutinability, expressed as the reciprocal of the lectin concentration required to give half-maximal agglutination, is roughly in direct proportion to the saturation density (Pollack and Burger, 1969). Table V shows the comparisons of these two parameters for a series of Balb-3T3 lines transformed by RNA or DNA viruses. Other evidence that the surface changes measured by interaction with these lectins actually reflect changes in control of cell growth are: (i) the resumption of growth in stationary confluent monolayers of 3T3 cells triggered by mild protease treatment (Burger, 1970); (ii) the lowering of the saturation density of polyoma-transformed cells with monovalent concanavalin A (Burger and Noonan, 1970); and (iii) the lowered saturation density and partial "normalization" of morphology displayed by concanavalin A revertants of transformed cells, selected as being

TABLE V

SATURATION DENSITIES AND AGGLUTINABILITIES OF NORMAL AND VIRUS TRANSFORMED CELL LINES

Cell line	Saturation density (3% serum) (No. cells/cm² × 10^{-4})	Half-maximal concentrations (μg/ml)	
		Wheat germ agglutinin	Concanavalin A
Swiss-3T3	1.4	1000	1500
Swiss-Py-3T3	17.2	150–200	100–200
Balb-3T3	2.1	1000	1500
Py-Balb-3T3	–	–	100–200
SV-Balb-3T3	9.4	100–200	100–200
MSV-Balb-3T3	10.0	100–150	100–150
RSV-Balb-3T3	14.7	100–150	200–300

resistant to the toxic effect of the lectin (Ozanne and Sambrook, 1971). Exceptions to the usual correlations of agglutinability with transformed cell behavior have nevertheless been noted (Burger, 1968, 1973; Sakiyama and Robbins, 1973).

2. TRANSPORT CHANGES

Enhanced rates of uptake of sugars and amino acids by transformed compared to normal cells have been noted in several laboratories (Hatanaka *et al.*, 1969, 1970a; Foster and Pardee, 1969; Isselbacher, 1972; Venuta and Rubin, 1973). Typically, cells on coverslips or petri dishes are washed and preincubated for 30–45 minutes in a salt buffer to deplete internal pools, then incubated for short periods of time (2–20 minutes) with radioactively labeled sugar or amino acid. Nonmetabolized analogs such as 2-deoxyglucose or 3-*O*-methylglucose and aminoisobutyric acid or cycloleucine have been used preferentially to avoid problems of internal utilization. When the initial rates of uptake are measured in this way, normalized to the amount of cell protein, and compared between normal and transformed cells, the reported values vary from 2- to 4-fold up to 10- to 15-fold increases in transformed, compared to normal, cells. In normal nondividing cells, glucose uptake may be stimulated either by trypsin or fresh serum (Sefton and Rubin, 1971). Continuous lines of fibroblasts, such as 3T3 and BHK and their SV40 or polyoma transformants, have been compared, as well as primary or secondary fibroblasts and their sarcoma virus-transformed derivatives. The patterns of increase show some selectivity among different sugars or amino acids, although the data pooled from different laboratories are not entirely in agreement.

In most instances, concentration and kinetic studies have shown increases in V_{max} with little or no change in K_m, except possibly in the case of a lowered K_m for 2-deoxyglucose uptake by cells transformed by the sarcoma viruses (Hatanaka *et al.*, 1970a,b). It is not clear whether the changes in rate of uptake of 2-deoxyglucose reflect qualitative changes in the cell membrane as well as increases in the number of transport sites. Only quantitative changes (increase in the number and/or mobility of carrier sites) appear to be involved in the enhanced uptake of 3-*O*-methylglucose in Rous sarcoma virus-transformed versus normal chick fibroblasts (Venuta and Rubin, 1973).

Events at the membrane, such as the binding of lectins, can alter transport rates (Isselbacher, 1972). Evidence from studies of conditionally lethal mutants of Rous sarcoma virus shows effects of a virus–gene product on sugar transport changes and agglutinability (see Section XI), and it would seem plausible that the primary effect be localized in the membrane. However, membrane involvement may be controlled by intracellular events that are the primary alterations; for example, rates of glucose utilization can affect the rates of glucose transport into the cell, and inhibitors of glycolysis have been shown to slow the rate of 2-deoxyglucose uptake (Rubin and Fodge, 1973).

VII. Specific Properties of Transformed Cells

The properties described in the preceding section are generally characteristic of transformed cells, including cells transformed by DNA or RNA viruses, by radiation or chemicals, or spontaneously. In addition to these general properties, virus-transformed cells possess a number of virus-specific properties not found in other types of transformed cells; these properties are related directly or indirectly to the presence of viral genetic material in transformed cells, and they can therefore be used to distinguish between cells transformed by different but closely related viruses of the same group. Procedures for detecting virus-specific properties are based on nucleic acid hybridization experiments and immunological techniques.

A. Detection of Viral Genetic Material

Various kinds of DNA–RNA and DNA–DNA hybridization experiments have been applied to the detection and quantitation of viral genomes in transformed cells. In general, the resolution in such experiments depends on the purity and specific radioactivity of the viral nucleic acid to be used in detecting homologous sequences in cell nucleic acids. With papova, adeno, and herpes viruses, viral DNAs can readily be labelled, isolated, and purified. The DNAs can be used directly in hybridization experiments or used as primer for the DNA-dependent RNA polymerase of *Escherichia coli* to make a specific complementary RNA of high specific radioactivity. With RNA tumor viruses the virion-associated RNA-dependent DNA polymerase ("reverse transcriptase"; Temin and Mizutani, 1970; Baltimore, 1970) can be utilized to make a specific DNA copy—either the single-strand complement of the viral RNA or the double-stranded copy. Labeled 70S viral RNA has also been used.

Some of the applications of hybridization with viral nucleic acids are summarized in Table VI. They include detection of viral genomes in transformed and revertant cells, quantitation of numbers of viral genomes in transformed and revertant cells, localization of viral genomes in fractions of cellular DNA, detection of covalent association between viral and cell DNAs during permissive and nonpermissive infections, detection of sequence homology between normal cell DNA and C-type viral RNA (i.e., detection of purported inherited provirus), and studies of transcription patterns in lytically infected or transformed cells which show the extent of DNA transcription, the processing of transcripts, and the relative amounts of different sequence transcripts. The experimental techniques center around two basic procedures: (1) the nitrocellulose filter technique of Gillespie and Spiegelman (1965), variations of which can be used for simple detection of hybrids or quantitative saturation studies, and (2) renaturation kinetics (Britten and Kohne, 1968; Wetmur and Davidson, 1968) and hydroxyapatite chromatography.

Papova viral DNAs can be obtained in highly purified form, either from density-gradient purified virus (Crawford, 1969) or by selective extraction from infected cells (Hirt, 1967). Purification is achieved by centrifugation in solutions of CsCl (density 1.55 gm/ml) containing ethidium bromide (100–200 μg/ml) which separates closed circular duplex DNA from open (nicked) circular or linear DNA (Radloff *et al.*, 1967). Viral DNA prepared in this way is free of linear host DNA encapsidated into pseudovirions (Michel *et al.*, 1967; Winicour, 1967, 1968, 1969). It may, however, contain host sequences covalently integrated into viral DNA. The latter occurs to a variable extent depending, at least in part, on the passage history of the virus; successive high multiplicity (undiluted) passage of polyoma or SV40 leads to the accumulation of defective virus containing deleted and substituted DNAs (Blackstein *et al.*, 1969; Yoshiike *et al.*, 1972; Tai *et al.*, 1972; Lavi and Winicour, 1972); such DNA would give a spurious background of homology between viral and host DNAs. Electron microscopic methods for determining contour lengths of native DNAs (Kleinschmidt, 1968) and

TABLE VI
HYBRIDIZATION EXPERIMENTS WITH VIRAL NUCLEIC ACIDS

Virus group	Purpose	Materials (methods)	References
Papova	1. Detection of virus-specific RNA in infected or transformed cells	Viral DNA and cell RNA (nitrocellulose filters)	Benjamin (1966)
	2. Nuclear vs cytoplasmic location of viral DNA in transformed or tumor cells	(a) *In vitro* complementary RNA and chromosomal DNA (nitrocellulose filters)	Sambrook *et al*. (1968)
		(b) Viral DNA and mitochondrial DNA (nitrocellulose filters)	Benjamin (1968)
	3. Detection and quantitation of genomes in transformed cells	Reassociation kinetics of viral and transformed cell DNAs (hydroxyapatite chromatography)	Gelb *et al*. (1971)
	4. Association of viral and cell DNA in nonpermissive infection	Viral complementary RNA and infected cell DNA (nitrocellulose filters)	Hirai and Defendi (1972), Defendi and Hirai (1973)
	5. Strand specificity and patterns of transcription in infected or transformed cells	Strand separation of DNA by complementary RNA hybridization; cell RNA and isolated single strand of viral DNA hydroxyapatite chromotography	Khoury *et al*. (1972, 1973), Sambrook *et al*. (1972), Lindstrom and Dulbecco (1972)
Adeno	1. Detection of virus-specific RNAs in transformed cells; patterns of transcription and relatedness of transcripts in cells infected or transformed by viruses of different subgroups	Cell RNA and viral DNA (nitrocellulose filters)	Green *et al*. (1968, 1971)
Herpes	1. Detection of viral DNA sequences and virus-specific RNA in tumor cells; (HSV-2)	Viral DNA and cell DNA (reassociation kinetics)	Frenkel *et al*. (1972)
		Viral DNA and cell RNA (solution hybridization)	Frenkel *et al*. (1972)
	2. Detection of viral DNA sequences in tumor cells; (EBV)	(a) Viral DNA and cell DNA (nitrocellulose filters)	zur Hausen *et al*. (1970a,b)
		(b) Viral complementary RNA and cell DNA (nitrocellulose filters)	Nonoyama and Pagano (1971)
C-type (avian)	1. Detection of DNA sequences in normal and transformed cells homologous to viral RNA	(a) Viral DNA (double stranded) and cell DNA (reassociation kinetics)	Varmus *et al*. (1972)
		(b) Viral RNA and cell DNA (nitrocellulose filters)	Baluda (1972)
	2. Detection of viral RNA in uninfected cells	Viral DNA (single stranded) and cell RNA (reassociation kinetics)	Hayward and Hanafusa (1973)

sequence homologies in denatured-renatured DNAs (Davis *et al*., 1971) can be of value in assessing the degree of homogeneity in viral DNAs, apart from or in conjunction with hybridization with cellular nucleic acids.

Maximal purity in viral RNAs from C-type viruses is obtained by isopycnic banding of whole virus particles in 20 to 50% sucrose gradients (density of virus $\simeq$ 1.16 gm/cc) followed by extractions of viral RNA (see references in Table VI); velocity sedimentation of the extracted RNA through sucrose gradients allows 70S viral RNA to be isolated from 4 to 5S RNA which may be, in part, host cell transfer RNA (Erikson, 1969).

B. Detection of Virus-Specific Antigens

Intracellular antigens are found in transformed or tumor cells induced by the following major virus groups: papova, adeno, Herpes avian C-type, and murine C-type. In addition, in most if not all the virus–cell systems examined, there are virus-specific antigens on the cell surface which can immunize against transplantation of tumor or transformed cells induced by the corresponding virus (Sjögren, 1965). Table VII lists the major classes of antigen for each virus group and the methods used for their detection.

1. Papova Viruses

The intranuclear tumor or T antigen is distinct for polyoma and SV40; cells of different species transformed by the same virus show the same T antigen specificity, while cells of a given species transformed by the two different viruses show correspondingly different T antigens. The same virus-specific T antigen found in transformed cells is also found in cells lytically infected by these viruses. In addition to the SV40 T antigen, a perinuclear or cytoplasmic antigen ("U" antigen) has been described in cells infected by adeno 2-SV40 hybrid viruses (Lewis and Rowe, 1971; Lewis *et al*., 1973); sera from SV40 tumor-bearing hamsters usually contain both T and U specificities, indicating the SV40-specific nature of the latter antigen. Little is known about the chemical nature of the T antigens; strong circumstantial, but no direct, evidence suggests that they are virus-coded proteins or specifically modified cellular antigens.

Cell surface antigens of papova virus-transformed cells and tumor cells have been detected by transplantation techniques; as in the case of the T antigens, transplantation antigens (TSTAs) are virus, not cell, species specific. Considerable variation in the specificity of the TSTAs induced by different strains of polyoma have been noted, however (Hare, 1967; Jarrett, 1966). Cell surface antigens can also be detected by cytotoxic tests

TABLE VII

VIRUS—SPECIFIC ANTIGENS IN TUMOR OR TRANSFORMED CELLS

Virus group	Class of antigen	Methods for detection	References
Papova	1. Tumor (T) antigens	(a) Complement fixation	(a) Black *et al.* (1963); Habel (1965, 1966)
		(b) Immunofluorescence using sera from tumor-bearing animal	(b) Pope and Rowe (1964)
	2. Transplantation (TSTA) antigens	(a) Rejection of cells by animal specifically immunized with homologous virus or other cells transformed by the homologous virus	(a) and (b) Habel (1961); Sjögren *et al.* (1961); Defendi (1963, 1968)
		(b) Immunogenicity of cells in inducing specific transplantation resistance against a test cell transformed by the homologous virus	
		(c) Cytotoxic and colony inhibition tests	(c) Hellström and Sjögren (1966); Tevethia and Rapp (1965)
	3. Surface (S) antigen (s)	(a) Immunofluorescence	(a) Tevethia *et al.* (1965); Malmgren *et al.* (1968); Irlin (1967)
		(b) Mixed hemadsorption	(b) Häyry and Defendi (1970)
Adeno-2-SV40 hybrid	1. U-antigen	(a) Immunofluorescence	(a) and (b) Lewis and Rowe (1971); Lewis *et al.* (1973)
		(b) Complement fixation with sera from SV40 tumor-bearing hamster	
Adeno	1. Tumor (T) antigen	(a) Complement fixation	(a) and (b) Huebner (1967)
		(b) Immunofluorescence	
	2. Transplantation antigens(s)	(a) Graft rejection	(a) Hellström and Sjögren (1967)
		(b) Cytotoxicity	(b) Ankerst and Sjögren (1969)
		(c) Colony inhibition	(c) Ankerst and Sjögren (1970)
Herpes (EBV)	1. Membrane antigen of tumor cells	Immunofluorescence	Klein (1972)
C-type			
(A) Avian	1. Group-specific antigens	(a) Complement fixation	(a) Sarma *et al.* (1964)
		(b) Radioimmune assay	(b) Fritz and Qualtiere (1973)
	2. Transplantation antigen	(a) Transplantation	(a) Jonsson (1966)
		(b) Cytotoxicity	(b) Kurth and Bauer (1972)
(B) Murine	1. Group-specific antigens	(a) Complement fixation	(a) Geering *et al.* (1966); Hartley *et al.* (1970)
		(b) Radioimmune assay	(b) Scolnick *et al.* (1972)
	2. Transplantation antigen	(a) Transplantation	(a) Ting (1968)
		(b) Colony inhibition	(b) Sjögren and Hellström (1966)

(release of radioactive chromium from target cells exposed to immune sera) and colony inhibition tests (loss of colony-forming ability in target cells exposed to immune sera or immune lymph node cells in the presence of complement) (see references in Table VII); the antigens detected in these tests are likely, though not certain, to be the same as the TSTAs. While the TSTAs may depend on the action of a viral gene, the surface or S antigen(s) detected by immunofluorescence and hemadsorption appear to be cell-specified and not dependent on persisting viral genes (Häyry and Defendi, 1970; Levin *et al.*, 1969; Levine *et al.*, 1970).

There is no antigenic relatedness between T and TSTA, nor between either of these antigens and structural components of the virus. Anti-T antisera (usually obtained as sera from an animal carrying a virus-free tumor) should be checked for absence of antiviral antibodies before being used in immunofluorescence or complement fixation tests in order to avoid confusion with viral capsid antigen(s), which also appear in nuclei of infected cells.

2. Adenoviruses

Both intranuclear T antigen(s) and transplantation antigen(s) have been described in cells of different species transformed by various human adenoviruses (see references in Table VII). A classification of these viruses into three groups has been based on guanine-cytosine content of the DNAs, extent of nucleic acid homology (both DNA–DNA and DNA-transformed cell RNA), and relative oncogenicity in hamsters (see Green *et al.*, 1968). The degree of relatedness in the T antigens induced by various human adenoviruses determined by complement fixation follows the same grouping based on other criteria (Huebner, 1967).

3. C-Type Viruses

Group-specific (gs) antigens are known for the avian and murine groups of leukemia and sarcoma viruses. The gs antigens are multiple by physical as well as immunological criteria (see Nowinsky *et al.*, 1972); there are both species-specific components and component(s) common to more than one species—e.g., the murine gs antigens contain an interspecific component shared by C-type viruses of all other mammalian species. Gs-antigens form an integral part of the internal ("core") structure of the virion; they are present not only in infected or transformed cells, but also in normal cells of birds and mice where its appearance follows Mendelian inheritance (Payne and Chubb, 1968; Hilgers *et al.*, 1972). Unlike the intracellular T antigens of papova and adenoviruses, the presence of gs antigens is not indicative of viral infection or transformation.

VIII. Transformation by Papova Viruses

A. Role of Cell Species

In contrast to cell interactions with C-type RNA viruses—in which virus production and cell transformation may occur in the same cell—papova viruses interact with susceptible cells in a mutually exclusive manner: in a given cell, the infection leads either to virus production and cell death (productive infection), or to abortive or stable transformation with no virus growth (nonproductive infection). The species of host cell is the major determinant of the outcome of infection with polyoma or SV40. Table VIII shows the commonly used cells and species of origin for productive and nonproductive infection by these two viruses. Some virus–cell combinations, such as the SV40–human, have been characterized as "semipermissive" since both types of interactions occur.

Usually the species of origin determines the majority response without playing an absolute role; the minority type of response can often be found in a given virus–cell system. With special precautions, rare transformants, such as SV40 transformed monkey cells or polyoma-transformed mouse cells, can be isolated from productively infected cultures (see below).

Nonproductive infection by polyoma or SV40 is accompanied by expression of early viral functions seen in productively infected cells (Benjamin, 1972; Sambrook, 1972); these include synthesis of T antigens, increased levels of several enzymes concerned with DNA synthesis, induction of cellular DNA synthesis, and cell surface alterations. No detectable syntheses occur of either viral DNA or viral structural proteins. The difference between nonproductive and productive infection thus lies in the extent of viral gene expression. The host cell factor(s) that play a role in governing

TABLE VIII

HOSTS FOR PRODUCTIVE AND NONPRODUCTIVE INFECTIONS BY POLYOMA AND SV40

Virus	Productive	Nonproductive
Polyoma	Mouse: Early passage mouse embryo fibroblasts; baby mouse (12–15 days) kidney primary (epithelial cells); established lines (3T3, Swiss or Balb)	Hamster: early passage hamster embryo fibroblasts; established lines (BHK, Nil) Rat: early passage rat embryo fibroblasts; established lines
SV40	Monkey: African green monkey kidney primary; established lines (BSC-1, Vero, CV-1) Human	Mouse: 3T3 cells Hamster: Chinese or Syrian hamster embryo fibroblasts Human

this expression are not known; however, the response of hamster (BHK)-mouse (3T3) hybrid cells to polyoma infection lies on the side of 3T3; i.e., the productive response seems to be dominant (Basilico *et al*., 1970). This suggests a requirement for a dominant cellular factor(s) to allow late viral gene expression rather than a dominant repressor of late viral functions in the nonproductive host. In order for transformation to occur, the positive cellular factor(s) which allow virus production clearly must be absent, since the latter event inevitably entails the death of the cell.

B. Quantitative Assays

Cells of either short-term culture or established lines can be transformed by papova viruses, although the latter are generally used for reasons of greater efficiency, shorter times, and better reproducibility. The most widely used assays for quantitative studies of transformations are: (1) the agar suspension method for polyoma with BHK as the target cell, and (2) the colony morphology method for SV40 with 3T3 as the target cell. These two assays have been described in detail (see chapters by Macpherson and Todaro in Habel and Salzman, 1969). Various laboratories depart in one or another aspect of these procedures; a general outline of these assays is given below with mention of qualifying factors that affect the efficiencies.

1. Agar Suspension Method

A polyoma–BHK system is used (see Macpherson in Habel and Salzman, 1969; Macpherson and Montagnier, 1964; refer also to Section VI,D).

a. Infection. BHK cells (or any fibroblast susceptible to transformation) may be infected either in monolayer culture or first trypsinized and resuspended and then infected in suspension. In either case, cells in exponential growth are required for optimal efficiency. Cells seeded 24 hours before infection at 5×10^5 to 1×10^6 cells per 60-mm petri dish are satisfactory. For infecting cells on the monolayer, the cultures are washed briefly with warm adsorption buffer (any balanced salt buffer, such as phosphate-buffered saline, with 2% calf serum is satisfactory; the pH may be adjusted downward to 6.0 to improve virus adsorption; standard growth media with 2% calf serum can also be used); the buffer is removed and 0.1–0.2 ml of virus diluted in adsorption buffer is added, and the cells are incubated for 1 hour at 37°C in a 5 to 10% CO_2 humidified incubator. The dishes are rocked after 20 and 40 minutes to spread the virus and prevent the monolayer from drying. At the end of the adsorption period, the monolayers are washed with adsorption buffer to remove the virus inoculum. Five milliliters of growth medium is added, and the cells returned to the incubator for at least 1 hour to allow the virus to penetrate; the next step of plating the infected cells in agar can be done between 1 and 6 hours after the end of the virus adsorption period; longer times, allowing cells to divide before being plated, may introduce uncertainties in estimating the efficiency of transformation. For infecting cells in suspension, exponentially growing cells are trypsinized, washed, and resuspended in adsorption buffer or medium containing virus in capped tubes (1 to 2×10^6 cells/ml and 10^8 or more PFU of virus per milliliter). The tubes are incubated in a 37°C water bath for 1 hour, with occasional gentle shaking; the cells are then spun down and resuspended in growth media for 1 hour before being plated in agar. A thin coat of paraffin on the inside of the tube is helpful to avoid cells sticking to the glass during the adsorption period; polystyrene tubes instead of glass can also be used to prevent sticking of cells.

b. Preparation of Bottom Agar Layer. Six milliliters of bottom agar containing 0.6% Difco Bacto agar and growth medium are pipetted into 60-mm petri dishes and allowed to harden. This is done most conveniently by mixing equal parts of 2× growth medium brought to 37°C and melted agar at 1.2% cooled to 45°C; the plates with bottom agar should be prepared the same day they are used. The number of plates to be prepared depends on the purpose of the experiment and on how much of the infected cell suspension is to be sampled; usually no more than 2 to 4×10^5 cells can be plated on a single dish without getting crowding and cross-feeding effects. Control experiments should be run beforehand to determine the backgrounds of clone size and frequencies of spontaneous transformants with a given cell line.

c. Plating of Cells in Top Agar. After adsorbing the virus and allowing for penetration, the cells are suspended in growth medium and an aliquot counted in a hemacytometer. At this time the presence of cell clumps should be noted and the suspension gently pipetted until the cells are dispersed; false-positive results are to be expected either from initially large clumps or from small clumps in which cells may grow through cross-feeding. Clumping is rarely a problem when infection is done on subconfluent monolayers, but may be encountered when the infection is done in suspension. Longer periods of incubation after adsorption may help to reduce the numbers of clumps; RDE (receptor-destroying enzyme) has also been used (Stoker and Abel, 1962). Cells, 1.5 to 2.0 ml, in 0.33% agar are gently layered on the solidified bottom layer; this is conveniently done by mixing 2 volumes of growth medium containing 0.5% agar with 1 volume of cell suspension; the agar-growth medium initially constituted at 45°C can be further cooled in a 37°C waterbath for a few minutes before the cells are added.

After 10 to 15 minutes at room temperature the plates are returned to the incubator. A well humidified incubator is necessary to avoid drying of the top layer; a few drops of water or of a 1:1 mixture of water and medium can be added to the dishes after 3 or 4 days of incubation, or as often as necessary, to maintain the loose consistency characteristic of the initially formed top layer. One or 2 ml of fresh growth media can also be added to the plates when kept for longer times.

d. Scoring. Assays may be counted as early as 5 to 7 days after plating, although scoring should also be done at later times. Large clones of transformed cells can be counted macroscopically or with a low-power microscope (see Fig. 2C). Control plates with uninfected cells should be scanned first to determine the upper limit of clone size to be considered as background as well as the number of spontaneous transformants. The cutoff point between normal and transformed clone size is usually easily determined on inspection. Precise measurements of clone size and clone size distributions can be obtained by using a linear micrometer scale (Stoker, 1968) or split-image eyepiece (Macpherson in Habel and Salzman, 1969). Uninfected cells should remain largely as single cells, with a small proportion undergoing 1 or 2, or occasionally 3, divisions. In infected cultures, 30 to 50% of the cells may undergo abortive transformation (see below) giving intermediate size clones (up to around 5 divisions), and a few percent become stably transformed and grow into large clones up to several hundred cells in size. Stable transformants can be distinguished from abortive transformants on the basis of their increasing size after the first 4 or 5 days. Clones can be picked with a Pasteur pipette from the top agar layer, dispersed by shaking or pipetting in a few milliliters of liquid medium, and plated to determine whether on further growth they display normal or transformed morphologies and growth patterns (see Fig. 1).

e. Abortive Transformation. Precisely the same procedure is followed as in the standard agar suspension assay, except that the top layer consists of cells suspended in "Methocel" medium instead of agar; a final concentration of 1.2% methyl cellulose in growth medium is used (Stoker, 1968). Clones are easily isolated and put back into liquid culture to analyze growth morphologies and to determine the proportions of abortive and stable transformation among clones of different size. Delayed transformation (Stoker, 1963) may be reflected in the agar suspension assay as stable transformation arising in what are initially abortively transformed clones.

f. Qualifying Factors and Efficiencies. A single-hit dose response curve holds over the range of input multiplicities of less than 1 up to about 500 PFU/cell (Macpherson and Montagnier, 1964). In this range, the efficiency of transformation by the virus is lower by a factor of roughly 10^4 compared to its plaque-forming ability. At multiplicities greater than 10^3 PFU/cell, the dose-response curve shows a plateau in the proportion of cells transformed; the maximum percent transformed is between 5 and 10% at a multiplicity of around 10^3 PFU/cell, whether the assays are conducted using the selective conditions of the agar suspension assay or the non-selective colony morphology test (see Fig. 2A; Macpherson and Stoker, 1962; Stoker and Abel, 1962). Thus, although both assays give linear responses over the same range of virus concentrations, the agar suspension test is preferable on the basis of its allowing up to a thousand times more infected cells to be sampled per culture dish.

The reasons for the low efficiency of virus in transformation are not known. An intrinsically low probability of the transforming event or sequence of events seems likely; heterogeneity of virus at the level of DNA length or base sequence may also be a factor.

The ceiling in cell transformation efficiency with saturating amounts of virus does not appear to be due to genetic heterogeneity in susceptibility in the uninfected cell population (Macpherson and Stoker, 1962; Sachs *et al.*, 1962; Black, 1964); however, some morphologically normal segregants from mixed clones of normal and transformed cells appear to have increased sensitivity to transformation (Stoker and Smith, 1964).

The following factors affecting the efficiency of transformation have been described. (1) X-irradiation of the cells (500 R) prior to or anytime within the first 2 days after infection enhances the frequency of transformed clones by a factor of 4–5 in the colony morphology assay, while lowering the overall colony formation (i.e., cell survival) almost 100-fold. Cloned survivors of X-irradiation do not exhibit higher transformation frequencies (Stoker, 1963, 1964). (2) pH levels around 7.6 to 7.8 result in 2- to 3-fold higher frequencies of transformation compared to pH levels in the range 6.9 to 7.1 in the colony morphology assay (Kisch and Fraser, 1964; Kisch and Subak-Sharpe, 1966). (3) Incubating cells at 24°C for 24 hours prior to or just after infection enhances the frequency of transformation (Medina and Sachs, 1963). (4) High Mg^{2+} concentrations (0.025 M) during the first 30 minutes post-infection (Stoker and Abel, 1962) gives a 2-fold enhancement. (5) Dimethyl sulfoxide, 0.2–1.0% present from the time of infection onward gives a variable (up to 5- or 10-fold) enhancement in the colony morphology assay but not in the agar suspension assay (Kisch, 1969). (6) An enhancement of up to 2-fold occurs when cells are infected in G-2 compared to G-1, using the agar suspension assay (Basilico and Marin, 1966). (7) Dibutyryl cyclic AMP, 0.6 mM, enhances 3- to 10-fold when present from 8 hours post-infection onward, in both types of assay (Smith *et al.*, 1973).

2. Colony Morphology Method

An SV40–3T3 system is used (see Todaro in Habel and Salzman, 1969; Todaro and Green, 1966a; refer also to Sections VI,A and B).

As in the polyoma–BHK system, transformation of 3T3 cells by SV40 can be assayed either by agar suspension (Black, 1966) or by colony morphology

(Todaro and Green, 1966a). Whereas the two methods have comparable efficiencies for polyoma, the colony morphology assay is approximately 200 times more efficient for SV40 (Black, 1966).

Proper handling of 3T3 cells is important, since spontaneous variants, altered in morphology, growth pattern, or saturation density, will lead to high and uncertain backgrounds. Factors in the routine handling of the cells which should be followed are: (1) use of Dulbecco's modified Eagle's medium with 10% calf serum, (2) incubation at 37°C in a 10% CO_2 humidified atmosphere, (3) routine subculture of the cells with high dilutions on transfer (1:1000 or 1:10,000), (4) occasional subcloning of cells with selection of uniform low density clones, (5) freezing of aliquots of cells and routine thawing every 4–6 months.

a. Adsorption. Cells may be infected either in monolayer or in suspension. Sixty-millimeter petri dishes with 5×10^5 to 10^6 cells/dish are washed with serum-free medium; 0.5 ml of virus (usually 10^6 PFU/ml or higher) in complete medium is added, and the plates incubated at 37°C for 3 hours with rocking every 20 minutes. Alternatively, one million freshly trypsinized cells may be suspended in medium with virus and incubated in a 37°C water bath for 1 or 2 hours with occasional shaking. At the end of the adsorption period, the cells on the monolayer are washed and fed with medium. One to several hours later the cells can be trypsinized, counted, diluted, and inoculated into new cultures. Cells infected in suspension are similarly plated out for growth of colonies immediately after the adsorption period.

b. Plating of Cells. Infected cells are inoculated at different dilutions into 60-mm dishes with 6 ml of fresh medium. Several dilutions with inocula covering the range of 50 to 10,000 cells/plate are made with several plates per dilution. Cultures are fluid changed twice a week, and colonies scored after 10 to 14 days.

c. Scoring. Plates may be scored at low power with an inverted or dissecting microscope; the distinction between normal and transformed colonies is easily made on the basis of the individual morphologies of the cells and the thickness or degree of multilayerings in the clone. A rapid scoring can be done directly on living cultures if they are to be incubated further; at the end of the assay, usually after 12 to 14 days, the monolayers are washed briefly with phosphate-buffered saline, fixed in 10% formalin for 10 minutes, stained with 1% hematoxylin, washed, and counted. Control plates from uninfected or mock-infected cultures are scored first to establish the background of variation in colony morphology as well as the cloning efficiency of the cells. The latter is calculated from the plates receiving the lowest number of cells (usually 50), and should be close to 50% for both control and infected cells. The frequency of transformation is based on counts taken from the series of plates with 5 to 50 transformed colonies per plate; counts from plates with more than 50 transformed colonies per plate are subject to error due to overlapping of adjacent colonies. The transformation frequency is calculated as the percentage of all colonies (normal and transformed) which are transformed, or as the percentage of cells plated which grew into transformed colonies.

A basically similar procedure has been used to isolate doubly transformed cells, beginning with polyoma-transformed 3T3 cells, infecting with SV40 and isolating clones which display mixed morphologies (Todaro and Green, 1965; Todaro *et al.*, 1965a).

d. Qualifying Factors. The percentage of cells transformed increases linearly with the multiplicity of infection over several logs of input virus multiplicities; the efficiency of transformation by the virus in the linear range is roughly 1 per 10^3 plaque-forming units, depending in part on the strain of virus (Todaro and Green, 1964a, 1966b; Black, 1966). At very high multiplicities of infection (above 10^3 PFU/cell), 50% of the cells may be transformed.

Other factors that affect the overall efficiency include the following: (1) infection of cells in suspension gives 2- to 4-fold higher efficiencies than infection on the monolayer (Todaro and Green, 1966a); (2) exponentially growing cells, infected in suspension, give 2-fold higher efficiencies than nongrowing cells; (3) X-irradiation of cells administered either prior to or after infection increases the frequency of transformation among the survivors; doses as high as 1200 rads give 7-fold higher frequencies while lowering the overall colony-forming ability 50- to 100-fold (Pollock and Todaro, 1968); (4) incorporation into cellular DNA of the halogenated pyrimidines (BUdR or IUdR in the medium at concentrations from 5 to 100 μg/ml, any time from 24 hours before to 96 hours after infection) gives enhancements of 2- to 9-fold (Todaro and Green, 1964b); (5) a minimum of 5 to 6 cell divisions (i.e., at least a 1:64 dilution of cells from a confluent infected monolayer) is necessary for full expression of transformation by all the potential transformants; studies on the effects of different dilutions on the transformation frequency have shown that potential transformants are gradually lost if the infected cells are not replated, but left in a confluent monolayer; a single cell division (1:2 dilution) is sufficient to fix irreversibly the transformed state, but another 3 to 5 divisions are required for the full expression of the transformed phenotype (Todaro and Green, 1966b); and (6) dibutyryl cyclic AMP (6 mM present from 0 to 24 hours after infection) has been shown to increase the frequency of transformation of Chinese hamster embryo cells by SV40 (Smith *et al.*, 1973).

e. Transformation to Serum Independence and Abortive Transformation (Refer to Section VI,C). When 3T3 cells undergo morphological transformation and lose density-dependent inhibition of growth (properties on which the colony morphology assay depends), they also acquire the ability to grow in media depleted of serum growth factor(s) required by normal 3T3 cells. The latter property may be acquired and selected for independently of the other two properties (Smith *et al.*, 1971; Todaro *et al.*, 1971). When the colony morphology assay is done using growth factor-free medium, the number of cell divisions (cumulative, in clones of all sizes) is directly proportional to the virus input. At least three classes of colonies are initiated under these conditions, as shown by further test: (1) those that cease dividing after 3 to 6 cell divisions, which grow to low saturation density in complete medium, and which lack SV40 T antigen (abortive and cryptic transformants), (2) those that continue to grow in factor-free medium and have SV40 T antigen, but maintain low (near normal) saturation densities in complete medium (so called stable serum-independent transformants), and (3) those that continue to grow in factor-free medium, have T antigen and grow to high saturation density in complete medium (stable "complete" transformants).

3. Transformation by Viral DNA

Both the closed circular form (component I) and the nicked or open circular form (component II) of viral DNA has transforming ability (Crawford *et al.*, 1964). Highest efficiencies are obtained using DEAE—dextran treatment of cells to enhance uptake of viral DNA (see Pagano in Habel and Salzman, 1969). The frequencies of transformation, however, are lower by a factor of several hundred to several thousand compared to whole virus.

C. Transformation of Cells in Short-Term Culture

The changes brought about in cells of established lines such as BHK and 3T3 by polyoma or SV40 are rapid; crucial events occur within the first cell doubling time after infection which fix the ultimate transformed state, and the full expression of transformed cell properties occurs after only a few subsequent cell doublings. The number of virus-cell interactions leading to transformation can easily be scored within about a week using assays that depend on the cloning of cells immediately after infection. The interactions of polyoma and SV40 with nonpermissive cells of short-term culture produce ultimately similar cellular alterations, but with lower efficiencies and over longer periods of time.

When secondary cultures of hamster embryo fibroblasts are infected by polyoma and followed through successive low dilution passages, there is a gradual evolution of a fully transformed phenotype. Two distinct stages in this evolution have been noted (Vogt and Dulbecco, 1962, 1963); (1) "Early" transformed cells can be seen as focal areas of refractile cells showing crisscross growth and variable tendencies to pile up; such clones appear at low but increasing frequencies in sparse (high dilution) cultures derived from early low-dilution passages. Focal lines derived from these altered cells have low efficiencies of cloning and are of low tumorigenicity. (2) After 2 or 3 weeks, sparse cultures initiated from later low dilution passages show increasing numbers of thick foci containing elongated and highly refractile cells, with a greater tendency toward multilayering than in the early thin foci. These cells also have a high cloning efficiency and are more tumorigenic in a "time to tumor" test. Early thin focal lines on further passage give rise to subclones of the thick late type. This progression is paralleled by the occurrence of abnormal mitoses. The frequency of anaphase bridges and chromosome breaks is high in the early focal lines; numerous nonviable cells are produced, and eventually the late fully transformed lines are selected; these are most often aneuploid and show fewer anaphase bridges. The primary action of the virus is seen as the initiation of the early focal lines. The need for further selection and progression is characteristic only of nonestablished cells, these steps perhaps being bypassed or already undergone by cells of established lines (Vogt and Dulbecco, 1962, 1963).

Focus assays can be carried out with polyoma virus using primary or secondary rat (Williams and Till, 1964) or hamster embryo cells (Stoker and Macpherson, 1961; Stanners, 1963). Factors that aid in the detection of transformation in these systems are: (1) use of X-irradiated normal cells as "feeders" (2000 to 7000 R; 1 to 2 $\times$ 10^5 cells/60-mm petri dish) to overcome the low plating efficiencies of the infected cells, and (2) use of 2 to 3% serum, rather than the usual 10%, which gives a relative growth advantage to the transformed cells (Stanners *et al.*, 1963; Stanners, 1963). The efficiency of transformation of nonestablished cells is usually lower by a factor of 10 or more compared to established cells. Although the number of foci increases with virus concentration over a wide range of virus input, the increase is not always linear (Williams and Till, 1964).

A similar time course and latent period has been observed for SV40 transformation of cells derived from hamster kidney (Black and Rowe, 1963) and human kidney cells (Shein and Enders, 1962). Not all of the eventual established tumorigenic lines show evidence of a persisting SV40 genome (Diamandopoulos and Enders, 1965). Enhanced frequencies of transformation by SV40 are observed with nonestablished human cells derived from individuals with certain inherited diseases (Todaro *et al.*, 1966; Todaro and Martin, 1967).

D. Transformation in Permissive Hosts

In the nonpermissive systems described above, transformation occurs without selection imposed by cytocidal interactions. Papova viruses will occasionally transform hosts which are permissive or semipermissive for virus growth (see Table VIII); such transforming events may be due to defective viruses that cannot complete productive infection, but this is not necessarily the case.

The following factors may be considered to enhance either the relative frequency of occurrence of transformation compared to productive infection, or the ease of detection of transformants, or both, in such systems: (1) use of partially inactivated virus—since the plaque-forming ability is nearly twice as sensitive as transforming ability to inactivation by radiation or chemicals (Benjamin, 1965; Basilico and DiMayorca, 1965; Latarjet *et al.*, 1967), a high-titer virus stock partially inactivated will undergo relatively more noncytocidal interactions leading to transformation than cytocidal ones; (2) irradiation of cells prior to or just after infection, or treatment with IUdR or BUdR, or other treatments, as noted above (Section IX,B), increases the frequency of the rare transforming events, probably with little or no effect on the frequency of the predominant cytocidal interaction; (3) in some instances, the rare transformants can be protected from being killed as a result of reinfection by virus produced from other cells; use of antiviral antiserum to neutralize the virus produced in the culture, frequent fluid changes and/or subculturing of the cells to dilute out the virus, and use of receptor-destroying enzyme (effective against polyoma but not SV40) to strip the rare emerging transformants of virus receptors, thereby making them insusceptible to infection (Benjamin, 1970).

IX. Transformation by Other DNA Viruses

A. Adenoviruses

Transformation of cells in short-term culture from a variety of species by human and simian adenoviruses has been reported (see Black, 1968; Casto, 1973a). In general, the efficiencies are low and the times of incubation long compared to papova virus systems. Human adenovirus 12 and simian adenovirus 7 are among the more efficient adenoviruses as transforming agents. Quantitative assays of transformation by these two viruses can be done using primary hamster embryo cells in a focus type of assay (Casto, 1968, 1973a). Monolayers are infected with high titer virus, and the cells are replated at 2–4 $\times$ 10^5 cells per 60 mm petri dish. The cultures are incubated for 48 hours in regular growth medium, and then switched to the same medium but containing lower amounts of calcium (0.1 m*M* instead of 18 m*M*). The calcium-deficient medium selects for adenovirus-transformed cells and against normal cells (Freeman *et al.*, 1967; Casto, 1968). Discrete foci, in numbers linearly related to the virus input, can be counted after 12 days.

Agar suspension has been used to assay transformation of Nil-2, a continuous line of hamster cells, by human adenovirus 12 (McAllister and Macpherson, 1968); Nil-2 and rat embryo cells have also been used to study transformation by other human adenoviruses (McAllister *et al.*, 1969).

Enhancement of adenovirus transformation by various physical and chemical agents have been studied as models for cocarcinogenesis. The following treatments cause substantial increases in the frequencies of transformation: X-irradiation (Coggin, 1969), halogenated pyrimidines (Casto, 1973a,b), steriod hormones (both enhancing and inhibitory, Milo *et al.*, 1972), ultraviolet light (Casto, 1973a,b), polycyclic hydrocarbons (Casto, 1973b; Casto *et al.*, 1973), and nitroquinoline oxide derivatives (Stich *et al.*, 1972).

B. Adenovirus–SV40 Hybrids

After the demonstration of oncogenicity of human adenovirus-7–SV40 hybrid virus in hamsters (Huebner *et al.*, 1964), the ability of this virus to cause transformation of various cultured cells has been reported, including hamster embryo or kidney cells (Black and Todaro, 1965; Black, 1966; Wells *et al.*, 1966; Black and White, 1967), human cells (Black and Todaro, 1965), and established hamster cells BHK and Nil-2 (Diamond, 1967). These studies are based largely on focal or colony morphology types of experiments; the evolution of morphologically transformed cells, even in the case of the established lines, is slow. Transformation of weanling hamster kidney cells can also be brought about by Adeno-2–, Adeno-3–, and Adeno-7–SV40 hybrid viruses (Black and White, 1967; Igel and Black, 1967). Use of low-calcium medium in focus assays, as originally used in transformation studies with human and simian adenoviruses, permits more efficient and rapid quantitation of transformation of hamster embryo fibroblasts by Adeno-7–SV40 virus (Duff and Rapp, 1970). Cells transformed by these hybrid viruses generally possess SV40 antigens, and some show both SV40 and adeno antigens (Rapp *et al.*, 1966; Igel and Black, 1967; Duff and Rapp, 1970). In permissive monkey cells, transformants may be obtained using partially UV-inactivated Adeno-7–SV40 hybrid virus (Shiroki and Shimojo, 1971).

C. Herpesviruses

Despite the known or suspected etiological role of several herpesviruses in various cancers of mammals, birds, and amphibians, little work has been done with these viruses as transforming agents *in vitro*. Different strains of HSV-1 are known to cause different morphological changes and changes in social behavior of infected cells, although the infected cells are ultimately killed (Ejercito *et al.*, 1968); the changes in social behavior induced by different strains is correlated with the synthesis of different viral and cell membrane glycoproteins (Keller *et al.*, 1970; Roizman, 1972). Herpes (HSV-2), partially inactivated by ultraviolet light, causes morphological transformation of hamster embryo fibroblasts and these cells are tumorigenic (Duff and Rapp, 1971; Rapp and Duff, 1972).

In a different context, UV-irradiated Herpes virus (HSV-1) has been shown to carry out a specific form of "genetic transformation" or "transduction"; the viral coded gene for thymidine kinase can be transferred by a strain of mouse L cells which itself lacks the kinase (Munyon *et al.*, 1971).

X. Transformation by RNA Viruses

The following brief description of interactions of C-type viruses with susceptible cells is based largely on work with avian viruses. With few exceptions to be noted below, the biology of the murine viruses, though less well studied, is closely analogous to that of the avian group. For comprehensive reviews and references covering both major virus groups, see Temin (1971, 1973).

A. Virus–Cell Interactions

1. The Infectious Cycle

The steps of adsorption and penetration of virus are not understood in detail, except that subgroup-specific receptors on the cell surface are involved in the avian system (see below). The first intracellular step leading to virus development and transformation is formation of the provirus by means of the virion-associated reverse transcriptase. The DNA product, presumed to be the direct precursor of the provirus, is then integrated in some manner into the host chromosomes, after which its further replication is presumably taken over by the cell machinery. There is no direct evidence pertaining to the intermediates in provirus formation or integration. The existence of the DNA provirus has been inferred on the basis of the sensitivity at early times after infection to various DNA inhibitors and also from the sensitivity to light inactivation of cells infected and incubated in the presence of BUdR. Subsequent to provirus formation, there is no expression of transformed cell properties and no production of progeny virus unless cells are allowed to divide [see Nakata and Bader (1968) and Bather and Leonard (1970) for examples of similar findings in the murine system]. The requirement for one round of cell division for the expression of viral functions has been shown in a variety of ways, including DNA synthesis inhibitors, serum starvation of cells, and X-ray inactivation studies comparing the ability of cells to form colonies with their ability to register as virus-producing cells. For expression of transformed cell properties, more than one cell division may be required. (These facts are similar to those in the SV40–3T3 system concerning cell division requirements for fixation and expression of transformation; see Section VIII,B.) Steps subsequent to the formation and integration of provirus and which lead to the expression of genes for virus production and transformation are not understood in detail. Transcription from the provirus appears to give rise only to viral RNA strands, which presumably serve as messenger RNA; no RNA complementary to viral RNA is found. Group-specific antigens appear (but see Section X,A,4 below) as well as virus-specific cell surface antigens; if the infected cell is susceptible to transformation, other surface changes also become apparent (agglutinability, transport changes) along with morphological changes. Progeny virus matures by budding of ribonucleoprotein particles (nucleoids) through modified areas of the cell surface membrane. The time course of these events is variable depending on culture conditions; under optimal conditions, morphological conversion can be complete in as little as 24 hours in a majority of cells (Hanafusa, 1969), with the first progeny virus appearing after the same period of time or less, depending on the stage of the cell cycle at the time of the infection.

In general, virus production and transformation occur side by side in infected cells; however, depending on both viral and cellular factors, infected cells may become virus producers without being transformed, they may be transformed without producing virus, or they may be transformed and produce defective virus or virus with altered host range. Multiple factors underlie the various outcomes of infection by C type viruses; these can be understood by considering the three topics discussed below: host genetic factors affecting virus susceptibility, viral factors affecting host range, and the presence in normal cells of C-type provirus.

2. Host Genetic Factors

a. Avian System. A series of unlinked autosomal genes in chickens, presumably specifying viral receptors on the cell surface, determine the pattern of susceptibility to avian C-type viruses. A separate locus is involved for each virus subgroup (see below), the allele for susceptibility being dominant in each case (Payne, 1972; Payne *et al.*, 1973). Patterns of susceptibility of other avian species (duck and quail) to certain virus strains are known as well as a definite but limited susceptibility of mammalian cells.

b. Murine System. A single locus in mice (FV-1) is the major determinant of susceptibility to murine leukemia and sarcoma viruses both *in vitro* and *in vivo* (Pincus *et al.*, 1971a,b; Ware and Axelrod, 1972; Hartley *et al.*, 1970). Two alleles are known, "n" and "b", which in the homozygous recessive condition confer susceptibility to "N-" and "B-" tropic strains of murine leukemia-sarcoma viruses, respectively. The degree of exclusion conferred by resistant genotypes in the mouse system is several orders of magnitude less, and is more readily overcome by increasing the multiplicity of infection than in the avian system. Also, consistent with these differences, the block in infection of resistant cells appears not to be at the cell surface but at some intracellular step in replication, as shown by the host ranges of phenotypically mixed particles between vesicular stomatitis virus and murine leukemia viruses of either N or B types (Huang *et al.*, 1973).

3. VIRAL FACTORS

a. Avian System. Three properties which are specified by viral envelope glycoproteins—host range, antigenicity, and viral interference—form the bases of the subgroup classification of avian C-type viruses (Vogt, 1965; Vogt *et al.*, 1966; Vogt and Ishizaki, 1966). Thus, viruses belonging to the same subgroup resemble one another in their abilities to recognize genetically defined cell surface receptors, in their type-specific antigenicities and patterns of cross-neutralization, and in their ability to exclude one another but not members of different subgroups in superinfection experiments (Steck and Rubin, 1966a,b; Ishizaki and Vogt, 1966; Vogt and Ishizaki, 1965). At least 6 subgroups, A–F, are known.

Incompatibility of viral envelope and cell surface membranes constitutes the first barrier to successful infection. For incompatible virus–cell systems, low levels of infection can sometimes be achieved by coinfecting with UV-inactivated Sendai virus (Robinson, 1967; Hanafusa *et al.*, 1970a). For genetically compatible virus–cell combinations, in both avian and murine systems, the efficiency of adsorption is often increased by using DEAE-dextran (5 to 40 μg/ml) during or before the adsorption period (Duc-Nguyen, 1968; Somers and Kirsten, 1968; Vogt, 1967a; Hartley *et al.*, 1970). Enhancement is also found by preinfection with a different subgroup leukosis virus (Hanafusa and Hanafusa, 1967).

Virus–virus interactions leading to phenotypic mixing is another important factor in determining virus host range (Vogt, 1967b). Certain strains of Rous sarcoma virus normally consist of a mixture of two or more viruses; the Bryan high-titer strain, for example, consists of at least three types of virus—two defective viruses with genes for transformation but which lack essential genes for replication (RSV_{α} lacking the reverse transcriptase and RSV_{β} lacking an envelope glycoprotein) and an avian leukosis virus. The latter virus, which replicates in but does not transform chick embryo fibroblasts, is present in a large excess (on the order of 10:1) over the transforming viruses, with the result that through coinfection the sarcoma virus particles are "helped" or completed by incorporating products specified by the leukosis genome. The host range and other properties are thus determined by the leukosis helper. When cells are infected at high virus dilution, singly infected cells are transformed and produce defective virus particles (the use of the term "defective" is not always accurate; in the case of RSV_{β}, for example, the virus can infect certain kinds of chicken cells as well as Japanese quail cells, and is therefore more accurately described as a virus with altered and relatively restricted host range). The Schmidt-Ruppin strain of Rous sarcoma virus, on the other hand, contains transforming virus which gives rise to nondefective particles of unaltered host range.

b. Murine System. Three common strains of mouse sarcoma virus (Moloney, Harvey, and Kirsten) contain genomes which are competent for transformation but defective in virus production, analogous in broad terms to the Bryan strain in the avian system. Production of infectious sarcoma virus therefore depends on phenotypic mixing with murine leukemia virus which is usually present in sarcoma virus stocks in a several hundred- to thousand-fold excess (Hartley and Rowe, 1966). From cultures of 3T3 cells infected at end-point dilutions for focus formation, transformed cells can be isolated which fail to produce virus particles detectable by physical, chemical, or biological methods (Aaronson and Rowe, 1970); such "nonproducer" transformed clones can be infected with murine leukemia virus leading to the "rescue" and production of infectious sarcoma virus with host range and antigenic markers of the superinfecting leukemia virus. The host range is either N-, B-, or NB depending on whether they grow in $FV\text{-}1^{n/n}$, $FV\text{-}1^{b/b}$, or both types of cells (Pincus *et al.*, 1971a,b); an isolate is also known which grows poorly in either type of host cell (Levy and Pincus, 1970). Transformed mouse cells producing various kinds of defective particles have also been described (Bassin *et al.*, 1971a; Peebles *et al.*, 1972; Somers and Kit, 1971), and these also can be superinfected by leukemia virus to give infectious sarcoma virus.

Murine sarcoma virus passaged in rat cells *in vitro* (Somers and Kit, 1971) leads to transformation with production of virus with altered host range, infecting rat but not mouse cells. Tumors induced in hamsters by murine sarcoma virus produce a hamster tropic sarcoma virus due apparently to the presence of a hamster C-type helper virus (Kelloff *et al.*, 1970); a similar set of findings has been described for passage of the murine virus in rats (Ting, 1967, 1968; Aaronson, 1971).

Interference among mammalian C-type viruses has been described (Somers and Kirsten, 1969; Sarma *et al.*, 1967; Sarma and Log, 1973), although it has not been as extensively investigated as in the avian system. Host range and antigenicity are not closely correlated in the murine viruses as they are in the avian viruses (Hartley *et al.*, 1970).

4. Presence of C-Type Provirus

Two lines of evidence support the conclusion that both avian and murine cells contain a complete provirus for C-type viral particles. (1) Cells of certain strains of inbred mice and chickens possess group-specific antigen(s) of their indigenous C-type viruses, without necessarily showing evidence of virus particles. In crosses between antigen negative and positive strains, the ability to produce the antigen is transmitted to the offspring in a Mendelian fashion, indicating that at least these internal viral components can be synthesized under host genetic control (Payne and Chubb, 1968; Hilgers *et al.*, 1972). In addition, another host genetic factor in the avian system has been shown to be involved in the synthesis of a viral envelope component (T. Hanafusa *et al.*, 1970; H. Hanafusa *et al.*, 1970; Weiss, 1969; Scheele and Hanafusa, 1971). The expression of this factor in chickens correlates with expression of avian group-specific antigens (T. Hanafusa *et al.*, 1972; Weiss and Payne, 1971). (2) The ability of cells to produce complete infectious (though perhaps not oncogenic) C-type virus has been shown by treatment with BUdR or IUdR or other chemicals; such experiments have been done on cloned populations of cells and under conditions which minimize the chance for accidental viral infection (Lieber *et al.*, 1973; Lieber and Todaro, 1973; Todaro, 1972; Lowy *et al.*, 1971; Weiss *et al.*, 1971; Aaronson *et al.*, 1971). C-type virus is also induced in mixed lymphocyte cultures with cells from histoincompatible mice, as well as *in vivo* after graft versus host stimulation (Hirsch *et al.*, 1970, 1972). Finally, crosses between mice with high and low spontaneous leukemia demonstrate the inherited capacity, determined by two discrete loci, to produce complete infectious leukemia virus (Rowe, 1972a).

Thus, the presence of group-specific antigens in cells cannot be used to determine either that cells have been infected by a C-type virus, or that they are producing such a virus (see Section VII,B). Furthermore, the findings on the cell genetic factor giving envelope specificity to infecting C-type viruses add another dimension to phenotypic mixing; as in the case of mixed viral infections, the use of viral components specified by a partially expressed endogenous viral genome by an infecting virus can lead to alterations in host range (Weiss, 1969; H. Hanafusa *et al.*, 1970; T. Hanafusa *et al.*, 1972); endogenous viruses may be fully induced by such infection (Weiss, 1969; T. Hanafusa *et al.*, 1970b). Interactions between endogenous and exogenous viral genomes may also occur by recombination (Weiss *et al.*, 1973). Phenotypic mixing and recombination could be involved in the numerous examples of altered host range properties when C-type viruses from one species are passed through cells of different species. A generalized view on the origins and evolution of C-type viruses has been given by Temin (1973).

B. Assays for Avian Viruses

Standard focus assays for Rous sarcoma virus on secondary cultures of chick embryo fibroblasts (see Fig. 2B) are employed in different laboratories with only minor variations in procedures. Vogt (1969) has given a thorough description of the techniques involved in the preparation of cells, assaying for interfering leukosis viruses, infection, incubation, and counting of foci; no attempt is made here to restate or summarize these procedures. Parts of the discussion in the following section on the murine viruses, pertaining to the dose-response relationships in focus assays, is also generally relevant here.

Transformation of chick cells by Rous sarcoma virus can also be carried out, though with lower efficiency, using agar suspension cultures (Rubin, 1966; Kawai and Yamamoto, 1970; Wyke, 1973a). Rous transformants of established hamster cells can also be isolated in this way (Macpherson, 1965, 1966).

Transformation of chick embryo fibroblasts by Rous sarcoma virus can be followed by observing morphological changes in the infected cells; under optimal conditions of infection, a rapid and synchronous transformation can be seen in up to 90% of the cells within 24 hours after infection (Hanafusa, 1969). The key factors in obtaining the high efficiency and short latency for expression of transformation are use of DEAE-dextran to enhance viral adsorption, and the state of the cells, being freshly plated and presumably very close to the G-1/S border; the establishment of a block to DNA-synthesis by excess thymidine, and the removal of the block prior to infection can also be used to maximize the degree of synchrony and overall efficiency. The ability to grow under agar is acquired by the cells along with the morphological conversion. These experimental conditions permit a wide variety of biochemical investigations to be directed at the events accompanying the early stages of infection and the emergence of the transformed phenotype; overall increases in DNA and RNA but not protein synthesis occur during and immediately after the morphological conversion (Hanafusa, 1969); specific changes in transport of sugars also occur in parallel with the morphological changes (Hatanaka and Hanafusa, 1970; see Section VI,F).

1. Assays with Avian Leukemia Virus

Traditional assays for avian leukosis viruses are based either on their interference properties (Rubin, 1960), or on their ability to rescue infectious sarcoma virus from nonproducing transformed cells. Some leukosis viruses are able to produce morphological alterations in cultured cells (Langlois *et al.*, 1967; Moscovici *et al.*, 1969). Leukosis viruses of certain subgroups can also be titered in plaque-type assays (Kawai and Hanafusa, 1972; Graf, 1972); one of these assays (Kawai and Hanafusa, 1972) appears to depend on complementation with a temperature-sensitive sarcoma genome.

C. Assays for Murine Viruses

1. Focus Assays

Focus assays have been developed for quantitating transformation by murine sarcoma viruses; minor variations in techniques have been employed by different groups of investigators using different virus strains and host cells; NIH-3T3 cells (FV-$1^{n/n}$), Balb-3T3 cells (FV-$1^{b/b}$) of mouse origin, and the NRK line of normal rat kidney cells are commonly used established lines; mouse embryo secondary cultures may also be used.

Some examples of focus assay methods are by Aaronson *et al.* (1970a), Aaronson and Rowe (1970), Bassin *et al.* (1968), and Somers and Kirsten (1969). Typically, cells are seeded in 60-mm petri dishes 24 hours before infection with 2 to 5 $\times$ 10^5 cells per dish. Monolayers are infected [DEAE-dextran treatment, 25 μg/ml, prior to or during virus adsorption may be used to enhance infection (Duc-Nguyen, 1968; Somers and Kirsten, 1968)] and incubated in appropriate growth media; fluid is changed after 4 or 5 days, and the foci are counted macroscopically after 7 to 10 days.

The dose response curves for focus formation are not always of the single-hit type. Several factors are involved:

a. Virus Spread. The development of foci may occur either through proliferation of transformed cells established by the infecting virus inoculum or through such proliferation accompanied by "recruitment" of neighboring cells by virus produced from transformed cells. In the former case, the dose-response curve would show true first-order kinetics for transformation. In the latter case, the response would require the initially infected cell to be doubly infected by a sarcoma virus particle and a leukosis virus helper. The kinetics would be pseudo first order at low virus dilutions, where the leukosis virus excess ensures double infection, and second order at higher dilutions. The range of dilution over which the virus gives a single-hit response is therefore dependent on the leukosis: sarcoma ration. Addition of leukemia virus at the higher dilutions, sufficient to infect all the cells, leads to the restoration of first-order kinetics. Log-log plots of virus titer (number of foci times dilution) versus virus dilution provide a simple way to analyze the data in dose-response experiments (Hartley and Rowe, 1966; Aaronson *et al.*, 1970a).

b. Time of Incubation. The process of focus formation by virus spread represents a substantial amplification, giving more rapid development and earlier detection of foci. Thus, assays on Balb-3T3 cells counted at early times, 3 days after infection, show a two-hit pattern since only rapidly developing foci due to virus spread are evident; the same assay counted at 7 days or later shows one-hit kinetics since singly infected non-virus-producing transformed cells have been able to establish visible foci through cell proliferation. Assays held for long times may show secondary foci due to released virus (Aaronson *et al.*, 1970b, 1971).

c. Cell Culture and Growth Conditions. Nonestablished cells appear to favor detection of rapidly appearing foci due to virus spread, and involve a risk of not detecting non-virus-producing transformants (Hartley and Rowe, 1966). Cell growth conditions must be favorable enough to allow the fixation and expression of transformation (see Section X, A), yet not so rich (in serum of other factors) that normal cells grow as well as the non-producing transformants (Fischinger and O'Connor, 1969). Established lines tend to have more uniform low-density backgrounds and can easily be used to detect both nonproducing and producing transformed cells (Jainchill *et al.*, 1969; Aaronson *et al.*, 1971).

d. Virus Aggregates. The tendency for C-type virus to form aggregates under certain conditions leads to another kind of "pseudo" first-order kinetics when conditions are such that virus spread is essential for detection. Pelleting and freezing of mixtures of sarcoma and leukemia viruses result in aggregate formation, and dose-response curves with such preparations show variable kinetics averaging between 1 and 2 hits (O'Connor and Fischinger, 1968a; Fischinger and O'Connor, 1969). Virus stocks prepared directly from tumors also contain aggregates (O'Connor and Fischinger, 1968b). Aggregates may be disrupted by sonication, or removed by velocity sedimentation or Millipore filtration (0.22–0.45 μm).

e. Cell and Virus Genotype. Incompatibility between viral host range and cell genotype at the FV-1 locus results in reduced infectivity and multihit titration curves (Rowe, 1972b).

2. Agar Suspension Assays

Transformation of cells by mammalian sarcoma viruses leads to the loss of anchorage dependence of growth (Section VI,D) as in the case of transformation by papova viruses (Section VIII, B). Assays based on suspension in soft agar have been used in studies of transformation of 3T3 by mouse sarcoma virus (Bassin *et al.*, 1970) and of BHK by hamster sarcoma virus (Zavada and Macpherson, 1970). Under certain conditions of cell concentration in the top agar layer, single-hit kinetics of transformation are observed; the efficiency, however, is far below that obtained with the same cells and virus in a focus assay (Bassin *et al.*, 1970).

3. Morphological Assay

Under optimal conditions, 80% of 3T6 cells infected by mouse sarcoma virus undergo progressive morphological transformations manifested 24 hours after infection and continuing for 3 or 4 days; the percentage of cells becoming agglutinable with concanavalin A increases over the same time period (Salzberg *et al.*, 1973). Such an assay, like the analogous one with Rous sarcoma virus (Hanafusa, 1969), affords a useful system in which to study the biochemical processes accompanying the establishment and expression of transformation.

4. Assays with Murine Leukemia Viruses

Titrations with murine leukemia viruses may be done by the XC syncitium assay (Rowe *et al.*, 1970), or by the $S^+ L^-$ lytic focus assay (Bassin *et al.*, 1971b). The former assay detects fibroblasts infected and producing leukemia virus without morphological alteration, and the latter detects what appear to be degenerative morphological changes in cells harboring an incomplete virus with sarcoma functions; both assays give similar estimates of virus titers. Leukosis viruses can also be titered by end-point dilution (measuring the ability to rescue sarcoma virus from nonproducing transformed cells) or by interference (measuring the ability to interfere with transformation by sarcoma virus of the homologous leukosis virus pseudotype); these assays are more laborious than the XC or S^+L^- assays but give comparable results (Bassin *et al.*, 1971b). Another type of assay for murine leukemia virus involving morphological transformation without degeneration is the focus assay on UCI-B cells (Hackett and Sylvester, 1972). Transformation appears to involve complementation between the infecting leukosis virus and a defective virus carried by the host cell.

XI. Viral and Cellular Factors in the Transformed Phenotype

A. Viral Factors

The process of virus-induced neoplastic transformation of cells is accompanied by the establishment of a stable association between viral and cellular genes, giving rise to the stable inheritance of the transformed phenotype. How viral genes become integrated into the cellular genome, and what role they may have in directing various aspects of the transformed phenotype are questions being approached through genetic studies of tumor viruses. Such studies have been initiated with papova viruses, adenoviruses, and avian and mammalian sarcoma viruses; the results on polyoma virus and Rous sarcoma virus are the most extensive.

The overall goal of genetic studies of tumor viruses has two parts: (1) characterization of the viral genome through the isolation of conditional lethal mutants, usually of the temperature-sensitive type; complementation and other experiments are carried out pertinent to the question of how many discrete gene functions are involved in virus replication or transformation; (2) physiological experiments in which cells infected by various mutants are studied under conditions both permissive and nonpermissive for the functional expression of the viral gene in question; nontransforming viral mutants are used as biochemical tools in efforts to determine the nature of the viral gene product and its manner of interaction with the cell. The brief summary given below outlines the main conclusions from such studies; the genetics of polyoma has recently been reviewed (Benjamin, 1972); recent reviews by Temin and others [see Temin (1973) for references] on the sarcoma viruses include genetic studies.

Based on studies of polyoma virus, at least two discrete viral functions are required for transformation. The first is *an initiation function* whose role appears to be the establishment of a stable association between viral and cellular genes; the molecular mechanisms involved in the expression of this viral gene are not understood, although it is clear that it is involved in viral DNA replication during productive infection, and appears likely to be involved in stable integration into chromosomal DNA in nonproductive infection. The action of this gene is required only transiently: once cells become transformed by allowing expression of this gene under permissive conditions, they remain transformed under nonpermissive conditions. The second is a *maintenance function* whose action is required continuously to maintain at least certain aspects of the transformed phenotype. Cells transformed by a temperature-sensitive mutant in a maintenance function revert toward a normal phenotype at the nonpermissive temperature. Properties that revert in this way, and therefore depend directly or indirectly on the mutated viral gene product, are morphology, growth pattern, and lectin agglutinability. A nontransforming mutant of the host-range class, presumably also defective in a maintenance function, is also unable to bring about the cell surface alteration detected by lectins (Benjamin and Burger, 1970). Temperature-sensitive mutants of SV40 have also been isolated and characterized; one of these is nontransforming and appears to be of the initiation type (Tegtmeyer and Ozer, 1971; Kimura and Dulbecco, 1973).

The abortive transformation response (see Section VIII, B) is important as a test to distinguish between the two classes of nontransforming mutants: initiation-defective mutants retain the ability to abortively transform cells (Stoker and Dulbecco, 1969) as well as to induce cell DNA-synthesis and the cell surface change (Eckhart *et al.*, 1971), whereas maintenance-defective mutants are unable to carry out abortive transformation and the cell surface change (Benjamin and Norkin, 1972; Norkin and Benjamin, 1973).

The size of the papova virus genome is approximately 3 million daltons, comprising an estimated 5–7 genes. Fewer viral genes are required for transformation than for productive infection; this conclusion is based on radiation inactivation studies which show the target size of the virus to be smaller as a transforming agent than as a plaque-forming agent, and also on genetic studies in which temperature-sensitive mutants in at least two, and most likely three, viral genes are unable to grow but are able to transform cells at the nonpermissive temperature. These genes are essential for late functions of productive infection, being required for syntheses of viral structural proteins but not of viral DNA (see Benjamin, 1972).

The overall picture of transformation based on studies of temperature-sensitive and other kinds of defective mutants of Rous sarcoma virus, and to some extent of mouse sarcoma virus, is basically similar to that obtained from the papova virus studies despite the enormous differences in the structure and composition between these two groups. Mutants of Rous sarcoma virus which lack the reverse transcriptase may properly be called initiation mutants, since the enzyme is presumably required to establish the stable virus-cell association in the form of the DNA provirus. Phenotypically mixed virus particles, containing an avian leukosis virus-derived reverse transcriptase and a defective sarcoma genome unable to code for the enzyme, can infect and transform normal cells which then produce non-infectious particles (Hanafusa and Hanafusa, 1971; H. Hanafusa *et al.*, 1972); such mutants are therefore unaffected in the transforming genes per se. Mutants in maintenance functions have been isolated and characterized by several groups; as in the case of the papova virus mutants, cells infected by these mutants become reversibly transformed and show changes in the cell surface as the ultimate site of alteration. Thus, with shifts of temperature, cells reverse their behavior with respect to morphology, growth patterns, growth in agar, lectin agglutinabilities, and 2-deoxyglucose uptake (Toyoshima and Vogt, 1969; Martin, 1970; Biquard and Vigier, 1970; Kawai and Hanafusa, 1971; Martin *et al.*, 1971; Scolnick *et al.*, 1972; Vogt, 1972; Burger and Martin, 1972).

The number of discrete sarcoma viral genes involved in maintenance of various aspects of the transformed phenotype is not known. Nontransforming mutants of Rous sarcoma virus are of two types—those which are unaffected in their ability to produce virus particles at the nonpermissive temperature, and those which are simultaneously affected in transformation and virus reproduction. At least four complementation groups are found among mutants of the first type and two among those of the second type (Wyke and Linial, 1973; Wyke, 1973a,b; Kawai *et al.*, 1972). It is not known whether as many as six separate genes are involved in maintaining transformed properties, or whether the number of complementation groups is larger than the number of genes due to intragenic complementation or other factors (see Wyke, 1973a,b).

The size of the genome of the sarcoma viruses is not known, but lies in the range of 2.5–10 million daltons of RNA. The complexity of the virion is substantially greater than for papova viruses; for the lower estimate of genome size, a large portion of the coding capacity of the virus would have to be concerned with syntheses of virus structural proteins. The presence in normal cells of the inherited provirus capable of being induced to form complete C-type virus particles (see Section X, A) adds a further dimension of complexity to the genetic studies of sarcoma and leukemia viruses.

Considering the many ways in which viral transformed cells differ from their normal counterparts (see Section VI and Table II), and the relatively restricted amount of genetic information in tumor viruses, it appears necessary to adopt the view that either the few viral gene products operative in maintaining transformed cell properties exert multiple, direct, "pleiotropic" effects, or that they play a regulatory role affecting the expression of numerous cellular functions (Benjamin, 1972). This general view is supported further by the considerations outlined below on cellular factors in transformation.

B. Cellular Factors

Cells in culture may become neoplastically transformed spontaneously or after exposure to various chemical carcinogens or to radiation; the properties of such transformed cells share many of the basic features of cells transformed by tumor viruses (see Section VI). These facts make it clear that essential aspects of the transformed phenotype are coded for by the cell and do not necessarily depend on added viral genetic information.

Certain of the general properties of transformed cells are variably expressed by normal cells according to physiological conditions. As discussed earlier, the saturation densities of normal cells are sensitive to serum concentration and to the pH of the medium. Other treatments can also trigger changes in normal cells which mimic in a transient or reversible way the properties of transformed cells; for example, mild treatments with proteolytic enzymes cause normal 3T3 cells in a confluent monolayer to initiate DNA synthesis and divide (Burger, 1970); the same treatment causes normal cells to agglutinate in the presence of low concentrations of concanavalin A or wheat germ agglutinin in a manner similar to viral or other types of transformed cells (Burger, 1969); similarly, the addition of 200 mM urea to the medium in cultures of normal cells causes loss of contact inhibition of movement resulting in increased amounts of nuclear overlap, loss of density inhibition of DNA synthesis, and increases in agglutinability by concanavalin A (Weston and Hendricks, 1972). Another pertinent observation is the regulated exposure during mitosis of sites for binding of wheat germ agglutinin (Fox *et al.*, 1971); thus, normal cells appear to regulate a change on their surface which transformed cells fail to regulate.

Various physiological and genetic facts are known to affect viral transformation at different levels. Some of the genetic factors affecting cell responses to tumor viruses have already been discussed (Sections VIII and X); these include genes which determine susceptibilities to productive versus nonproductive infection by papova viruses, and susceptibility to avian sarcoma viruses via cell surface receptors and to murine viruses at a presumably early but unknown intracellular stage of infection. In addition, the species and tissue origin of genetically susceptible cells play a role in determining morphological properties after viral transformation; cells of different species or tissue origin transformed by either the same or different strains of Rous sarcoma virus have detectably different morphologies (see Section VI, A). Other examples are known of tissue specificity in transformation by avian or mammalian sarcoma viruses; thus, avian myeloblastosis virus replicates in chick embryo fibroblasts without transforming them, while morphological transformation does occur in some cells derived from yolk sac, muscle, and other tissues (Baluda and Goetz, 1961; Moscovici *et al.*, 1969); the spleen focus-forming component of Friend virus (Lilly, 1972) as well as the various avian and murine leukosis viruses, may be cited as further examples of viruses with tissue-specific target cells for transformation, although *in vitro* studies have not been done to confirm this.

A final group of related observations can be cited to show that normal cellular genes can exert control over the transformed phenotype directed by an integrated viral genome. Phenotypically normal revertants of virus transformed cells have been isolated in a number of ways; in general the procedures are based on negative selection in which cells growing under conditions which restrict growth of normal cells are selectively killed. Revertants, usually of SV40-transformed 3T3 cells, have been isolated which display lowered saturation densities, normal or near normal morphologies, increased dependence on serum for growth, inability to grow in agar, decreased agglutinability, and increased resistance to the toxic effects of Concanavalin A. Methods for isolating and characterizing these types of revertants have recently been described (Vogel and Pollack, 1973).

A general finding is that the SV40 viral genome is still present in the revertants, as judged from the presence of SV40 T antigen; in some instances, SV40-specific RNA has been looked for and found (Ozanne *et al.*, 1973); SV40 virus can be recovered from some but not all of the revertants by fusion with permissive monkey cells; whenever tests have been made, the recovered virus appears to be wild type with respect to its transforming ability. Similar observations have been made on a temperature-sensitive SV40-transformed 3T3 cell (Renger and Basilico, 1972, 1973; Noonan *et al.*, 1973). The conclusions drawn from these studies are that the cell can exert control over the transformed phenotype following the establishment of a viral genome in the transformed cell, and that this most likely occurs by bypassing the effects of the integrated viral genome rather than by a direct action on the viral genome itself.

The mechanisms of these phenotypic-reversions are not clear; however, variations in the number and kind of normal cell chromosomes has been cited as a possible factor (Pollack *et al.*, 1970; Hitotsumachi *et al.*, 1971; Yamamoto *et al.*, 1973). An independent series of observations also suggests that the presence of an excess of normal cellular chromosomes can suppress the effects of viral transformation; somatic cell hybrids formed by fusion of various normal, nonmalignant cells with various virus-transformed, malignant cells behave normally, i.e., are nonmalignant (Harris *et al.*, 1969; Harris and Klein, 1969; Klein *et al.*, 1971; Bregula *et al.*, 1971; Weiner *et al.*, 1971). Such nontumorigenic hybrids may become malignant upon losing chromosomes, presumably of the normal parent (Klein *et al.*, 1971). Similarly, in the *in vitro* studies cited above, increases in chromosome number accompany reversion while losses accompany re-reversion (Pollack *et al.*, 1970). Efforts have been made to identify the chromosomes responsible for the suppressive effects (Hitotsumachi *et al.*, 1971; Yamamoto *et al.*, 1973). Further efforts must be made to understand how normal growth control mechanisms behave in a dominant way over the virus-triggered events leading to the loss of growth control.

References

Aaronson, S. (1971). *Virology* **44**, 29.
Aaronson, S., and Rowe, W. (1970). *Virology* **42**, 9.
Aaronson, S., and Todaro, G. (1968). *Science* **162**, 1024.
Aaronson, S., Jainchill, J., and Todaro, G. (1970a). *Proc. Nat. Acad. Sci. U. S.* **66**, 1236.
Aaronson, S., Todaro, G., and Huebner, R. (1970b). *Lepetit Colloquium, 2nd, Paris, 1969*. 138.
Aaronson, S., Todaro, G., and Scolnick, E. (1971) *Science* **174**, 157.
Abercrombie, M. (1962) *Cold Spring Harbor Symp. Quant. Biol.* **27**, 427.
Abercrombie, M., and Ambrose, E. (1958). *Exp. Cell Res.* **15**, 322.
Abercrombie, M., and Heaysman, J. (1954). *Exp. Cell Res.* **6**, 293.

Ankerst, J., and Sjögren, H. (1969). *Int. J. Cancer* **4**, 279.
Ankerst, J., and Sjögren, H. (1970). *Int. J. Cancer* **6**, 84.
Baltimore, D. (1970). *Nature* (*London*) **226**, 1209.
Baluda, M. (1972). *Proc. Nat. Acad. Sci. U.S.* **69**, 576.
Baluda, M., and Goetz, I. (1961). *Virology* **15**, 185.
Basilico, C., and DiMayorca, G. (1965). *Proc. Nat. Acad. Sci. U.S.* **54**, 125.
Basilico, C., and Marin, G. (1966). *Virology* **28**, 429.
Basilico, C., Matsuya, Y., and Green, H. (1970). *Virology* **41**, 295.
Bassin, R., Simons, P., Chesterman, F., and Harvey, J. (1968). *Int. J. Cancer* **3**, 265.
Bassin, R., Tuttle, N., and Fischinger, P. (1970). *Int. J. Cancer* **6**, 95.
Bassin, R., Phillips, L., Kramer, M., Haapala, D., Peebles, P., Nomura, S., and Fischinger, P. (1971a). *Virology* **44**, 29.
Bassin, R., Tuttle, N., and Fischinger, P. (1971b). *Nature* (*London*) **229**, 564.
Bather, R., and Leonard, A. (1970). *J. Nat. Cancer Inst.* **40**, 551.
Benjamin, T. (1965). *Proc. Nat. Acad. Sci. U.S.* **54**, 121.
Benjamin, T. (1966). *J. Mol. Biol.* **16**, 359.
Benjamin, T. (1968). *Virology* **36**, 685.
Benjamin, T. (1970). *Proc. Nat. Acad. Sci. U.S.* **67**, 394.
Benjamin, T. (1972). *Curr. Top. Microbiol. Immunol.* **59**, 107.
Benjamin, T., and Burger, M. (1970). *Proc. Nat. Acad. Sci. U.S.* **67**, 929.
Benjamin, T., and Norkin, L. (1972). *In* "Molecular Studies in Viral Neoplasia: 25th Annual Symposium Fundamental Cancer Research." (In press.)
Berwald, Y., and Sachs, L. (1965). *J. Nat. Cancer Inst.* **35**, 641.
Biggs, P., de-Thé, G., and Payne, L., eds. (1972). *IARC Sci. Publ.* n.2.
Biquard, J., and Vigier, P. (1970). *C. R. Acad. Sci.* (*Paris*) **271**, 2430.
Black, P. (1964). *Virology* **24**, 179.
Black, P. (1966). *Virology* **28**, 760.
Black, P. (1968). *Annu. Rev. Microbiol.* **22**, 391.
Black, P., and Rowe, W. (1963). *Proc. Soc. Exp. Biol. Med.* **114**, 721.
Black, P., and Todaro, G. (1965). *Proc. Nat. Acad. Sci. U.S.* **54**, 374.
Black, P., and White, B. (1967). *J. Exp. Med.* **125**, 629.
Black, P., Rowe, W., Turner, H., and Huebner, R. (1963). *Proc. Nat. Acad. Sci. U.S.* **50**, 1148.
Blackstein, M., Stanners, C., and Farmilo, A. (1969). *J. Mol. Biol.* **42**, 301.
Borek, C., and Sachs, L. (1966). *Nature* (*London*) **210**, 276.
Borek, C., and Sachs, L. (1967). *Proc. Nat. Acad. Sci. U.S.* **57**, 1522.
Borek, C., and Sachs, L. (1968). *Proc. Nat. Acad. Sci. U.S.* **59**, 83.
Bregula, U., Klein, G., and Harris, H. (1971). *J. Cell. Sci.* **8**, 673.
Britten, R., and Kohne, D. (1968). *Science* **161**, 529.
Burger, M. (1968). *Nature* (*London*) **219**, 499.
Burger, M. (1969). *Proc. Nat. Acad. Sci. U.S.* **62**, 994.
Burger, M. (1970). *Nature* (*London*) **227**, 120.
Burger, M. (1972). *Methods Enzymol.* **23**.
Burger, M. (1973). *Fed. Proc., Fed. Amer. Soc. Exp. Biol.*, **32**, 91.
Burger, M., and Martin, G. (1972). *Nature New Biol.* **237**, 9.
Burger, M., and Noonan, K. (1970). *Nature* (*London*) **228**, 512.
Burk, R. (1966). *Nature* (*London*) **212**, 1261.
Casto, B. (1968). *J. Virol.* **2**, 376.
Casto, B. (1973a). *Progr. Exp. Tumor Res.* **18**, 166.
Casto, B. (1973b). *Cancer Res.* **33**, 402.
Casto, B., Pieczynski, W., and DiPaolo, J. (1973). *Cancer Res.* **331**, 819.

Ceccarini, C. (1971). *Nature New Biol.* **233**, 271.

Ceccarini, C., and Eagle, H. (1971). *Proc. Nat. Acad. Sci. U. S.* **68**, 229.

Clarke, G., and Stoker, M. (1971). *Ciba Found. Symp. Cell Cult.* 17–32.

Coggin, J. (1969). *J. Virol.* **3**, 458.

Crawford, L. (1969). *In* "Fundamental Techniques in Animal Virology" K., Habel and N. Salzman, eds.), pp. 75–81. Academic Press, New York.

Crawford, L., Dulbecco, R., Fried, M., Montagnier, L., and Stoker, M. (1964). *Proc. Nat. Acad. Sci. U. S.* **52**, 148.

Cunningham, D., and Pardee, A. (1969). *Proc. Nat. Acad. Sci. U. S.* **64**, 1049.

Davis, R., Simon, M., and Davidson, N. (1971). *Methods Enzymol.* **21**, 383.

Defendi, V. (1963). *Proc. Soc. Exp. Biol. Med.* **113**, 12.

Defendi, V. (1968). *Transplantation* **6**, 642.

Defendi, V., and Hirai, K. (1973). *In* "Control of Proliferation in Animal Cells." May, 1973. Cold Spring Harbor, New York.

Defendi, V., and Stoker, M., eds. (1967). *Wistar Inst. Symp. Monogr.* **7**.

Defendi, V., Lehman, J., and Kraemer, P. (1963). *Virology* **19**, 592.

Diamandopoulos, G., and Enders, J. (1965). *Proc. Nat. Acad. Sci. U. S.* **54**, 1092.

Diamond, L. (1967). *Inst. J. Cancer* **2**, 143.

Duc-Nguyen, H. (1968). *J. Virol.* **2**, 643.

Duff, R., and Rapp, F. (1970). *J. Virol.* **5**, 568.

Duff, R., and Rapp, F. (1971). *Nature (London)* **233**, 48.

Dulbecco, R. (1970). *Nature (London)* **227**, 802.

Dulbecco, R., and Stoker, M. (1970). *Proc. Nat. Acad. Sci. U. S.* **66**, 204.

Eagle, H. (1971). *Science* **174**, 500.

Eagle, H., Foley, G., Koprowski, H., Lazarus, H., Levine, E., and Adams, R. (1970). *J. Exp. Med.* **131**, 863.

Eckhart, W., Dulbecco, R., and Burger, M. (1971). *Proc. Nat. Acad. Sci. U. S.* **68**, 283.

Edwards, J., and Campbell, J. (1971). *J. Cell. Sci.* **8**, 53.

Edwards, J., Campbell, J., and Williams, J. (1971). *Nature New Biol.* **231**, 147.

Ejercito, P., Kieff, E., and Roizman, B. (1968). *J. Gen. Virol.* **3**, 357.

Ephrussi, B., and Temin, H. (1960). *Virology* **11**, 547.

Erikson, R. (1969). *Virology* **37**, 124.

Fenner, F., ed. (1968). "The Biology of Animal Viruses." Academic Press, New York.

Fischinger, P., and O'Connor, T. (1969). *J. Nat. Cancer Inst.* **43**, 487.

Foster, D., and Pardee, A. (1969). *J. Biol. Chem.* **244**, 2675.

Fox, T., Sheppard, J., and Burger, M. (1971). *Proc. Nat. Acad. Sci. U. S.* **68**, 244.

Freeman, A., Black, P., Wolford, R., and Huebner, R. (1967). *J. Virol.* **1**, 362.

Freeman, A., Price, P., Igel, H., Young, J., Maryak, J., and Huebner, R. (1970). *J. Nat. Cancer Inst.* **44**, 65.

Freeman, A., Price, P., Bryan, R., Gordon, R., Gilden, R., Kelloff, G., and Huebner, R. (1971). *Proc. Nat. Acad. Sci. U. S.* **68**, 445.

Frenkel, N., Roizman, B., Cassai, E., and Nahmias, A. (1972). *Proc. Nat. Acad. Sci. U. S.* **69**, 3874.

Fritz, R., and Qualtiere, L. (1973). *J. Virol.* **11**, 736.

Geering, G., Old, L., and Boyse, E. (1966). *J. Exp. Med.* **124**, 753.

Gelb, L., Kohne, D., and Martin, M. (1971). *J. Mol. Biol.* **57**, 129.

Gillespie, D., and Spiegelman, S. (1965). *J. Mol. Biol.* **12**, 829.

Graf, T. (1972). *Virology* **50**, 567.

Green, H. (1973). *In* "Control of Proliferation in Animal Cells." May, 1973. Cold Spring Harbor, New York.

Lindstrom, D., and Dulbecco, R. (1972). *Proc. Nat. Acad. Sci. U. S.* **69**, 1517.
Lipton, A., Klinger, I., Paul, D., and Holley, R. (1971). *Proc. Nat. Acad. Sci. U. S.* **68**, 2799.
Lowy, D., Rowe, W., Teich, N., and Hartley, J. (1971). *Science* **174**, 155.
Macpherson, I. (1965). *Science* **148**, 1731.
Macpherson, I. (1966). *Recent Results Cancer Res.* **6**, 1.
Macpherson, I. (1969). *In* "Fundamental Techniques in Virology" (K. Habel and N. Salzman, eds.), pp. 212–219. Academic Press, New York.
Macpherson, I., and Montagnier, L. (1964). *Virology* **20**, 291.
Macpherson, I., and Russell, W. (1966). *Nature (London)* **210**, 1343.
Macpherson, I., and Stoker, M. (1962). *Virology* **16**, 147.
Malmgren, R., Takemoto, K., and Carney, P. (1968). *J. Nat. Cancer Inst.* **40**, 263.
Maroudas, N. (1972). *Exp. Cell Res.* **74**, 337.
Martin, G. (1970). *Nature (London)* **227**, 1021.
Martin, G. (1971). *Lepetit Colloquium, 2nd, Paris, 1970*, p. 320.
Martin, G., Venuta, S., Weber, M., and Rubin, H. (1971). *Proc. Nat. Acad. Sci. U. S.* **68**, 2739.
Martinez-Palomo, A., and Brailovsky, C. (1968). *Virology* **34**, 379.
Martz, E. (1972). *J. Cell. Physiol.* **81**. 39.
Martz, E., and Steinberg, M. (1972). *J. Cell. Physiol.* **79**, 189.
Martz, E., and Steinberg, M. (1973). *J. Cell. Physiol.* **81**, 25.
McAllister, R., and Macpherson, I. (1968). *J. Gen. Virol.* **2**, 99.
McAllister, R., and Macpherson, I. (1969). *J. Gen. Virol.* **4**, 29.
McAllister, R., Nicolson, G., Reed, G., Kern, J., Gilden, R., and Huebner, R. (1969). *J. Nat. Cancer Inst.* **43**, 917.
Medina, D., and Sachs, L. (1963). *Virology* **19**, 127.
Michel, M., Hirt, B., and Weil, R. (1967). *Proc. Nat. Acad. Sci. U. S.* **58**, 1381.
Milo, G., Schaller, J., and Yohn, D. (1972). *Cancer Res.* **32**, 2338.
Montagnier, L. (1968). *C. R. Acad. Sci. (Paris).* **267**, 921.
Montagnier, L. (1971). *Ciba Found. Symp. Growth Cont. Cell Cult.*, 33–41.
Moscovici, C., Moscovici, M., and Zanetti, M. (1969). *J. Cell. Physiol.* **73**, 105.
Munyon, W., Kraiselburd, E., and Davis, D. (1971). *J. Virol.* **7**, 813.
Nakata, Y., and Bader, J. (1968). *J. Virol.* **2**, 1255.
Nicolson, G. (1971). *Nature New Biol.* **233**, 244.
Nicolson, G. (1972). *Nature New Biol.* **239**, 193.
Nilhausen, K., and Green, H. (1965). *Exp. Cell. Res.* **40**, 166.
Nonoyama, M., and Pagano, J. (1971). *Nature New Biol.* **233**, 103.
Noonan, K., Renger, H., Basilico, C., and Burger, M. (1973). *Proc. Nat. Acad. Sci. U. S.* **70**, 347.
Norkin, L., and Benjamin, T. (1973). (In preparation.)
Nowinsky, R., Sarkar, N., and Fleissner, E. (1972). *Methods Cancer Res.* **8**.
O'Connor, T., and Fischinger, P. (1968a). *Science* **159**, 325.
O'Connor, T., and Fischinger, P. (1968b). *J. Nat. Cancer Inst.* **43**, 487.
Ozanne, B., and Sambrook, J. (1971). *Lepetit Colloquium, 2nd, Paris, 1970*, p. 248.
Ozanne, B., Sharp, P., and Sambrook, J. (1973). *J. Virol.* (in press).
Paul, D., Lipton, A., and Klinger, I. (1971). *Proc. Nat. Acad. Sci. U. S.* **68**, 645.
Payne, L. (1972). *In* "RNA Viruses and Host Genome in Oncogenesis" (P. Emmelot and O. Bentelgen, eds.), pp. 93–116.
Payne, L., and Chubb, R. (1968). *J. Gen. Virol.* **3**, 379.
Payne, L. (1973). *Lepetit Colloquium, 4th, 1972*.
Payne, L. N., Crittenden, L. B., and Weiss, R. A. (1973). Possible epipsomes in eukaryotes. In preparation.
Peebles, P., Haapala, D., and Gadzar, A. (1972). *J. Virol.* **9**, 488.

Pincus, T., Hartley, J., and Rowe, W. (1971a). *J. Exp. Med.* **133**, 1219.
Pincus, T., Rowe, W., and Lilly, F. (1971b). *J. Exp. Med.* **133**, 1234.
Pollack, R., and Burger, M. (1969). *Proc. Nat. Acad. Sci. U. S.* **63**, 1074.
Pollack, R., and Teebor, G. (1969). *Cancer Res.* **29**, 1770.
Pollock, A., and Todaro, G. (1968). *Nature (London)* **219**, 520.
Pollack, R., Green, H., and Todaro, G. (1968). *Proc. Nat. Acad. Sci. U. S.* **60**, 126.
Pollack, R., Vogel, A., and Wollman, L. (1970). *Nature (London)* **228**, 938.
Pope, J., and Rowe, W. (1964). *J. Exp. Med.* **120**, 121.
Radloff, R., Bauer, W W., and Vinograd, J. (1967). *Proc. Nat. Acad. Sci. U.S.* **57**, 1514.
Rapp, F., and Duff, R. (1972). *IARC Publ.* **2**, pp. 447–450.
Rapp, F., Butel, J., Tevethia, S., Katz, M., and Melnick, J. (1966). *J. Immunol.* **97**, 833.
Renger, H., and Basilico, C. (1972). *Proc. Nat. Acad. Sci. U. S.* **69**, 109.
Renger, H., and Basilico, C. (1973). *J. Virol.* **11**, 702.
Risser, R. (1973). *In* "Control of Proliferation in Animal Cells." May, 1973. Cold Spring Harbor, New York.
Robinson, H. (1967). *Proc. Nat. Acad. Sci. U. S.* **57**, 1655.
Roizman, B. (1972). *IARC* 2, pp. 1–17.
Rosenblith, J., Ukena, T., Yin, H., Berlin, R., and Karnovsky, M. (1973). *Proc. Nat. Acad. Sci. U. S.* **70**, 1625.
Rowe, W. (1972a). *J. Exp. Med.* **136**, 1272.
Rowe, W. (1972b). *Perspect. Virol.* **10**, 782.
Rowe, W., Hartley, J., and Pugh, W. (1970). *Virology* **42**, 1136.
Rubin, H. (1960). *Proc. Nat. Acad. Sci. U. S.* **46**, 1105.
Rubin, H. (1966). *Exp. Cell Res.* **41**, 138.
Rubin, H. (1970). *Science* **167**, 1271.
Rubin, H., and Fodge, D. (1973). *In* "Control of Proliferation in Animal Cells." May, 1973. Cold Spring Harbor, New York.
Sachs, L., Medina, D., and Berwald, Y. (1962). *Virology* **17**, 491.
Sakiyama, H., and Robbins, P. (1973). *Fed. Proc., Fed. Amer. Soc. Exp. Biol.* **32**, 86.
Saksela, E., and Moorhead, P. (1963). *Proc. Nat. Acad. Sci. U. S.* **50**, 390.
Salzberg, S., Robbin, S., and Green, M. (1973). *Virology* **53**, 186.
Sambrook, J. (1972). *Advanc. Cancer* **16**, 141.
Sambrook, J., Westphal, H., Srinivasan, P., and Dulbecco, R. (1968). *Proc. Nat. Acad. Sci. U. S.* **60**, 1288.
Sambrook, J., Sharp, P., and Keller, W. (1972). *J. Mol. Biol.* **57**, 71.
Sanders, F., and Burford, B. (1964). *Nature (London)* **201**, 786.
Sanders, F., and Smith, J. (1970). *Nature (London)* **227**, 513.
Sanford, K., Barker, B., Woods, M., Parshad, R., and Law, L. (1967). *J. Nat. Cancer Inst.* **39**, 705.
Sarma, P., and Log, S. (1973). *Virology* **54**, 160.
Sarma, P., Turner, W., and Huebner, R. (1964). *Virology* **23**, 313.
Sarma, P., Cheong, M., Hartley, J., and Huebner, R. (1967). *Virology* **33**, 180.
Scheele, C., and Hanafusa, H. (1971). *Virology* **45**, 401.
Scher, C., and Nelson-Rees, W. (1971). *Nature New Biol.* **233**, 263.
Scolnick, E., Parks, W., and Livingston, D. (1972). *J. Immunol.* **109**, 570.
Sefton, B., and Rubin, H. (1970). *Nature (London)* **227**, 843.
Sefton, B., and Rubin, H. (1971). *Proc. Nat. Acad. Sci. U. S.* **68**, 3154.
Shein, H., and Enders, J. (1962). *Proc. Nat. Acad. Sci. U. S.* **48**, 1350.
Shiroki, K., and Shimojo, H. (1971). *Virology* **45**, 163.
Shodell, M. (1972). *Proc. Nat. Acad. Sci. U. S.* **69**, 1455.

Sjögren, H. (1965). *Progr. Exp. Tumor Res.* **6**, 289.
Sjögren, H., and Hellström, I. (1966). *Subviral Carcinog. Int. Symp. Tumor Viruses, 1st, 1966,* pp. 207–219.
Sjögren, H., Hellström, I., and Klein, G. (1961). *Cancer Res.* **21**, 329.
Smith, H., Scher, C., and Todaro, G. (1971). *Virology* **44**, 359.
Smith, H., Gelb, L., and Martin, M. (1972). *Proc. Nat. Acad. Sci. U. S.* **69**, 152.
Smith, B., Defendi, V., and Wigglesworth, N. (1973). *Virology* **51**, 230.
Somers, D., and Kirsten, W. (1968). *Virology* **36**, 155.
Somers, D., and Kirsten, W. (1969). *Int. J. Cancer* **4**, 697.
Somers, D., and Kit, S. (1971). *Virology* **46**, 774.
Stanners, C. (1963). *Virology* **21**, 464.
Stanners, C., Till, J., and Siminovitch, L. (1963). *Virology* **21**, 448.
Steck, F., and Rubin, H. (1966a). *Virology* **29**, 628.
Steck, F., and Rubin, H. (1966b). *Virology* **29**, 642.
Stich, H., Hammerberg, O., and Casto, B. (1972). *Can. J. Genet. Cytol.* **15**, 911.
Stoker, M. (1963). *Nature* (*London*) **200**, 756.
Stoker, M. (1964). *Virology* **24**, 123.
Stoker, M. (1968). *Nature* (*London*) **218**, 234.
Stoker, M., and Abel, P. (1962). *Cold. Spring Harbor Symp. Quant. Biol.* **27**, 375.
Stoker, M., and Dulbecco, R. (1969). *Nature* (*London*) **223**, 317.
Stoker, M., and Macpherson, I. (1961). *Virology* **14**, 359.
Stoker, M., and Rubin, H. (1967). *Nature* (*London*) **215**, 171.
Stoker, M., and Smith, A. (1964). *Virology* **24**, 175.
Stoker, M., O'Neill, M., Berryman, C., and Waxmann, V. (1968). *Int. J. Cancer* **3**, 683.
Stoker, M., Thornton, M., Riddle, P., Birg, F., and Meyer, G. (1972). *Int. J. Cancer* **10**, 613.
Tai, H., Smith, C., Sharp, P., and Vinograd, J. (1972). *J. Virol.* **9**, 317.
Taylor-Papadimitriou, J., Stoker, M., and Riddle, P. (1971). *Int. J. Cancer* **7**, 269.
Tegtmeyer, P., and Ozer, H. (1971). *J. Virol.* **8**, 517.
Temin, H. (1960). *Virology* **10**, 182.
Temin, H. (1966). *J. Nat. Cancer Inst.* **37**, 167.
Temin, H. (1967). *Wistar Inst. Symp Monogr.* **7**, 103.
Temin, H. (1968). *Int. J. Cancer* **3**, 771.
Temin, H. (1970). *J. Cell. Physiol.* **75**, 107.
Temin, H. (1971). *Annu. Rev. Microbiol.* **25**, 609.
Temin, H. (1973). *Adv. Cancer Res.* **19** (in press).
Temin, H., and Mizutani, S. (1970). *Nature* (*London*) **226**, 1211.
Temin, H., Pierson, R., and Dulak, N. (1971). *In* "Nutrition and Metabolism of Cells in Culture" V. Cristofalo and G. Rothblat, (eds.). Academic Press, New York.
Tevethia, S., and Rapp, F. (1965). *Proc. Soc. Exp. Biol. Med.* **120**, 455.
Tevethia, S., Katz, M., and Rapp, F. (1965). *Proc. Soc. Exp. Biol. Med.* **119**, 896.
Ting, R. (1967). *Proc. Soc. Exp. Biol. Med.* **126**, 778.
Ting, R. (1968). *J. Virol.* **2**, 865.
Todaro, G. (1972). *Nature* (*London*) **240**, 157.
Todaro, G., and Green, H. (1963). *J. Cell Biol.* **17**, 299.
Todaro, G., and Green, H. (1964a). *Virology* **23**, 117.
Todaro, G., and Green, H. (1964b). *Virology* **24**, 393.
Todaro, G., and Green, H. (1965). *Science* **147**, 513.
Todaro, G., and Green, H. (1966a). *Virology* **28**, 756.
Todaro, G., and Green, H. (1966b). *Proc. Nat. Acad. Sci. U. S.* **55**, 302.
Todaro, G., and Martin, G. (1967). *Proc. Soc. Exp. Biol. Med.* **124**, 1232.
Todaro, G., Green, H., and Goldberg, B. (1964). *Proc. Nat. Acad. Sci. U. S.* **51**, 66.

Todaro, G., Habel, K., and Green, H. (1965a). *Virology* **27**, 179.
Todaro, G., Lazar, G., and Green, H. (1965b). *J. Cell. Comp. Physiol.* **6**, 325.
Todaro, G., Green, H., and Swift, M. (1966). *Science* **153**, 1252.
Todaro, G., Matsuya, Y., Bloom S., Robbins, A., and Green, H. (1967). *Wistar Inst. Symp. Monogr.* **7**, 87.
Todaro, G., Scher, C., and Smith, H. (1971). *Ciba Found. Symp. Contr. Cell Cult.* pp. 151–167.
Tooze, J., ed. (1973). "The Molecular Biology of Tumor Viruses" Cold Spring Harbor Laboratory, New York.
Toyoshima, K., and Vogt, P. (1969). *Virology* **39**, 930.
Tsai, R., and Green, H. (1973). *Nature New Biol.* **243**, 168.
Varmus, H., Weiss, R., Früs, R., Levinson, W., and Bishop, J. (1972). *Proc. Nat. Acad. Sci. U. S.* **69**, 20.
Venuta, S., and Rubin, H. (1973). *Proc. Nat. Acad. Sci. U. S.* **70**, 653.
Vogel, A., and Pollack, R. (1973). This volume.
Vogt, P. (1965). *Advan. Cancer Res.* **11**, 293.
Vogt, P. (1967a). *Virology* **33**, 175.
Vogt, P. (1967b). *Virology* **32**, 708.
Vogt, P. (1969). *In* "Fundamental Techniques in Virology" (K. Habel, and N. Salzman, eds.); pp. 198–211. Academic Press, New York.
Vogt, P. (1972). *J. Nat. Cancer Inst.* **48**, 3.
Vogt, M., and Dulbecco, R. (1962). *Cold Spring Harbor Symp. Quant. Biol.* **27**, 367.
Vogt, M., and Dulbecco, R. (1963). *Proc. Nat. Acad. Sci. U. S.* **49**, 171.
Vogt, P., and Ishizaki, R. (1965). *Virology* **26**, 664.
Vogt, P., Ishizaki, R., and Duff, R. (1966). *Subviral Carcinog. Int. Symp. Tumor Viruses, 1st, 1966*, pp. 297–310.
Ware, L., and Axelrod, A. (1972). *Virology* **50**, 339.
Weiler, E. (1959). *Exp. Cell Res., Suppl.* **7**, 244.
Weiner, F., Klein, G., and Harris, H. (1971). *J. Cell. Sci.* **8**, 681.
Weiss, R. (1969). *J. Gen. Virol.* **5**, 511.
Weiss, R., and Payne, L. (1971). *Virology* **45**, 508.
Weiss, R., Früs, R., Katz, E., and Vogt, P. (1971). *Virology* **46**, 290.
Weiss, R., Mason, W., and Vogt, P. (1973). *Virology* **52**, 535.
Wells, S., Rabson, A., Malmgren, R., and Ketcham, A. (1966). *Cancer* (*Philadelphia*) **19**, 1411.
Weston, J., and Henricks, K. (1972). *Proc. Nat. Acad. Sci. U. S.* **69**, 3727.
Wetmur, J. G., and Davidson, N. (1968). *J. Mol. Biol.* **31**, 349.
Williams, J., and Till, J. (1964). *Virology* **24**, 505.
Winicour, E. (1967). *Virology* **31**, 15.
Winicour, E. (1968). *Virology* **34**, 571.
Winicour, E. (1969). *Advan. Virus Res.* **14**, 153.
Wolstenholme, G., and Knight, J., eds. (1971) *Ciba Found. Symp. Growth Contr. Cell Cult.*
Wyke, J. (1973a). *Virology* **52**, 587.
Wyke, J. (1973b). *Virology* **54**, 28.
Wyke, J., and Linial, M. (1973). *Virology* **53**, 152.
Yamamoto, T., Rabinowitz, Z., and Sachs, L. (1973). *Nature New Biol.* **243**, 247.
Yoshiike, K., Furono, A., and Suzuki, K. (1972). *J. Mol. Biol.* **70**, 415.
Zavada, J., and Macpherson, I. (1970). *Nature* (*London*) **225**, 24.
Zetterberg, A., and Auer, G. (1970). *Exp. Cell. Res.* **62**, 262.
zur Hausen, H., and Schulte-Holthausen, H. (1970a). *Nature* (*London*) **227**, 245.
zur Hausen, H., Schulte-Holthausen, H., Klein, G., Henle, W., Henle, G., Clifford, P., and Santesson, L. (1970b). *Nature* (*London*) **228**, 1056.

Author Index

Numbers in italics refer to the pages on which the complete references are cited.

H

M

P

T

X

Y

Subject Index

C

D

A 4
B 5
C 6
D 7
E 8
F 9
G 0
H 1
I 2
J 3